FISH PHYSIOLOGY

Volume I

Excretion, Ionic Regulation, and Metabolism

CONTRIBUTORS

FRANK P. CONTE

EDWARD M. DONALDSON

ROY P. FORSTER

LEON GOLDSTEIN

CLEVELAND P. HICKMAN, JR.

P. W. HOCHACHKA

F. G. T. HOLLIDAY

W. N. HOLMES

ARTHUR M. PHILLIPS, JR.

BENJAMIN F. TRUMP

FISH PHYSIOLOGY

Edited by
W. S. HOAR
DEPARTMENT OF ZOOLOGY
UNIVERSITY OF BRITISH COLUMBIA
VANCOUVER, CANADA

and
D. J. RANDALL
DEPARTMENT OF ZOOLOGY
UNIVERSITY OF BRITISH COLUMBIA
VANCOUVER, CANADA

Volume I

Excretion, Ionic Regulation, and Metabolism

 Academic Press   New York and London   1969

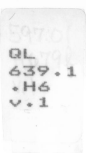
COPYRIGHT © 1969, BY ACADEMIC PRESS, INC.
ALL RIGHTS RESERVED
NO PART OF THIS BOOK MAY BE REPRODUCED IN ANY FORM,
BY PHOTOSTAT, MICROFILM, RETRIEVAL SYSTEM, OR ANY
OTHER MEANS, WITHOUT WRITTEN PERMISSION FROM
THE PUBLISHERS.

ACADEMIC PRESS, INC.
111 Fifth Avenue, New York, New York 10003

United Kingdom Edition published by
ACADEMIC PRESS, INC. (LONDON) LTD.
Berkeley Square House, London W1X 6BA

LIBRARY OF CONGRESS CATALOG CARD NUMBER: 76-84233

PRINTED IN THE UNITED STATES OF AMERICA

LIST OF CONTRIBUTORS

Numbers in parentheses indicate the pages on which the authors' contributions begin.

FRANK P. CONTE (241), *Department of Zoology, Oregon State University, Corvallis, Oregon*

EDWARD M. DONALDSON (1), *Vancouver Laboratory, Fisheries Research Board of Canada, Vancouver, British Columbia, Canada*

ROY P. FORSTER (313), *Department of Biological Sciences, Dartmouth College, Hanover, New Hampshire*

LEON GOLDSTEIN* (313), *Department of Physiology, Harvard Medical School, Boston, Massachusetts*

CLEVELAND P. HICKMAN, JR. (91), *Department of Biology, Washington and Lee University, Lexington, Virginia*

P. W. HOCHACHKA (351), *Department of Zoology, University of British Columbia, Vancouver, British Columbia, Canada*

F. G. T. HOLLIDAY (293), *Biology Department, University of Stirling, Stirling, Scotland*

W. N. HOLMES (1), *Department of Biology, University of California, Santa Barbara, California*

ARTHUR M. PHILLIPS, JR. (391), *U.S. Department of the Interior, Fish and Wildlife Service, Bureau of Sport Fisheries and Wildlife, Cortland, New York*

BENJAMIN F. TRUMP (91), *Department of Pathology, Duke University, Durham, North Carolina*

---

* Present address: Division of Biological and Medical Sciences, Brown University, Providence, Rhode Island.

PREFACE

More than a decade has passed since "The Physiology of Fishes" was published under the editorship of M. E. Brown. During this period, an increasing number of physiologists and biochemists have chosen to work on fishes. These investigators have opened up many additional areas of research, have developed new concepts to explain previously conflicting phenomena, and, at the same time, have raised many engaging questions which will only be answered by further study. It is possible that some of the impetus for this burst of activity can be attributed to "The Physiology of Fishes."

"Fish Physiology," a six-volume treatise, attempts to review recent advances in selected areas of fish physiology, to relate these advances to the existing body of earlier literature, and to delineate useful areas for further study. It is published with the hope that it will serve biologists of the 1970's as "The Physiology of Fishes" served its readers throughout the 1960's. Margaret Brown (Varley) found it impossible to undertake the editorial work associated with the production of this treatise, and, therefore, we agreed to assume the task.

The increase in the number of volumes from the two of "The Physiology of Fishes" is not only a reflection of the rapid increase of interest in this group of animals but of their physiological diversity as well. Since the term "fishes" includes the Agnatha, Chondrichthyes, Actinopterygii, and Choanichthyes, the treatise deals with physiological mechanisms whose vertebrate phylogeny covers an expanse of 500 million years. It considers adaptive processes associated with successful living in a full range of aquatic habitats extending from the tropics to the frigid zones; it also describes primitive air-breathing systems of great diversity, the physiology of most of the vertebrate organ systems, and numerous curious devices for protection and communication. While discussing these many functional processes, the authors have referred to a wealth of comparative material so that the treatise has become more than an account of the physiology of fishes; it contains many fundamental concepts and principles important in the broad field of comparative animal physiology. It is our hope that "Fish Physiology" will prove as valuable in fisheries research laboratories as in university reference libraries and that it will be a rich source of detailed information for the comparative physiologist and the zoologist as well as the specialist in fish physiology.

Taxonomists may quarrel with the lack of uniformity in the scientific terminology used throughout the treatise. We have bowed to the author's choice in all cases and have not attempted to impose any particular classification. This decision was made after lengthy discussion and consultation with taxonomists who felt that all groups and species could be readily identified in standard reference books; this seems to be the essential requirement for the physiologist.

Volume I deals with water and electrolyte balance, excretion, and some aspects of metabolism. Succeeding volumes will consider the endocrine system, reproduction, development, luminescence, chromatophores and venoms, the circulatory and respiratory systems, the nervous system and the sense organs, several aspects of fish behavior, and special adaptations to environmental change.

June, 1969   W. S. HOAR

D. J. RANDALL

CONTENTS

CONTENTS OF OTHER VOLUMES

1

THE BODY COMPARTMENTS AND THE
DISTRIBUTION OF ELECTROLYTES

W. N. HOLMES and EDWARD M. DONALDSON

I. INTRODUCTION

The cells contained within the integument of all but the simplest multicellular organisms are bathed in a relatively stable fluid medium. Basing his judgment on observations on the composition of blood, Claude Bernard perceived that the constancy of this medium, which he termed the *"milieu interieur,"* was an essential condition of free and independent life. Water is the solvent of this fluid medium and in it are dissolved numerous inorganic and organic solutes. The organic solutes are mainly nutrients and the products of metabolism while the inorganic solutes

1

consist of oxygen, carbon dioxide, and electrolytes occurring in ratios similar to those found in dilute seawater. From this medium the cells take up oxygen and nutrients and into it they discharge carbon dioxide and the products of metabolism.

Separating the extracellular environment from the internal environment of the cells is a highly organized functional boundary, the cell membrane. This membrane is approximately 75 Å thick and consists of a bimolecular lipoprotein layer which severely restricts the passive movement of solutes from the extracellular fluid into the cell. Again, water is the major solvent of the intracellular fluid but, although the number of solvent particles per unit volume of intracellular and extracellular fluid is roughly equal, the composition in terms of the relative abundance of particle species in each fluid is strikingly different. In the extracellular fluid the concentrations of Na^+, Cl^-, HCO_3^-, and Ca^{2+} are relatively high and, with the exception of blood plasma, the protein concentration is relatively low when compared to the intracellular fluid. In contrast, the concentrations of K^+, PO_4^{3-}, Mg^{2+} and protein in the intracellular fluid are relatively high when compared to the extracellular fluids. Since some of the constituents of both the extracellular and the intracellular fluids possess electrical charges and some of these charged particles are either selectively accumulated in one of the fluids or are nondiffusible, the intracellular fluid becomes negatively charged with respect to the extracellular fluid. Such a charge tends to aid or mitigate against the passive movement of charged particles into or out of the cell body.

There exist, therefore, regulatory mechanisms, presumed to be situated in the cell membrane, which are repsonsible for the maintenance of the differential distribution of solutes between the extracellular and intracellular fluids. These mechanisms involve the expenditure of energy in order to move particle species against the concentration and/or the electrochemical gradient. Thus any tendency to change the concentrations of solutes in these fluids either by passive diffusion, albeit often slow, or by changes in the metabolic state of the individual are counteracted.

By definition such mechanisms are termed "active transport" mechanisms. Probably the mechanism associated with the transport of Na^+ out of the cell and K^+ into the cell is the one best understood at this time. Energy derived from metabolism is used to sustain these ion fluxes and, on the average, these active movements of Na^+ and K^+ just balance the diffusion of Na^+ into and K^+ out of the cell. The potential which exists across the membrane is believed to be related to the fact that K^+ tends to permeate the membrane more rapidly than does Na^+.

Among the Vertebrata we find animals which have adapted to a wide range of ecological niches and even within the class Agnatha and the various classes of gnathostomatous fishes there exist forms which have adapted to freshwater, brackish water, and marine environments. Indeed, others such as certain members of the Salmonoidei and Anguilloidei have adopted life cycles involving sojourns in both freshwater and seawater. Others, such as the lungfishes, have even become adapted to the terrestrial habitat for at least part of their lives. In general, however,

Table I

The Approximate Steady State Ion Concentrations and Potentials Existing between the Muscle Cells and Their Interstitial Fluid in the Various Physiological and Environmental States of the Eel, *Anguilla anguilla*[a]

		Ion concentrations (mmole/liter)			
		Interstitial fluid (ion$_o$)	Intra-cellular fluid (ion$_i$)	$\dfrac{[\text{ion}_o]}{[\text{ion}_i]}$	ε_{ion}(mV)
Ion					
Parietal muscle					
Freshwater yellow eel	Na$^+$	143.2	20.3	7.0542	+47.1
	K$^+$	2.26	129	0.0175	−99.3
Freshwater silver eel	Na$^+$	150.1	26.7	5.6217	+42.3
	K$^+$	1.75	112.0	0.0156	−102.1
Seawater yellow eel	Na$^+$	164.2	18.9	8.6878	+53.0
	K$^+$	3.35	146.0	0.0229	−92.7
Seawater silver eel	Na$^+$	183.3	22.4	8.1830	+51.6
	K$^+$	3.22	143.0	0.0225	−93.1
Tongue muscle					
Freshwater yellow eel	Na$^+$	143.2	15.4	9.2987	+54.7
	K$^+$	2.26	122.0	0.1852	−97.9
Freshwater silver eel	Na$^+$	150.1	21.3	7.0469	+47.9
	K$^+$	1.75	120.0	0.0146	−103.7
Seawater yellow eel	Na$^+$	164.2	14.1	11.6454	+60.2
	K$^+$	3.35	159	0.0211	−94.7
Seawater silver eel	Na$^+$	183.3	22.6	8.1106	+51.4
	K$^+$	3.22	154	0.0209	−94.9

[a] The membrane potentials were calculated using the Nernst equation for univalent ions:

$$\varepsilon_{\text{ion}} = \frac{R \cdot T}{F \cdot Z} \log_e \frac{[\text{ion}_o]}{[\text{ion}_i]} \quad \text{V}$$

where R is the gas constant (8.31 joules/mole-deg absolute), T is the temperature (deg absolute), F is the Faraday constant (96,500 C/mole), and Z is the valency (1+). By converting to $\log_{10}$, expressing ε in millivolts, and assuming the temperature of the tissues to be the same as that of the environment (12°C), the equation becomes $\varepsilon_{\text{ion}} = 56.5 \log_{10}[\text{ion}_o]/[\text{ion}_i]$ mV. Calculated from Chan *et al.* (1967).

despite the range of ecological niches which have been occupied by the vertebrates, a remarkable constancy in the compositions of the extracellular and intracellular fluids exists. Furthermore, among the Agnatha and the gnathostomatous fishes, only the members of the order Myxiniformes and the class Chondrichthyes show exceptions to the generalized pattern of electrolyte distribution between the two major body compartments. An approximation of the steady state ion concentrations and the attendant potential differences expected to occur in fish may be visualized from an examination of Table I.

So far the concept of body compartments has only been dealt with in general terms, and we must now proceed to consider more rigorous definitions of these spaces.

II. THE TOTAL BODY VOLUME

The total body volume is of course self-evident, but as a physiological parameter in most vertebrates it is somewhat meaningless when the distribution and movement of solutes are under consideration. Large portions of the total body volume are occupied by structures having very low turnover rates of metabolites and solutes. The integument and skeleton of many fishes for instance cannot be considered to be solute pools having significant short-term exchanges of water and solutes with the surrounding tissues and fluids. For this reason, therefore, the total body water content is more usefully related to the body compartments since it is the common solute of, and is apportioned between, the major compartments and their subdivisions. Although water is passively distributed according to the disposition of the solutes, it delineates the compartments of organisms in which the metabolic reactions may take place.

A. The Intracellular Compartment

The intracellular compartment of any tissue, organ, or organism may be defined as the sum of the cellular volumes contained within the limits of the cell membrane. This is of course an oversimplification of the actual state of affairs obtaining within each individual cell. The cells are structurally extremely complex and contain structures such as the nucleus, the nucleoli, the mitochondria, the endoplasmic reticulum, and so on.

Some of the organelles are themselves bounded by membranes, e.g., the nucleus and the mitochondria, and consequently they represent compartments within the cell body itself. Many of these membranes almost certainly possess specific active transport properties, and substances appear to be selectively accumulated within these organelles.

Probably, the most intensely studied organelle from this standpoint is the mitochondrion. Isolated mitochondria have been observed to actively take up K^+ from the surrounding medium. Respiratory substrates are necessary for this process, but in contrast to the Na^+ and K^+ active transport mechanisms in the plasma membrane, it is not inhibited by the presence of the cardiac glycoside, ouabain. Thus, the system can be clearly differentiated from that occurring in the cell membrane. Also isolated kidney mitochondria have been observed to show a 50-fold increase in their Ca^{2+} content during respiration *in vitro* (Vasington and Murphy, 1962). This active uptake of Ca^{2+} into the mitochondrion *in vitro* is dependent upon the presence of Mg^{2+}, inorganic phosphate, and adenosine triphosphate (ATP) in the medium. Furthermore, the influx of inorganic phosphate approximates the simultaneous influx of Ca^{2+}. The quantities of inorganic phosphate and Ca^{2+} entering the mitochondrion *in vitro*, however, far exceeds the solubility of the possible calcium salts (Lehninger *et al.*, 1963). These authors concluded that it would be necessary for at least one salt, possibly hydroxyapatite, to precipitate within the mitochondrion; such areas of precipitation may be associated with the dense osmophilic granules within the mitochondria (see Lehninger, 1965). The *in vivo* accumulation of Ca^{2+} within the mitochondrion is by no means so striking and indeed appears to be considerably less than that observed *in vitro*.

Surprisingly little consideration, however, has been given to the possible physiological implications of intramitochondrial accumulation of ions. Several years ago it was postulated that the mitochondria evolved from bacteria which had originally parasitized and ultimately become symbiotic within the aerobic cell. Such speculation considerably complicates our concept of the intracellular compartment. Nevertheless, the role of the mitochondrial membrane appears to be, at least in part, associated with maintaining a constant *intramitochondrial milieu* in which the enzyme systems may function. Further the active transport mechanisms in the membrane of the mitochondrion may function to regulate the solute composition of the hyaloplasm. Such a mechanism would of course only achieve a temporary sequestration of ions within the mitochondrion, but it may serve to temporarily regulate any localized concentrations of solutes. The existing membranes of the bacteria and their associated active transport mechanisms could well have been adapted to these purposes.

The foregoing serves to illustrate the complexity of the intracellular compartment in even the simplest cell type. Much of the experimental data concerning the role of the mitochondrion within the cell has been derived from mammalian tissues, but there is no reason to believe that the data derived from these tissues do not equally apply to the tissues of fishes. Indeed, the mammalian studies do, in fact, indicate the directions of future studies in fishes. For a precise and elegant survey of the role of the mitochondrion in cellular regulation the reader is referred to Lehninger (1965).

It is clear, therefore, that one must recognize the presence of islands, rich in specific ions and cell solutes, which occur within the intracellular compartment. Further, the compartment is not in reality a homogeneous solution within the aqueous phase of the cell. Even so, it is convenient and meaningful to consider the intracellular compartment as a single aqueous phase when discussing the movement of solutes, particularly electrolytes, between the cell and its surrounding medium.

B. The Extracellular Compartment

The extracellular compartment is that space which exists outside the plasma membranes of the cells, and it contains the fluid and the inclusions surrounding these cells. Anatomically the extracellular compartment can be divided into several subcompartments.

One group of extracellular spaces is anatomically characterized by the presence of a continuous layer of epithelial cells separating it from the remainder of the extracellular compartment. These spaces are collectively termed the "transcellular space." Fluids passing through the epithelial cell boundary into the transcellular compartment are invariably modified. In vertebrates generally they include the gastrointestinal and biliary secretions, the cerebrospinal, intraoccular, pericardial, peritoneal, synovial, and pleural fluids; the luminal fluid of the thyroid gland; the cochlear endolymph, the secretions of the sweat and other glands and the contents of the renal tubules and urinary tract.

The remainder of the extracellular compartment is composed of the *intravascular fluid* or blood plasma, the *interstitial fluid* and the lymph.

The intravascular fluid is circulated through a closed system consisting of arteries and veins which are connected by the capillary network. The capillary network is a system of narrow vessels, 10–20 μ in diameter, with walls composed of a single layer of flattened endothelial cells. During the passage of blood through the capillary network a certain "leakage" of water and plasma solutes take place. This net fluid movement out of the capillaries occurs, according to the Starling hypothesis, as the resultant of the forces of filtration and the forces of absorption along the

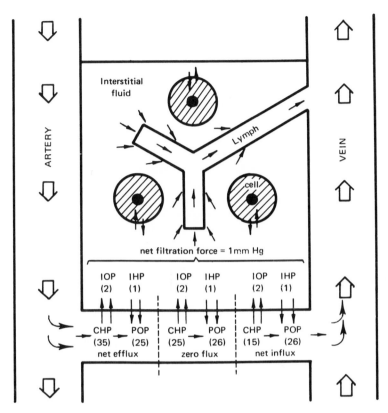

Fig. 1. A diagrammatic representation of the transcapillary movement of ultra-filtrate into the interstitial space and the return of this filtrate via the lymphatic system to the intravascular space. CHP = capillary hydrostatic pressure, IHP = interstitial hydrostatic pressure, POP = plasma oncotic pressure, and IOP = interstitial oncotic pressure. All values in parentheses present hypothetical pressures in mm Hg.

inside and outside of the capillaries. The forces tending to filter fluid out of the capillaries are greater at the arterial end of the capillary and consist of the capillary hydrostatic pressure (CHP) and the oncotic pressure of the interstitial fluid (IOP). Conversely, the forces tending to absorb fluid from the interstitial space are greater at the distal or venous end of the capillary and they include the hydrostatic pressure of the interstitial fluid (IHP) and the oncotic pressure of the plasma (POP). Thus, the net force of filtration $(+ve)$ or absorption $(-ve)$ at any point along the capillary $= (CHP + IOP) - (IHP + POP)$. Values for the forces of filtration and absorption in fish have not been determined, but typical values taken from the mammalian literature are shown in Fig. 1.

Clearly, a net filtration of fluid out of the capillary system into the

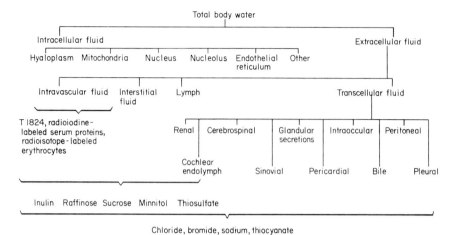

Fig. 2. A summary of the major body compartments in the vertebrates. The approximate volumes of distribution of some of the more common indicator substances are bracketed.

interstitial space occurs during the passage of blood through the capillary network. It is this continuous transcapillary efflux of water and solutes out of the blood which constitutes the intersitial fluid.

By virtue of the limited permeability of the capillary wall, much of the protein remains in the capillary and an ultrafiltrate of plasma emerges into the interstitial space. This phenomenon is illustrated by the progressively increasing oncotic pressure of the capillary blood as it passes from the arterial end to the venous end of the capillary loop. Further, the progressive reduction in plasma volume as the filtration continues along the length of the capillary results in a progressive decrease in the capillary hydrostatic pressure.

The existence of drainage channels, which start as blind tubules surrounded by interstitial fluid, ensure that the interstitial fluid does not remain stagnant. These endothelial tubules collect into larger vessels, known as "lymphatic ducts," which finally empty into a large vein after passing through the *lymph nodes*. In the lymph nodes the fluid is filtered through a trabeculum of cells, and lymphocytes are added before it rejoins the venous circulation. This system of tubules, beginning with the blind ducts and ending in the large trunk vessels, is known as the "lymphatic system" and the fluid contained in the system is known as "lymph." Since it is derived from interstitial fluid, lymph may also be considered to be an ultrafiltrate of plasma. However, although the inorganic constituents occur at approximately the same concentrations as those found in plasma, the protein composition is quite variable and even the largest molecules such as fibrinogen may be present in the lymph at some point in the system.

A summary and classification of these body compartments are illustrated in Fig. 2.

III. METHODS FOR THE DETERMINATION
OF BODY COMPARTMENTS

With the exception of total body water, the compartmental volumes of an individual cannot be measured directly, and therefore indirect methods have been devised. These methods often involve the application of the dilution principle to some nontoxic indicator which is rapidly and homogeneously distributed throughout the compartment to be measured. The volume in which a known amount of this substance is homogeneously distributed may then be calculated after the concentration of the substance in the compartmental fluid has been determined. The general equation for this relationship is as follows:

$$V = Q/C \tag{1}$$

where V is the volume of distribution, Q is the quantity administered, and C is the concentration of the indicator in the compartmental fluid. Should this solute leave the compartment as a result of excretion or metabolism, then a correction must be applied for this loss. The amount lost from the compartment is subtracted from the amount administered and the general formula becomes:

$$V = (Q - E)/C \tag{2}$$

where E is the quantity excreted. The volume of distribution is therefore defined as the volume which would be necessary to accommodate all the indicator substances in the body at a specific time if it were distributed in the compartment at the observed plasma concentration of the substance at that time.

Two principal dilution methods exist for the determination of compartmental volumes. These are the *infusion–equilibrium method* and the *kinetic method*. The infusion–equilibrium method is most suited to use with indicator substances which are rapidly excreted from the organism, e.g., inulin. A priming dose, sufficient to saturate the tissues, is administered intravenously, and the substance is then slowly infused until a constant plasma concentration is achieved. At this point the infusion is stopped and excretory products are continuously collected until the plasma concentration of the indicator has declined to zero. The amount of indicator collected in the excretory products represents the amount which was present in the body (Q) at the time the infusion was stopped. By dividing this quantity (Q) by the plasma concentration at the end of infusion (C) the volume of distribution of the indicator is obtained [Eq. (1)].

This method has serious drawbacks to its application in studies on fish. First, since the excretion of the indicator may extend over a considerable period, it is necessary to retain an intravenous cannula throughout the collection period. Such techniques are difficult in fish, particularly if the individuals are maintained in a free-swimming state. Second, many species of fish possess both renal and extrarenal pathways of excretion, and therefore both pathways must be monitored to determine the amount of indicator excreted. Third, in instances where the indicator is an inorganic substance such as chloride, bromide, deuterium oxide, or tritium oxide, the possibility exists that these substances may, after having been excreted by the kidney, reenter the circulation from the environment via the extrarenal uptake mechanisms in the gill epithelium. Therefore, cognizance must be taken of these possible sources of error when volumes of distribution in fish are determined according to the infusion–equilibrium method.

In the kinetic method a series of plasma samples are analyzed following the intravenous administration of a single dose of indicator substance. The log concentration of the indicator substance is then plotted against time. Initially the semilogarithmic decline in the plasma concentration of indicator substance is nonlinear, but thereafter it is linear. By extrapolation of the linear portion of the curve to zero time a theoretical value for the plasma concentration at this time is obtained. This value represents what the equilibrium concentration of indicator substance in the plasma would have been if instantaneous distribution had occurred following the injection. The volume of distribution is then obtained by dividing the amount of indicator administered (Q) by the zero-time equilibrium concentration (C) [Eq. (1)]. The volume of distribution may also be determined at any other time after injection by measuring the plasma concentration at this time and the amount of indicator substance excreted up to this time. These values, together with the amount of indicator initially injected, are then substituted in Eq. (2). Again, when the latter technique is applied to fish, the excretion via both the renal and extrarenal pathways must be monitored and consideration, in some instances, must be given to the possible reentry of excreted indicator into the fish.

We will now consider the methods available to the physiologist for the determination of specific compartmental volumes.

A. Total Body Water

The total body water content of any animal is easily and most accurately determined by a comparison of the wet and dry body weight of the animal following desiccation to constant weight. The terminal nature of this method does, of course, limit its usefulness.

Other less drastic and indirect techniques have therefore been devised; but in each case their accuracy must be determined by simultaneous comparison with the desiccation method.

Each of these methods involves the application of the dilution principle to indicator substances which become rapidly and homogeneously distributed throughout the total body water, including the transcellular spaces. The chemicals commonly used for this purpose include urea, thiourea, sulfanilamide, antipyrine, 4-acetyl-4-aminoantipyrine (NAAP), deuterium oxide, and tritium oxide. It is assumed that these chemicals become evenly distributed throughout the body, and the accuracy of each method is largely determined by the extent to which this assumption is true.

The use of urea is limited owing to the endogenous production of this substance, and thiourea has been shown to distribute unevenly in the body water (Winkler et al., 1943). Sulfanilamide, at least in some mammals, may become conjugated in the liver although values derived from the use of this substance in the dog are in good agreement with desiccation values. Antipyrine is rapidly distributed throughout the body water, but it is also rapidly metabolized and excreted in the urine (Brodie, 1951). Although the metabolism and excretion of antipyrine appears to occur at a uniform rate (Soberman, 1950) and the appropriate corrections may be applied [Eq. (2)], the use of the compound is probably more appropriate in acute experiments. However, the use of NAAP would seem to be more suitable under experimental and environmental circumstances where the rates of urine flow in teleosts are likely to vary (Holmes and McBean, 1963; Holmes and Stainer, 1966). This compound is only slowly metabolized by the tissues and is excreted extremely slowly. When deuterium oxide and tritium oxide are used to determine the total body water content of fish maintained in a closed environment, the possible reentry of the excreted compounds into the body must be examined. Studies on mammals also indicate that approximately 5% of the labile hydrogen atoms from both compounds will exchange with unlabeled hydrogen atoms of substances other than those in the body water (Elkinton and Danowski, 1955).

B. The Extracellular Volume

Since whole blood is itself a tissue consisting of the blood cells suspended in their extracellular environment the total *intravascular volume* may be determined by establishing (1) the total blood cell volume or (2) the plasma volume. In each case an estimate of the total intravascular volume may be obtained as follows if a simultaneous measurement of the hematocrit is made:

$$\text{Intravascular volume} = \text{plasma volume} \times \frac{100}{100 - \text{hematocrit}} \qquad (3)$$

$$\text{Intravascular volume} = \frac{\text{cell volume}}{\text{hematocrit} \times 100} \qquad (4)$$

Plasma volume may be determined according to the kinetic dilution method after the intravenous administration of known amounts of radio-iodinated serum albumin (RISA), or T 1824 (Evans blue) which binds to the plasma proteins. Since serum albumin tends to leak out of the vascular space during passage through the capillaries, the estimates of plasma volume using RISA or T 1824 tend to be high. By using a larger protein molecule exclusively as the indicator, such as radioiodine-labeled fibrinogen, the transcapillary leakage is minimized and the plasma volume values obtained are some 2–12% lower than those obtained with albumin and other small protein molecules.

The alternative method of determining total intravascular volume depends upon the selective accumulation of certain compounds within the blood cells, principally the erythrocytes. When isolated erythrocytes are incubated in the presence of radioisotopes such as ^{52}Fe, ^{55}Fe, ^{32}P, ^{42}K and thorium B, the radioisotopes penetrate the cells and become bound to the hemoglobin or some other protein within the cell. The more firmly bound the isotope becomes, the lower the rate of loss of the radioactive indicator from the cells after it has been injected into the animal. In this regard ^{51}Cr, which becomes firmly bound to the globin portion of the hemoglobin molecule, shows the lowest rate of loss from the system. No appreciable loss occurs during the first 24 hr after injection; thus, only one blood sample is necessary to establish the dilution (Sterling and Gray, 1950; Gray and Sterling, 1950). On the other hand, ^{32}P is less firmly bound and up to 6% per hour may be lost from the circulating cells. Many blood samples must therefore be taken in order to establish the dilution curve for ^{32}P labeled cells (Gregersen and Rawson, 1959). A unique technique for the estimation of blood cell volume has recently been developed in the Pacific hagfish, *Eptatretus stoutii*, by McCarthy (1967). This method involves labeling the red blood cells of the hagfish with L-methionine-methyl-^{14}C. Blood from donor animals is incubated *in vitro* in the presence of the L-methionine-methyl-^{14}C in a rotary incubator for 2 hr at 10°C. At the end of the incubation period the blood–isotope mixture is gently and rapidly centrifuged, the plasma supernatant is removed, and the cells are resuspended three times in cold physiological saline. The cells are finally suspended in a small amount of physiological saline and intravenously injected into the experimental animals. Approximately 3% per hour of the zero-hour blood concentration of isotope is lost from the hagfish circulation.

When plasma volumes or total intravascular volumes are calculated from dilution studies using labeled red blood cells, the values tend to be lower than those obtained by labeled plasma dilution techniques. This discrepancy arises from two sources. First, the red blood cells are not homogeneously distributed throughout the intravascular compartment, and therefore the hematocrit is not constant for all blood vessels. This is particularly true of the hagfish (Johansen *et al.*, 1962). Second, a certain amount of plasma is always trapped between the blood cells when they are separated from the plasma (Gregersen and Rawson, 1959). The most accurate estimation of intravascular volume therefore is obtained by simultaneous measurement of plasma and cell volumes (Armin *et al.*, 1952).

The distribution of several substances is purported to measure the combined intravascular, interstitial, and lymphatic spaces. These substances include inulin, raffinose, sucrose, mannitol, and thiosulfate, and the value which each substance gives increases in that order. The molecular weights of these substances, which are in a reverse order, largely determine the rates at which they diffuse throughout the various extracellular subcompartments, particularly the connective tissue and the transcellular spaces. For this reason the volume of distribution of the compound is often cited rather than the specific volume it is purported to measure. Inulin and sucrose appear to show two phases of distribution: A rapidly equilibrating phase is followed by a second phase where it is believed that these substances become selectively accumulated in the macrophages. These substances are not, therefore, distributed homogeneously at this point and consequently erroneously high values may be obtained. Mannitol does not have this characteristic and is a more reliable index in chronic experiments.

An estimation of the volumes of distribution of body Na$^+$ and Cl$^-$ may also be used to determine extracellular volume. Neither Na$^+$ nor Cl$^-$ is homogeneously distributed throughout the extracellular compartment of any vertebrate. They are preferentially distributed between certain transcellular spaces, e.g., gastric secretion and renal tubular fluid; and some, albeit small, amounts of both ions occur within the intracellular compartment. Nevertheless, surprisingly accurate estimations of extracellular volume may be made if corrections are applied for the heterogeneous distribution of the ions. Indeed the methods are particularly useful when applied to tissue samples such as skeletal muscle. A refined method for calculating the Cl⁻ space of a tissue or organism was described by Manery (1954) as follows:

$$\text{Cl}^- \text{ space } (\text{H}_2\text{O})_{E^{\text{Cl-}}} = \frac{\text{Cl}_t^- \times r_{\text{Cl}-} \times \text{H}_2\text{O}_p}{\text{Cl}_p^-} \quad \text{g water/kg wet weight} \quad (5)$$

where Cl_t^- = tissue Cl⁻ concentration in millimoles per kilogram wet weight tissue,

H_2O_p = plasma water content in milliliter per kilogram wet weight plasma,

Cl_p^- = plasma Cl⁻ concentration in millimoles/liter plasma, and

$r_{\text{Cl}-}$ = Gibbs-Donnan ratio for Cl⁻.

The following analogous equation may also be derived for calculating the Na⁺ space of an organism or tissue:

$$\text{Na}^+ \text{ space } (\text{H}_2\text{O})_{E^{\text{Na+}}} = \frac{\text{Na}_t^+ \times \text{H}_2\text{O}_p}{\text{Na}_p^+ \times r_{\text{Na}^+}} \quad \text{g water/kg tissue} \quad (6)$$

where Na_t^+ = tissue Na⁺ concentration in millimoles per kilogram wet weight tissue,

H_2O_p = plasma water content in milliliter per kilogram wet weight plasma,

Na_p^+ = plasma Na⁺ concentration in millimoles per liter plasma, and

r_{Na^+} = Gibbs-Donnan ratio for Na⁺.

The Gibbs-Donnan ratios for Na⁺ and Cl⁻ used in Eqs. (5) and (6) have not been estimated for lower vertebrate tissues. Therefore, the values obtained from mammalian studies must be applied as approximations and these values are $r_{\text{Na}^+} = 0.942$ and $r_{\text{Cl}-} = 0.977$. For a critical evaluation of the methods used in the determination of extracellular volume from the distributions of Na⁺ and Cl⁻, the reader is referred to reviews by Manery (1954) and Cotlove and Hogben (1962).

C. The Intracellular Volume

The intracellular volume cannot be measured directly. As pointed out above, it is an extremely complex and certainly not a homogeneous compartment of the body. An estimate of intracellular volume may be estimated by subtracting simultaneous values for the total body water and the extracellular volume, but this estimate is clearly dependent upon the accuracy of the methods used for the determination of body water and extracellular space.

A summary of the classification of the body compartments is included in Fig. 2, and the range of distribution of some of the indicator substances is outlined.

IV. COMPARTMENTAL SPACES IN FISH

A. Class Agnatha

ORDER MYXINIFORMES AND ORDER PETROMYZONTIFORMES

Until 1959 only one value for a compartmental volume in the Agnatha appeared in the literature. This value was the blood volume determined in a single specimen of the sea lamprey, *Petromyzon marinus*, by Welcker in 1858. The animal taken from the sea had a body weight of 1094 g, and the blood volume was reported to be 4.16% of the body weight. This value is extremely low when compared to the subsequent data obtained for the freshwater form of this species by Thorson (1959). Although Thorson's data (Table II) are undoubtedly more reliable, the discrepancy is nevertheless large. According to Thorson (1959) the difference probably reflects an inverse relationship between relative blood volume and body size similar to that which has been demonstrated in the elasmobranchs (Martin, 1950). The possibility also exists that there may be a difference in the distribution of body compartments between the freshwater and marine forms of this species. The blood volume of the Pacific hagfish, however, is even higher than that of the freshwater form of *Petromyzon marinus* (Table II).

The relative intravascular or blood volume of the Pacific hagfish, *Eptatretus stoutii*, is the highest reported for any vertebrate species. This high value is because of both a high plasma volume and a high red blood cell volume in this species when compared to *Petromyzon marinus* or other groups of fishes. Furthermore, the total extracellular volume of *Eptretatus stoutii* is similar to that of *Petromyzon marinus* (Table II), and the increase in intravascular volume appears to have occurred at the expense of the interstitial space (McCarthy and Conte, 1966; McCarthy, 1967).

Table II

Summary of the Available Data on Compartmental Volumes in Two Species of Cyclostomes[a]

Parameter	Pacific hagfish, *Eptatretus stoutii*, (McCarthy, 1967) (ml/100 g body wt)	Method of determination	Sea lamprey, *Petromyzon marinus*, (Thorson, 1959) (g/100 g body wt)	Method of determination
Total body water	74.6 ± 3.4 (5)	Desiccation	75.6 ± 0.15 (12)	Desiccation
Extracellular water	25.9 ± 5.1 (5)	Inulin-carboxyl-^{14}C	23.9 ± 0.23 (12)	Sucrose space
Interstitial water	10.3 ± 1.6 (5)	Inulin space minus T 1824 space	18.4 ± 0.19 (12)	Sucrose space minus T 1824 space
Intravascular volume	16.9 ± 2.1 (5)	T 1824 space plus red blood cell volume	8.5 ± 0.12 (12)	T 1824 space and hematocrit
Plasma volume	13.8 ± 3.1 (10)	T 1824 space	5.5 ± 0.09 (12)	T 1824 space
Red blood cell volume	4.9 ± 2.0 (5)	L-Methionine-methyl-^{14}C space	3.0 (12)	Intravascular space minus plasma volume
Intracellular volume	48.7 ± 1.7 (5)	Intravascular volume minus inulin space	51.7 ± 0.26 (12)	Total body water minus sucrose space

[a] The values taken from Thorson (1959) were reported by the author in g/100 g body weight. These values may be converted to ml/100 g body weight by using the reported specific gravity values of 1.018 and 1.040 for plasma and blood, respectively. The numbers in parentheses indicate the number of individual fish used for the determination. All means ± S.E.

B. Class Chondrichthyes

SUBCLASS ELASMOBRANCHII

An earlier set of data reporting the blood volumes of elasmobranchs (Table III; Martin, 1950) showed somewhat lower mean values than those which were later reported by Thorson (1958) (Table IV). The lower values found by Martin may, however, be attributable to the considerably larger fish used in his study. A single specimen of *Raja binoculata*, which has not been included in the synopsis of his data (Table III), had a body weight of 4750 g, which was within the body

Table III

Summary of the Data on Blood Volumes in Chondrichthian Fishes[a,b]

Species		Body wt (g)	Hematocrit (% cells)	Blood vol (% body wt)	p value
Chondrichthyes					
Ratfish		1225 ± 53	18.0 ± 0	2.6 ± .49	NS[c]
Chimaera colliei					
Dogfish		2168 ± 816	18.0 ± 2.5	9.0 ± 2.3	<0.02
Squalus sucklii					
Skate		3225 ± 751	18.5 ± 1.8	5.0 ± .66	<0.02
Raja rhina					
Skate	♂	1468 ± 129	18.3 ± 1.9	4.1 ± .35	<0.01
Raja binoculata	♀	3334 ± 337	17.1 ± 1.1	3.4 ± .23	NS[c]
				3.7 ± .21	<0.01
Osteichthyes					
Lingcod		4435 ± 896	35 ± 4.7	2.8 ± .22	—
Ophiodon elongatus					

[a] From Martin (1950).

[b] These values are compared to the blood volumes of an osteichthyan, *Ophiodon elongatus*. (All values are means ± S.E.; p values represent the significance with respect to the value for *Ophiodon elongatus*.)

[c] NS indicates not significant.

weight range of this species studied by Thorson. The blood volume for this individual was 7.3%; a value close to that reported by Thorson for this species (8.0%). Furthermore, Martin (1958) found that the best fit for his data from the heavier skate was obtained when the blood weights were plotted against the 0.75 exponent of the body weight. This correction would bring the mean blood weight values of the larger skate closer to those found in the smaller individuals by Thorson (cf. Tables III and IV).

Table IV

A Summary of the Body Compartmental Volumes in Chondrichthian Fishes[a-c]

	Marine species						Freshwater species
	Long-nosed skate, *Raja rhina*	Big skate, *Raja binoculata*	Dogfish, *Squalus acanthias*	Lemon shark, *Negaprion brevirostris*	Nurse shark, *Ginglymostoma cirratum*	Ratfish, *Hydrolagus colliei*	Lake shark, *Carchashinus nicaraguensis*
Body weight (g)	4933 (12) (1400–16550)	8184 (4) (2646–18100)	2631 (33) (1120–6350)	6400 (9) (3180–12270)	1603 (5) (1136–2270)	1067 (16) (520–1573)	48070 (10) (27700–57200)
Hematocrit (% cells)	16.8 (11) (12–21)	17.5 (4) (15–20)	18.2 (25) ±0.55	21.5 (9) ±0.37	17.4 (5) ±1.3	20.2 (12) (15–25)	22.8 (10) ±1.4
Pulse (beats/min)	11.1 (10) (6–18)	16.0 (4) (12–22)	31.0 (14) (18–40)	26 (9) (20–32)	22 (5) (16–24)	31.8 (11) (18–42)	12.2 (10) (8–18)
Plasma volume (T 1824 space)	5.9 (8) ±0.53	6.5 (4) ±0.60	5.5 (24) ±0.29	5.4 (9) ±0.11	5.7 (5) ±0.27	4.2 (8) ±0.32	5.1 (10) ±0.22
Blood volume (T 1824 space and hematocrit)	7.2 (8) (4–9.5)	8.0 (4) (6.5–9.9)	6.8 (24) ±0.37	7.0 (9) ±0.14	6.8 (5) ±0.33	5.2 (8) (4.1–7.4)	6.8 (10) ±0.35
Extracellular fluid (inulin)	11.8 (8) ±0.67	13.2 (2) ±1.6	12.7 (13) ±0.35	—	—	10.6 (8) ±0.25	—
Extracellular fluid (sucrose)	—	—	21.2 (3) ±1.4	21.2 (8) ±0.47	21.9 (4) ±1.2	—	19.7 (8) ±0.54
Interstitial fluid (inulin or sucrose space minus T 1824 space)	5.5 (6) (4–8.6)	7.9 (2) (5.6–10.2)	15.7 (3)	15.8 (8)	16.2 (4)	6.7 (6) (5.9–7.7)	14.6 (8)
Intracellular fluid (total water minus extracellular space)	68.3	67.5	50.5	49.9	49.8	59.7	52.4
Total body water (desiccation)	82.0 (3) ±0.41	82.6 (13) ±0.40	71.7 (16) ±0.48	71.1 (6) ±0.14	71.7 (3) ±0.22	71.4 (13) ±0.47	72.1 (4) ±0.37

[a] From Thorson (1958).
[b] All volumes are expressed as percentages of mean body weight.
[c] The range of measurements or the S.E. of the mean is given under each mean value. The number of individual fish used for each determination is given in parentheses after the mean value.

A comparison of the blood volumes of several elasmobranch species with the blood volume of a single osteichthyan species, the lingcod (*Ophiodon elongatus*), also showed that the blood volumes of the elasmobranchs were significantly greater than those of the lingcod (Martin, 1960) (Table III). This trend has been further documented by Thorson (1958, 1961, 1962; Tables VIII and IX).

As in the case of many vertebrate species, the values obtained for the extracellular volume of the elasmobranchs depend upon the chemical indicator used for the determinations. The degree of penetration, and the resulting volume of distribution, is inversely related to the molecular weight of inulin, raffinose, and sucrose (Table V). On the other hand,

Table V

Comparison of Inulin, Raffinose, and Sucrose Spaces of *Squalus acanthias*[a]

Indicator	No. of specimens	Av body weight (g)	Volume (% body weight)	Range
Inulin	13	2818	12.7	11.4–14.4
Raffinose	2	2605	15.2	15.1–15.4
Sucrose	3	2183	21.2	18.5–24.3

[a] Data from Thorson (1958).

sodium thiocyanate appears to be readily excreted via both the renal and extrarenal pathways and consequently, when no correction is made, gives inordinately low values for the volume of distribution (Thorson, 1958).

Owing to the different degrees of penetration observed for inulin and sucrose, the calculated interstitial fluid spaces (inulin or sucrose space minus plasma space) of the elasmobranchs tend to be lower or higher depending upon whether the volume of distribution of inulin or sucrose is used in the calculation (Table IV). Conversely, the estimates of intracellular volume (total body water minus inulin or sucrose space) tend to be higher or lower for the same reason (Table IV).

The compartmental spaces of only one freshwater elasmobranch have been measured. This species was the Lake Nicaragua shark, *Carcharhinus nicaraguensis*, and none of the compartmental spaces, including total body water, differed significantly from the corresponding values in a variety of marine species (Table IV).

A limited amount of data is available on the transcellular spaces in elasmobranchs (Table VI). The relatively large volumes of peritoneal and cerebrospinal fluid in the long-nosed skate, *Raja rhina*, probably is reflected in the high values for total body water of this species (cf. Tables IV and VI). Since neither sucrose nor inulin was observed to penetrate

Table VI

The Fluid Volumes of Some Transcellular Spaces
in Three Species of Chondrichthian Fishes[a,b]

Indicator	Hydrolagus colliei	Raja rhina	Squalus acanthias
Peritoneal fluid	Trace (7)	1.0 (6) (0.52–1.48)	0.33 (6) (0.11–0.51)
Cerebrospinal fluid	0.25 (7) (0.17–0.40)	0.77 (6) (0.58–1.0)	0.40 (5) (0.32–0.52)
Ocular fluid	0.85 (5) (0.78–0.99)	0.17 (3) (0.14–0.20)	0.38 (5) (0.30–0.57)

[a] From Thorson (1958).
[b] The values in parentheses below the mean values indicate the ranges, and the numerals in parentheses after the mean values indicate the number of fishes used for the determination.

the transcellular spaces measured, these high values for some of the transcellular spaces are therefore not reflected in the estimates of extracellular volumes (Thorson, 1958).

C. Class Osteichthyes

1. CLASSES SARCOPTERYGII, BRACHIOPTERYGII, AND ACTINOPTERYGII

The plasma and blood volumes of osteichthyan fishes have long been known to be low when compared to those of mammalian species. Until 1961, however, the available data were rather scant; indeed no data were recorded in the literature between 1858 and 1934. Most of the values reported by various investigators up to 1964 are recorded in Table VII. A more detailed taxonomic analysis of the blood volumes in osteichthyan fishes, however, was published by Thorson in 1961, and these data are summarized in Table VIII. The more primitive freshwater chondrostean and holostean fishes tend to have somewhat higher plasma and blood volumes than the more advanced teleostean fishes from freshwater. Comparison of the mean values from most of the chondrostean and holostean species examined are in fact statistically higher than the corresponding values from the individual freshwater teleostean species. Furthermore, when the values from the plasma volumes from all the species of chondrostean fishes are pooled, the mean value is significantly greater than the pooled value for the freshwater teleost species (Table IX, $p < 0.001$). This relationship does not hold, however, when the pooled mean value for the holostean species is compared to the mean pooled value for the freshwater teleost species. Also, a comparison of the plasma volumes of the freshwater and marine teleost species does not reveal any significant difference (Table IX).

Values obtained for the measurement of extracellular fluid volumes in the Osteichthyes vary according to the degree of penetration of the indicator substance, the usual inverse relationship being found between the molecular weight of the indicator and the recorded volume of distribution (Table X). Again the extracellular compartment, as indicated by the sucrose space, tends to be smaller in the more advanced teleosts.

The mean pooled extracellular volumes of both the chondrostean and the holostean species are significantly greater than that of the freshwater teleost species (Table IX, $p < 0.001$ and < 0.05, respectively). Also, the marine teleosts tend to have slightly greater extracellular volumes than do the freshwater teleosts (Table IX, $p < 0.05$).

The total body water composition of the chondrostean fishes is significantly higher than that of the freshwater teleost species (Table IX, $p < 0.02$), but there appears to be no significant difference between the body water content of the holosteans, freshwater teleosts, or the marine teleosts (Table IX). As a consequence of the extracellular space and the total water composition, the intracellular volumes of the more primitive members of the class Osteichthyes tend to be lower than intracellular volumes estimated for the teleost species (Table IX).

2. Blood Volume Changes Associated with the Evolution of the Fishes

Among the classes Agnatha, Chondrichthyes, and Osteichthyes there appears to be a correlation between the blood volume and the degree of primitiveness of the individual fish (Thorson, 1961), the trend being toward lower red blood cell (RBC) and plasma volumes in the Chondrichthyes and Osteichthyes. Although large RBC and plasma volumes may be, in general, considered to be primitive characteristics among the aquatic vertebrates, the reasons why this should be so are not immediately apparent. One is tempted to suggest that the circulatory system in the primitive fish may be less efficient. Unfortunately, there are no comparative data on the cardiac outputs of fish to substantiate this hypothesis. Indeed the available data on pulse and respiration rates of the agnathans, chondrichthyans, and osteichthyans do not seem to show any phlogenetic trend (Thorson, 1961; Table VIII). An examination of the cyclostome hemoglobins, however, suggests that they may represent the early stages in the phylogeny of oxygen transport. In *Lampetra fluviatilis* the hemoglobin molecule contains a single peptide chain having an amino acid

Table VII

A Summary of the Data on Blood Volumes Presented by Authors Other than Thorson (1961)[a]

Species[b]	No. of fish	Body weight (g)	Blood vol (ml)	Blood wt (g)	Blood wt (% body wt)	Reference and method
Cyprinus tinea	1	269.5	4.81	5.04	1.87	Welcker (1858) (bleeding)
Perch	1	122.7	1.26	1.32	1.07	Welcker (1858) (bleeding)
Perca fluviatilis						
Perch	1	98.2	1.26	1.32	1.34	Welcker (1858) (bleeding)
Perca fluviatilis						
Tautog	3	—	—	—	1.5	Derrickson and Amberson (1934) (bleeding)
Tautoga onitis						
Bullhead	6	171.4 ± 34.7	2.16 ± 0.33	—	1.76 ± 0.26	Prosser and Weinstein (1950) (T 1824 and hematocrit)
Ameiurus natalis						
Lingcod	8	4435 ± 896	202 ± 50	211 ± 52	2.8 ± 0.22	Martin (1950) (T 1824 and hematocrit)
Ophiodon elongatus						
Rock fish	1	2150	57	60	2.8	Martin (1950) (Vital red and hematocrit)
Sebastodes sp.						
Sculpin	3	4020 ± 189	89.3 ± 10.2	93.7 ± 10.2	2.3 ± 0.19	Martin (1950) (Vital red and hematocrit)
Cottidae sp.						
Goldfish	—	—	—	—	2.5-3.0	Korzhuyev and Nikolskaya (1951) (Hemoglobin washout)
Carassius carassius						
Common sole	—	—	—	—	4.0-5.9	(Hemoglobin washout)
Solea vulgaris						

Species	n					Reference
Common sucker *Catostomus commersoni*	23	200–999	—	—	1.5	Lennon (1954) (bleeding)
Rainbow trout—FW *Salmo gairdneri*	10	8–29	—	—	2.25	Schiffman and Fromm (1959) (T 1824 and hematocrit)
Rainbow trout—FW	10	—	—	—	3.5 ± 0.9	Conte et al. (1963) (T 1824 and hematocrit)
Rainbow trout—FW	5	—	—	—	3.3 ± 0.9	Albumin-^{131}I and hematocrit
Rainbow trout—FW	9	—	—	—	2.9 ± 0.8	Simultaneous T 1824 and albumin-^{131}I and hematocrit
Rainbow trout—FW	6	—	—	—	2.8 ± 1.0	T 1824 or albumin-^{131}I and ^{51}Cr-labeled red blood cell
Steelhead trout—SW *Salmo gairdneri*	13	553 ± 127	—	—	6.9 ± 1.8	Smith (1966) (T 1824 and hematocrit)
Rainbow trout—FW *Salmo gairdneri*	4	548 ± 49	—	—	2.4 ± 0.4	Smith (1966) (bleeding)
Sockeye salmon—FW *Oncorhynchus nerka*	8	1814	—	—	4.0 ± 0.6	Smith (1966) (T 1824 and hematocrit)
Coho salmon—SW *Oncorhynchus kisutch*	8	930 ± 104	—	—	4.5 ± 1.5	Smith (1966) (T 1824 and hematocrit)
Pink salmon—SW *Oncorhynchus gorbuscha*	6	1012 ± 120	—	—	2.3 ± 0.4	Bleeding
		—	—	—	7.8	Smith and Bell (1964) (T 1824 and hematocrit)
Atlantic cod *Gadus morhua*	—	—	—	—	2.4	Ronald et al. (1964) (Fluorescein and hematocrit)

[a] All mean values are recorded $\pm$ S.E.

[b] Here FW indicates freshwater and SW seawater.

Table VIII

Parameter	Freshwater Chondrostei (order Acipenseriformes)		Freshwater Holostei (order Lepisosteiformes)		Freshwater Teleostei	
	Lake sturgeon, *Acipenser fulvescens*	Paddlefish, *Polyodon spatula*	Bowfin, *Amia calva*	Short-nosed gar, *Lepisosteus platostomum*	Common sucker, *Catostomus commersoni*	Carp, *Cyprinus carpio*
Weight (g)	3058 (8)	4679 (5)	1963 (6)	1185 (7)	617 (2)	2412 (7)
	(2275–4530)	(3740–5910)	(1020–3265)	(855–1730)	(580–655)	(1585–3190)
Pulse	49 (8)	22 (5)	20 (6)	19 (6)	55 (2)	28 (6)
(beats/min)	(44–52)	(16–28)	(14–28)	(14–24)	(47–64)	(18–39)
Respiration	53 (8)	17 (5)	14 (6)	28 (5)	47 (2)	27 (3)
(per min)	(40–72)	(10–26)	(9–20)	(24–36)	(45–50)	(19–38)
Hematocrit	22 (8)	30 (5)	32 (6)	42 (7)	39 (2)	33 (7)
(% cells)	(19–29)	(24–37)	(32–34)	(33–50)	(39–40)	(23–24)
Specific gravity, plasma	1.016 (3)	1.017 (3)	1.018 (3)	1.016 (3)	1.016 (3)	1.019 (3)
	(all 1.016)	(1.016–1.018)	(1.0175–1.0185)	(1.015–1.017)	(1.015–1.017)	(1.018–1.0495)
Specific gravity, blood	1.036 (3)	1.040 (3)	1.045 (3)	1.051 (3)	1.041 (3)	1.040 (3)
	(1.033–1.040)	(1.039–1.041)	(1.044–1.047)	(1.050–1.052)	(1.040–1.042)	(1.039–1.0495)
Plasma volume (T 1824 space)	2.8 ± 0.18 (8)	2.2 ± 0.09 (5)	2.2 ± 0.19 (6)	2.1 ± 0.14 (7)	1.2 ± 0.14 (2)	1.8 ± 0.10 (7)
Blood volume	3.7 (8)	3.0 (5)	3.4 (6)	3.8 (7)	2.2 (2)	3.0 (7)
	(2.8–4.9)	(2.4–3.6)	(2.9–5.0)	(3.0–5.2)	(1.8–2.7)	(2.4–3.5)
Extracellular fluid (sucrose space)	20.1 ± 1.4 (8)	15.6 ± 0.42 (5)	18.9 ± 1.4 (6)	13.6 ± 0.33 (7)	12.2 ± 0.32 (2)	15.5 ± 1.3 (7)
Interstitial fluid (sucrose space minus plasma)	17.3	13.4	16.7	11.5	11.0	13.7
Total body water	72.7 ± 0.25 (8)	74.0 ± 0.48 (6)	74.5 ± 0.48 (6)	66.7 ± 0.70 (7)	74.4 ± 0.45 (2)	71.4 ± 0.45 (7)
Intracellular fluid (total water minus sucrose space)	52.6	58.4	55.6	53.1	62.2	55.9

[a] From Thorson (1961).
[b] Values are indicated as means ± S.E. or with the ranges reported in parentheses below.
[c] All volumes expressed as percentage of body weight.

sequence more closely related to that of mammalian myoglobin than the α or β chains of mammalian hemoglobin (Braunitzer *et al.*, 1964). The molecular weight of this hemoglobin is 17,000; it is monomeric and it does not show any of the heme–heme interaction properties of the dimeric and tetrameric forms of hemoglobin. Furthermore, the molecule probably has an oxygen dissociation curve which is hyperbolic or sigmoidal with a flattened zone over the range of tissue and environmental oxygen partial pressures in the lamprey. Hemoglobin of this type would probably release relatively small amounts of oxygen over the range of oxygen partial pressures to which it was exposed. In contrast the heme–heme interaction properties of the tetrameric, and possibly the dimeric, forms of hemoglobin would result in sigmoidal oxygen dissociation curves with progressively smaller slopes over their intermediate ranges. These molecules would unload more oxygen per unit change in oxygen partial

A Summary of the Data on Compartmental Volumes in Several Species of Osteichthyan Fishes[a,b,c]

	Marine Teleostei							
	Bigmouth buffalo fish, Ictiobus cyprinellus	Tiger rockfish, Mycteroperca tigris	Nassau grouper, Epinephelus striatus	Red snapper, Lutianus campechanus	Gray snapper, Lutianus griseus	Green moray, Gymnothorax funebris	Great barracuda, Sphyraena barracuda	Rainbow parrot fish, Pseudoscarus guacamaia
	3395 (8)	5885 (1)	1270 (2)	3765 (2)	3711	4062 (6)	2204 (11)	4607 (19)
	(1980–5440)		(930–1610)	(3130–4400)	(1900–4680)	(3050–4815)	(1432–4575)	(1650–6830)
	8 (8)	51 (1)	52 (1)	58 (2)	54 (3)	70 (5)	68 (11)	48 (16)
	(19–70)			(48–68)	(44–66)	(60–88)	(32–98)	(30–98)
	22 (4)	—	—	—	41 (4)	22 (3)	41 (7)	40 (14)
	(20–24)				(30–48)	(18–28)	(30–54)	(22–64)
	29 (8)	28 (1)	28 (2)	36 (2)	35 (6)	26 (6)	31 (11)	30 (27)
	(18–40)		(28–29)		(28–40)	(24–28)	(25–36)	(20–40)
	1.016 (3)	Used av	Used av	Used av	Used av	Used av	Used av	Used av
	(all 1.016)	(1.017)	(1.017)	(1.017)	(1.017)	(1.017)	(1.017)	(1.017)
	1.042 (3)	Used av	Used av	Used av	Used av	Used av	Used av	Used av
	(1.041–1.043)	(1.042)	(1.042)	(1.042)	(1.042)	(1.042)	(1.042)	(1.042)
	1.9 (8) ± 0.22	2.3 (1)	1.8 ± 0.18 (2)	1.3 ± 0.04 (2)	1.3 ± 0.04 (2)	1.6 ± 0.26 (6)	1.9 ± 0.09 (10)	2.4 ± 0.11 (16)
	2.8 (8)	3.3 (1)	2.6 (2)	2.2 (2)	2.0 (6)	2.2 (6)	2.8 (10)	3.6 (16)
	(1.8–4.1)							
	13.2 (8) ± 0.45	12.5 (1)	14.5 ± 1.0 (2)	14.0 (2)	14.0 ± 0.40 (6)	15.8 ± 1.1 (6)	15.9 ± 0.73 (8)	16.6 ± 0.57 (8)
	11.3	10.2	12.7	12.7	12.7	14.2	14.0	14.2
	70.6 (8) ± 1.2	71.1 (1)	71.7 ± 0.95 (2)	71.3 ± 0.11 (2)	72.3 ± 0.42 (6)	63.7 ± 2.4 (6)	70.6 ± 0.65 (9)	73.1 ± 0.32 (14)
	57.4	58.6	57.2	57.3	58.3	47.9	54.7	56.5

pressure. Therefore, compared to the number of monomeric hemoglobin molecules per unit volume of blood in the cyclostome, it would seem that fewer monomeric units of hemoglobin would be necessary to release the same volume of oxygen to the tissues if these units were arranged in dimers having heme–heme interaction properties. In the tetrameric form still fewer hemoglobin units would be required to release this volume of oxygen.

We feel therefore that a broader knowledge of the molecular forms of hemoglobin occurring in the fishes is necessary to fully explain the decreasing blood volumes in the more advanced forms. These studies should also include the characterization of the oxygen carrying and releasing properties of the hemoglobins at the tissue and environmental temperatures and at the oxygen tensions present in the tissues and environment of the organism.

Table IX

A Summary of the Mean Compartmental Spaces in the Various Taxonomic Groups of the Osteichthyes[a,b]

Parameter	Osteichthyes			
	Freshwater Chondrostei	Freshwater Holostei	Freshwater Teleostei	Marine Teleostei
Weight (g)	3681 (13)	1544 (13)	2664 (17)	3710 (47)
Pulse (beats/min)	39 (13)	20 (12)	37 (16)	57 (39)
Respiration (per min)	47 (13)	20 (11)	29 (9)	38 (28)
Hematocrit (% cells)	25 (13)	38 (13)	32 (17)	30 (55)
Specific gravity, plasma	1.0185 (6)	1.0185 (6)	1.017 (9)	—
Specific gravity, blood	1.048 (6)	1.048 (6)	1.041 (9)	—
Plasma volume (T 1824 space)	2.5 ± 0.11^{c} (13)	2.1 ± 0.12^{d} (13)	1.8 ± 0.11 (17)	1.9 ± 0.06^{d} (43)
Blood volume	3.5 (13)	3.6 (13)	2.8 (17)	2.9 (43)
Extracellular fluid (sucrose space)	18.4 ± 0.87^{c} (13)	16.0 ± 0.65^{e} (13)	14.0 ± 0.56 (17)	15.4 ± 0.31^{e} (33)
Interstitial fluid (sucrose space minus plasma)	15.9	13.9	12.2	13.5
Total body water	73.2 ± 0.39^{f} (13)	70.3 ± 0.43^{d} (13)	71.4 ± 0.60 (17)	70.8 ± 0.41^{d} (40)
Intracellular fluid (total water minus sucrose space)	54.8	54.3	57.4	55.4

[a] From Thorson (1961). All volumes are expressed as percentage of body weight.

[b] Values are reported as means ± S.E., and numerals in parentheses indicate the number of individual determinations.

[c] $p < 0.001$.

[d] Not significant with respect to the corresponding value for the group of freshwater teleost species.

[e] $p < 0.05$.

[f] $p < 0.02$.

Table X

Comparison of Inulin, Raffinose, and Sucrose Spaces of *Pseudoscarus guacamaia*[a]

Indicator	No. of specimens	Av body weight (g)	Volume (% body weight)	Range
Inulin	8	5451	11.4	9.2–14.5
Raffinose	4	4096	14.4	12.7–16.4
Sucrose	8	4696	16.6	14.3–18.9

[a] Data from Thorson (1961).

3. CHANGES IN THE EXTRACELLULAR COMPARTMENTS OF EURYHALINE SPECIES

Hatchery reared steelhead trout, *Salmo gairdneri*, maintained at seasonal temperatures and photoperiod, show significant declines in their plasma Ca^{2+}, Cl^-, and water concentrations and the concentrations of Ca^{2+} and Cl^- in muscle during the period of growth from 25 to 110 g body weight. This trend also occurred, but at a much slower rate, during the period of growth from 110 to 250 g body weight (Houston, 1959; Table XI). Estimation of the Cl^- space [Eq. (5)] in the muscle of these fish indicated that the extracellular volume per unit wet weight of muscle also declined quite markedly during the growth of the smaller trout and declined much less rapidly during the growth period of the larger individual fishes (Table XI). At the same time the relative intracellular volumes (total tissue water minus extracellular volume) of muscle samples from the smaller weight range of trout increased with increases in body weight, while in the larger fish the relative intracellular volume remained unchanged (Table XI).

The period of growth and development represented in the steelhead trout studied by Houston (1959) included the period of parr–smolt transformation in this species. This stage of development in all salmonid fishes is a period of profound physiological change and is associated with the preadaptation of the individual fishes to the marine environment (Huntsman and Hoar, 1939; Parry, 1958, 1961). Later studies by Houston (1960) and Houston and Threadgold (1963) were directed toward an elucidation of the changes occurring in the composition and distribution of body fluids during the parr–smolt transformation of the Atlantic salmon, *Salmo salar*. They found that the plasma Cl^- concentration of this species declined sharply with the onset of smoltification, but later recovered to a level somewhat higher than that observed in the nonsmolting parr. Some decrease was observed in the muscle Cl^- concentration at the onset of smoltification, but no secondary increase was observed as the meta-

Table XI

Variation of Plasma and Tissue Chloride and Calcium, Plasma Water, Chloride Space, and Cellular Space (Tissue Water–Chloride Space) with Weight in Freshwater Steelhead Trout[a]

Parameter	June–July, 1957			February–March, 1958		
	Sample size	Weight range (g)	Regression[b] ($Y = a + bX$)	Sample size	Weight range (g)	Regression[b] ($Y = a + bX$)
Plasma chloride	39	25–70	$Y = 152.8 - 0.30X$	23	76–250	$Y = 141.0 - 0.04X$
Plasma water	39	25–70	$Y = 958.4 - 0.22X$	19	76–250	$Y = 934.3 - 0.016X$
Plasma calcium	24	47–118	$Y = 2.71 - 0.004X$			
Tissue chloride	35	30–70	$Y = 31.9 - 0.18X$	23	80–250	$Y = 14.9 - 0.02X$
Tissue calcium	23	47–118	$Y = 3.24 - 0.015X$			
Chloride space	33	32–70	$Y = 217.1 - 1.28X$	19	80–250	$Y = 103.4 - 0.156X$
Cellular space	33	32–70	$Y = 628.3 + 0.43X$	19	80–250	$Y = 619.2 - 0.023X$

[a] Data from Houston (1959).

[b] $Y = a + bX$, where Y = plasma or tissue electrolyte and water concentrations (mEq/liter, mEq/kg wet weight, g H_2O/kg wet weight), a = the ordinate intercept, b = the slope, and X = body weight (g).

morphosis progressed. Changes in the Cl⁻ space of the Atlantic salmon suggest that the parr–smolt transformation process is characterized by a shift in the distribution of the body fluids in this species also. A sharp decline occurred in the extracellular volume (Cl⁻ space) of muscle during the silvery parr stage of development. This corresponded to the growth period from 20 to 35 g body weight. With the onset of the full smolting condition the extracellular volume of muscle stabilized at a value which was approximately 20% less than that of the presmolting fish. Concomitant with these changes in the muscle extracellular volume, increases in the intracellular volume (total muscle water minus Cl⁻ space) were observed.

Houston and Threadgold (1963) suggested that the changes in the compartmental distribution of electrolytes and water during the smoltification of the Atlantic salmon were consistent with possible changes in the pattern of renal excretion. We now know, at least in the trout, *Salmo gairdneri,* that this is indeed true. During the period of onset of smolting in this species the rates of urine flow and electrolyte excretion declined to approximately one-half of the presmolt values (Holmes and Stainer, 1966). This reduction was entirely attributable to a reduction in the glomerular filtration rate. If the trout were retained in freshwater until they eventually lost their overt characteristics of smolting, the glomerular filtration rate and the renal excretory pattern returned to those found in the presmolting fish (Holmes and Stainer, 1966). Therefore, since the smolting salmonid shows (1) a reduction in the extracellular volume and an increase in the intracellular volume of the muscle, (2) a reduction in the muscle water content, and (3) a reduction in the rate of water and electrolyte excretion via the kidneys, the new steady state must be presumed to be accompanied by a concomitant decrease in the extrarenal influx of water and electrolytes. Furthermore, it is possible that an increase in the extrarenal efflux of electrolytes also occurs at this time.

These data do not establish whether the changes in compartmental volumes occur as a direct result of the process of smoltification per se or whether they are merely manifestations of changes occurring due to the growth of the organism. Decreases in the extracellular compartment are characteristic of periods of rapid growth in several vertebrate species. Fellers *et al.* (1949) have demonstrated decreases of 79 and 57% in the thiocyanate and Na⁺ spaces, respectively, in humans between infancy and maturity. Similar changes have also been reported for the rat and the chicken (Barlow and Manery, 1954; Medway and Kare, 1959).

Nevertheless, the physiological changes which occur in the smolting salmonid, whether they result from the smolting process per se or coincidental changes in growth rate, certainly predispose the individual fish to life in the marine environment (Gordon, 1959a,b; Houston, 1959, 1960,

1963; Parry, 1961). The reduced glomerular filtration and urine flow rates of the trout, *Salmo gairdneri* (Holmes and Stainer, 1966), may be interpreted as part of the preadaptation to a marine environment. Upon adaptation to seawater, this species shows even larger decreases in the rates of urine flow (R. M. Holmes, 1961) and glomerular filtration (Holmes and McBean, 1963). Furthermore, Houston (1960) was able to demonstrate that Atlantic salmon in the full smolting condition were able to adapt to seawater much more readily than fish in the early stages of the parr–smolt transformation. Nonsmolting parr, on the other hand, were invariably unable to withstand an abrupt transfer from freshwater to seawater.

Following the abrupt transfer of salmonid fishes from freshwater to seawater, a series of physiological changes take place. These responses may be divided into two phases. An acute adaptive phase occurs immediately following transfer of the fish to seawater, and this is followed by a chronic regulative phase which ultimately results in the establishment of a new steady state with respect to the tissue water and electrolyte composition of the fish. Several recent studies involving a variety of salmonid species have been devoted to an examination of the changes which occur in the inorganic ion and water composition of the fish during adaptation to seawater (e.g., Gordon, 1959a,b; Houston, 1959; Parry, 1961). Typical of these studies are the data derived from the trout, *Salmo gairdneri*, and included in Fig. 3 (W. N. Holmes, unpublished data). Of

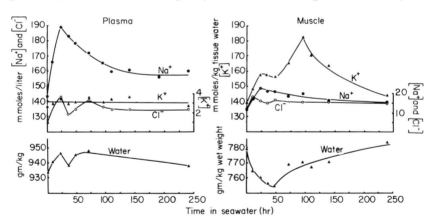

Fig. 3. Changes in the electrolyte and water concentrations of plasma and muscle from the trout, *Salmo gairdneri*, following abrupt transfer to 60% standard seawater (284 mM Na$^+$, 6 mM K$^+$) at approximately 5°C. The response may be divided into two phases: an acute adaptive phase immediately following transfer, followed by a chronic regulative phase which ultimately establishes a new steady state of electrolyte distribution (W. N. Holmes, unpublished data).

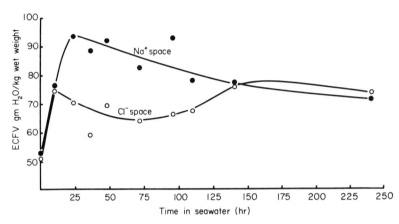

Fig. 4. Changes in the Na⁺ and Cl⁻ spaces of skeletal muscle from trout (*Salmo gairdneri*) following abrupt transfer to 60% seawater (284 mM Na⁺, 6 mM K⁺) at approximately 5°C. (ECFV = extracellular fluid volume.) Cl⁻ space and Na⁺ space as given in Eqs. (5) and (6). (From W. N. Holmes, unpublished data.)

particular interest are the oscillatory changes which occur in some parameters between 24 and 72 hr and between 72 and 140 hr in the regulatory phase (Fig. 3). Similar oscillations were reported in this species by Houston (1959). Estimations of the Na⁺ and Cl⁻ spaces [Eqs. (5) and (6), respectively] in the skeletal muscle of the freshwater trout before transfer to seawater indicated essentially similar values for the extracellular fluid volume (W. N. Holmes, unpublished data; Fig. 4) These values were 53.0 ± 1.8 and 50.8 ± 0.5 g/kg wet weight of muscle for the Na⁺ and the Cl⁻ spaces, respectively. In the same species, Houston (1959) reported somewhat higher values for the muscle Cl⁻ space (63–72 g/kg wet weight) in fish of the same size range (200–250 g body weight). During the first 10 hr after transfer to seawater both the Cl⁻ and Na⁺ spaces in muscle showed rapid increases to approximately 75 g/kg wet weight muscle (Fig. 3). After 10 hr, however, the Cl⁻ space commenced to decline but the Na⁺ space continued to rise until at 24 hr after transfer it was almost twice the freshwater value (Fig. 4). Thereafter, the Na⁺ space also declined and between 140 and 240 hr after transfer to seawater the Na⁺ and Cl⁻ spaces were essentially constant at a level which was approximately 45% higher than the freshwater value. Between 72 and 140 hr after transfer, however, a secondary rise occurred in the Cl⁻ space of the muscle; a similar phenomenon was also reported by Houston (1963) for this species, but the reason for its occurrence remains obscure.

Clearly, the estimated volume of distribution in muscle of either Na⁺ or Cl⁻ is not a reliable index of extracellular fluid volume during the early regulative phase of adaptation to seawater. Only the simultaneous estimations of inulin and/or sucrose space, together with measurements of the tissue water and electrolyte compositions, will elucidate this problem.

V. ELECTROLYTE COMPOSITION

A. Class Agnatha

1. ORDER MYXINIFORMES

Early determinations of the blood freezing point of the myxinoids indicated that the blood was slightly hypertonic to seawater (Dekhuyzen, 1904; Greene, 1904). Borei (1935) found the plasma of *Myxine glutinosa* to be considerably hypotonic to seawater, but this finding has never been substantiated. In more recent studies the serum of *Myxine glutinosa* was found to be virtually isosmotic to seawater by Robertson (1954). The earlier findings of hypertonic blood were explained by McFarland and Munz (1958) when they discovered a relationship between serum osmotic pressure and the degree of handling which the fish had received. Slime production resulted in a 1–3% hypertonicity which lasted an hour or more. It is now generally agreed that the Myxiniformes are isosmotic with seawater (Morris, 1960; Chester-Jones *et al.*, 1962; Robertson, 1963; McFarland and Munz, 1965). Morris (1965), however, still maintains that the blood is slightly hypertonic with respect to the environmental seawater and that for this reason the animal produces a small amount of hypotonic urine.

Exposure of *Polistotrema stoutii* to 80% seawater resulted in rapid weight gain with a return to normal after 7 days; exposure to 122% seawater resulted in weight loss with no return to initial weight. These findings are explained by a high permeability to water and a lack of regulatory mechanisms for sodium chloride (McFarland and Munz, 1965).

Although the osmotic pressure of myxinoid plasma is very similar to that of the environment, its ionic composition is dissimilar. The Na^+ concentration in both *Myxine glutinosa* and *Polistotrema stoutii* is higher than in seawater (Table XII). The plasma Cl^- concentration in *Myxine glutinosa* has been found to be higher than in seawater by Bellamy and Chester-Jones (1961) and Morris (1965) and to be lower than in seawater by Robertson (1954, 1966). In *Polistotrema stoutii,* Urist (1963) and McFarland and Munz (1958, 1965) have found the plasma concentration of Cl^- to be lower than in seawater. The plasma concentration of K^+ is similar to that in seawater. The divalent ions Ca^{2+}, Mg^{2+}, and SO_4^{2-} are all found at lower concentrations in the plasma than in seawater. In *Myxine glutinosa* plasma Ca^{2+} ranges in concentration from 53–67% of its concentration in seawater, Mg^{2+} varies from 24–61%, and SO_4^{2-} varies from 12–87% (Cole, 1940; Robertson, 1954; Bellamy and Chester-Jones, 1961; Morris, 1965; Robertson, 1966). In contrast to a urea concentration of 60 mmoles/liter determined by Borei (1935) the serum urea content of *Myxine glutinosa* was found to be 4 mmoles/liter by Cole (1940) and 2–4 mmoles/kg of water by Robertson (1954, 1966); thus, urea plays no significant osmoregulatory role in the hagfish.

McFarland and Munz (1965) have shown that although there is probably no active transport of Na^+ across the gut, gills, or skin, the low Na^+ content of the slime secretion probably serves to maintain the high plasma Na^+ concentrations. In contrast to a low Na^+ content, the slime has a high content of Ca^{2+}, Mg^{2+}, and K^+. In addition, Mg^{2+}, K^+, SO_4^{2-}, and phosphate are secreted into the glomerular filtrate by the mesonephric duct cells and appear in the urine at higher concentrations than in the plasma (Munz and McFarland, 1964; McFarland and Munz, 1965). The lack of a renal mechanism for the reabsorption of sodium chloride or water supports the theory of the marine origin of the Myxiniformes.

2. Order Petromyzontiformes

Unlike the hagfish which have an exclusively marine habitat the lampreys invariably breed in freshwater. Lampreys spend the first part of the life cycle in freshwater as amoecoete larvae and after metamorphosis they either remain in freshwater (*Lampetra planerii*), migrate into estuarine or coastal waters (*Lampetra fluviatilis*), or migrate into the open sea (*Petromyzon marinus* and *Lampetra tridentata tridentata*). In the Great Lakes there is a potodramous population of *Petromyzon marinus* which gained access to their present habitat when the Welland Canal, which bypasses Niagara Falls, was opened in 1827 (Urist, 1963). This population spawns in the freshwater streams which drain into the Great Lakes.

Although the anadromous lampreys spend much of their life cycle in salt water the only analysis so far available from this habitat is that of Burian (1910) who showed the osmotic pressure of the blood of a single specimen of *Petromyzon marinus* to be 317 mOsm/liter compared with 1236 mOsm/liter for Mediterranean seawater. Thus, with respect to its osmotic concentration, the blood of the marine lamprey closely resembles that of the marine teleost and is approximately four times more dilute than that of the hagfish. Urist (1963) transferred metamorphosed poto-

Table XII
Cyclostomata, Myxiniformes—Blood Chemistry

Fish or medium	Na	K	Mg	Ca	NH₄	Cl	HCO₃	PO₄	SO₄	Urea	Protein (g%)	Total ions	Units	Osmotic pressure (mOsm/liter)	Comments	Author	Date
Myxine glutinosa	—	—	—	—	—	—	—	—	—	—	—	—	—	962ᵃ	—	Dekhuyzen	1904
Seawater	—	—	—	—	—	—	—	—	—	—	—	—	—	930ᵃ	—	—	—
Polistotrema stouti	—	—	—	—	—	—	—	—	—	—	—	—	—	1059ᵃ	—	Greene	1904
Seawater	—	—	—	—	—	—	—	—	—	—	—	—	—	1032ᵃ	—	—	—
Polistotrema stouti	—	—	—	—	—	344	—	—	—	0	—	—	mmoles/liter	—	Blood chloride varies linearly with sea-water chloride	Bond *et al.*	1932
Seawater diluted to 2.5% salinity	—	—	—	—	—	384	—	—	—	0	—	—	mmoles/liter	—	—	—	—
Polistotrema stouti	—	—	—	—	—	414	—	—	—	—	—	—	mmoles/liter	—	—	—	—
Seawater diluted to 3.0% salinity	—	—	—	—	—	467	—	—	—	0	—	—	mmoles/liter	—	—	—	—
Polistotrema stouti	—	—	—	—	—	471	—	—	—	—	—	—	mmoles/liter	—	—	—	—
Seawater	—	—	—	—	—	530	—	—	—	—	—	—	mmoles/liter	—	—	—	—
Polistotrema stouti	—	—	—	—	—	570	—	—	—	0	—	—	mmoles/liter	—	—	—	—
Seawater concentrated to 4.0%	—	—	—	—	—	626	—	—	—	—	—	—	mmoles/liter	—	—	—	—
Myxine glutinosa	—	—	—	—	—	325	—	—	—	60	—	—	mmoles/liter	806ᵃ	Urea value has proved to be erroneous	Borei	1935
Seawater	—	—	—	—	—	520	—	—	—	—	—	—	—	1021ᵃ	—	—	—
Myxine glutinosa	402	9.1	22.5	5.3	—	448	3.7	—	6.0	4	—	901	mmoles/liter	—	—	Cole	1940
Seawater	416	9.1	50.2	9.4	—	483	2.2	—	30.4	—	—	1000	—	—	—	—	—
Myxine glutinosa	558	9.6	19.4	6.3	—	576	—	12.5ᵇ	6.7	3	—	1183	mmoles/kg	—	Serum is isotonic within 1%	Robertson	1954
Seawater	506	10.7	58	11.1	—	592	—	—	30.6	—	—	1209	—	—	—	—	—
Polistotrema stouti	370	—	—	—	—	408	—	—	—	—	—	—	mmoles/liter	859ᵃ	30 hr immersion	McFarland and Munz	1958
85% Seawater	338	—	—	—	—	444	—	—	—	—	—	—	—	852ᵃ	—	—	—
Polistotrema stouti	428	—	—	—	—	483	—	—	—	—	—	—	—	1022ᵃ	30 hr immersion	—	—
Seawater	450	—	—	—	—	498	—	—	—	—	—	—	—	1003ᵃ	—	—	—
Polistotrema stouti	505	—	—	—	—	522	—	—	—	—	—	—	—	1200ᵃ	30 hr immersion	—	—
116% Seawater	583	—	—	—	—	608	—	—	—	—	—	—	—	1165ᵃ	—	—	—

34

	Na	K	Ca	Mg		Cl						Osmotic conc.	Units	Osmotic conc.	Remarks	Reference	Year
Myxine glutinosa	549	11.1	189	5.1	—	563	—	5.0	—	—	—	1152	mmoles/liter	—	—	Bellamy and Chester-Jones	1961
Seawater	470	12.2	49.7	8.4	—	550	—	—	—	—	—	1067	mmoles/liter	—	—	—	—
Myxine glutinosa	535	9.1	—	—	—	—	—	—	—	—	—	—	mmoles/liter	—	—	Chester-Jones *et al.*	1962
Seawater	489	10.2	—	—	—	—	—	—	—	—	—	—	—	—	—	—	—
Myxine glutinosa	355	5.2	—	—	—	—	—	—	—	—	—	—	—	—	Serum sodium remains higher than seawater sodium	—	—
60% Seawater	287	5.5	—	—	—	—	—	—	—	—	—	—	—	—	—	—	—
Myxine glutinosa	371	7.8	—	—	—	—	—	—	—	—	—	—	—	—	—	—	—
73% Seawater	344	7.7	—	—	—	—	—	—	—	—	—	—	—	—	—	—	—
Myxine glutinosa	1136	14.5	—	—	—	—	—	—	—	—	—	—	—	—	Water loss increases serum sodium	—	—
165% Seawater	776	14.8	—	—	—	—	—	—	—	—	—	—	—	—	—	—	—
Polistotrema stouti	544	7.7	10.4	5.4	—	446	5.2	1.0	4.4	—	4.2	1026.4	mmoles/liter	—	—	Urist	1963
Seawater	509	30.0	47.5	10.0	—	540	2.0	0.0	30.0	—	—	1168.5	—	—	—	—	—
Polistotrema stouti	570	7.0	12.0	4.5	—	547	—	3.7	0.9	—	—	—	mmoles/kg	1034	—	Munz and McFarland	1964
Seawater	496	10.3	51.6	10.9	—	543	—	0.0	25.7	—	—	—	—	1029	—	—	—
Polistotrema stouti	522	10.9	13.8	3.9	—	501	—	—	—	—	—	—	mmoles/kg	954	2 days' immersion	McFarland and Munz	1965
100% Seawater	456	11.3	50.8	8.9	—	528	—	—	—	—	—	—	—	953	—	—	—
Polistotrema stouti	405	10.3	12.8	4.4	—	371	—	—	—	—	—	—	—	743	2 days' immersion	—	—
75% Seawater	363	8.4	40.4	6.9	—	402	—	—	—	—	—	—	—	740	—	—	—
Polistotrema stouti	439	7.6	10.8	3.6	—	407	—	—	—	—	—	—	—	706	33 days' immersion	—	—
73% Seawater	336	7.9	39.6	7.9	—	371	—	—	—	—	—	—	—	700	—	—	—
Myxine glutinosa	529	10.4	25.6	6.4	—	534	—	—	18.3	—	—	1123.4	mmoles/kg	—	Running seawater	Morris	1965
Seawater	455	9.4	52.6	9.8	—	524	—	—	27.3	—	—	1078.1	—	—	—	—	—
Myxine glutinosa	471	12.1	50.7	4.9	—	500	—	—	19.7	—	—	1058.6	—	1038[a]	Still seawater high magnesium an effect of urethan	—	—
Seawater	463	9.5	52.3	9.8	—	535	—	—	27.4	—	—	1096.7	—	1011[a]	—	—	—
Myxine glutinosa	486	8.2	11.9	5.1	—	508	7.2	2.1	3.0	2.8	—	1035	mmoles/kg	—	—	Robertson	1966
Seawater	439	9.3	50.0	9.6	—	513	2.2	—	26.4	—	—	1050	—	—	—	—	—

[a] Derived from freezing point depression.

[b] Expressed in mEq/kg water as $H_2PO_4^-$ and HPO_4^{2-} with a valency of 1.84.

Table XIII

Cyclostomata, Petromyzontiformes—Blood Chemistry

Fish or medium	Na	K	Mg	Ca	NH$_4$	Cl	HCO$_3$	PO$_4$	SO$_4$	Urea	Protein (g%)	Total ions	Units	Osmotic pressure (mOsm/liter)	Comments	Author	Date
Lampetra fluviatilis	—	—	—	—	—	—	—	—	—	—	—	—	—	258[a]	—	Dekhuyzen	1904
Petromyzon marinus	—	—	—	—	—	—	—	—	—	—	—	—	—	317[a]	In seawater	Burian	1910
Seawater	—	—	—	—	—	—	—	—	—	—	—	—	mmoles/liter	1236[a]	—	—	—
Petromyzon marinus	—	—	—	—	—	—	—	—	—	—	—	—	—	290[a]	—	Fontaine	1930
Freshwater	—	—	—	—	—	—	—	—	—	—	—	—	—	11[a]	—	—	—
Petromyzon marinus	—	—	—	—	—	119	—	—	—	—	—	—	—	301[a]	—	—	—
Dilute seawater	—	—	—	—	—	—	—	—	—	—	—	—	—	263[a]	—	—	—
Petromyzon marinus	—	—	—	—	—	115	—	—	—	—	—	—	—	355[a]	—	—	—
Dilute seawater	—	—	—	—	—	—	—	—	—	—	—	—	—	435[a]	—	—	—
Petromyzon marinus	—	—	—	—	—	127	—	—	—	—	—	—	—	414[a]	—	—	—
Dilute seawater	—	—	—	—	—	—	—	—	—	—	—	—	—	537[a]	—	—	—
Petromyzon marinus	—	—	—	—	—	147	—	—	—	—	—	—	—	581[a]	—	—	—
Dilute seawater	—	—	—	—	—	—	—	—	—	—	—	—	—	1054[a]	—	—	—
Petromyzon marinus	—	—	—	—	—	240	—	—	—	—	—	—	mmoles/liter	247[a]	—	—	—
Seawater	—	—	—	—	—	121	—	—	—	—	—	—	—	280[a]	In tap water	Galloway	1933
Lampetra fluviatilis	—	—	—	2.7	—	—	—	—	—	—	—	—	—	306[a]	22 hr in seawater	—	—
Lampetra fluviatilis	—	—	—	—	—	—	—	—	—	—	—	—	—	301[a]	5.5 hr in seawater	—	—
One-third seawater	—	—	—	—	—	—	—	—	—	—	—	—	—	462[a]	—	—	—
Lampetra fluviatilis	—	—	—	—	—	—	—	—	—	—	—	—	—	403[a]	7 hr in seawater	—	—
One-half seawater	—	—	—	—	—	—	—	—	—	—	—	—	—	935[a]	—	—	—
Lampetra fluviatilis	—	—	—	—	—	—	—	—	—	—	—	—	—	253[a]	—	—	—
Seawater	—	—	—	—	—	—	—	—	—	—	—	—	mmoles/kg	—	—	Wikgren	1953
Lampetra fluviatilis	119.6	3.2	2.1	2.0	0.4	95.9	6.4	12.8[b]	2.7	—	3.6	239	mmoles/liter	—	—	Robertson	1954
Lampetra fluviatilis	—	—	—	—	—	58	—	—	—	—	—	—	—	220	Ammoecoete larva	Hardisty	1956
Lampetra planeri	—	—	—	—	—	101	—	—	—	—	—	—	—	226	Adult	—	—
Lampetra planeri	—	—	—	—	—	—	—	—	—	—	—	—	—	—	—	—	—

Species											Units	Osmolality	Description	Reference	Year
Lampetra planeri	—	—	—	—	61.0	—	—	—	—	—	—	—	Ammocoete control	—	—
Lampetra planeri	—	—	—	—	54.1	—	—	—	—	—	—	—	Ammocoete 14 days' distilled water	—	—
Lampetra planeri	—	—	—	—	71.0	—	—	—	—	—	—	—	Ammocoete 14 days' tap water	—	—
Lampetra fluviatilis	—	—	—	—	113	—	—	—	—	—	—	286	November adult	—	—
Lampetra fluviatilis	—	—	—	—	118	—	—	—	—	—	—	272	March adult	—	—
Lampetra fluviatilis	—	—	—	—	96.8	—	—	—	—	—	mmoles/liter	—	November freshwater	Morris	1956
Lampetra fluviatilis	—	—	—	—	100.5	—	—	—	—	—	mmoles/liter	—	January seawater	Morris	—
Lampetra fluviatilis	—	—	—	—	—	—	—	—	—	—	—	247[a]	Freshwater	Morris	1958
Lampetra fluviatilis	—	—	—	—	—	—	—	—	—	—	—	296[a]	In 33% seawater	—	—
33% Seawater	—	—	—	—	—	—	—	—	—	—	—	355[a]	—	—	—
Lampetra fluviatilis	—	—	—	—	—	—	—	—	—	—	—	306[a]	In 50% seawater 4 out of 18 fresh run fish	—	—
50% seawater	—	—	—	—	—	—	—	—	—	—	—	522[a]	—	—	—
Lampetra fluviatilis	85	33	—	—	—	—	—	—	—	—	mmoles/liter	—	Tap water	Bentley and Follett	1963
Lampetra fluviatilis	89	37	—	—	—	—	—	—	—	—	—	—	100 mmoles/liter NaCl	—	—
Petromyzon tridentata	87.0	6.1	1.4	2.8	80.1	3.1	4.7	0.5	3.6	186.1	mmoles/liter	—	Migrating anadromous	Urist	1963
Petromyzon marinus	139.0	6.2	1.9	2.4	113.0	5.2	1.4	0.9	—	278.9	mmoles/liter	—	Migrating potodramous	—	—
Petromyzon marinus	136.0	5.1	1.8	2.7	112.0	5.2	1.3	0.7	3.9	273.9	mmoles/liter	—	Spawning potodramous	—	—
Lake Huron water	0.02	0.05	0.25	0.9	0.05	1.75	0.003	2.3	—	5.32	mmoles/liter	—	—	—	—
Lampetra planeri	99.4	6.4	—	—	81.2	—	—	—	—	—	mmoles/kg	227	November	Bull and Morris	1967
Ammocoete larva	97.4	7.8	—	—	80.3	—	—	—	—	—	—	240	July	—	—
Petromyzon marinus	103.0	3.4	1.6	2.4	91.0	6.0	1.3	0.1	2.7	208.8	—	—	Ammocoete larva	Urist and Van de Putte	1967
Petromyzon marinus	134.0	4.0	1.8	2.4	122.1	5.0	1.3	0.1	3.2	270.6	—	—	Metamorphosed downstream	—	—
Petromyzon marinus	137.0	3.3	2.0	2.2	122.1	5.0	1.4	0.1	3.5	272.1	—	—	Parasitic adult	—	—
Petromyzon marinus	124	—	—	1.8	105	—	2.7	—	0.17	—	—	—	Metamorphosed freshwater	—	—
Petromyzon marinus	184	—	—	3.1	166	—	3.3	—	0.16	—	—	—	Metamorphosed 2 hr in seawater	—	—
Petromyzon marinus	212	—	—	3.5	173	—	4.3	—	0.15	—	—	—	Metamorphosed 4 hr in seawater	—	—

[a] Converted from freezing point.

[b] Expressed in mEq/kg water present as $H_2PO_4^-$ and HPO_4^{2-} with a valence of 1.84.

dramous lampreys, *Petromyzon marinus*, into artificial seawater and observed the changes in serum composition after 2 and 4 hr of immersion. In 4 hr the plasma Ca^{2+} concentration increased from 1.8 to 3.5 mmoles/ liter, inorganic phosphorous from 2.7 to 4.3 mmoles/liter, Cl^- from 105 to 173 mmoles/liter, and Na^+ from 124 to 173 mmoles/liter, but urea nitrogen remained constant at 0.17 to 0.15 mmoles/liter. Thus, there was an increase in inorganic ions during the 4-hr adaptation but no urea retention, suggesting that the mechanism of urea retention is of more recent origin than the cyclostomes. Other workers have captured anadomous lampreys during their spawning migration in freshwater and transferred them back into seawater in order to gain a knowledge of the ionic composition of their blood in saltwater (Fontaine, 1930; Galloway, 1933; Hardisty, 1956; Morris, 1956, 1958). It appears, however, that the osmoregulatory ability of the seawater lamprey is rapidly lost upon entry into freshwater, and therefore these experiments have been only partially successful.

Migrating *Petromyzon marinus* (Fontaine, 1930) and *Lampetra fluviatilis* (Galloway, 1933) (Table XIII) were found to show an increased blood osmotic pressure on transfer to dilute seawater. Immersion in full seawater was fatal in both cases. Morris (1958) caught maturing *Lampetra fluviatilis* at an early stage of their migration up the River Trent in England and transferred them to 50% seawater. Of the 18 animals tested only 4 were able to osmoregulate on the basis of maintaining a relatively constant body weight and a plasma osmotic pressure well below that of the environment. The mean osmotic pressure of these *Lampetra fluviatilis* was 306 mOsm/liter compared to 317 mOsm/liter for *Petromyzon marinus* captured in the Mediterranean, thus these two species may have similar blood concentrations in the marine habitat. Morris (1960) has proposed that there are three factors which contribute to the inability of maturing lampreys to osmoregulate in freshwater: First, an increase in the water permeability of the external surface which may lead them to migrate into water of lower salinity, second, a reduction in the swallowing rate related to a decrease in the diameter of the alimentary canal, and, third, a decrease in the abundance of Cl^- excretory cells in the gill epithelium.

A complete analysis of the plasma constituents of freshwater *Lampetra fluviatilis* by Robertson (1954) (Table XIII) showed it to be very similar in ionic composition to the plasma of the freshwater teleost, *Coregonus clupoides* (Table XVI). The only notable differences were that the Na^+, Cl^-, and HCO_3^- concentrations are lower in the lamprey. Urist (1963) and Urist and Van de Putte (1967) have analyzed the sera of the potodramous *Petromyzon marinus* in the Great Lakes and the anadromous *Petromyzon tridentata tridentata* of Oregon (Table XIII). The amoecoete larvae of

Petromyzon marinus were found to have a low serum ionic concentration of only 209 mmoles/liter. After metamorphosis and downstream migration there was an increase in serum Na^+ and Cl^- concentrations to give a total ionic concentration of 271 mmoles/liter. During parasitic adult life, upstream migration, and spawning there was no significant change in the total ion concentration, but an increase in the K^+ and SO_4^{2-} concentrations occurred (Table XIII). Anadromous *P. tridentata tridentata* migrating up the Willamette River in Oregon had a much lower concentration of serum ions (186 mmoles/liter) than the potodramous *Petromyzon marinus* at the same stage in the life cycle. Urist (1963) and Urist and Van de Putte (1967) showed that this was the result of a loss of Na^+ and Cl^- in *P. tridentata tridentata* owing to a breakdown of osmoregulatory ability during its spawning migration.

Transfer of *Petromyzon marinus* from freshwater to Ca^{2+}-deficient freshwater for 2 hr resulted in a small drop in serum Ca^{2+} concentration from 2.6 to 2.2 mmoles/liter and an increase in the serum inorganic phosphate concentration from 3.0 to 4.8 mmoles/liter (Urist, 1963). The increase in inorganic phosphate is similar to the response in bony vertebrates with hypocalcemia. The finding of only a small decrease in Ca^{2+} concentration indicates that the lamprey is able to regulate Ca^{2+} by means of its gill membranes and mucosal skin despite the lack of a skeletal Ca^{2+} reservoir (Urist, 1963).

Hardisty (1956) recorded a seasonal variation in the total Cl^- content of the amoecoete larvae of *Lampetra planeri*. Bull and Morris (1967) have shown that this change was related to the nutritional status of the animal and that there is no difference in the ionic composition of the serum in animals sampled in July and November. The concentrations of Na^+ and Cl^- observed by Bull and Morris are lower than those found by Robertson (1954) in adult *Lampetra fluviatilis* but similar to those found by Urist and Van de Putte (1967) in the amoecoete larva of *Petromyzon marinus* indicating that low Na^+ and Cl^- concentrations may be a general characteristic of amoecoete larvae. The serum K^+ concentration obtained by Bull and Morris, on the other hand, was higher than those obtained by either Robertson (1954) or Urist and Van de Putte (1967) except in the case of migrating and spawning *Petromyzon marinus* (Table XIII).

B. Class Chondrichthyes

1. Subclass Elasmobranchii (Marine Species)

a. Plasma Composition. Although the marine Elasmobranchii resemble the Myxiniformes in having a plasma osmotic pressure similar to

Table XIV

Chondrichthyes Blood Chemistry

Fish or medium	Na	K	Mg	Ca	Cl	HCO3	CO2	PO4	SO4	Urea	TMAO	Protein (g%)	Total ions	Units	Osmotic pressure (mOsm/liter)	Comments	Author	Date
							Elasmobranchii Marine											
Mustelus canis	—	—	—	—	—	—	—	—	—	—	—	—	—	—	1011[a]	—	Garrey	1905
Seawater	—	—	—	—	—	—	—	—	—	—	—	—	—	—	978[a]	—	—	—
Raja stabuloforis	255	4.9	2.8	3.8	241	—	6.1	1.0	—	453	—	—	—	—	—	—	Smith	1929[a]
Seawater	416	9.1	50.0	9.4	483	—	2.2	—	30.4	—	—	—	507	mmoles/liter	—	—	—	—
Raja diaphenes	237	6.8	3.5	5.1	227	—	5.6	1.4	3.1	377	—	—	1090	—	—	—	—	—
Seawater	206	4.6	22.6	4.5	225	—	2.0	—	13.2	—	—	—	501	—	—	—	—	—
Carcharias littoralis	267	5.5	2.3	5.5	235	—	10.3	3.0	0.5	381	—	—	504	—	—	—	—	—
Mustelus canis	270	5.0	3.0	5.5	234	—	11.6	3.5	2.2	381	—	—	544	—	—	—	—	—
Seawater	445	10.6	50.0	14.5	517	—	2.4	0.2	30.6	—	—	—	541	—	—	—	—	—
Scyllium canicula	156	—	—	4.3	299	—	—	—	—	—	—	3.0	1162	—	1107[a]	♂	Pora	1936[b]
	186	—	—	6.7	267	—	—	—	—	—	—	4.0	—	—	1123[a]	♂ Elevated oxygen 48 hr	—	—
Raja undulata	192	—	—	3.3	277	—	—	—	—	—	—	3.2	—	—	1098[a]	♀	—	—
	207	—	—	5.1	248	—	—	—	—	—	—	4.0	—	—	1124[a]	♀ Elevated oxygen 48 hr	—	—
Squatina angelus	—	—	—	4.1	—	—	—	—	—	—	—	4.2	—	—	1097[a]	♂	Pora	1936[c]
	—	—	—	3.6	255	—	—	—	—	—	—	2.3	—	—	1125[a]	♀	—	—
Torpedo marmorata	—	—	—	4.9	369	—	—	—	—	—	—	5.3	—	—	1102[a]	♂	Pora	1936[d]
Raja clavata	—	—	—	—	285	—	—	—	—	—	—	2.5	—	—	1098[a]	♀	—	—
Seawater	—	—	—	3.6	—	—	—	—	—	—	—	2.5	—	—	1095[a]	♂	—	—
Raja erinacea	—	—	—	—	—	—	—	—	—	—	—	—	—	—	1074[a]	♀	—	—
Seawater	—	—	—	—	—	—	—	—	—	—	—	—	—	—	968[a]	Osmotic pressure is of whole blood	Chaisson and Friedman	1935
Raja erinacea	254	8.0	2.5	6.0	255	—	—	—	—	320	—	—	—	—	925	—	—	—
Rhinobatus percellens	143	12.8	1.0	3.7	144	—	—	—	—	349	—	—	—	mmoles/liter	—	—	Hartman et al.	1941
Narcine brasiliensis	134	7.0	1.5	6.0	159	—	—	—	—	209	—	—	—	mmoles/liter	—	—	Pereira and Sawaya	1957
Seawater	354	9.8	37.6	8.0	—	—	—	—	—	—	—	—	—	—	—	—	—	—
Squalus acanthias	—	—	—	—	—	—	—	—	—	—	71	—	—	mmoles/liter	—	—	Cohen et al.	1958

Species															Units	Remarks	Osmolality	Reference	Year
Mustelus canis	—	—	—	—	275	—	—	—	—	—	—	—	—	—	mmoles/kg	—	970	Davson and Grant	1960
Mustelus canis	288	8	—	3	275	270	6	2	—	3	342	97	2.85	1037	mmoles/kg	—	962	Doolittle et al.	1960
Aprionodon isodon	238	7.0	—	—	252	—	—	—	—	—	—	—	—	—	mmoles/liter	—	—	Sulya et al.	1960
Carcharhinus limbatus	258	10.0	—	—	241	—	—	—	—	—	—	—	—	—	—	—	—	—	1960
Sphyrna tiburo	289	12.5	—	—	254	—	—	—	—	—	—	—	—	—	—	—	—	—	—
Mustelus canis	—	—	—	—	—	—	—	—	—	—	—	—	—	—	mmoles/kg	—	981.3	Bloete et al.	1961
Raja clavata	289	4.0	—	—	311	—	—	—	—	—	444	—	—	—	mmoles/kg	—	995a	Murray and Potts	1961
Seawater	—	—	—	—	—	—	—	—	—	—	—	—	—	—	mmoles/kg	—	—	—	—
Platyrnoidoidis triseriata	234	11.4	4.6	5.3	208	3.2	2.1	—	—	—	—	—	—	—	mmoles/liter	—	973a	Urist	1961
Carcharhinus leucas leucus	223.4	9.0	2.9	4.5	236	5.1	2.0	0.6	—	—	333	2.4	484	—	mmoles/liter	—	—	Urist	1962a
Squalus acanthias	255	6.6	—	—	239	—	—	—	—	—	333	—	—	—	mmoles/liter	—	—	Maren	1962a
Triakus semifasciatus	235	10.0	3.0	5.0	230	5.0	1.2	0.5	—	—	333	3.5	—	—	mmoles/liter	—	973/kg	Urist	1962b
Seawater	509	30.0	47.5	10.0	540	2.0	0.0	30.0	—	—	—	—	—	—	—	—	—	—	—
Raja stabuliforis	182	4.2	3–5	3–5	220	—	—	1	—	—	—	—	—	—	mmoles/liter	—	958/kg	Maren et al.	1963
Raja ocellata	285	3.5	3–5	3–5	255	—	—	1	—	—	—	—	—	—	—	—	928	—	—
Raja erinacea	260	3.5	3–5	3–5	253	—	—	1	285	—	—	—	—	—	—	—	917	—	—
Raja clavata	285	4.0	—	—	240	—	—	—	—	—	—	—	—	—	—	—	—	Enger	1964
Squalus acanthias	240	3.6	—	—	259	—	—	—	—	—	—	—	—	—	mmoles/liter	—	—	Robin et al.	1964
Squalus acanthias	—	—	—	—	234.6	—	—	—	—	—	—	—	—	—	mmoles/liter	—	997	Burger	1965
Squalus acanthias	—	—	—	—	242.6	—	—	—	—	—	—	—	—	—	mmoles/liter	Rectal gland removed 21 days	1000	—	—
Seawater	—	—	50	9	500	—	—	—	—	—	—	—	—	—	—	—	932	—	—
Dasyatis americana	251	18.8	1.9	11.6	256	—	5.9	—	—	—	351	3.55	—	—	mmoles/liter	—	864	Bernard et al.	1966
Dasyatis say	256	20.6	1.6	9.8	262	—	7.8	—	—	—	382	2.60	—	—	mmoles/liter	—	840	—	—
Negaprion brevirostris	307	5.5	—	—	277	—	—	—	—	—	—	—	—	—	mmoles/liter	—	—	Oppelt et al.	1966
Ginglymostoma cirratum	291	4.2	3.1	—	287	6	—	—	—	—	357	4.0	—	—	—	—	—	—	—
Squalus acanthias	263	4.1	—	6.6	249	—	—	—	—	—	—	—	—	—	mmoles/liter	—	1007	Murdaugh and Robin	1967
Raja eglanteria	—	—	—	—	195	—	—	—	—	—	320	—	—	—	mmoles/liter	24 hr immersion	737a	Price and Creaser	1967
Dilute seawater	—	—	—	—	211	—	—	—	—	—	—	—	—	—	—	—	608a	—	—
Raja eglanteria	—	—	—	—	291	—	—	—	—	—	344	—	—	—	—	24 hr immersion	823a	—	—
Dilute seawater	—	—	—	—	378	—	—	—	—	—	—	—	—	—	—	—	703a	—	—
Raja eglanteria	—	—	—	—	222	—	—	—	—	—	368	—	—	—	—	—	844a	—	—
Seawater	—	—	—	—	421	—	—	—	—	—	—	—	—	—	—	—	817a	—	—
Raja eglanteria	243	11	—	5.0	249	—	1.2	0.5	—	—	366	—	—	—	mmoles/liter	—	—	Price	1967
Heterodontus triseriata	235	10.0	3.0	5.0	230	5.0	—	—	—	—	338	—	—	—	mmoles/liter	—	—	Urist and Van de Putte	1967

41

Table XIV (Continued)

Fish or medium	Na	K	Mg	Ca	Cl	HCO₃	CO₂	PO₄	SO₄	Urea	TMAO	Protein (g %)	Total ions	Units	Osmotic pressure (mOsm/ liter)	Comments	Author	Date
Elasmobranchii Freshwater																		
Pristis microdon	—	—	—	—	170	—	—	3.1	—	130	—	—	—	mmoles/liter	548[a]	—	Smith	1931
Dasyatis uarnak	—	—	—	—	212	—	—	—	—	104	—	—	—	—	548[a]	—	—	—
Carcharhinus melanop	—	—	—	—	158	—	—	—	—	103	—	—	—	—	484[a]	—	—	—
Hypolopus sephen	—	—	—	—	146	—	—	—	—	81	—	—	—	—	—	—	—	—
C. Leucas nicaraguensis	200.1	8.2	2.0	3.0	180.5	6.0	—	4.0	0.5	132	—	3.4	404.3	mmoles/liter	—	—	Urist	1961
Freshwater San Juan River	0.7	0.1	0.2	0.8	0.8	1.7	—	0.001	0.7	—	—	—	5.1	—	—	—	—	—
Pristis perotteti	216.6	6.5	0.9	4.2	193.1	—	—	—	—	—	—	3.7	—	mmoles/liter	—	—	T. B. Thorson	1967
Carcharhinus leucas	245.8	6.4	1.6	4.5	219.3	7.0	—	4.4	0.7	180	—	3.0	—	—	—	—	—	—
Holocephali																		
Hydrolagus colliei	—	—	—	—	—	—	—	—	—	—	—	—	—	—	801[a]	—	Nicol	1950
Chimaera monstrosa	363	10.2	—	—	380	—	—	—	—	266	—	—	—	mmoles/liter	—	—	Fänge and Fugelli	1962
Hydrolagus colliei	268	6.9	1.5	4.8	272	1.7	—	2.2	0.6	303	—	—	557.7	mmoles/liter	—	—	Urist	1966

[a] Value converted from freezing point.

42

that of seawater they differ in that urea is responsible for a considerable portion of the plasma osmotic pressure (Table XIV). Urea was first identified in the blood of both the Rajiformes and Squaliformes by Staedeler and Frerichs (1858). Rodier (1899) found that the freezing point of elasmobranch blood was a little lower than that of seawater and that urea was responsible for approximately one-third of its osmotic pressure. These findings were subsequently confirmed by Fredericq (1904), Garrey (1905), and Bottazzi (1906).

The first thorough investigation of the inorganic constituents of marine elasmobranch plasma was carried out by Smith (1929a) who examined the plasma of the sharks, *Carcharias littoralis* and *Mustelus canis,* and the skates, *Raja stabuliforis* and *Raja diaphenes* (Table XIV). No striking differences were apparent between the concentrations of ions and urea in the sharks and in the skates. The mean Na$^+$ concentration in plasma was 257 mmoles/liter, while K$^+$ was 5.5 mmoles/liter, Ca^{2+} 5.0 mmoles/liter, and Cl$^-$ 234 mmoles/liter. The concentrations observed by Smith (1929a) were similar to those of McCallum (1926) who examined the sera of *Acanthias vulgaris* and *Carcharias littoralis* except that Smith observed a greater variability in the K$^+$ and Ca^{2+} levels. McCallum (1926) concluded that the high plasma urea concentrations which he observed in sharks were a result of an inability of the kidney to remove them; whereas Smith (1929a, 1936), noting that both urea and trimethylamine oxide (TMAO) are present in the urine at a lower concentration than in the plasma and that the uremia persists even when the fish is in a state of inanition, argued that urea and trimethylamine oxide are actively reabsorbed from the glomerular filtrate. Thus, urea and trimethylamine oxide form a significant component of the plasma osmotic solutes, the gills and integument being relatively impermeable. Hartman et al. (1941) extended the findings of Smith (1936) on the reciprocal relationship between plasma Na$^+$ and urea concentrations. During inanition there was a decrease in the plasma urea concentration and an increased plasma Na$^+$ concentration in *Raja erinacea*. This change was partially offset by feeding. Intramuscular injection of urea resulted in an increased plasma urea concentration and a decreased plasma Na$^+$ concentration. Neither interrenalectomy nor removal of the rectal gland was found to have a significant effect on the plasma constituents of the skate (Hartman et al., 1944; Burger, 1965). Since the investigations of Smith (1929a) and Hartman et al. (1941), several workers have examined the serum composition of other elasmobranchs, and these data are listed in Table XIV. Urist and Van de Putte (1967) report ranges for marine elasmobranchs of 480–490 mmoles/liter for total serum ion concentration, 225–250 mmoles/liter for Na$^+$, 4.0–5.0 mmole/liter for Ca^{2+}, and 250–330 mmoles/liter for urea.

Pora (1936a,b) investigated the effect of hyperoxygenation on *Scyllium canicula* and found increases in the serum Na^+, Ca^{2+}, protein, and total osmotic concentrations and a decrease in the serum Cl^- concentration of both males and females (Table XIV). Pora (1936c) also observed a sex difference in the serum composition of *Raja undulata*, the osmotic pressure being higher in the female and serum protein and Ca^{2+} concentrations higher in the male (Table XIV). In *Scyllium canicula* the male was found to have the higher serum osmotic pressure. In contrast to the findings of a sex difference in serum Ca^{2+} concentration by Pora (1936c), Hess *et al.* (1928) reported that there was no sex difference for serum Ca^{2+} concentration in the dogfish; the concentration was high in both male and female. This was confirmed by Smith (1929a) and Hartman *et al.* (1941, 1944). Urist (1961) examined the serum Ca^{2+} concentration in several species of elasmobranch both male and female, mature and immature, gravid and nongravid, and found it to be high in all cases (5 mmoles/liter), a situation quite different from that in the teleost (Hess *et al.*, 1928).

The effects of immersion in dilute seawater on the plasma Cl^-, urea, and osmotic concentrations of the skate, *Raja eglanteria*, were investigated by Price and Creaser (1967) (Table XIV). Immersion in water of low salinity resulted in a loss of serum Cl^- and urea. Introduction of skates adapted to low salinity water (21‰) into high salinity water (up to 31‰) resulted in an increase in serum urea and Cl^-. The time taken to reach osmotic equilibrium in *Raja eglanteria* was 48 hr after a change of 2.4‰ and 70 hr after a change of 10.0‰ in the external salinity. Thus, skate in living in esturine conditions where the salinity changes with each tide probably never reach osmotic equilibrium. Although Price and Creaser (1967) found a lowering of serum Cl^- concentration in low salinity water in the laboratory, Price (1967) was unable to show a significant relationship between serum Cl^- concentration and external salinity in captured skate.

b. Composition of Other Fluids. i. Pericardial and perivisceral fluids. Smith (1929a) found the pH of the pericardial fluid to be lower than that of the plasma with a correspondingly lower concentration of HCO_3^-. The Ca^{2+} concentration was also found to be lower while the K^+ concentration was higher (Table XV). Perivisceral fluid was also found to be more acidic than plasma and to have a lower Ca^{2+} concentration. Unlike the pericardial fluid the concentration of K^+ in the perivisceral fluid was similar to that in the serum and the concentrations of SO_4^{2-} and Mg^{2+} were higher than those in the serum. Pericardial and perivisceral fluid urea concentrations were the same or lower than the plasma (Smith, 1929a) (Table XV).

In *Raja erinacea* the perivisceral fluid was found to have a higher concentration of Mg^{2+}, K^+, Cl^-, and urea and a lower concentration of Ca^{2+} than the plasma (Hartman *et al.*, 1941) (Table XV). Rodnan *et al.* (1962) and Murdaugh and Robin (1967) compared plasma and coelomic fluid concentrations in *Squalus acanthias* and found the Na^+, Mg^{2+}, H^+, and Cl^- concentrations to be higher and the Ca^{2+}, HCO_3^-, and protein concentrations to be significantly lower in the coelomic fluid (Table XV). Although NH_4^+ was not measured, Rodnan *et al.* (1962) suggest that it may make up some of the cation deficit in the coelomic fluid.

The most recent measurements of batoid pericardial and perivisceral fluids have been carried out on the stingray, *Dasyatis americana*, by Bernard *et al.* (1966) (Table XV); Ca^{2+}, Mg^{2+}, and protein were less concentrated, and the osmotic pressure was lower in both pericardial and perivisceral fluid. In pericardial fluid K^+ was less concentrated, and H^+, Cl^-, and NH_4^+ were elevated in both pericardial and perivisceral fluids. These data were interpreted by Bernard *et al.* (1966) as indicating that communication between the perivisceral fluid and seawater via the abdominal pores and between the pericardial cavity and the abdominal cavity via the pericardioperitoneal canals must be nonexistent or limited.

ii. Cerebrospinal and cranial fluids. The composition of the cranial fluid of several elasmobranchs was found by Smith (1929a) to approximate that of a plasma dialysate (Table XV). Davson and Grant (1960) compared subdural fluid (cranial fluid) to true cerebrospinal fluid (CSF) in *Mustelus canis* and found CSF to be isosmolar and similar in Cl^- concentration to plasma while the subdural fluid had a lower osmotic pressure but a similar Cl^- concentration (Table XV). In *Raja clavata* cranial fluid had a significantly lower concentration of K^+ than the plasma; concentrations of sodium chloride and urea were not different (Murray and Potts, 1961) (Table XV). Maren (1926a) examined the true cerebrospinal fluid of 200 *Squalus acanthias* and found the Cl^- concentration to be 7% higher than in the plasma. In the CSF, Na^+ and K^+ were slightly or questionably higher (Table XV). The carbonic anhydrase inhibitor acetazolamide abolished the normal Cl^- excess in the CSF indicating that Cl^- is being actively secreted into the CSF (Maren, 1926b). Acetazolamide also changed the CSF–plasma ratio for carbon dioxide from 1.15 to 1.72 suggesting that carbonic anhydrase may be responsible for the removal of metabolic carbon dioxide from the central nervous system (Maren and Frederick, 1958). Cranial fluid and plasma were found to have similar

Table XV
Chondrichthyes— Chemistry of Other Body Fluids

Fish and fluid	Na	K	Mg	Ca	Cl	HCO₃	CO₂	PO₄	SO₄	Urea	TMAO	Protein (g%)	Total ions	Units	Osmotic pressure (mOsm/liter)	Comments	Author	Date
Elasmobranchii Marine																		
Raja stabuloforis	—	—	—	—	—	—	—	—	—	—	—	—	—	—	—	—	Smith	1929a
Cranial fluid	264	4.1	0.9	4.9	263	—	5.55	1.0	—	436	—	—	—	mmoles/liter	—	—	—	—
Periviseral fluid	283	5.8	17.9	2.4	309	—	0.35	0.7	7.2	450	—	—	—	mmoles/liter	—	—	—	—
Raja stabuloforis	—	—	—	—	—	—	—	—	—	—	—	—	—	—	—	—	—	—
Pericardial fluid	335	20.2	2.5	0.6	369.5	—	0.4	0.7	—	364	—	—	—	mmoles/liter	—	—	—	—
Raja diaphenes	—	—	—	—	—	—	—	—	—	—	—	—	—	—	—	—	—	—
Cranial fluid	—	—	—	—	—	—	4.6	0.9	0.4	198	—	—	—	—	—	—	—	—
Periviseral fluid	155	6.2	14.0	2.1	188	—	—	0.7	8.0	254	—	—	—	mmoles/liter	—	—	—	—
Pericardial fluid	—	—	—	—	—	—	—	—	—	—	—	—	—	—	—	—	—	—
Carcharias littoralis	—	—	—	—	—	—	—	—	—	—	—	—	—	—	—	—	—	—
Cranial fluid	—	—	—	—	—	—	10.5	—	—	317	—	—	—	mmoles/liter	—	—	—	—
Periviseral fluid	276	8.9	21.4	5.1	306	—	—	0.6	1.27	355	—	—	—	mmoles/liter	—	—	—	—
Pericardial fluid	290	9.3	2.7	2.8	—	—	—	—	—	—	—	—	—	—	—	—	—	—
Mustelus canis	—	—	—	—	—	—	—	—	—	—	—	—	—	—	—	—	—	—
Cranial fluid	270	6.0	1.0	4.0	260	—	8.8	1.7	1.0	300	—	—	—	mmoles/liter	—	—	—	—
Periviseral fluid	—	—	—	—	—	—	—	—	—	—	—	—	—	—	—	—	—	—
Pericardial fluid	—	—	—	—	—	—	—	—	—	—	—	—	—	—	—	—	—	—
Raja erinacea	—	—	—	—	—	—	—	—	—	—	—	—	—	—	—	—	Hartman et al.	1944
Periviseral fluid	—	10	21	—	—	—	3	—	—	—	—	—	—	mmoles/liter	—	—	—	—
Acanthias vulgaris	—	—	—	—	300	—	—	—	—	—	—	—	—	mmoles/liter	—	—	Jensen and Vilstrup	1954
Endolymph	280	60	—	—	—	—	—	—	—	—	—	—	—	—	—	—	—	—
Perilymph	247	37	—	—	—	—	—	—	—	—	—	—	—	—	—	Potassium value erroneous	—	—
Squalus acanthias	—	—	—	—	—	—	—	—	—	—	—	—	—	—	—	—	Maren and Frederick	1958
Plasma	—	—	—	—	—	—	7.4	—	—	—	—	—	—	mmoles/liter	—	—	—	—
Aqueous humor	—	—	—	—	—	—	8.9	—	—	—	—	—	—	mmoles/liter	—	Control	—	—
Aqueous humor	—	—	—	—	—	—	5.3	—	—	—	—	—	—	mmoles/liter	—	Acetazolamide treated	—	—

Fluid												Units	Osmolality	Notes	Reference	Year
Mustelus canis	—	—	—	—	—	—	—	—	—	—	—	—	—	—	Davson and Grant	1960
Aqueous humor	—	—	—	—	245	—	—	—	—	—	—	mmoles/kg	947	Hypotonic to plasma	—	—
Cerebrospinal fluid	—	—	—	—	282	—	—	—	—	—	—	—	964	Isotonic to plasma	—	—
Subdural fluid	—	—	—	—	273	—	—	—	—	—	—	—	944	Similar to aqueous humor	—	—
Mustelus canis	—	—	—	—	—	—	—	—	—	—	—	—	—	—	Doolittle et al.	1960
Aqueous humor	279	7	3	3	256	15	1	320	85	22	980	mmoles/kg	935	—	—	—
Mustelus canis	—	—	—	—	—	—	—	—	—	—	—	—	960.7/kg	—	Bloete et al.	1961
Anterior aqueous humor	—	—	—	—	—	—	—	—	—	—	—	—	—	—	—	—
Posterior aqueous humor	—	—	—	—	—	—	—	—	—	—	—	—	952.8	—	—	—
Vitreous humor	—	—	—	—	—	—	4	—	—	—	—	—	968	—	—	—
Raja clavata	—	—	—	—	—	—	—	—	—	—	—	—	—	—	Murray and Potts	1961
Cranial fluid	280	3.3	—	—	311	—	—	437	—	—	—	—	—	—	—	—
Perilymph	281	3.5	—	—	321	—	—	447	—	—	—	—	—	—	—	—
Endolymph jelly	295	63.4	—	—	391	—	—	381	—	—	—	—	—	—	—	—
Lorenzini jelly	443	12.5	—	—	581	—	—	75	—	—	—	—	—	—	—	—
Squalus acanthias	—	—	—	—	—	—	—	—	—	—	—	—	—	—	Maren	1962a
Cerebrospinal fluid	271	7.8	—	—	264	—	—	—	—	—	—	mmoles/liter	986/kg	—	—	—
Aqueous humor	276	7.5	—	—	251	—	—	—	—	—	—	—	979	—	—	—
Squalus acanthias	—	—	—	—	—	—	—	—	—	—	—	—	—	—	Rodnan et al.	1962
Coelomic fluid	296	4.4	3.7	2.2	328	0	—	415	—	0.02	—	mmoles/liter	1005	—	—	—
Raja clavata	—	—	—	—	—	—	—	—	—	—	—	—	—	—	Enger	1964
Cranial fluid	286	4.6	—	—	255	—	—	—	—	—	—	mmoles/liter	—	—	—	—
Endolymph	287	58.7	—	—	322	—	—	—	—	—	—	—	—	—	—	—
Cetorhinus maximus	—	—	—	—	—	—	—	—	—	—	—	—	—	—	—	—
Cranial fluid	248	3.9	—	—	238	—	—	—	—	—	—	—	—	—	—	—
Endolymph (saccular cavity)	276	56.0	—	—	319	—	—	—	—	—	—	—	—	—	—	—
Endolymph semicircular canal	286	86.9	—	—	378	—	—	—	—	—	—	—	—	—	—	—
Perilymph	259	3.1	—	—	237	—	—	—	—	—	—	—	—	—	—	—
Dasyatis americana	—	—	—	—	—	—	—	—	—	—	—	—	—	—	Bernard et al.	1966
Perivisceral fluid	255	20.4	2.6	0.3	310	—	—	365	—	0.111	—	mmoles/liter	827	—	—	—
Pericardial fluid	262	11.9	0.9	0.7	308	—	—	340	—	0.031	—	mmoles/liter	828	—	—	—
Dasyatis say	—	—	—	—	—	—	—	—	—	—	—	mmoles/liter	—	—	—	—
Perivisceral fluid	236	20.5	3.9	1.5	289	—	—	356	—	0.153	—	mmoles/liter	813	—	—	—
Pericardial fluid	258	11.7	1.9	0.8	295	—	—	354	—	0.040	—	mmoles/liter	810	—	—	—

Table XV (*Continued*)

Fish and fluid	Na	K	Mg	Ca	Cl	HCO3	CO2	PO4	SO4	Urea	TMAO	Protein (g%)	Total ions	Units	Osmotic pressure (mOsm/liter)	Comments	Author	Date
Seawater	346	21.4	30	20.1	399	—	—	—	—	—	—	—	—	mmoles/liter	709	—	—	—
Napaprion brevirostris	—	—	—	—	—	—	—	—	—	—	—	—	—	—	—	—	Oppelt *et al.*	1966
Ventricular fluid	317	5.2	—	—	285	—	8.4	—	—	—	—	—	—	mmoles/liter	—	—	—	—
Extrabrain fluid	313	5.3	—	—	290	—	9.4	—	—	—	—	—	—	—	—	—	—	—
Ginglymostoma cirratum	—	—	—	—	—	—	—	—	—	—	—	—	—	—	—	—	—	—
Ventricular fluid	332	5.5	—	—	310	—	8.3	—	—	—	—	—	—	—	—	—	—	—
Squalis acanthias	—	—	—	—	—	—	—	—	—	—	—	—	—	—	—	—	Cserr and Rall	1967
Plasma	—	4.1	—	—	—	—	—	—	—	—	—	—	—	—	—	—	—	—
Plasma	—	6.0	—	—	—	—	—	—	—	—	—	—	—	—	—	—	—	—
Cerebrospinal fluid	—	3.5	—	—	—	—	—	—	—	—	—	—	—	—	—	Anoxic 15 min	—	—
Cerebrospinal fluid	—	5.8	—	—	—	—	—	—	—	—	—	—	—	—	—	—	—	—
Extradural fluid	—	3.4	—	—	—	—	—	—	—	—	—	—	—	—	—	Anoxic 15 min	—	—
Extradural fluid	—	3.2	—	—	—	—	—	—	—	—	—	—	—	—	—	Anoxic 15 min	—	—
Elasmobranchii Freshwater																		
Pristis microdon	—	—	—	—	—	—	—	—	—	—	—	—	—	—	—	—	Smith	1931
Coelemic fluid	—	—	—	—	204	—	—	—	—	150	—	—	—	mmoles/liter	538	—	—	—
Pericardial fluid	—	—	—	—	201	—	—	—	—	98	—	—	—	—	—	—	—	—
Dasyatis uarnak	—	—	—	—	—	—	—	—	—	—	—	—	—	—	—	—	—	—
Coelemic fluid	—	—	—	—	278	—	—	—	—	—	—	—	—	—	—	—	—	—
Pericardial fluid	—	—	—	—	216	—	—	—	—	—	—	—	—	—	—	—	—	—
Carcharhinus leucas	—	—	—	—	—	—	—	—	—	—	—	—	—	—	—	—	Murdaugh and Robin	1967
Cranial fluid	247.4	4.7	1.8	3.0	219.3	6.6	—	0.18	0.5	—	—	1.3	—	mmoles/liter	—	—	—	—
Periviseral fluid	211.0	5.8	2.2	0.8	217.4	—	—	—	0.025	—	—	Trace	—	—	—	—	—	—
Pericardial fluid	249.6	10.9	1.8	1.4	254.7	—	—	—	Trace	—	—	Trace	—	—	—	—	—	—

48

concentrations of Cl-, Na+, and K+ in *Raja clavata* and *Cetorhinus maximus* by Enger (1964) (Table XV). Oppelt *et al.* (1966) (Table XV) have compared the composition of CSF and extrabrain fluid (cranial fluid) in the lemon shark, *Nagaprion brevirostris*. Sodium chloride and total carbon dioxide concentrations were higher in CSF than in plasma. Extrabrain fluid was closer in composition to CSF than to plasma even though it has been regarded as being similar to plasma in inorganic composition. No connection has been found between the CSF and the extrabrain fluid, the former fluid having a relatively rapid turnover rate compared to the latter (Oppelt *et al.*, 1966).

Cserr and Rall (1967) have questioned the finding by Maren (1962a) and Oppelt *et al.* (1966) that the K+ concentrations in CSF and plasma are similar. Rapidly sampled *Squalus acanthias* CSF had a K+ concentration of 3.5 mmoles/kg (Table XV) which was relatively independent of the plasma K+ concentration and similar to the concentration observed in mammalian CSF.

iii. Ear fluids and Lorenzini jelly. In the mammal the perilymph is regarded as being similar in composition to the cerebrospinal fluid, while the endolymph contains a high concentration of K+ balanced by a low Na+ concentration (Potts and Parry, 1964). The first analyses of these fluids in elasmobranchs were carried out by Kaieda (1930) (*Scolioidontus laticandus*) and Jensen and Vilstrup (1954) (*Acanthias vulgaris*), and in both cases the K+ concentration in the endolymph was less than twice that in the perilymph. More recently, Murray and Potts (1961) (Table XV) have examined *Raja clavata* and found the perilymph to be similar in composition to the serum while the endolymph K+ concentration was 19 times as concentrated as that in the perilymph. This compares to a 30-fold differential reported for the mammal. Also, Na+ and Cl- were higher in the endolymph while urea was present at a lower concentration. The jelly occupying the tubes and ampullae of Lorenzini was found to be 5% hypertonic to seawater and to contain more Na+, K+, and Cl- and considerably less urea than plasma (Murray and Potts, 1961) (Table XV). Also, K+ and Cl- were found to be present at higher concentrations in endolymph in *Raja clavata* than in plasma by Enger (1964), but in this report no difference was shown in the Na+ concentrations (Table XV).

iv. Eye fluids. A complete analysis of aqueous humor and plasma in the smooth dogfish, *Mustelus canis,* has been published by Doolittle *et al.* (1960) (Table XV). Urea, trimethylamine oxide, Na+, Cl-, protein, and osmotic concentrations were lower in the aqueous humor while HCO_3^- was present at a concentration of 15 mmoles/kg compared to 6 mmoles/kg in the plasma. Davson and Grant (1960) also found low osmotic and Cl^- concentrations in the aqueous humor of this fish. Maren and Frederick (1958) found the ratio between carbon dioxide concentration in aqueous humor and plasma to be 1.2. This was lowered to 0.7 after acetazolamide and carbonic anhydrase activity was located in the ciliary process, iris, and retina. Thus, as in the mammalia, carbonic anhydrase may be responsible for the transfer of carbon dioxide into the aqueous humor. Bloete et al. (1961) measured the osmotic concentrations of the eye fluids of Mustelus canis and found them to be in the order arterial plasma > anterior aqueous > posterior aqueous > vitreous. This sequence suggests that water probably does not pass through the cornea from the hypotonic seawater.

In contrast to the earlier work, Maren (1962a) (Table XV) found the carbon dioxide equilibrium and osmotic pressure to be the same in the plasma and aqueous humor of Squalis acanthias. The observation that Na^+ was definitely more concentrated and that K^+ was slightly more concentrated in the aqueous humor led Maren (1962a) to suggest that Na^+ secretion may be the major factor in aqueous humor formation. It is interesting to note, however, that Maren (1967) recently referred to his 1962 data (Table XV) as indicating a slightly greater amount of HCO_3^- in aqueous humor than in plasma in Squalus acanthias.

2. SUBCLASS ELASMOBRANCHII (FRESHWATER SPECIES)

a. Plasma Composition. Four Malaysian elasmobranchs collected in freshwater were examined by Smith (1931). These were the shark, Carcharhinus melanopterus, the sawfish, Pristis microdon, and two rays, Dasyatis uarnak and Hypolophus sephen. Most of the experiments were carried out on Pristis microdon caught from the Perak River at least 25 miles upstream from the last traces of seawater. Plasma osmotic concentration was 548 mOsm/liter in Pristis microdon. This value is much lower than that found in marine elasmobranchs but higher than that found in freshwater teleosts. The blood urea concentration was only 30%, and the Cl^- concentration only 75% of the concentrations observed in marine elasmobranchs (Table XV).

Although elasmobranchs are widely distributed in freshwater (Herre, 1955), recent studies have been confined to those occurring in Lake Nicaragua and the Rio San Juan which connects it to the ocean (Urist, 1961; Thorson, 1967) (Table XIV). Urist (1961) compared the serum composition of Carcharhinus leucas nicaraguensis from Rio San Juan with those of the teleost, Megalops atlanticus, from the same location and Carcharhinus leucas leucas from a marine habitat. The total ion concentration of the serum in the freshwater *Carcharhinus* was 83%, the Ca^{2+} 66%, and the urea 30% of the values observed in the marine *Carcharhinus* (Table XIV). The serum of *Megalops atlanticus* had a quite different composition having a much lower Na^+ concentration and a minimal concentration of urea (Table XVI). The freshwater sharks and sawfish of the Rio San Juan and Lake Nicaragua are now believed to have migrated freely from the Atlantic Ocean and thus to be identical to those occurring in the ocean and capable of osmoregulating both in the seawater and in freshwater (Thorson, 1967). Serum concentrations in *Pristis perotteti* and *Carcharhinus leucas* from Lake Nicaragua were similar to those observed for *Carcharhinus leucas nicaraguensis* (Urist, 1961) except that the Na^+ and urea concentrations were somewhat higher in the *Carcharhinus leucas* from Lake Nicaragua (Thorson, 1967) (Table XIV).

b. Composition of Other Fluids. Early work by Smith (1931) showed that urea is present in both the perivisceral and pericardial fluids of the freshwater elasmobranchs, the concentration being lowest in the pericardial fluid as in the marine elasmobranchs (Table XV). The pericardial fluid of *Carcharhinus leucas* from Nicaragua has the expected high K^+ concentration; however, the perivisceral fluid of this fish does not have the elevated Cl^- concentration found in some marine elasmobranchs (Thorson, 1967) (Table XV). The cranial fluid of *Carcharhinus leucas* from Lake Nicaragua is similar in composition to the serum except that it has a lower phosphate and protein concentration (Thorson, 1967) (Table XV).

3. SUBCLASS HOLOCEPHALI

The first determination of the osmotic pressure of the blood of the ratfish, *Hydrolagus colliei*, indicated that it was slightly hypoosmotic to seawater (801 mOsm/liter) (Nicol, 1950). Fänge and Fugelli (1962) examined the blood of *Chimaera monstrosa* and concluded that it was probably isosmotic with seawater. It differed from the marine elasmobranchs in having higher Na^+ and Cl^- concentrations and a lower urea concentration (Table XIV). The observations of Fänge and Fugelli have since been confirmed by Urist (1966) who observed a similar situation with respect to Na^+, Cl^-, and urea in *Hydrolagus colliei* (Table XIV).

C. Class Osteichthyes

1. SUBCLASS SARCOPTERYGII

a. Order Coelacanthiformes. Morphological studies have shown that the coelacanths evolved from a group of fish lying close to the ancestral

Table XVI
Osteichthyes—Blood Chemistry

Fish or medium	Na	K	Mg	Ca	NH₄	Cl	HCO₃	CO₂	PO₄	SO₄	Urea	Protein (g%)	Total ions	Units	Osmotic pressure (mOsm/liter)	Comments	Author	Date
Coelacanthiformes																		
Latimeria chalumnae	181	51.3	14.4	3.5	—	199	4.7	—	—	—	355	5.1	—	mmoles/liter	1181	Fish frozen, blood hemolyzed	Pickford and Grant	1967
Sarcopterygii																		
Seawater	—	—	—	—	—	—	—	—	—	—	—	—	—	—	1090			
Dipteriformes																		
Proptopterus aethiopicus	99.0	8.2	Trace	2.1	—	44.1	—	35.0	1.0	Trace	0.6	—	—	mmoles/liter	238	Nonestivating male	Smith	1930
Actinopterygii																		
Acipenseriformes																		
Acipenser stellatus	—	—	—	—	—	96	—	—	—	—	3.5	—	—	mmoles/liter	—	Seawater	Korjuev	1938
Acipenser stellatus	—	—	—	—	—	92	—	—	—	—	3.4	—	—	—	—	Freshwater	—	—
Acipenser sturio	—	—	—	—	—	92	—	—	—	—	3.3	—	—	—	—	Seawater	—	—
Acipenser sturio	—	—	—	—	—	93	—	—	—	—	3.5	—	—	—	—	Freshwater	—	—
Acipenser oxyrhynchus	150.6	2.67	0.9	1.9	—	112.9	—	—	3.1	—	—	—	—	mmoles/liter	—	Young fish freshwater	Magnin	1962
Acipenser oxyrhynchus	164.9	2.84	1.3	1.5	—	132.9	—	—	2.5	—	—	—	—	—	—	Young fish seawater	—	—
Acipenser sturio	163.6	4.65	1.57	2.1	—	126.4	—	—	4.6	0.5	—	—	—	—	343ᵃ	Adult seawater	—	—
Acipenser sturio	155.8	4.3	1.47	2.3	—	119.7	—	—	5.0	0.7	—	—	—	—	318ᵃ	Adult freshwater	—	—
Acipenser fulvescens	143	4.2	1.25	1.28	—	107	—	—	2.6	—	—	—	—	—	—	Ottawa River	—	—
Acipenser fulvescens	143	3.56	0.85	1.25	—	104	—	—	2.4	—	—	—	—	—	—	St. Lawrence River	—	—
Acipenser fulvescens	148	4.46	1.6	1.4	—	113	—	—	2.7	—	—	—	—	—	—	Lake Nipissing (mature)	—	—
Acipenser transmontanus	130	2.5	2.1	1.7	—	115.1	5.2	—	2.9	0.4	1.0	2.6	259.9	mmoles/liter	—	Seawater males	Urist and Van de Putte	1967
Acipenser transmontanus	129	2.7	2.0	1.8	—	111.0	6.0	—	3.3	0.5	0.9	2.5	256.3	—	—	Freshwater males and females	—	—
Acipenser transmontanus	123	2.0	1.1	4.6	—	116.0	5.0	—	4.1	0.2	1.1	4.0	256.0	—	—	Freshwater mature females	—	—
Amiiformes																		
Amiatus calva	132.5	2.0	0.4	5.3	—	119.5	—	4.3	4.1	2.2	—	—	—	mmoles/liter	—	—	Smith	1929b

Species										Units		Reference	Year
Lepisosteiformes													
Lepisosteus osseus	140	2.7	0.3	6.1	118	—	9.1	3.9	2.4	mmoles/liter	—	Smith	1929b
Lepisosteus productus	149	5.0	—	—	141	—	—	—	—	mmoles/liter	Low salinity species	Sulya et al.	1960
Lepisosteus spatula	153	6.2	—	—	129	—	—	—	—	—	Euryhaline species	—	—
Lepisosteus osseus	159	4.2	—	—	133	—	—	—	—	—	Low salinity species	—	—
Teleostei—Marine													
Syngnathus	212	16.5	9.7	3.4	192	—	4.7	—	20.5	mmoles/liter	—	Edwards and Condorelli	1928
Hippocampus	162	—	—	—	146	—	8.0	—	44.1	mmoles/liter	—	—	—
Lophius	229	6.6	3.7	2.3	162	—	3.1	—	9.4	mmoles/liter	—	—	—
Mureana	253	—	—	3.9	177	—	3.1	—	14.9	—	—	—	—
Lophius piscatorius	198	7.4	5.8	2.9	186	6.6	5.3	—	2.7	mmoles/liter	—	Smith	1929b
Gadus callarias	180	4.9	3.8	5.0	158	4.8	5.3	—	1.0	mmoles/liter	—	—	—
Spheroides maculatus	—	3.2	4.5	8.3	162	8.0	2.4	—	—	mmoles/liter	—	—	—
Conger vulgaris	—	—	—	—	186	—	2.7	4.5	— (430[a])	mmoles/liter	—	Boucher-Firley	1934
Mureana helena	—	—	—	—	187	—	—	5.6	— (441[a])	mmoles/liter	—	—	—
Conger vulgaris	180	—	—	—	161	—	—	5.6	— (425[a])	—	—	Pora	1936d
Lophius piscatorius	211.8	5.1	2.5	2.8	196	8.0	9.8	3.9	5.7 (452[a])	mmoles/kg	—	Brull and Nizet	1953
Mureana helena	564	1.95	2.43	3.87	188.4	2.07	9.6[b]	8.0	34.0	mmoles/kg	—	Robertson	1954
Seawater	—	12.0	64.4	12.4	659	8.0	—	—	1.4	—	—	—	—
Lophius americanus	198	3.4	0.8	2.2	177	3.6	5.2	4.3	— (396)	mmoles/liter	Average of fish 11 and 12	Forster and Berglund	1956
Scomberomorus maculatus	188	9.8	—	—	167	—	—	3.5	— (386)	mmoles/liter	—	Becker et al.	1958
Thunnus thynnus	190	26.8	—	—	181	—	—	6.5	— (437)	—	—	—	—
Mycteroperca venenosa	190	6.4	—	—	181	—	—	2.3	— (467)	—	—	—	—
Sphyraena barracuda	215	6.4	—	—	189	—	—	3.3	— (476)	—	—	—	—
Mycteroperca bonasi	228	7.9	—	—	208	—	—	2.8	— (461)	—	—	—	—
Promicrops itaiara	200	5.8	—	—	166	—	—	5.6	— (384)	—	—	—	—
Seawater	445	9.8	—	—	537	3.6	—	—	1.2 (1070)	—	—	—	—
Brevoortia patronus	178	11.8	—	—	154	—	—	—	—	mmoles/liter	Euryhaline species	Sulya et al.	1960
Dorosoma petenese	166	14.3	—	—	163	—	—	—	—	—	Low salinity species	—	—
Dorosoma cepedianum	163	2.8	—	—	—	—	—	—	—	—	Low salinity species	—	—
Elops saurus	176	9.8	—	—	106	—	—	—	—	—	Euryhaline species	—	—
Galeichthys felis	187	10.0	—	—	178	—	—	—	—	—	Euryhaline species	—	—
Bagre marina	185	10.8	—	—	189	—	—	—	—	—	Euryhaline species	—	—
Mugil cephalus	177	6.3	—	—	153	—	—	—	—	—	Euryhaline species	—	—
Mugil curema	179	5.6	—	—	151	—	—	—	—	—	Euryhaline species	—	—
Oligophites saurus	182	7.0	—	—	167	—	—	—	—	—	High salinity species	—	—

Table XVI (*Continued*)

Fish or medium	Na	K	Mg	Ca	NH₄	Cl	HCO₃	CO₂	PO₄	SO₄	Urea	Protein (g%)	Total ions	Units	Osmotic pressure (mOsm/liter)	Comments	Author	Date
Caranx hippos	201	3.0	—	—	—	170	—	—	—	—	—	—	—	—	—	Euryhaline species	—	—
Coryphaena hippurus	192	9.2	—	—	—	170	—	—	—	—	—	—	—	—	—	High salinity species	—	—
Cynoscion nebulosus	195	7.8	—	—	—	251	—	—	—	—	—	—	—	—	—	Euryhaline species	—	—
Cynoscion arenarius	219	5.8	—	—	—	—	—	—	—	—	—	—	—	—	—	Euryhaline species	—	—
Micropogon undulatus	179	7.6	—	—	—	160	—	—	—	—	—	—	—	—	—	Euryhaline species	—	—
Sciaenops ocellata	189	8.6	—	—	—	—	—	—	—	—	—	—	—	—	—	Euryhaline species	—	—
Leiostomus xanthurus	176	3.1	—	—	—	—	—	—	—	—	—	—	—	—	—	Euryhaline species	—	—
Pogonias cromis	156	6.8	—	—	—	—	—	—	—	—	—	—	—	—	—	Euryhaline species	—	—
Lagodon rhomboides	183	8.3	—	—	—	174	—	—	—	—	—	—	—	—	—	Euryhaline species	—	—
Archosargus probatocephalus	163	7.6	—	—	—	157	—	—	—	—	—	—	—	—	—	Euryhaline species	—	—
Scomberomorus maculatus	181	8.8	—	—	—	178	—	—	—	—	—	—	—	—	—	High salinity species	—	—
Paralabrax clathratus	180	5.0	1.5	3.0	—	147	10.0	—	1.3	0.5	0.1	3.5	—	mmoles/liter	—	—	Urist	1962b
Gadus morhua	185	3.7	—	—	—	157	—	—	—	—	—	—	—	mmoles/liter	—	—	Enger	1964
Cottus scorpius	184	3.8	—	—	—	173	—	—	—	—	—	—	—	—	—	—	—	—
Salmo irideus	170	1.4	—	—	—	135	—	—	—	—	—	—	—	—	—	—	—	—
Serranus scriba	189	—	—	—	—	159	—	—	—	—	—	—	—	mmoles/liter	—	Seawater	Motais *et al.*	1965
Serranus scriba	136	—	—	—	—	109	—	—	—	—	—	—	—	—	—	Freshwater 3 hr		
Sphaeroides maculatus	98	17.1	—	3.4	—	185	—	—	—	—	—	—	—	mmoles/liter	—	Control	Eisler and Edmunds	1966
Sphaeroides maculatus	107	23.4	—	3.6	—	183	—	—	—	—	—	—	—	mmoles/liter	—	1 ppm Endrin 96 hr		
Seawater	175	6.8	—	7.3	—	406	—	—	—	—	—	—	—	mmoles/liter	—	—	—	—
Teleostei—Freshwater																		
Tinca vulgaris	140.9	—	—	—	—	106	—	—	—	—	—	—	—	—	280ª	—	Keys and Hill	1934
Esox lucius	—	—	—	—	—	—	—	—	—	—	—	—	—	—	274ª	—	—	—
Coregonus clupoides	128	3.81	1.69	2.67	0.33	116.8	10.58	—	8.7ᵇ	2.29	0.7	4.2	—	mmoles/liter	—	—	Robertson	1954
Micropterus dolomieue	128	2.5	—	—	—	123	—	—	1.68	—	—	2.6	—	mmoles/liter	—	June 24	Shell	1959
	128	2.3	—	3.4	—	111	—	—	1.32	—	—	2.2	—	—	—	July 24	—	—
	140	3.9	—	2.5	—	128	—	—	1.28	—	—	2.3	—	—	—	August 27	—	—
	140	2.1	—	3.0	—	116	—	—	0.99	—	—	2.7	—	—	—	October 7	—	—
	140	2.3	—	4.9	—	110	—	—	0.97	—	—	4.5	—	—	—	November 18	—	—

54

Species																	
Megalops atlanticus	101	6.2	1.4	2.5	—	140	10.1	—	5.0	0.40	6.0	6.0	266.6	mmoles/liter	—	Anadromous taken in freshwater	Urist	1961
Salmo gairdnerii irideus	139.2	4.7	—	2.2	—	128.1	—	—	1.0	—	—	3.7	—	mmoles/liter	—	1–4 days after capture	Huggel *et al.*	1963
Squalus cephalus	115.5	5.1	—	1.8	—	98.1	—	—	1.2	—	—	4.3	—	—	—	12 days after capture	—	—
Carassius auratus	140.7	4.2	—	2.2	—	121.1	—	—	0.8	—	—	4.5	—	—	—	Caudal plasma	Levine and Musallam	1964
	—	—	—	—	—	—	—	—	—	—	—	—	—	—	289[a]	—	—	—
Salvelinus namaycush	—	—	—	—	—	117.3	6.40	6.65	—	—	—	—	—	mmoles/kg	298	Control	Hoffert and Fromm	1966
Cyprinus carpio	130	2.93	1.23	2.12	—	111.5	8.47	8.96	—	—	—	—	—	—	—	Diamox	Houston and Madden	1968
	—	—	—	—	—	125.2	—	—	0.39	—	—	2.2	—	mmoles/liter	274	17°C	—	—

[a] Derived from freezing-point depression.
[b] Expressed in mEq/kg water present as $H_2PO_4^-$ and HPO_4^{2-} with a valency of 1.84.

stock of gnathostomes, the Osteolepids (Rhipidistia), which gave rise to the lungfishes (Dipnoi), and also to the ancestral amphibia (Young, 1950). The Chondrichthyes, Sarcopterigii, and Amphibia differ biochemically from the Actinopterygii in that the three former groups are able to form urea by the ornithine urea cycle (Brown and Brown, 1967). Although the blood chemistry of only one specimen of the living coelacanth, *Latimeria chalumnae*, has been examined it is evident that these fish possess a blood urea concentration within the range found in the marine elasmobranchs and that the serum osmolarity is close to that of seawater (Pickford and Grant, 1967) (Table XVI). Blood from the renal and hepatic portal veins was hyperosmotic to seawater while blood from the heart was hyposmotic. If the value for heart blood turns out to be the correct one then *Latimeria* must drink seawater and excrete the excess ions ingested. But, if the blood is indeed hyperosmotic then *Latimeria* resembles an elasmobranch. The serum Na^+ concentration was lower than that of marine elasmobranchs and similar to that of marine teleosts. The measured values for blood K^+ and Mg^{2+} were both high owing to hemolysis. The Mg^{2+} concentration was also high in the aqueous humor (Table XVII), but in the elasmobranchs the Mg^{2+} concentration of this fluid is similar to that of plasma. Similar high Mg^{2+} concentrations had previously only been found in stressed *Lophius piscatorius* (Brull and Cuypers 1954). In contrast to the normal situation in teleosts, *Latimeria* had a blood Cl^- concentration higher than that for Na^+. This may, however, be explained by the complete hemolysis which occurred in the blood sample.

b. Order Dipteriformes. The lungfishes normally have a very low concentration of urea in their plasma (Table XVI). During the dry season, the shallow freshwaters which they inhabit dry up and the fish estivate in the underlying mud. During this period of water and nutritional deprivation the fish relies on tissue protein as an energy source and the amino acid nitrogen is metabolized to urea. To conserve the water, which would be necessary for urea excretion, the urea is allowed to accumulate in the blood and muscle until the end of the estivation period (Smith, 1930; Janssens, 1964). The inorganic composition of the blood of the normal lungfish was determined by Smith (1930). The plasma Cl^- concentration which he observed appears to be very low when compared to that in teleosts.

2. SUBCLASS ACTINOPTERYGII

a. Order Acipenseriformes. Although the sturgeons have an almost wholly cartilaginous skeleton and bear other resemblances to the elasmo-

branchs, it is now accepted that they are of actinopterygian descent
(Young, 1950). Physiological uremia does not occur in *Acipenser* either
in freshwater or in saltwater (Korjuev, 1938) (Table XVI). An investigation has been carried out into the changes in blood chemistry which occur
during the migration of young sturgeon and spawned adults from freshwater into saltwater and of the migration of prespawning adults in the
reverse direction by Magnin (1962). When immature *Acipenser oxyrhynchus* migrated from freshwater to saltwater there were increases
in the plasma concentrations of Na^+, Mg^{2+}, and Cl^- and decreases in the
concentrations of Ca^{2+} and phosphate (Table XVI). Migration of prespawning *Acipenser sturio* from the sea into freshwater resulted in a
significantly lower plasma osmolar, Na^+ and Cl^- concentration. Plasma
SO_4^{2-} was elevated in freshwater. Magnin (1962) noted a greater degree
of variability in the ionic composition of the mature fish and concluded
that gonadal muturation results in a perturbation of the metabolism of
water and electrolytes. In the third species which he investigated,
Acipenser fulvescens, Magnin compared specimens captured in the
Ottawa and St. Lawrence rivers with those from Lake Nipissing. The
mean plasma concentrations of Na^+, Cl^-, and Mg^{2+} were higher in the lake
fish, a consequence of the majority of these fish being sexually mature
(Table XVI). Magnin concluded that the homeostatic mechanisms of
the sturgeon like those of the salmon did not completely regulate the
osmotic pressure and ionic composition of the blood when the fish passed
from one osmotic environment to another. Instead there were small but
significant changes in the plasma concentrations of the ions.

In a recent investigation, the plasma of *Acipenser transmontanus* was
found to have only a slightly lower total concentration of ions in freshwater than in seawater (Urist and Van de Putte, 1967) (Table XVI) indicating the presence of a very effective homeostatic mechanism in this
species. The low concentration of Ca^{2+} in the plasma of this and other
species of sturgeon (Table XVI) is unique and may be related to the
progressive loss of bone during the evolution of the Acipenseriformes
(Urist and Van de Putte, 1967). An exception to this generalized hypocalcemia are the mature female *Acipenser transmontanus* which show the
hypercalcemia and hyperphosphotemia characteristic of mature female
teleosts.

b. Orders Amiformes and Lepisosteiformes. Representatives of each
of these two holostean orders are found in the Great Lakes. Of the two,
the long-nosed gar, *Lepidosteus osseus,* is the more primitive, while
Amiatus calva diverged more recently from the teleostean stock (Young,
1950). Blood analyses by Smith (1929b), however, revealed no major

58   W. N. HOLMES AND EDWARD M. DONALDSON

differences between the two species or between them and the freshwater teleosts except that the two holosteans had approximately double the Ca^{2+} and one quarter the Mg^{2+} found in *Coregonus clupoides* (Robertson, 1954) (Table XVI). Three species of gars including *Lepidosteus osseus* examined from the Gulf of Mexico (Sulya *et al.*, 1960) (Table XVI) had higher concentrations of Na^+, K^+, and Cl^- in the plasma than were reported by Smith, but this may be explained by the presumably higher environmental salinity.

3. GROUP "TELEOSTI"

a. Blood Chemistry. Although many analyses of the blood chemistry of marine teleosts were carried out in the first half of this century, Robertson (1954) appears to be the first worker to have carried out an analysis of the environmental seawater for comparison. In his studies, Na^+ comprised 92% and K^+ less than 1% of the cations in plasma while Cl^- accounted for 87% of the anions. Only HCO_3^- and phosphate ions were present at higher concentrations in the blood plasma than in seawater. The abundances of the other ions in plasma with respect to their concentrations in seawater were in the following order: $Na^+ > Ca^{2+} > Cl^- > SO_4^{2-} > K^+ > Mg^{2+}$, where plasma Na^+ was 38% of the Na^+ concentration in seawater and plasma Mg^{2+} was 4% of the concentration of this ion in seawater. The total concentration of ions plus urea was 32% of the seawater ionic concentration (Robertson, 1954) (Table XVI). In an earlier analysis of the plasma of *Mureana* and three other teleosts, erroneously high urea concentrations were obtained (Edwards and Condorelli, 1928) (Table XVI).

Analysis of the plasma composition of the freshwater teleost, *Coregonus clupoides*, showed it to have a total ion concentration which was only 66% of that in *Mureana helena*, thus the freshwater teleost does not maintain ionic concentrations as high as those in the marine teleost. All ions except K^+ and HCO_3^- were lower in *Coregonus* with the major ions of sodium and chloride showing the greatest differences (Robertson, 1954) (Table XVI). Apart from the data of Robertson other complete analyses have been presented for the marine teleosts, *Lophius piscatorius* (Brull and Nizet, 1953), *Lophius americanus* (Forster and Berglund, 1956), and *Paralabrax clathratus* (Urist, 1962b) and for the teleost *Megalops atlanticus* (Urist, 1961) captured in freshwater (Table XVI). Other investigators have covered a greater number of different species but measured only Na^+, K^+, and Cl^- (Becker *et al.*, 1958; Sulya *et al.*, 1960) (Table XVI). One observation of note from the former paper is the high plasma K^+ concentration in the bluefin tuna, *Thunnus thynnus*

(Table XVI). However since only one fish was examined it is not possible to determine whether this is a characteristic of the species as a whole. The only other plasma K^+ concentration which approached that in *Thunnus thynnus* was reported by Eisler and Edmunds (1966) for *Sphaeroides maculatus* treated with the insecticide Endrin (Table XVI).

An interesting study was carried out by Shell (1959) to determine whether seasonal changes occur in the electrolyte concentrations of the freshwater teleost, *Micropterus dolomieue*. Although the investigation was only carried out between June and November in one year, some seasonal differences are apparent (Table XVI). This observation may indicate the presence of some cyclic metabolic patterns. These cycles were related by Shell to a presumed change in the endocrine status of the bass during the period of the study. Experiments have shown that the stress of capture and transportation on *Salmo gairdnerii irideus* resulted in increased plasma Na^+ and Cl^- concentrations and decreased plasma K^+ and protein concentrations. These concentrations returned to normal after 12 days (Huggel *et al.*, 1963) (Table XVI).

b. Euryhaline Species. Many teleosts are stenohaline and thus limited to habitats having only a narrow range of osmolarity such as freshwater rivers and lakes or the open sea. Other species are euryhaline and can accommodate themselves by means of homeostatic mechanisms to changes in external ionic concentrations. The euryhaline species may be divided into two groups, one of which inhabits the interface between the true freshwater and the true seawater environments, e.g., in an estuary. The second group of euryhaline fishes spend part of their life cycle in freshwater and part of it in seawater and may be divided into catadromous species which migrate downstream to spawn in the sea, e.g., *Anguilla rostrata*, and anadromous species which migrate upstream from the sea to spawn in freshwater, e.g., *Oncorhynchus nerka*. Much scientific effort has been expended in investigations of the changes and the reasons for the changes which occur in the plasma composition of these euryhaline species. We will first deal with some of the coastal euryhaline species and then, in more detail, some members of the orders *Anguilliformes* and *Salmoniformes*.

Intertidal blennies have been studied by Raffy (1949), House (1963), Gordon *et al.* (1965), and Evans (1967a). The plasma Na^+ and Cl^- concentrations in *Xiphister atropurpureus*, which has a salinity tolerance of from 10 to 100% seawater, were found to be stable over the range of 31 to 100% seawater but to fall by about 15% when transferred to salinities equivalent to between 31 and 10% seawater. Mature fish showed lower concentrations of Na^+ and Cl^- than the nonreproductive fish in 10% sea-

Table XVII

Fish or medium	Na	K	Mg	Ca	NH$_4$	Cl	HCO$_3$	CO$_2$	PO$_4$	SO$_4$	Urea	TMAO	Protein (g %)
				Teleostei—Euryhaline—General									
Alosa pseudoharengus	—	—	—	—	—	168	—	—	—	—	—	—	5.9
Alosa pseudoharengus	—	—	—	—	—	151	—	—	—	—	—	—	5.3
Fundulus kansae	160	3.9	—	2.5	—	—	—	—	—	—	—	—	—
Fundulus kansae	231	2.0	—	4.6	—	—	—	—	—	—	—	—	—
Fundulus kansae	157	1.2	—	3.0	—	—	—	—	—	—	—	—	—
Seawater	440	14.0	—	10.5	—	—	—	—	—	—	—	—	—
Pleuronectes flesus	193.7	5.4	—	—	—	166.1	—	—	—	—	—	0.7	—
Pleuronectes flesus	156.8	5.1	—	—	—	113.7	—	—	—	—	—	0.4	—
Platichthyes flesus	168	—	—	—	—	156	—	—	—	—	—	—	—
Platichthyes flesus	162	—	—	—	—	148	—	—	—	—	—	—	—
Platichthyes flesus	141.7	3.4	—	3.3	—	168.1	—	—	—	—	—	—	—
Platichthyes flesus	123.9	2.9	—	2.7	—	131.7	—	—	—	—	—	—	—

Fish	Na	K	Mg	Ca	NH$_4$	Cl	HCO$_3$	CO$_2$	PO$_4$	SO$_4$	Urea	Cholesterol (mg %)	Protein (g %)
				Teleostei—Euryhaline—Salmonoidei									
Oncorhynchus keta	—	—	—	1.55	—	89.5	—	—	—	—	5.33	—	—
Oncorhynchus keta	—	—	—	1.50	—	86.8	—	—	—	—	4.06	—	—
Oncorhynchus keta	—	—	—	1.17	—	80.8	—	—	—	—	20.9	—	—
Oncorhynchus keta	—	—	—	1.03	—	71.8	—	—	—	—	11.4	—	—
Oncorhynchus gorbushka	—	—	—	1.50	—	84.0	—	—	—	—	3.46	—	—
Oncorhynchus gorbushka	—	—	—	1.65	—	76.3	—	—	—	—	2.7	—	—
Oncorhynchus gorbushka	—	—	—	1.40	—	77.0	—	—	—	—	12.5	—	—
Oncorhynchus gorbushka	—	—	—	1.36	—	72.2	—	—	—	—	6.33	—	—
Oncorhynchus nerka	—	—	0.33	2.68	—	—	—	—	—	1.9	2.14	570	—
Oncorhynchus nerka	—	—	0.18	3.41	—	—	—	—	—	2.0	2.21	572	—
Oncorhynchus nerka	—	—	0.77	2.48	—	—	—	—	—	0.86	1.53	436	—
Oncorhynchus nerka	—	—	0.59	2.97	—	—	—	—	—	0.61	1.32	499	—
Oncorhynchus nerka	—	—	0.42	1.56	—	—	—	—	—	1.5	1.74	394	—
Oncorhynchus nerka	—	—	0.45	1.94	—	—	—	—	—	1.53	1.50	202	—
Salmo trutta	155	5.1	0.45	1.55	—	121	—	—	1.5	0.04	—	—	—
Salmo trutta (brown trout)	150	2.7	—	—	—	133	—	—	—	—	—	—	—
Salmo trutta (brown trout)	—	—	—	—	—	185	—	—	—	—	—	—	—
Salmo trutta (brown trout)	163	3.1	—	—	—	150	—	—	—	—	—	—	—
Salmo trutta (sea trout)	155	5.3	—	—	—	124	—	—	—	—	—	—	—
Salmo trutta (sea trout)	205	0.9	—	—	—	191	—	—	—	—	—	—	—
Salmo trutta (sea trout)	166	3.5	—	—	—	138	—	—	—	—	—	—	—
Salmo trutta (sea trout)	144	>10	—	—	—	122	—	—	—	—	—	—	—
Oncorhynchus masou	—	—	—	—	—	108	—	—	—	—	—	—	—
Oncorhynchus masou	—	—	—	—	—	109	—	—	—	—	—	—	—
Oncorhynchus masou	—	—	—	—	—	101	—	—	—	—	—	—	—
Oncorhynchus tschawytscha	187	1.8	—	—	—	—	—	—	—	—	2.7	501	6.6
Oncorhynchus tschawytscha	162	1.0	—	—	—	—	—	—	—	—	—	530	4.6
Oncorhynchus tschawytscha	160	0.8	—	—	—	—	—	—	—	—	3.1	754	5.8
Oncorhynchus tschawytscha	145	0.7	—	—	—	—	—	—	—	—	2.9	365	4.1
Oncorhynchus tschawytscha	165	2.0	—	—	—	—	—	—	—	—	3.1	126	1.7
Salmo salar—parrs	117	2.19	—	2.33	—	130	—	—	—	—	—	—	—
Salmo salar—smolts	131	3.03	—	2.00	—	183	7.3	—	—	—	—	—	—
Salmo salar—post-smolts	156	3.28	—	—	—	133	—	—	—	—	—	—	—
Salmo salar—smolts	159	3.62	—	—	—	166	—	—	—	—	—	—	—
Salmo salar—adults	212	3.15	—	3.43	—	157	11.0	—	—	—	—	—	—

Osteichthyes—Blood Chemistry

Total ions	Units	Osmotic pressure (mOsm/liter)	Comments	Author	Date
			Teleostei—Euryhaline—General		
—	mmoles/liter	—	Prespawners seawater	Sindermann and Mairs	1961
—	—	—	Postspawners freshwater	—	—
—	mmoles/liter	181	Freshwater	Stanley and Fleming	1964
—	—	242	Seawater 3 days	—	—
—	—	190	Seawater 20 days	—	—
—	—	—	—	—	—
—	mmoles/liter	364	Seawater 10 days	Lange and Fugelli	1965
—	—	304	Freshwater 10 days	—	—
—	mmoles/liter	—	Seawater	Motais et al.	1965
—	—	—	Freshwater 3 hr	—	—
—	—	297	Seawater	Lahlou	1967
—	—	240	Freshwater	—	—

Total ions	Units	Osmotic pressure (mOsm/liter)	Comments	Author	Date
			Teleostei—Euryhaline—Salmonoidei		
—	mmoles/liter	—	Male Estuary	Lysaya	1951
—	—	—	Female Estuary	—	—
—	—	—	Male Spawning grounds	—	—
—	—	—	Female Spawning grounds	—	—
—	—	—	Male Estuary	—	—
—	—	—	Female Estuary	—	—
—	—	—	Male Spawning grounds	—	—
—	—	—	Female Spawning grounds	—	—
—	mmoles/liter	—	Male Lummi Island	Idler and Tsuyuki	1958
—	—	—	Female Lummi Island	—	—
—	—	—	Male Lillooet	—	—
—	—	—	Female Lillooet	—	—
—	—	—	Male Forfar Creek	—	—
—	—	—	Female Forfar Creek	—	—
—	mmoles/liter	—	Freshwater	Phillips and Brockway	1958
—	mmoles/liter	326	Freshwater	Gordon	1959a
—	—	335	½ seawater 240 hr, seawater 48 hr	—	—
—	—	342	½ seawater 240 hr, seawater 64 days	—	—
—	—	351	Freshwater	—	—
—	—	430	Seawater 24 hr	—	—
—	—	356	Seawater 5 months	—	—
—	—	358	Brackish indefinite	—	—
—	mmoles/liter	387	Inshore or estuarine waters	Kubo	1961
—	—	387	Lower reaches	—	—
—	—	349	Upper reaches	—	—
—	—	—	Sea	Robertson et al.	1961
—	—	—	Migrating fall run	—	—
—	—	—	Migrating spring run	—	—
—	—	—	Spawning fall run	—	—
—	—	—	Spawning spring run	—	—
—	mmoles/kg	328	Freshwater	Parry	1961
—	—	312	Freshwater	—	—
—	—	349	Freshwater	—	—
—	—	—	Seawater 2 weeks	—	—
—	—	344	Seawater	—	—

Table XVII

Fish	Na	K	Mg	Ca	NH$_4$	Cl	HCO$_3$	CO$_2$	PO$_4$	SO$_4$	Urea	Cholesterol (mg %)	Protein (g %)
Salmo salar—adults spawning	176	3.03	—	3.45	—	172	—	—	—	—	—	—	—
Salmo salar—adult kelts	294	3.44	—	3.98	—	—	—	—	—	—	—	—	—
Salmo gairdnerii	145	2.33	—	—	—	—	—	—	—	—	—	—	—
Salmo gairdnerii	160	2.42	—	—	—	—	—	—	—	—	—	—	—
Salmo gairdnerii	144	6.0	—	2.65	—	151	—	—	—	—	—	—	—
Salmo gairdnerii	162	3.81	—	2.02	—	124.6	—	—	1.27	—	—	—	—
Oncorhynchus kisutch	146	—	—	—	—	132	—	—	—	—	—	—	—
Oncorhynchus kisutch	147	—	—	—	—	131	—	—	—	—	—	—	—
Oncorhynchus kisutch	182	—	—	—	—	149	—	—	—	—	—	—	—
Oncorhynchus kisutch	205	—	—	—	—	182	—	—	—	—	—	—	—
Oncorhynchus tschawytscha	159	0.4	1.8	2.9	—	112	4.0	—	4.5	0.3	13.6	—	6.7
Oncorhynchus tschawytscha	161	0.3	1.2	2.7	—	114	4.1	—	4.0	0.5	10.3	—	5.1
Oncorhynchus tschawytscha	160	1.3	1.0	2.3	—	137	4.7	—	4.7	0.3	7.5	—	4.9
Oncorhynchus tschawytscha	179	1.0	0.9	1.0	—	139	5.0	—	4.1	0.3	7.1	—	4.7

Fish	Na	K	Mg	Ca	NH$_4$	Cl	HCO$_3$	CO$_2$	PO$_4$	SO$_4$	Urea	TMAO	Protein (g %)
				Teleostei—Euryhaline—Anguilloidei									
Anguilla anguilla	143.5	1.49	—	3.32	—	—	—	—	—	—	—	—	—
Anguilla anguilla	112	4.13	—	7.65	—	—	—	—	—	—	—	—	—
Anguilla anguilla	150	2.75	—	—	—	105	—	—	—	—	—	—	—
Anguilla anguilla	175	3.15	—	—	—	154.5	—	—	—	—	—	—	—
Anguilla anguilla	143	3.4	—	—	—	70	—	—	—	—	—	—	—
Anguilla anguilla	181	2.8	—	—	—	148	—	—	—	—	—	—	—
Anguilla anguilla	98.8	6.2	—	—	—	—	—	—	—	—	—	—	—
Anguilla anguilla	75.7	9.0	—	—	—	—	—	—	—	—	—	—	—
Anguilla anguilla	128.0	3.5	—	—	—	—	—	—	—	—	—	—	—
Anguilla anguilla	149.4	3.52	—	—	—	96.6	—	—	—	—	—	—	—
Anguilla anguilla	49.1	2.77	—	—	—	47.3	—	—	—	—	—	—	—
Anguilla anguilla	162.8	3.49	—	—	—	107.1	—	—	—	—	—	—	—
Anguilla anguilla	159.9	4.50	—	—	—	134.0	—	—	—	—	—	—	—
Anguilla anguilla	178.0	4.20	—	—	—	141.0	—	—	—	—	—	—	—
Anguilla anguilla	143.2	2.26	3.04	2.31	—	—	—	—	1.30	—	—	—	—
Anguilla anguilla	150.1	1.75	2.13	2.29	—	88.25	—	—	1.81	—	—	—	—
Anguilla anguilla	164.2	3.35	4.18	2.75	—	—	—	—	2.02	—	—	—	—
Anguilla anguilla	183.3	3.22	3.74	2.37	—	139.6	—	—	1.75	—	—	—	—
Anguilla anguilla	133.8	2.75	4.08	2.90	—	—	—	—	1.86	—	—	—	—
Anguilla anguilla	111.7	2.38	2.11	2.89	—	—	—	—	—	—	—	—	—
Anguilla anguilla	110.4	3.34	2.48	3.53	—	—	—	—	2.10	—	—	—	—
Anguilla anguilla	124.7	3.47	2.69	3.30	—	55.9	—	—	1.81	—	—	—	—
Anguilla anguilla	185.0	3.45	7.93	3.73	—	—	—	—	2.30	—	—	—	—
Anguilla anguilla	194.0	4.33	9.63	4.03	—	207.7	—	—	1.50	—	—	—	—
Anguilla anguilla	66.6	3.64	2.74	1.64	—	—	—	—	—	—	—	—	—
Anguilla anguilla	119.7	2.35	1.69	2.33	—	—	—	—	—	—	—	—	—
Anguilla anguilla	122.6	1.82	1.88	1.79	—	—	—	—	1.94	—	—	—	—
Anguilla anguilla	195.4	3.15	10.31	3.05	—	—	—	—	2.23	—	—	—	—
Anguilla anguilla	222.4	3.24	10.27	4.23	—	214.4	—	—	1.93	—	—	—	—
Anguilla anguilla	111.5	3.10	1.78	3.47	—	—	—	—	—	—	—	—	—
Anguilla anguilla	84.9	4.05	—	2.57	—	53.0	—	—	—	—	—	—	—
Anguilla anguilla	205.1	4.08	10.75	4.65	—	181.4	—	—	—	—	—	—	—

(Continued)

otal ions	Units	Osmotic pressure (mOsm/liter)	Comments	Author	Date
—	—	328	Freshwater	—	—
—	—	371	Freshwater + 2 days seawater	—	—
—	—	—	Freshwater	Holmes and McBean	1963
—	—	—	80% seawater 10 days	—	—
—	mmoles/liter	—	Freshwater	Fromm	1963
—	mmoles/liter	—	—	Hickman et al.	1964
—	mmoles/liter	295	Freshwater—control	Conte	1965
—	—	291	Freshwater—X ray	—	—
—	—	331	Seawater—control	—	—
—	—	372	Seawater—X ray	—	—
80.9	mmoles/liter	—	Spawning freshwater—male	Urist and Van de Putte	1967
87.8	—	—	Spawning freshwater—female	—	—
11.4	—	—	Marine habitat—male	—	—
31.9	—	—	Marine habitat—female	—	—

Total ions	Units	Osmotic pressure (mOsm/liter)	Comments	Author	Date
			Teleostei—Euryhaline—Anguilloidei		
—	mmoles/liter	—	Control	Fontaine	1964
—	—	—	Stanniectomized	—	—
—	mmoles/liter	328	Silver eel—freshwater	Sharratt et al.	1964
—	—	377	Silver eel—seawater	—	—
—	—	—	Yellow eel—freshwater	—	—
—	—	—	Yellow eel—seawater	—	—
—	mmoles/liter	—	Distilled water—6 weeks	Chester-Jones et al.	1965
—	—	—	Distilled water—12 weeks	—	—
—	—	—	Aldactone—6 weeks	—	—
—	mmoles/liter	—	Freshwater controls (silver)	Butler	1966
—	—	—	Freshwater hypox—3 weeks (silver)	—	—
—	—	—	Freshwater controls (yellow)	—	—
—	—	—	Seawater controls (silver)	—	—
—	—	—	Seawater hypox—3 weeks (silver)	—	—
—	mmoles/liter	—	Control—yellow eel—freshwater	Chan et al.	1967
—	—	—	Control—silver eel—freshwater	—	—
—	—	—	Control—yellow eel—4-5 weeks in seawater	—	—
—	—	—	Control—silver eel—4-5 weeks in seawater	—	—
—	—	—	Control—yellow eel—4 weeks in distilled water	—	—
—	—	—	Control—silver eel—4 weeks in distilled water	—	—
—	—	—	Stanniectomized—yellow eel—3 weeks freshwater	—	—
—	—	—	Stanniectomized—silver eel—3 weeks freshwater	—	—
—	—	—	Stanniectomized—yellow eel—3 weeks seawater	—	—
—	—	—	Stanniectomized—silver eel—3 weeks seawater	—	—
—	—	—	Stanniectomized—yellow eel—3 weeks distilled water	—	—
—	—	—	Adrenal insufficient—yellow eel—3 weeks freshwater	—	—
—	—	—	Adrenal insufficient—silver eel—3 weeks freshwater	—	—
—	—	—	Adrenal insufficient—yellow eel—3 weeks seawater	—	—
—	—	—	Adrenal insufficient—silver eel—3 weeks seawater	—	—
—	—	—	Stanniectomized + adrenal insufficient—yellow eel—1 week freshwater	—	—
—	—	—	Stanniectomized + adrenal insufficient—silver eel—2 weeks freshwater	—	—
—	—	—	Stanniectomized + adrenal insufficient—silver eel—2 weeks seawater	—	—

Table XVII

Fish	Na	K	Mg	Ca	NH$_4$	Cl	HCO$_3$ CO$_2$	PO$_4$	SO$_4$	Urea	TMAO	Protein (g%)
Anguilla anguilla	147.1	—	—	—	—	130.4	— —	—	—	—	—	—
Anguilla anguilla	181.5	—	—	—	—	167.9	— —	—	—	—	—	—
Anguilla anguilla	139.8	—	—	—	—	137.3	— —	—	—	—	—	—
Anguilla anguilla	187.7	—	—	—	—	175.0	— —	—	—	—	—	—
Anguilla anguilla	136.6	—	—	—	—	133.4	— —	—	—	—	—	—
Anguilla anguilla	136.3	—	—	—	—	130.0	— —	—	—	—	—	—
Anguilla anguilla	136.7	—	—	—	—	121.0	— —	—	—	—	—	—
Anguilla rostrata	153.4	5.0	—	—	—	103.6	— —	—	—	—	—	—
Anguilla rostrata	146.0	5.3	—	—	—	103.2	— —	—	—	—	—	—
Anguilla rostrata	132.5	4.1	—	—	—	72.8	— —	—	—	—	—	—
Anguilla rostrata	147.1	4.7	—	—	—	89.8	— —	—	—	—	—	—
Anguilla rostrata	153.6	4.74	—	—	—	113.7	— —	—	—	—	—	—
Anguilla rostrata	143.4	6.00	—	—	—	103.3	— —	—	—	—	—	—
Anguilla rostrata	137.0	6.10	—	—	—	98.4	— —	—	—	—	—	—
Anguilla rostrata	138.1	2.83	1.26	3.33	—	70.48	— —	—	—	—	—	—
Anguilla rostrata	138.0	2.95	1.23	3.29	—	67.31	— —	—	—	—	—	—
Anguilla rostrata	132.0	2.94	0.89	3.26	—	70.94	— —	—	—	—	—	—
Anguilla rostrata	151.3	2.23	1.28	3.00	—	109.60	— —	—	—	—	—	—
Anguilla rostrata	144.0	2.39	1.40	2.98	—	85.54	— —	—	—	—	—	—
Anguilla rostrata	135.4	2.65	1.20	2.72	—	80.20	— —	—	—	—	—	—
Anguilla rostrata	150.7	2.77	1.21	3.19	—	111.11	— —	—	—	—	—	—
Anguilla rostrata	143.3	2.60	1.22	3.11	—	77.35	— —	—	—	—	—	—
Anguilla rostrata	131.6	2.90	1.07	2.80	—	82.88	— —	—	—	—	—	—

water. Inability to lower its permeability to Cl⁻ may determine the lower end of the salinity range of this fish (Evans, 1967b). Transfer of the flounder, *Pleuronectes flesus*, from seawater to freshwater results in a decrease in plasma osmolarity and in the concentration of Na⁺, K⁺, and Cl⁻, and, to a lesser extent, trimethylamine oxide (Lange and Fugelli, 1965) (Table XVI). This lowering of the plasma osmotic pressure would be expected to result in an increase in the volume of the cells from the osmotic influx of water. However, a loss of intracellular free ninhydrin-positive substances and trimethylamine oxide occurs, and consequently the size of the cells remain constant. When the euryhaline, *Platichthys flesus*, together with the stenohaline marine fish, *Serranus scriba*, were transferred from seawater to freshwater for 3 hr it was found that there was only a small drop in the plasma Na⁺ and Cl⁻ concentrations in the former fish but a considerable drop, leading to death, occurred in the plasma concentrations of these ions in the stenohaline species (Table XVI). Transfer of the flounder to freshwater was found to result in an immediate 85% decrease in the rate of Na⁺ and Cl⁻ efflux. This did not occur in the stenohaline fish (Motais *et al.*, 1965). Hickman (1959) transferred the starry flounder, *Platichthys stellatus*, from 25% seawater to 5.4% seawater and freshwater. The plasma osmolarity was reduced, reaching

(Continued)

Total ions	Units	Osmotic pressure (mOsm/liter)	Comments	Author	Date
—	mmoles/liter	—	Seawater—sham	Mayer *et al.*	1967
—	—	—	Seawater—adx.	—	—
—	—	—	2 hr freshwater—sham + cortisol	—	—
—	—	—	2 hr freshwater—adx.	—	—
—	—	—	2 hr freshwater—adx. + cortisol	—	—
—	—	—	12 hr seawater—sham	—	—
—	—	—	12 hr seawater—adx.	—	—
—	mmoles/liter	—	Controls	Butler and Langford	1967
—	—	—	Mock adx. 3 weeks steel sutures	—	—
—	—	—	Adx. 3 weeks silk sutures	—	—
—	—	—	Adx. 3 weeks steel sutures	—	—
—	mmoles/liter	—	Intact	Butler	1967
—	—	—	Hypox—3 weeks controls	—	—
—	—	—	Hypox—3 weeks prolactin	—	—
—	mmoles/liter	271.6	Freshwater silver intact controls (March)	Butler *et al.*	1969b
—	—	266.7	Freshwater silver—sham adx. 3 weeks (March)	—	—
—	—	265.6	Freshwater silver—adx. 3 weeks (March)	—	—
—	—	307.9	Freshwater yellow intact controls (June)	—	—
—	—	287.4	Freshwater yellow—sham adx. 3 weeks (June)	—	—
—	—	271.6	Freshwater yellow—adx. 3 weeks (June)	—	—
—	—	297.4	Freshwater yellow intact controls (July)	—	—
—	—	275.8	Freshwater yellow—sham adx. 3 weeks (July)	—	—
—	—	251.8	Freshwater yellow—adx. 3 weeks (July)	—	—

equilibrium in less than 24 hr in the former medium and between 1 and 3 days in the latter medium. Recently, the changes observed in the plasma composition of *Platichthys flesus* after transfer from seawater to freshwater have been reconfirmed by Lahlou (1967) (Table XVI).

Transfer of the euryhaline mud skipper, *Periophthalmus sobrinus,* from 1000 mOsm/liter seawater to 200 mOsm/liter seawater resulted in no significant change in the plasma osmotic, Na^+ or Cl^- concentrations. Direct transfer to freshwater, however, resulted in a sharp drop in these three parameters and a high rate of mortality. Fish adapted to 200 mOsm/ liter seawater over a period of 6 days on the other hand were able to withstand transfer to freshwater (Gordon *et al.*, 1965).

An attempt to show an effect of the steroids hydrocortisone, 9α-fluoro-hydrocortisone and *dl*-aldosterone on the changes in plasma electrolyte concentrations accompanying the transfer of the stenohaline wrasse, *Thalassoma dupperey,* the partially euryhaline mullet, *Mugil cephalus,* and the euryhaline tilapia, *Tilapia mossambica,* did not succeed (Edelman *et al.*, 1960).

On transfer from freshwater to seawater plasma Na^+, Ca^{2+}, and osmolar concentrations increased during the first 3 days and then returned to near normal levels after 20 days in *Fundulus kansae*. The plasma K⁺ concentration on the other hand was lower after 3 and 20 days in seawater (Stanley and Fleming, 1964). A remarkable observation in this fish was the finding of hypertonic urine between the second and tenth days after transfer to seawater (Stanley and Fleming, 1964; Fleming and Stanley, 1965). In relation to these findings it has been shown recently that the increase in permeability to Na⁺ to Cl⁻ after transfer of *Fundulus heteroclitus* from freshwater to saltwater takes many hours. In contrast, the reverse transfer results in a very rapid decrease in permeability (Potts and Evans, 1967). The plasma composition of an andromous species, the alewife, *Alosa pseudoharengus*, has been investigated in prespawning fish in saltwater and postspawning fish in freshwater. The latter group had lower serum protein and Cl⁻ concentrations than the former; however, it was not possible to distinguish between reproductive and migratory effects (Sindermann and Mairs, 1961).

c. Order Anguilliformes. The changes in the plasma electrolyte composition of eels in relation to their catadromous migration have been the subject of intensive study in the last 5 years. The major part of the life cycle of *Anguilla anguilla* is spent as a yellow eel in freshwater. When the eel becomes sexually differentiated prior to its catadromous migration, it stops feeding and the ventral surface turns a silvery white color. Transfer of silver eels from freshwater to seawater resulted in a 15% increase in the serum osmotic concentration. This change was mainly accounted for by increases in serum Na⁺ and Cl⁻. Freshwater yellow eels had a serum Cl⁻ concentration noticeably lower than the freshwater silver eels (Sharratt *et al.*, 1964, Table XVII). Removal of the corpuscles of Stannius from silver eels resulted in a significant decrease in the serum Na⁺ concentration and significant increases in the serum Ca²⁺ and K⁺ concentrations. Injection of a lyophilized preparation of the corpuscles into stanniectomized eels tended to lower serum Ca²⁺ and K⁺ concentrations and raise serum Na⁺ concentration, indicating a possible osmoregulatory role for the corpuscles of Stannius (Fontaine, 1964) (Table XVII). Recently, the effects of stanniectomy have been studied in the goldfish, *Carassius auratus* (Ogawa, 1968). Removal of the corpuscles in the goldfish maintained in freshwater resulted in a statistically nonsignificant fall in plasma osmolarity and sodium concentration, which was reversed by injection of angiotensin II. Maintenance of freshwater silver eels in distilled water for 6 or 12 weeks resulted in a progressive decrease in serum Na⁺ and a progressive increase in serum K⁺ concentrations. Treatment of the distilled-water eels over the 6-week period, with the antialdosterone compound, Aldactone, resulted in a higher serum Na⁺ and a lower serum K⁺ concentration (Chester-Jones *et al.*, 1965) (Table XVII). Hypophysectomy of freshwater silver eels, *Anguilla an-*

guilla, resulted in significant decreases in the serum Na^+, K^+, and Cl^- concentrations. Intact yellow eels had higher serum Na^+ and Cl^- concentrations than intact silver eels in the same environment. Hypophysectomized silver eels maintained in seawater had higher serum concentrations of Na^+ and Cl^- than intact silver eels in seawater (Butler, 1966) (Table XVII). Thus, the pituitary gland plays a role in serum electrolyte homeostasis in *Anguilla anguilla*. Hypophysectomy of freshwater *Anguilla rostrata* also resulted in decreased serum Na^+ and Cl^- concentrations, but the changes were not reversed by injection of ovine prolactin (Butler, 1967) (Table XVII). This lack of response may have resulted from the amount and dosage regimen used, or may have indicated that other pituitary factors play a major osmoregulatory role in this fish. Contrary to the findings of Butler (1966), Chan *et al.* (1967) (Table XVII) noted lower serum Na^+ and higher serum K^+ concentrations in freshwater yellow eels than in freshwater silver eels. Adaptation of silver and yellow eels to seawater resulted in increases in serum Na^+, K^+, Cl^-, Mg^{2+}, and phosphate concentrations (Table XVII). Stanniectomy of freshwater eels resulted in a decrease in serum Na^+ and Cl^- and an increase in serum K^+ and Ca^{2+} concentrations, thus corroborating the findings of Fontaine (1964). Stanniectomized seawater eels, also analyzed 3 weeks postoperatively, had increased serum concentrations of Na^+, Ca^{2+}, and Mg^{2+}. In addition, silver eels showed an increase in serum K^+ concentration (Table XVII). Adrenalectomy was followed by a lowering of serum Na^+, Ca^{2+} and Mg^{2+} concentrations in freshwater eels. Adrenalectomy of seawater silver eels caused an increase in the serum concentrations of Na^+, Ca^+, and Mg^{2+}, but it must be emphasized that data for sham-adrenalectomized eels were not presented. After both stanniectomy and adrenalectomy, freshwater yellow eels had lower serum Na^+ and Mg^{2+} concentrations, while freshwater silver eels had lower serum Na^+ and Cl^- concentrations and increased serum K^+ concentrations (Table XVII). Injection of cortisol lowered the serum Na^+, K^+, and phosphate concentrations in the freshwater yellow eels and lowered serum Na^+ concentration in freshwater silver eels. Aldosterone lowered the serum K^+ concentrations in both yellow and silver freshwater eels. Injection of the adrenal inhibitor metyrapone plus betamethasone caused a lowering of the serum Na^+ concentrations. Adrenocorticotropic hormone (mammalian) had no effect on serum electrolyte concentrations even when injected at the rate of 2 IU/day for 10 days (Chan *et al.*, 1967). These authors draw attention to the fact that while there are similarities between the effects of adrenalectomy and stanniectomy there are also differences (Table XVII). The changes following adrenalectomy were interpreted by Chan *et al.* as being a result of the removal of cortisol and possibly also an unidentified mineralocorticoid.

The reasons for the changes following stanniectomy, however, remain obscure (Chan et al., 1967). Additional work on *Anguilla anguilla* has further confirmed that cortisol is probably the Na^+ excreting factor in this fish (Mayer et al., 1967) (Table XVII).

Recently, Butler and Langford (1967) (Table XVII) showed that while adrenalectomized freshwater yellow eels with silk-wound closure had significantly lower serum Na^+ and Cl^- concentrations than intact control eels, fish with stainless steel wire-wound closure did not show serum electrolyte changes. These findings were interpreted as indicating that the silk-sutured fish had lost ions by diffusion owing to poor wound healing and that partial adrenalectomy as carried out in the experiment did not affect serum electrolyte composition. Since adrenalectomy by the above technique did not measurably lower the plasma cortisol concentration (Butler and Donaldson, 1967) the operation has been improved in such a way as to remove a greater portion of the interrenal tissue, and this we have found to result in plasma cortisol concentrations similar to those observed in hypophysectomized eels (Butler et al., 1969a,b). Adrenalectomy by the improved technique resulted in a lower plasma Mg^{2+} concentration in freshwater silver eels when compared to sham-operated fish. There was no significant difference between the serum composition of adrenalectomized and sham-adrenalectomized freshwater yellow eels in the June experiment. In the July experiment there were significant decreases in the serum Na^+, Ca^{2+}, and osmotic concentrations of adrenalectomized freshwater yellow eels relative to the sham-adrenalectomized fish (Butler et al., 1969b) (Table XVII). These changes observed in the July experiment are explained by the fact that additional interrenal tissue was removed from these fish. Although the removal of further interrenal tissue did not result in a detectable decrease in the plasma cortisol concentration it may have lowered the cortisol concentration below a critical threshold for electrolyte homeostasis. Experiments by Chan et al. (1968) point to a role for both cortisol and prolactin in the endocrine control of the electrolyte composition of the freshwater European eel, *Anguilla anguilla*.

d. Order Salmoniformes. Early work dealing with the changes in blood chemistry which occur in migrating salmon has been reviewed by Black (1951). A 2-year study of blood composition in the pink salmon, *Oncorhynchus nerka*, and chum salmon, *Oncorhynchus gorbushka*, of the Amur Estuary and Amgun River (Lysaya, 1951) showed the expected decrease in plasma Cl^- concentrations when male and female fish migrated from the estuary to the spawning grounds. The blood Ca^{2+} concentrations observed in this study, however, did not show the usual sex difference observed in other maturing fish (refer to Section V, C, 3, f below). The increase in the blood urea concentration during the anadromous migration of these two species was considerable (Table XVII) and remained unconfirmed prior to the recent work of Urist and Van de Putte (1967) on *Oncorhynchus tschawytscha* (Table XVII); however, the change was not as great in this latter study. Lysaya considered the increasing urea concentration during the anadromous migration to be a homeostatic mechanism which would tend to maintain a constant blood osmotic pressure in the face of the observed electrolyte depletion and that this uremia eventually leads to the death of the salmon. This thesis has yet to be substantiated. Idler and Tsuyuki (1958) in the course of their investigation of the changes in the blood of sockeye salmon, *Oncorhynchus nerka*, during its anadromous migration in the Fraser River system also measured plasma urea, but rather than finding an increase in urea concentration during the migration they actually observed a slight decline. These authors also noted a decline in the cholesterol, Mg^{2+}, and Ca^{2+} concentrations of the plasma as the fish migrated up the river (Table XVII). At Lummi Island at the start of the upstream migration the female fish had a higher plasma Ca^{2+} concentration, but the sex difference was not apparent during the migration, indicating that either the major mobilization of the calcium phosphoprotein occurred prior to the fluvial migration or that the rate of utilization of this material by the ovary during the migration was sufficient to prevent an increase in the plasma concentration.

Transfer of *Oncorhynchus nerka* at the end of their spawning migration up the Dalnee River back into full seawater or even isotonic or somewhat hypotonic media resulted in an increase in plasma osmolarity, loss of coordination and death after 4–5 hr (Zaks and Sokolova, 1961). In contrast, *Oncorhynchus nerka* collected at the end of their anadromous migration to Great Central Lake while showing poor survival when transferred directly to full seawater showed very good survival if the salinity was increased gradually (Donaldson, 1968). Retention of downstream migrating *Oncorhynchus nerka* in freshwater resulted in a slight decrease in plasma osmolarity after 5 days, while the plasma albumin concentration fell from its initial level of 8.7% to 5.7% after 24 days. These changes together with changes in kidney function were interpreted as indicating a tendency toward "freshwater" characteristics (Zaks and Sokolova, 1965). Rapid transfer of adult *Oncorhynchus nerka* from saltwater to freshwater resulted in a decline in plasma Na^+ concentration of 22 mmoles/liter and osmolarity of 63 mOsm/liter. As in the earlier study by these workers, transfer of fish at the end of their upstream migration back into seawater with a freezing-point depression greater than $-0.6°$ resulted in high mortality (Zaks and Sokolova, 1965).

An investigation of physiological changes in the blood of migrating adult *Oncorhynchus tschawytscha* (Robertson *et al.*, 1961) (Table XVII) showed that plasma Na^+ and especially K^+ concentrations fell during the upstream migration. The hypokalemia was especially noticeable in the fall run fish. The plasma cholesterol concentration rose during the migratory period and then fell in the spawning fish to levels which were lower than the concentration in the sea salmon. In this study, unlike those of Lysaya (1951) and Urist and Van de Putte (1967), there was no increase in plasma urea during the anadromous migration. Studies of the inshore migration of *Oncorhynchus masou* showed an increase in the blood osmotic and Cl^- concentration in the late spring at the beginning of the inshore migration followed by a decrease as the fish approached the estuary (Kubo, 1960). As the fish migrated up the river there was a decline in the blood osmolar and Cl^- concentration followed by a temporary rise during preparation for spawning and finally a marked drop just prior to spawning (Kubo, 1961) (Table XVII). Kubo has also investigated the changes occurring in the blood during the downstream migration. At first it seemed that the osmotic pressure of the blood became higher after the parr–smolt transformation when compared to parr or dark parr fish (Kubo, 1953). Later studies on the blood of *Oncorhynchus masou*, however, showed that when the smolts are about to descend into the sea the osmotic pressure of the blood became lower again (Kubo, 1955). Examination of *Oncorhynchus tschawytscha* in the marine habitat showed that they had a blood chemistry similar to that of marine teleosts. On the other hand, specimens of these fish sampled in freshwater showed a noticeable hypokalemia which was related to the prolonged starvation which the fish had undergone (Urist and Van de Putte, 1967) (Table XVII). Urist also pointed out that the plasma Ca^{2+} in the female salmon fell from a high of 6.0 mmoles/ liter down to 2.5 mmoles/liter when the ova were fully developed. Ionizing radiation was found to cause an increase in the plasma Cl^-, Na^+, and osmotic concentrations of juvenile coho salmon, *Oncorhynchus kisutch*, kept in saltwater but not in coho kept in freshwater. The irradiation appeared to have interfered with the extrarenal osmoregulatory systems (Conte, 1965) (Table XVII). The plasma osmotic concentration of premigratory juvenile *Oncorhynchus kisutch* increased from 295 to 400 mOsm/liter immediately after transfer to seawater but then tended to decline with time. In no case did the osmotic concentration exceed 415 mOsm/liter. Migratory and postmigratory fish were also able to osmoregulate satisfactorily (Conte *et al.*, 1966). These findings for the coho were in marked contrast to those for the steelhead trout, *Salmo gairdnerii*, which showed a regression of osmoregulatory activity after the migration period had passed (Conte and Wagner, 1965).

The first fairly complete analysis of the plasma of a fish from the genus *Salmo* was carried out on *Salmo trutta* (Phillips and Brockway, 1958) (Table XVII). In general the concentrations were similar to those determined in the carp by Field *et al.* (1943). A comparison of the osmoregulatory abilities of the brown trout and sea trout forms of *Salmo trutta* in the Aberdeen area indicated that they were virtually identical in this respect (Gordon, 1959a) (Table XVII) and also similar to American hatchery raised *Salmo trutta* (Gordon, 1959b). Plasma ionic concentrations were either unchanged or increased by less than 10% after the fish had been transferred from freshwater to seawater and acclimated (Gordon, 1959a) (Table XVII). A complete study of the major plasma ions at various points in the life cycle of the Atlantic salmon, *Salmo salar*, has been published by Parry (1961). There was an increase in the osmotic pressure of the plasma during development from the fry to the smolt stage in the life cycle. This was followed by a decline during the downstream migration of the smolt. This decrease in *Salmo salar* was not significant owing to the individual variability of the fish. Kubo (1955), however, attached more significance to his observed decrease in plasma osmolarity at this stage in *Oncorhynchus masou*. *Salmo salar* smolts which had been in seawater for 2 weeks had an elevated plasma osmotic concentration. Postsmolts which had been confined to freshwater for a year after they would normally have migrated also showed the same response. Adult prespawning fish in seawater had lower plasma osmotic pressures than the smolts in seawater and the osmotic concentration decreased further after entry into freshwater on the anadromous migration. Postspawned salmon (kelts) which had been in seawater for 2 days had the highest plasma osmotic pressure (Parry, 1961) (Table XVII). The hypokalemia noted in spawning *Oncorhynchus tschawytscha* (Urist and Van de Putte, 1967) was not present in the spawning *Salmo salar* (Table XVII) indicating that it may be related to the inevitable postspawning death of the Pacific salmon. Hickman *et al.* (1964) determined the inorganic composition of plasma from *Salmo gairdnerii* acclimated to 16° (Table XVII) and of trout transferred gradually to water at 6°. The most notable changes observed were the immediate significant declines in the plasma Cl⁻ and phosphate concentrations. Fromm (1963) has also reported on the inorganic composition of *Salmo gairdnerii* plasma (Table XVII).

An investigation of the role of adrenocorticosteroids in the osmoregulatory processes of *Salmo gairdnerii* has been carried out by Holmes and his co-workers. The injection of corticosteroids into saline-loaded freshwater *Salmo gairdnerii* resulted in increases in extrarenal Na⁺ excretion of 61% (deoxycorticosterone) and 89% (cortisol) (Holmes, 1959). Injection of cortisol, corticosterone, or aldosterone into nonsmolting fresh-water rainbow trout resulted in a decrease in the plasma Na^+ concentration, the response to aldosterone being the most dramatic. Plasma K^+ concentrations showed an initial rise after the first injection followed by a decline. These findings and those of Chester-Jones (1956) on *Salmo trutta* are in contrast to the normal pattern of response to corticosteroids in the mammal, i.e., an increase in the Na–K ratio (Holmes and Butler, 1963). As a part of a study devoted to determining the glomerular filtration rate of *Salmo gairdnerii*, Holmes and McBean (1963) measured plasma Na^+ and K^+ concentrations in freshwater-adapted fish and in fish after they had been transferred to seawater. A significant increase in the plasma Na^+ concentration ($p < 0.001$) occurred, but no change in the K^+ concentration was observed (Table XVII). Although differences have been observed in the urine electrolyte composition of *Salmo gairdnerii* smolts, presmolts and postsmolts maintained in freshwater, no significant differences have been observed in their plasma electrolyte composition (Holmes and Stainer, 1966). At the present time we have a fairly consistent picture of the changes in blood chemistry which occur in members of the Salmonoidei during their downstream and anadromous migrations, but our knowledge of the hormonal control of osmoregulation in these fish is incomplete. For example, there is some evidence that *Salmo gairdnerii* is able to survive in freshwater after hypophysectomy (Donaldson and McBride, 1967) longer than many other fish (Schreibman and Kallman, 1966) indicating that prolactin may not be obligatory for the maintenance of plasma osmolarity in these fish. On the other hand, we have evidence for the presence of prolactin in the pituitary of *Oncorhynchus tschawytscha* collected in freshwater (Donaldson *et al.*, 1968), and it may be that the 3-month freshwater survival of some hypophysectomized *Salmo gairdnerii* was a result of only a gradual loss of Na^+ similar to that which occurs in the goldfish, which also has a long, but not indefinite, survival period in freshwater following hypophysectomy (Yamazaki and Donaldson, 1968) and in the eel (Maetz *et al.*, 1967).

e. Effect of Low Temperature on Plasma Constituents. The first experiment to determine the effect of cold on teleost plasma was that of Platner (1950) who observed a 159% increase in serum Mg^{2+} in goldfish after 60 hr at 0–1°C. Anoxic goldfish at the same temperature showed an increase in plasma Mg^{2+} concentration of 627%. The source of the increased Mg^{2+} was probably the red blood cells. Arctic teleosts living in deep water have plasma freezing points in the range $-0.91°$ to $-1.01°$. The water in which they live has a year round temperature of $-1.73°$, and the plasma is thus in a supercooled state. Arctic teleosts living in surface waters on the other hand have plasma freezing points of $-0.80°$ to $-0.81°$ in the summer and $-1.47°$ to $-1.50°$ in the winter (Scholander *et al.*, 1957) (Table XVIII). Thus the plasma osmotic pressure of the surface species, which are subject to ice crystal seeding, is raised in the winter to prevent the plasma from freezing, while those species living in deeper water are not subject to the occurrence of ice seeding. Concentrations of Na^+, K^+, and Cl^- were found to be elevated in the plasma of the cod, *Gadus morhua*, caught in water below $2°$ in the winter, but this was not observed in cod caught below this temperature during the summer months. The data were interpreted to indicate an osmotic imbalance in those fish living at low temperatures during the winter months (Woodhead and Woodhead, 1958, 1959). Furthermore, these authors proposed that the increased plasma osmotic concentration of the arctic surface species examined by Scholander *et al.* (1957) may also have been in a state of osmotic imbalance. *Gadus callarias* caught at $-1.5°$ in the Bering Sea had a plasma freezing point of $0.797°$, a value which, although higher than the freezing point of plasma in these fish at $4.5°$, still indicated a considerable degree of supercooling similar to that found by Scholander *et al.* (1957) (Eliassen *et al.*, 1960) (Table XVIII). The plasma freezing point of *Cottus scorpius* was reduced from $-0.60°$ at a water temperature of $10°$ to $-0.80°$ to $-0.90°$ after being kept 6–8 weeks at $-1.5°$ regardless of the season (Eliassen *et al.*, 1960). These authors question the theory that there was a definite temperature threshold below which the fish enter a state of osmotic imbalance and regarded the uniform response to cold which they observed in each group of fish as indicative of an adjustment to a new state of balance (Eliassen *et al.*, 1960).

The winter flounder, *Pseudopleuronectes americanus*, also showed a lowered serum freezing point in cold water and, like the cod, Cl^- accounted for a lower percentage of the total plasma osmolarity during the winter months than during the summer months (Pearcy, 1961) (Table XVIII). Thus, organic components in addition to inorganic components may be responsible for the lowering of plasma freezing point at low temperatures. With this in mind Gordon *et al.* (1962) returned to Hebron Fjord (Scholander *et al.*, 1957) to reinvestigate the problem. After examining many plasma components (Table XVII), these workers concluded that the antifreeze components of *Gadus ogac* may be part of the nonprotein nitrogen (NPN) fraction, but this was not the case for the sculpin, *Myxocephalus scorpius*.

f. Calcium in the Female Teleost. The first data showing a sex dif-

Table XVIII

Teleostei—the Effect of Low Temperature on Blood Chemistry

Fish	Na	K	Mg	Ca	Cl	PO₄	NPN	Urea	Amino N	Protein (g%)	Units	Freezing point depression Plasma	Sea-water	Water temp.	Comments	Author	Date
Carassius auratus	—	—	0.8	—	—	—	—	—	—	—	mmoles/liter	—	—	20–25	Control	Platner	1950
Carassius auratus	—	—	3.0	—	—	—	—	—	—	—		—	—	0–1	Nonaerated	—	—
Gadassius auratus	—	—	1.5	—	—	—	—	—	—	—		—	—	0–1	Aerated	Scholander *et al.*	1957
Carus ogac	—	—	—	—	—	—	—	—	—	—		1.47	—	−1.73	Winter, depth 2–8 meters	—	—
Gadus ogac	—	—	—	—	—	—	—	—	—	—		0.80	—	4–7	Summer, depth 2–6 meters	—	—
Myxocephalus scorpius	—	—	—	—	—	—	—	—	—	—		1.50	—	−1.73	Winter, depth 2–8 meters	—	—
Myxocephalus scorpius	—	—	—	—	—	—	—	—	—	—		0.80	—	4–7	Summer, depth 2–6 meters	—	—
Salvelinus alpinus	—	—	—	—	—	—	—	—	—	—		0.81	—	4–7	Summer, depth 2–6 meters	—	—
Boreogadus saida	—	—	—	—	—	—	—	—	—	—		1.02	—	−1.73	Temperature constant depth 100–200 meters	—	—
Lycodes turneri	—	—	—	—	—	—	—	—	—	—		0.97	—	−1.73	100–200 meters	—	—
Liparis koefoedi	—	—	—	—	—	—	—	—	—	—		0.91	—	−1.73	100–200 meters	—	—
Gymnacanthus tricuspis	—	—	—	—	—	—	—	—	—	—		0.93	—	−1.73	100–200 meters	—	—
Icelus spatula	—	—	—	—	—	—	—	—	—	—		0.96	—	−1.73	100–200 meters	—	—

Gadus callarias	—	—	—	188	—	—	—	—	mmoles/kg	0.797	—	−1.5	—	Eliassen et al.	1960
Gadus callarias	—	—	—	178	—	—	—	—	—	0.693	—	4.5	—	—	—
Gadus callarias	—	—	—	162	—	—	—	—	—	0.64	—	15.0	—	—	—
Cyclopterus lumpus	—	—	—	218	—	—	—	—	—	0.88	—	−1.5	—	—	—
Cyclopterus lumpus	—	—	—	165	—	—	—	—	—	0.57	—	10.0	—	—	—
Anarhichas minor	—	—	—	195	—	—	—	—	—	0.80	—	−1.5	—	—	—
Drepanopsetta platessoides	—	—	—	210	—	—	—	—	—	0.93	—	−1.5	—	—	—
Pseudopleuronectes americana	—	—	—	—	—	—	—	—	—	1.15	—	—	Winter, [NaCl] = 57% of osmotic pressure	Pearcy	1961
Pseudopleuronectes americana	—	—	—	—	—	—	—	—	—	0.63	—	—	Summer, [NaCl] = 83% of osmotic pressure	—	—
Myxocephalus scorpius	216	4.3	—	234	0.55	92.5	14.3	10.0	mEq/liter	1.25[a]	−1.75	−1.7	Labrador spring	Gordon et al.	1962
Myxocephalus scorpius	276	6.4	—	184	—	121.5	—	—	—	0.80[a]	—	—	New Brunswick spring	—	—
Gadus ogac	216	5.5	—	243	0.72	289	24.9	17.8	—	0.94[a]	—	—	Labrador spring	—	—
Microgadus tomcod	246	8.3	—	166	—	92.5	—	—	—	0.98[a]	—	—	New Brunswick spring	—	—
Microgadus tomcod	231	5.1	—	142	—	71.5	—	—	—	0.82[a]	—	—	New Brunswick spring	—	—
Carassius auratus	—	—	—	123	—	—	—	—	mmoles/liter	—	—	21	—	Houston	1962
Carassius auratus	—	—	—	90	—	—	—	—	—	—	—	3	800 min immersion	—	—
Cyprinus carpio	127.9	2.74	1.19	115.6	0.39	—	—	1.8	mmoles/liter	0.515[a]	—	4	3 weeks immersion	Houston and Madden	1968

[a] Derived from osmotic pressure.

Table XIX

Osteichthyes—Chemistry of Other Body Fluids

Fish and fluid	Na	K	Mg	Ca	Cl	HCO₃	CO₂	PO₄	SO₄	Urea	TMAO and TMA	Protein (g %)	Total ions	Units	Osmotic pressure (mOsm/liter)	Comments	Author	Date
Lophius piscatorius	—	—	—	—	—	—	—	—	—	—	—	—	—	—	—	—	Smith	1929b
Spinal fluid	194	6.4	6.0	2.7	184	—	7.2	4.7	3.1	—	—	—	—	mmoles/liter	—	—	—	—
Pericardial fluid	197	5.9	8.2	2.6	197	—	9.0	—	3.2	—	—	—	—	mmoles/liter	—	—	—	—
Perivisceral fluid	195	6.1	6.8	3.0	192	—	8.8	4.4	4.1	—	—	—	—	mmoles/liter	—	—	—	—
Gadus callarias	—	—	—	—	—	—	—	—	—	—	—	—	—	—	—	—	—	—
Perivisceral fluid	165	4.6	3.1	5.1	151	—	11.0	3.3	—	—	—	—	—	mmoles/liter	—	—	—	—
Pericardial fluid	163	4.7	1.0	5.1	169	—	7.2	—	—	—	—	—	—	mmoles/liter	—	—	—	—
Protopterus aethiopicus	—	—	—	—	—	—	—	—	—	—	—	—	—	—	—	—	Smith	1930
Perivisceral fluid	91.0	4.2	Trace	2.1	46	—	36.0	1.6	—	—	—	—	—	mmoles/liter	222	—	—	—
Pericardial fluid	97.0	3.2	Trace	1.9	34	—	—	—	—	—	—	—	—	—	208	—	—	—
Lophius piscatorius	—	—	—	—	—	—	—	—	—	—	—	—	—	—	—	—	Derrien	1937
Vitreous body	—	—	—	—	152	—	—	—	—	—	—	—	—	mmoles/liter	—	Plasma Cl = 183	—	—
Lophius americanus	—	—	—	—	—	—	—	—	—	—	—	—	—	mmoles/liter	—	—	Forster and Berglund	1956
Pericardial fluid	200	4.3	1.7	2.3	187	—	—	4.5	1.1	—	—	—	—	—	—	Mean of Nos. 1, 9, and 15	—	—

									Reference	Year
Stenodomus versicolor	—	—	—	—	—	—	—	—	—	—
Aqueous and vitreous humor	—	—	—	—	—	—	364	Plasma osmotic pressure = 397	Davson and Grant	1960
Tautoga onitis	—	—	—	—	—	—	—	—	—	—
Aqueous and vitreous humor	—	—	—	—	—	—	321	Plasma osmotic pressure = 360	—	—
Gadus morhua	—	—	—	—	—	mmoles/liter	—	—	Enger	1964
Cranial fluid	193	3.2	—	167	—	—	—	—	—	—
Endolymph	136	72.9	—	166	—	—	—	—	—	—
Cottus scorpius	—	—	—	—	—	—	—	—	—	—
Cranial fluid	191	3.6	—	176	—	—	—	—	—	—
Endolymph	152	50.8	—	177	—	—	—	—	—	—
Salmo irrideus	—	—	—	—	—	—	—	—	—	—
Cranial fluid	151	5.0	—	143	—	—	—	—	—	—
Endolymph	104	72.9	—	—	—	—	—	—	—	—
Salvelinus namaycush	—	—	—	—	—	—	—	—	Hoffert and Fromm	1966
Aqueous humor	93.5	6.19	6.4	—	—	mmoles/liter	244	Control	—	—
Aqueous humor	127.4	4.79	4.96	—	—	—	—	Diamox	—	—
Latimeria chalumnae	—	—	—	—	—	—	—	—	Pickford and Grant	1967
Aqueous humor	—	3.7	—	—	303	mmoles/liter	952	—	—	—

ference for plasma calcium concentration in the teleost were provided by Hess et al. (1928). These workers found a range of 9–12.5 mg % in the male cod and 12.7–29 mg % in the female cod. The females showing the highest concentrations were those with large mature gonads. A similar situation was found in the puffer fish. Pora (1935, 1936e) examined the serum of male and female *Cyprinus carpio* and *Labrus bergylta*, confirmed the findings of Hess et al., and in addition noted a somewhat elevated protein concentration in the female. Estrogen treatment of male or female goldfish resulted in an increased serum concentration of nonultrafilterable calcium and phosphorus and in total protein. Some of the increased calcium was present as colloidal calcium phosphate, the remainder was bound to the serum phosphoprotein, vitellin (Bailey, 1957). Estrogenization of maturing sockeye salmon, *Oncorhynchus nerka*, resulted in increased serum levels of calcium, protein phosphorous, and total protein. Untreated maturing female salmon had higher serum concentrations of calcium and protein phosphorous than maturing male salmon (Ho and Vanstone, 1961). Female brook trout, *Salvelinus fontinalis*, examined between July and May showed a markedly increased total serum protein and calcium concentration during the spawning period of October to November. Male *Salvelinus* did not show similar changes (Booke, 1964). Oguri and Takada (1966) noted a fivefold increase in the serum calcium concentration of the snake headfish, *Channa argus*, after injection of estradiol. They also noted an increase in serum calcium in the goldfish after injection of estradiol and were able to relate the serum calcium concentration in normal female goldfish to the gonadosomatic index (Oguri and Takada, 1967). The target tissue in the maturing female fish which responds to estrogen is the liver which synthesizes a calcium phosphoprotein–phospholipid glycolipoprotein complex which is then transported via the bloodstream to the developing ova (Urist, 1964).

g. *Chemistry of Other Body Fluids. i. Perivisceral fluid.* The chemical composition of this fluid in the Osteichthyes appears only to have been examined by Smith (1929b, 1930). In *Lophius piscatorius* and *Gadus callarias* there were higher carbon dioxide concentrations in perivisceral fluid than in plasma indicating a greater alkalinity. In the elasmobranchs the reverse was true. Other ions were present in similar concentrations to those in the plasma, and only a trace of protein was measured. In the lungfish, *Protopterus aethiopicus*, Smith (1930) reported no differences between plasma and perivisceral fluid (Table XIX).

ii. *Pericardial fluid.* This fluid was also found to have a higher carbon dioxide content than plasma by Smith (1929b) (Table XIX). The ionic composition of the pericardial fluid of *Lophius americanus* showed no differences from that of plasma which cannot be explained in terms of Gibbs–Donnan effect and experimental error (Forster and Berglund, 1956) (Table XIX).

iii. Cranial fluid. This fluid which is collected from the brain case was found to be similar in ionic composition to the plasma in *Lophius piscatorius* by Smith (1929b) (Table XIX). Enger (1964) examined the cranial fluid of three species of teleost and compared it to plasma and endolymph. The cranial fluid was similar in composition to the plasma in the two marine teleosts examined, but in *Salmo irideus* there was a considerably higher K⁺ concentration in the cranial fluid (Table XIX). The composition of the cerebrospinal fluid does not appear to have been investigated in the Osteichthyes.

iv. Endolymph. This fluid has been isolated from the ear and analyzed in three teleosts. In all cases the fluid showed the characteristically high K⁺ concentration found throughout the vertebrates. The endolymph to cranial fluid ratio for Na⁺ in these teleosts was 0.7 and for Cl⁻ was 1.0, whereas in the elasmobranchs examined the ratios were 1.05 and 1.3, respectively (Enger, 1964) (Tables XV and XVIII).

v. Eye fluids. The vitreous body of *Lophius piscatorius* was found to have a lower concentration of Cl⁻ than that of plasma by Derrien (1937), while Davson and Grant (1960) reported that a combination of aqueous and vitreous humor is hypoosmotic with respect to plasma in both *Stenotomus versicolor* and *Tautoga onitis* (Table XIX). These findings have recently been substantiated by Hoffert and Fromm (1966) who found both the osmotic and Cl⁻ concentrations were lower in the aqueous humor. Treatment with the carbonic anhydrase inhibitor Diamox increased the Cl⁻ and lowered the total carbon dioxide and HCO_3^- concentrations in the aqueous humor, but the exact role of carbonic anhydrase in aqueous humor formation in the teleost remains to be clarified (Hoffert and Fromm, 1966; Maren, 1967b). The aqueous humor of the coelacanth, *Latimeria chalumnae,* appears to have a lower urea concentration and a higher osmolarity than heart blood from the same animal (Pickford and Grant, 1967) (Table XIX).

REFERENCES

Armin, J., Grant, R. T., Pels, H., and Reeve, E. B. (1952). The plasma cell and blood volumes of albino rabbits as estimated by the dye T 1824 and P-32 marked cell methods. *J. Physiol. (London)* **116,** 59–73.
Bailey, R. E. (1957). The effect of estradiol on serum calcium, phosphorus, and protein of goldfish. *J. Exptl. Zool.* **136,** 455–469.

Barlow, J. S., and Manery, J. F. (1954). The changes in electrolytes, particularly chloride, which accompany growth in chick muscle. *J. Cellular Comp. Physiol.* **43**, 165.

Becker, E. L., Bird, R., Kelly, J. W., Schilling, J., Solomon, S., and Young, N., (1958). Physiology of marine teleosts. I. Ionic composition of tissue. *Physiol. Zool.* **31**, 224–227.

Bellamy, D., and Chester-Jones, I. (1961). Studies on Myxine glutinosa. I. The chemical composition of the tissues. *Comp. Biochem. Physiol.* **3**, 175–183.

Bentley, P. J., and Follet, B. K. (1963). Kidney function in a primitive vertebrate, the cyclostome *Lampetra fluviatilis. J. Physiol. (London)* **169**, 902–918.

Bernard, G. R., Wynn, R. A., and Wynn, G. G. (1966). Chemical anatomy of the pericardial and perivisceral fluids of the stingray *Dasyatis americana. Biol. Bull.* **130**, 18–27.

Black, V. S. (1951). Some aspects of the physiology of fish. II. Osmotic regulation in teleost fishes. *Publ. Ontario Fisheries Res. Lab.* **71**, 53–89.

Bloete, M., Naumann, D. C., Frazier, H. S., Leaf, A., and Stone, W., Jr. (1961). Comparison of the osmotic activity of ocular fluids with that of arterial plasma in the dogfish. *Biol. Bull.* **121**, 383–384.

Bond, R. M., Cary, M. K., and Hutchinson, G. E. (1932). A note on the blood of the hagfish *Polistotrema stouti* (Lockington). *J. Exptl. Biol.* **9**, 12–14.

Booke, H. E. (1964). Blood serum protein and calcium levels in yearling brook trout. *Progressive Fish Culturist* **26**, 107–110.

Borei, H. (1935). Uber die Zusammensetzung der Korperflussigkeiten von *Myxine glutinosa* L. (Composition of blood of cyclostome, Myxine.) *Arkiv Zool.* **28B**, 1–5.

Bottazzi, F. (1906). Sulla regulazione della pressione osmotica negli organismi animali. *Arch. Fisiol.* **3**, 416–446.

Boucher-Firley, S. (1934). Sur quelques constituants chimiques du san de Congre et de Murene. *Bull. Inst. Oceanog.* **651**, 1–6.

Braunitzer, G., Hilse, K., Rudloff, V., and Hilschmann, N. (1964). *Advan. Protein Chem.* **19**, 1–65.

Brodie, B. B. (1951). *Methods Med. Res.* **4**, 31.

Brown, G. W., and Brown, S. G. (1967). Urea and its formation in Coelacanth liver. *Science* **155**, 570–573.

Brull, L., and Cuypers, Y. (1954). Quelques characteristiques biologiques de, *Lophius piscatorius* L. *Arch. Intern. Physiol.* **62**, 70–75.

Brull, L., and Nizet, E. (1953). Blood and urine constituents of *Lophius piscatorius. J. Marine Biol. Assoc. U. K.* **32**, 321–328.

Bull, J. M., and Morris, R. (1967). Studies on fresh water osmoregulation in the ammocoete larva of *Lampetra planeri* (Bloch). I. Ionic constituents, fluid compartments, ionic compartments and water balance. *J. Exptl. Biol.* **47**, 485–494.

Burger, J. W. (1965). Roles of the rectal gland and the kidneys in salt and water excretion in the spiny dogfish. *Physiol. Zool.* **38**, 191–196.

Burian, R. (1910). Funktion der Nierenglomeruli und Ultrafiltration. *Arch. Ges. Physiol.* **136**, 741–760.

Butler, D. G. (1966). Effect of hypophysectomy on osmoregulation in the European eel (*Anguilla anguilla* L.). *Comp. Biochem. Physiol.* **18**, 773–781.

Butler, D. G. (1967). Effect of ovine prolactin on tissue electrolyte composition of hypophysectomized fresh water eels (*Anguilla rostrata*). *J. Fisheries Res. Board Can.* **24**, 1823–1826.

Butler, D. G., and Donaldson, E. M. (1967). Unpublished data.

Butler, D. G., and Langford, R. W. (1967). Tissue electrolyte composition of the fresh water eel (Anguilla rostrata) following partial surgical adrenalectomy. Comp. Biochem. Physiol. 22, 309–312.

Butler, D. G., Donaldson, E. M., and Clarke, W. C. (1969a). Physiological evidence for a pituitary adrenocortical feedback mechanism in the eel (Anguilla rostrata). Gen. Comp. Endocrinol. 12, 173–176.

Butler, D. G., Clarke, W. C., Donaldson, E. M., and Langford, R. W. (1969b). Surgical adrenalectomy of a teleost fish (Anguilla rostrata Lesueur): Effect on plasma cortisol and tissue electrolyte and carbohydrate concentrations. Gen. Comp. Endocrinol. 12, 502–514.

Chaisson, A. F., and Freidman, M. H. F. (1935). The effect of histamine, adrenaline and destruction of the spinal cord on the osmotic pressure of the blood in the skate. Proc. Nova Scotian Inst. Sci. 18, 240–244.

Chan, D. K. O., Chester-Jones, I., Henderson, I. W., and Rankin, J. C. (1967). Studies on the experimental alteration of water and electrolyte composition of the eel (Anguilla anguilla L.). J. Endocrinol. 37, 297–317.

Chan, D. K. O., Chester-Jones, I., and Mosley, W. (1968). Pituitary and adrenocortical factors in the control of the water and electrolyte composition of the freshwater European eel (Anguilla anguilla L.). J. Endocrinol. 42, 91–98.

Chester-Jones, I. (1956). The role of the adrenal cortex in the control of water and salt electrolyte metabolism in vertebrates. Mem. Soc. Endocrinol. 5, 102–119.

Chester-Jones, I., Phillips, J. G., and Bellamy, D. (1962). Studies on water and electrolytes in cyclostomes and teleosts with special reference to Myxine glutinosa L. (the Hagfish) and Anguilla anguilla L., (the Atlantic Eel). Gen. Comp. Endocrinol. Suppl. 1, 36–47.

Chester-Jones, I., Henderson, I. W., and Butler, D. G. (1965). Water and electrolyte flux in the European eel (Anguilla anguilla). Arch. Anat. Microscop. Morphol. Exptl. 54, 453–468.

Cohen, J. J., Krupp, M. A., and Chidsey, C. A., III (1958). Renal conservation of trimethylamine oxide by the spiny dogfish, Squalus acanthias. Am. J. Physiol. 194, 229–235.

Cole, W. H. (1940). The composition of fluids and sera of some marine animals and of the sea water in which they live. J. Gen. Physiol. 23, 575–584.

Conte, F. P. (1965). Effects of ionizing radiation on osmoregulation in fish Oncorhynchus kisutch. Comp. Biochem. Physiol. 15, 292–302.

Conte, F. P., and Wagner, H. H. (1965). Development of osmotic and ionic regulation in juvenile steelhead trout Salmo gairdneri. Comp. Biochem. Physiol. 14, 603–620.

Conte, F. P., Wagner, H. H., Fessler, J., and Gnose, C. (1966). Development of osmotic and ionic regulation in juvenile coho salmon Oncorhynchus kisutch. Comp. Biochem. Physiol. 18, 1–15.

Cotlove, E., and Hogben, C. A. M. (1962). Mineral Metab. 2, Part B, 109–173.

Cserr, H., and Rall, D. P. (1967). Regulation of cerebrospinal fluid [K+] in the spiny dogfish Squalus acanthias. Comp. Biochem. Physiol. 21, 431–434.

Davson, H., and Grant, C. T. (1960). Osmolarities of some body fluids in the elasmobranch and teleost. Biol. Bull. 119, 293.

Dekhuyzen, M. C. (1904). Ergebnisse von osmotischen Studien, nametlich bei Knochenfischen, und der biologischen Station des Bergenser Museums. Bergens Museums Arbok. Naturv. Rekke 8, 3–7.

Derrickson, M. B., and Amberson, W. R. (1934). Determination of blood volume in the lower vertebrates by direct method. *Biol. Bull.* **67**, 329 (abstr.).

Derrien, Y. (1937). Repartition du chlorure de sodium et du glucose entre le plasma sanguin et le corps vitre de *Lophius piscatorium* L. *Compt. Rend. Soc. Biol.* **126**, 943–945.

Donaldson, E. M. (1968). Unpublished observations.

Donaldson, E. M., and McBride, J. R. (1967). The effects of hypophysectomy in the rainbow trout *Salmo gairdnerii* (Rich.) with special reference to the Pituitary Interrenal axis. *Gen. Comp. Endocrinol.* **9**, 93–101.

Donaldson, E. M., Yamazaki, F., and Clarke, W. C. (1968). Effect of hypophysectomy on plasma osmolarity in Goldfish and its reversal by ovine prolactin and a preparation of salmon pituitary "Prolactin." *J. Fisheries Res. Board Can.* **25**, 1497–1500.

Doolittle, R. F., Thomas, C., and Stone, W., Jr. (1960). Osmotic pressure and aqueous humor formation in dogfish. *Science* **132**, 36–37.

Edelman, I. S., Young, H. L., and Harris, J. B. (1960). Effects of corticosteroids on electrolyte metabolism during osmoregulation in teleosts. *Am. J. Physiol.* **199**, 666–670.

Edwards, J. G., and Condorelli, L. (1928). Electrolytes in blood and urine of fish. *Am. J. Physiol.* **86**, 383–398.

Eisler, R., and Edmunds, P. H. (1966). Effects of endrin on blood and tissue chemistry of a marine fish. *Trans. Am. Fisheries Soc.* **95**, 153–159.

Eliassen, E., Leivestad, H., and Moller, D. (1960). Effect of low temperature on the freezing point of plasma and on the potassium: Sodium ratio in the muscle of some boreal and subartic fishes. *Arbok. Univ. Bergen, Mat.-Nat. Ser.* No. 14, 1–24.

Elkinton, J. R., and Danowski, T. S. (1955). "The Body Fluids." Williams and Wilkins, Baltimore, Maryland.

Enger, P. S. (1964). Ionic composition of the cranial and labyrinthine fluids and saccular D.C. potentials in fish. *Comp. Biochem. Physiol.* **11**, 131–137.

Evans, D. H. (1967a). Sodium, chloride and water balance of the intertidal teleost, *Xiphister atropurpureus*. 1. Regulation of plasma concentration and body water content. *J. Exptl. Biol.* **47**, 513–518.

Evans, D. H. (1967b). Sodium, chloride and water balance of the intertidal teleost, *Xiphister atropurpureus*. 2. The role of the kidney and the gut. *J. Exptl. Biol.* **47**, 519–534.

Fänge, R., and Fugelli, K. (1962). Osmoregulation in chimaeroid fishes. *Nature* **196**, 689.

Fellers, F. X., Barnett, H. L., Hare, K., and McNamara, H. (1949). Changes in thiocyanate and sodium[24] spaces during growth. *Pediatrics* **3**, 622.

Field, J. B., Elveljem, C. A., and Juday, C. (1943). A study of blood constituents of carp and trout. *J. Biol. Chem.* **148**, 261–269.

Fleming, W. R., and Stanley, J. G. (1965). Effects of rapid changes in salinity on the renal function of a euryhaline teleost. *Am. J. Physiol.* **209**, 1025–1030.

Fontaine, M. (1930). Recherches sur le milieu intérieur de la lamproie marine (*Petromyzon marinus*). Ses variations en fonction de celles du milieu extérieur. *Compt. Rend.* **191**, 680–682.

Fontaine, M. (1964). Corpuscules de Stannius et regulation ionique (Ca, K, Na) du milieu intérieur de l'Anguille (*Anguilla anguilla* L.). *Compt. Rend.* **259**, 875–878.

Forster, R. P., and Berglund, F. (1956). Osmotic diuresis and its effect on total electrolyte distribution in plasma and urine of the aglomerular teleost, *Lophius americanus*. *J. Gen. Physiol.* **39**, 349–359.

Fredericq, L. (1904). Sur la concentration moléculaire du sang et des tissues chez les animaux aquatiques. *Arch. Biol.* (*Liege*) **20**, 709–730.

Fromm, P. O. (1963). Studies on renal and extra-renal excretion in a freshwater teleost, *Salmo gairdneri*. *Comp. Biochem. Physiol.* **10**, 121–128.

Galloway, T. M. (1933). Osmotic concentration of blood of Cyclostome *Petromyzon*. *J. Exptl. Biol.* **10**, 313–316.

Garrey, W. E. (1905). The osmotic pressure of sea water and of the blood of marine animals. *Biol. Bull.* **8**, 257–270.

Gordon, M. S. (1959a). Osmotic and ionic regulation in Scottish brown trout and sea trout (*Salmo trutta* L.). *J. Exptl. Biol.* **36**, 253–260.

Gordon, M. S. (1959b). Ionic regulation in the brown trout (*Salmo trutta* L.). *J. Exptl. Biol.* **36**, 227–252.

Gordon, M. S., Amdur, B. H., and Scholander, P. F. (1962). Freezing resistance in some northern fishes. *Biol. Bull.* **122**, 52–56.

Gordon, M. S., Boettius, J., Boettius, I., Evans, D. H., McCarthy, R., and Oglesby, L. C. (1965). Salinity adaptation in the mudskipper fish *Periophthalmus sobrinus*. *Hvalradets Skrifter Norske Videnskaps. Akad. Oslo* **48**, 85–93.

Gray, S. J., and Sterling, K. (1950). The tagging of red cells and plasma proteins with radioactive chromium. *J. Clin. Invest.* **29**, 1604–1613.

Greene, C. W. (1904). Physiological studies of the chinook salmon. *U.S. Bur. Fisheries, Bull.* **24**, 431–456.

Gregersen, M. I., and Rawson, R. A. (1959). Blood volume. *Physiol. Rev.* **39**, 307–342.

Hardisty, M. W. (1956). Some aspects of osmotic regulation in lampreys. *J. Exptl. Biol.* **33**, 431–447.

Hartman, F. A., Lewis, L. A., Brownell, K. A., Sheldon, F. F., and Walther, R. F. (1941). Some blood constituents of the normal skate. *Physiol. Zool.* **14**, 476–486.

Hartman, F. A., Lewis, L. A., Brownell, K. A., Angerer, K. A., and Sheldon, F. F. (1944). Effect of interrenalectomy on some blood constituents in the skate. *Physiol. Zool.* **17**, 228–238.

Herre, A. W. C. T. (1955). Sharks in fresh water. *Science* **122**, 417.

Hess, A. F., Bills, C. E., Weinstock, M., and Rivkin, H. (1928). Difference in calcium level of the blood between the male and female cod. *Proc. Soc. Exptl. Biol. Med.* **25**, 349–350.

Hickman, C. P., Jr. (1959). The osmoregulatory role of the thyroid gland in the starry flounder, *Platichthys stellatus*. *Can. J. Zool.* **37**, 997–1060.

Hickman, C. P., Jr., McNabb, R. A., Nelson, J. S., Van Breemen, E. O., and Comfort, D. (1964). Effect of cold acclimation on electrolyte distribution in rainbow trout (*Salmo gairdnerii*). *Can. J. Zool.* **42**, 577–597.

Ho, F. C. W., and Vanstone, W. E. (1961). Effect of estradiol monobenzoate on some serum constituents of maturing sockeye salmon (*Oncorhynchus nerka*). *J. Fisheries Res. Board Can.* **18**, 859–864.

Hoffert, J. R., and Fromm, P. O. (1966). Effect of carbonic anhydrase inhibition on aqueous humor and blood bicarbonate ion in the teleost (*Salvelinus namaycush*). *Comp. Biochem. Physiol.* **18**, 333–340.

Holmes, R. M. (1961). Kidney function in migrating salmonids. *Rept. Challenger Soc.* (*Cambridge*) **3**, No. 13, 23.

Holmes, W. N. (1959). Studies on the hormonal control of sodium metabolism in the rainbow trout (*Salmo gairdnerii*). *Acta Endocrinol.* **31**, 587–602.

Holmes, W. N., and Butler, D. G. (1963). The effect of adrenocortical steroids on the tissue electrolyte composition of the fresh water rainbow trout (*Salmo gairdneri*). *J. Endocrinol.* **25**, 457–464.

Holmes, W. N., and McBean, R. L. (1963). Studies on the glomerular filtration rate of rainbow trout (*Salmo gairdneri*). *J. Exptl. Biol.* **40**, 335–341.

Holmes, W. N., and Stainer, I. M. (1966). Studies on the renal excretion of electrolytes by the trout (*Salmo gairdneri*). *J. Exptl. Biol.* **44**, 33–46.

Holmes, W. N. Unpublished data.

House, C. R. (1963). Osmotic regulation in the brackish water teleost, *Blennius pholis*. *J. Exptl. Biol.* **40**, 87–104.

Houston, A. H. (1959). Osmoregulatory adaptation of steelhead trout (*Salmo gairdneri* Richardson) to sea water. *Can. J. Zool.* **37**, 729–748.

Houston, A. H. (1960). Variations in the plasma level of chloride in hatchery reared yearling Atlantic salmon during parr-smolt transformation and following transfer into sea water. *Nature* **185**, 632–633.

Houston, A. H. (1962). Some observations on water balance in the goldfish *Carassius auratus* L. during cold death. *Can. J. Zool.* **40**, 1169–1174.

Houston, A. H., and Madden, J. A. (1968). Environmental temperature and plasma electrolyte regulation in the carp *Cyprinus carpio*. *Nature* **217**, 969–970.

Houston, A. H., and Threadgold, L. T. (1963). Body fluid regulation in smolting Atlantic salmon. *J. Fisheries Res. Board Can.* **20**, 1355–1367.

Huggel, H., Kleinhaus, A., and Hamzehpour, M. (1963). The blood composition of *Salmo gairdneri irideus* and *Squalius cephalus* (Teleostei, Pisces). *Rev. Suisse Zool.* **70**, 286–290.

Huntsman, A. G., and Hoar, W. S. (1939). Resistance of Atlantic salmon sea water. *J. Fisheries Res. Board Can.* **4**, 409.

Idler, D. R., and Tsuyuki, H. (1958). Biochemical studies on sockeye salmon during spawning migration. I. Physical measurements, plasma cholesterol, and electrolyte levels. *Can. J. Biochem. Physiol.* **36**, 783–791.

Janssens, P. A. (1964). The metabolism of the aestivating African lungfish. *Comp. Biochem. Physiol.* **11**, 105–117.

Jensen, C. E., and Vilstrup, T. (1954). Determination of some inorganic substances in the labyrinthine fluid. *Acta Chem. Scand.* **8**, 697–698.

Johansen, K., Fänge, R., and Johannessen, M. W. (1962). Relations between blood, sinus fluid and lymph in *Myxine glutinosa* L. *Comp. Biochem. Physiol.* **7**, 23–28.

Kaieda, J. (1930). Biochemische Untersuchungen des Labyrinthwassers und der cerebrospinal Flussigkeit der Haifische. *Z. Physiol. Chem.* **188**, 193–202.

Keys, A., and Hill, R. M. (1934). The osmotic pressure of the colloids in fish sera. *J. Exptl. Biol.* **11**, 28–33.

Korjuev, P. A. (1938). Urea and chlorides of the blood of sea ganoids. *Bull. Biol. Med. Exptl. URSS* **6**, 158–159.

Kubo, T. (1953). On the blood of salmonid fishes of Japan during migration. I. Freezing point of blood. *Bull. Fac. Fisheries, Hokkaido Univ.* **4**, 138–148.

Kubo, T. (1955). Changes of some characteristics of the blood of the smolts of *Oncorhynchus masou* during seaward migration. *Bull. Fac. Fisheries, Hokkaido Univ.* **6**, 201–207.

Kubo, T. (1960). Notes on the blood of Masou salmon during inshore migration with special reference to the osmoconcentration. *Bull. Fac. Fisheries, Hokkaido Univ.* **11**, 15–19.

Kubo, T. (1961). Notes on the blood of Masou salmon (*Oncorhynchus masou*) during upstream migration for spawning with special reference to the osmoconcentration. *Bull. Fac. Fisheries, Hokkaido Univ.* **12**, 189–195.

Lahlou, B. (1967). Excretion renale chez un poisson euryhaline, le flet (*Platichthys flexus* L.): Characteristiques de l'urine normale en eau douce et en eau de mer et effects des changements de milieu. *Comp. Biochem. Physiol.* 20, 925–938.

Lange, R., and Fugelli, K. (1965). The osmotic adjustment in the euryhaline teleosts, the flounder, *Pleuronectes flesus* L. and the three-spined stickleback, *Gasterosteus aculeatus* L. *Comp. Biochem. Physiol.* 15, 283–292.

Lehninger, A. L. (1965). "The Mitochondrion." Benjamin, New York.

Lehninger, A. L., Rossi, C. S., and Greenwalt, J. W. (1963). Respiration-dependent accumulation of inorganic phosphate and Ca^{++} by rat liver mitochondria. *Biochem. Biophys. Res. Commun.* 10, 444.

Lennon, R. E. (1954). Feeding mechanism of the sea lamprey and its effect on host fishes. *U.S. Fish Wildlife Serv., Fishery Bull.* 56, 247–293.

Levine, L., and Musallam, D. A. (1964). A one-drop cryoscope: The tonicity of frog and goldfish sera. *Experientia* 20, 508.

Lysaya, N. M. (1951). Changes in the blood composition of salmon during the spawning migration. *Izv. Tikhookeansk. Nauchn.-Issled. Inst. Rybn. Khoz. i Okeanogr.* 35, 47–60.

McCallum, A. B. (1926). Paleochemistry of body fluids and tissues. *Physiol. Rev.* 6, 316–357.

McCarthy, J. E. (1967). Vascular and extravascular fluid volumes in the Pacific hagfish (*Eptatretus stoutii*, Lockington). M.A. Thesis, Oregon State University.

McCarthy, J. E., and Conte, F. P. (1966). Determination of the volume of vascular and extravascular fluids in the Pacific Hagfish, *Eptatretus stoutii*. *Am. Zoologist* 6, 605 (abstr.).

McFarland, W. N., and Munz, F. W. (1958). A re-examination of the osmotic properties of the Pacific hagfish (*Polistotrema stouti*). *Biol. Bull.* 114, 348–356.

McFarland, W. N., and Munz, F. W. (1965). Regulation of body weight and serum composition by hagfish in various media. *Comp. Biochem. Physiol.* 14, 383–398.

Maetz, J., Mayer, N., and Chartier-Baraduc, M. M. (1967). La balance minérale du sodium chez *Anguilla anguilla* en eau de mer, en eau douce et au cours de transfert d'un milieu a l'autre: Effets de l'hypophysectomie et de la prolactine. *Gen. Comp. Endocrinol.* 8, 177–188.

Magnin, E. (1962). Recherches sur la systématique et la biologie des acepenserides *Acipenser sturio, Acipenser oxyrhynchus,* et *Acipenser fulvescens.* Ch. 4. Quelques aspects de l'équilibre hydrominéral. *Ann. Stat. Centr. Hydrob. Appl.* 9, 170–242.

Manery, J. F. (1954). Water and electrolyte metabolism. *Physiol. Rev.* 34, 334–417.

Maren, T. H. (1962a). Ionic composition of cerebrospinal fluid and aqueous humor of the dogfish (*Squalus acanthias*). I. Normal values. *Comp. Biochem. Physiol.* 5, 193–200.

Maren, T. H. (1962b). Ionic composition of cerebrospinal fluid and aqueous humor of the dogfish *Squalus acanthias*. II. Carbonic anhydrase activity and inhibition. *Comp. Biochem. Physiol.* 5, 201–215.

Maren, T. H. (1967a). Special body fluids of the elasmobranch. In "Sharks, Skates and Rays" (P. W. Gilbert, R. F. Mathewson, and D. P. Rall, eds.), pp. 287–292. Johns Hopkins Press, Baltimore, Maryland.

Maren, T. H. (1967b). Carbonic anhydrase: Chemistry, physiology, and inhibition. *Physiol. Rev.* 47, 595–781.

Maren, T. H., and Frederick, A. (1958). Carbonic anhydrase inhibition in the elasmobranch: Effect on aqueous humor and cerebrospinal fluid CO_2. *Federation Proc.* 17, 391.

Maren, T. H., Rawls, J. A., Burger, J. W., and Myers, A. C. (1963). The alkaline (Marshall's) gland of the skate. *Comp. Biochem. Physiol.* **10,** 1–16.

Martin, A. W. (1950). Some remarks on the blood volume of fish. In "Studies Honoring Trevor Kincaid," pp. 125–140. Univ. of Washington Press, Seattle, Washington.

Mayer, N., Maetz, J., Chan, D. K. O., Forster, M., and Chester-Jones, I. (1967). Cortisol: A sodium excreting factor in the eel (*Anguilla anguilla* L.) adapted to sea water. *Nature* **214,** 1118–1120.

Medway, W., and Kare, M. R. (1959). Thiocyanate space in growing domestic fowl. *Am. J. Physiol.* **196,** 783.

Morris, R. (1956). The osmoregulatory ability of the lamprey *Lampetra fluviatilis* in sea water during the course of its spawning migration. *J. Exptl. Biol.* **33,** 235–248.

Morris, R. (1958). The mechanism of marine osmoregulation in the Lampern (*Lampetra fluviatilis* L.) and the causes of its breakdown during the spawning migration. *J. Exptl. Biol.* **35,** 649–664.

Morris, R. (1960). General problems of osmoregulation with special reference to cyclostomes. *Symp. Zool. Soc. London* **1,** 1–16.

Morris, R. (1965). Studies on salt and water balance in *Myxine glutinosa* (L.). *J. Exptl. Biol.* **42,** 359–371.

Motais, R., Romeu, F. G., and Maetz, J. (1965). Mechanism of eurhalinity—comparative investigation of *Platichthys* and *Serranus* following their transfer to fresh water. *Compt. Rend.* **261,** 801–804.

Munz, F. W., and McFarland, W. N. (1964). Regulatory function of a primitive vertebrate kidney. *Comp. Biochem. Physiol.* **13,** 381–400.

Murdaugh, H. V., and Robin, E. D. (1967). Acid-base metabolism in the dogfish shark. In "Sharks, Skates and Rays" (P. W. Gilbert, R. F. Mathewson, and D. P. Rall, eds.), pp. 249–264. Johns Hopkins Press, Baltimore, Maryland.

Murray, R. W., and Potts, W. T. W. (1961). Composition of the endolymph, perilymph, and other body fluids of elasmobranchs. *Comp. Biochem. Physiol.* **2,** 65–76.

Nicol, J. A. C. (1960). The autonomic nervous system of the Chimaeroid fish *Hydrolagus colliei. Quart. J. Microscop. Sci.* **91,** 379–400.

Ogawa, M. (1968). Osmotic and ionic regulation in goldfish following removal of the corpuscles of Stannius or the pituitary gland. *Can. J. Zool.* **46,** 669–676.

Oguri, M., and Takada, N. (1966). Effects of some hormonic substances on the urinary and serum calcium levels of the snake headfish *Channa argus. Bull. Japan. Soc. Sci. Fisheries* **32,** 28–31.

Oguri, M., and Takada, N. (1967). Serum calcium and magnesium levels of goldfish, with special reference to the gonadal maturation. *Bull. Japan. Soc. Sci. Fisheries* **33,** 161–166.

Oppelt, W. W., Adamson, R. H., Zubrod, C. G., and Rall, D. P. (1966). Further observations on the physiology and pharmacology of elasmobranch ventricular fluid. *Comp. Biochem. Physiol.* **17,** 857–866.

Parry, G. (1958). Size and osmoregulation in salmonid fishes. *Nature* **181,** 1218.

Parry, G. (1961). Osmotic and ionic changes in blood and muscle of migrating salmonids. *J. Exptl. Biol.* **38,** 411–427.

Pearcy, W. G. (1961). Seasonal changes in osmotic pressure of flounder sera. *Science* **134,** 193–194.

Pereira, R. S., and Sawaya, P. (1957). Contributions to the study of the chemical composition of the blood of certain selachians of Brazil. *Univ. Sao Paulo, Fac. Filsof., Cienc. Letras, Bol. Zool.* **21,** 85–92.

Phillips, A. M., Jr., and Brockway, D. R. (1958). The inorganic composition of brown trout blood. *Progressive Fish Culturist* **20**, 58–61.

Pickford, G. E., and Grant, F. B. (1967). Serum osmolarity in the coelacanth *Latimeria chalumnae* urea retention and ion regulation. *Science* **155**, 568–570.

Platner, W. S. (1950). Effects of low temperature on Mg content of blood, body fluid and tissue of goldfish and turtle. *Am. J. Physiol.* **161**, 399–405.

Pora, E. A. (1935). Differences minerales dans la composition du sang suivant le sexe, chez *Cyprinus carpio*. *Compt. Rend. Soc. Biol.* **119**, 373–375.

Pora, E. A. (1936a). De l'influence de l'oxgénation du milieu extérieur sur la composition du sang, chez (*Scyllium canicula*). *Compt. Rend. Soc. Biol.* **121**, 194–196.

Pora, E. A. (1936b). Influence des fortes oxygénations du milieu extérieur sur la composition du sang de *Scyllium canicula*. *Ann. Physiol. Physicochim. Biol.* **12**, 238.

Pora, E. A. (1936c). Sur les différences chimiques et physico-chimiques du sang des deux sexes des Sélaciens. *Compt. Rend. Soc. Biol.* **121**, 105–107.

Pora, E. A. (1936d). Quelques données analytiques sur la composition chimique et physico-chimique du sang de quelques invertebrés et vertebrés marins. *Compt. Rend.* **121**, 291–293.

Pora, E. A. (1936e). Sur les différences chimiques et physico-chimiques du sang, suivant les sexes chez *Labrus bergylta*. *Compt. Rend. Soc. Biol.* **121**, 102–105.

Potts, W. T. W., and Evans, D. H. (1967). Sodium and chloride balance in the killifish *Fundulus heteroclitus*. *Biol. Bull.* **133**, 411–425.

Potts, W. T. W., and Parry, G. (1964). "Osmotic and Ionic Regulation in Animals." Pergamon Press, Oxford.

Price, K. S. (1967). Fluctuations in two osmoregulatory components, urea and sodium chloride, of the clearnose skate *Raja eglanteria* Bosc 1802. II. Upon natural variation of the salinity of the external medium. *Comp. Biochem. Physiol.* **23**, 77–82.

Price, K. S., and Creaser, E. P. (1967). Fluctuations in two osmoregulatory components, urea and sodium chloride, of the clearnose skate *Raja eglanteria* Bosc 1802. I. Upon laboratory modification of external salinities. *Comp. Biochem. Physiol.* **23**, 65–76.

Prosser, C. L., and Weinstein, S. J. F. (1950). Comparison of blood volume in animals with open and closed circulatory systems. *Physiol. Zool.* **23**, 113–124.

Raffy, A. (1949). L'euryhalinité de *Blennius pholis* L. *Comp. Rend. Soc. Biol.* **143**, 1575–1576.

Robertson, J. D. (1954). The chemical composition of the blood of some aquatic chordates including members of the *Tunicata*, *Cyclostomata* and *Osteichthyes*. *J. Exptl. Biol.* **31**, 424–442.

Robertson, J. D. (1963). Osmoregulation and Ionic composition of cells and tissues. *In* "Biology of Myxine" (A. Brodal and R. Fänge, eds.), pp. 504–515. Oslo Univ. Press, Oslo.

Robertson, J. D. (1966). Osmotic constituents of the blood plasma and parietal muscle of *Myxine glutinosa* L. *In* "Some Contemporary Studies in Marine Science" (H. Barnes, ed.), pp. 631–644. Allen & Unwin, London.

Robertson, O. H., Krupp, M. A., Favour, C. B., Hane, S., and Thomas, S. F. (1961). Physiological changes occurring in the blood of the Pacific Salmon (*Oncorhynchus tshawytscha*) accompanying sexual maturation and spawning. *Endocrinology* **68**, 733–746.

Robin, E. D., Murdaugh, H. V., Jr., and Weiss, E. (1964). Acid-base, fluid and electrolyte metabolism in the elasmobranch. 1. Ionic composition of erythrocytes, muscle and brain. *J. Cellular Comp. Physiol.* **64**, 409–418.

Rodier, E. (1899). Observations et expériences comparatives sur l'eaux de mer, le sang et les liquides internes des animaux marins. *Trav. Lab. Soc. Sci. Stat. Zool. Arachon* pp. 103–123.

Rodnan, G. P., Robin, E. D., and Andrus, M. H. (1962). Dogfish coelomic fluid. 1. Chemical anatomy. *Bull. Mt. Desert Isl. Biol. Lab.* **4**, 69–70.

Schiffman, R. H., and Fromm, P. O. (1959). Measurement of some physiological parameters in rainbow trout (*Salmo gairdnerii*). *Can J. Zool.* **37**, 25–32.

Scholander, P. F., Van Dam, L., Kanwisher, J. W., Hammel, H. T., and Gordon, M. S. (1957). Supercooling and osmoregulation in arctic fish. *J. Cellular Comp. Physiol.* **49**, 5–24.

Schreibman, M. P., and Kallman, K. D. (1966). Endocrine control of fresh water tolerance in teleosts. *Gen. Comp. Endocrinol.* **6**, 144–155.

Sharratt, B. M., Chester Jones, I., and Bellamy, D. (1964). Water and electrolyte composition of the body and renal function of the eel (*Anguilla anguilla*). *Comp. Biochem. Physiol.* **11**, 9–18.

Shell, E. W. (1959). Chemical composition of the blood of small mouth bass. Ph.D. Thesis, Cornell University.

Sindermann, C. J., and Mairs, D. F. (1961). Blood properties of pre-spawning and post-spawning anadromous alewives (*Alosa pseudoharengus*). *U.S. Fish Wildlife Serv., Fishery Bull.* **183**, 145–151.

Smith, H. W. (1929a). The composition of the body fluids of elasmobranchs. *J. Biol. Chem.* **81**, 407–419.

Smith, H. W. (1929b). The composition of the body fluids of the goose fish (*Lophius piscatorius*). *J. Biol. Chem.* **82**, 71–75.

Smith, H. W. (1930). Metabolism of the lungfish *Protopterus aethiopicus*. *J. Biol. Chem.* **88**, 97–130.

Smith, H. W. (1931). The absorption and excretion of water and salts by the elasmobranch fishes. I. Fresh-water elasmobranchs. *Am. J. Physiol.* **98**, 279.

Smith, H. W. (1936). The retention and physiological role of urea in the *Elasmobranchii*. *Biol. Rev.* **11**, 49–92.

Soberman, R. J. (1950). Use of antipyrine in measurement of the total body water in animals. *Proc. Soc. Exptl. Biol. Med.*, **74**, 789–792.

Staedeler, G., and Frerichs, F. T. (1858). Uber das Vorkommen von Harnstoff' Taurin und Scyllit in den Organen der Plagiostomen. *J. Prakt. Chem.* **73**, 48–55.

Stanley, J. G., and Fleming, W. R. (1964). Excretion of hypertonic urine by a teleost. *Science* **144**, 63–64.

Sterling, K., and Gray, S. J. (1950). Determination of the circulating red cell volume in man by radioactive chromium. *J. Clin. Invest.* **29**, 1614–1619.

Sulya, L. L., Box, B. E., and Gunther, G. (1960). Distribution of some blood constituents in fish from the Gulf of Mexico. *Am. J. Physiol.* **199**, 1177–1180.

Thorson, T. (1958). Measurement of the fluid compartments of four species of marine Chondrichthyes. *Physiol. Zool.* **31**, 16–23.

Thorson, T. (1959). Partitioning of the body water in sea lamprey. *Science* **130**, 99–100.

Thorson, T. (1961). The partitioning of body water in Osteichthyes: Phylogenetic and ecological implications in aquatic vertebates. *Biol. Bull.* **120**, 238–254.

Thorson, T. (1962). Partitioning of body fluids in the Lake Nicaragua shark and three marine sharks. *Science* **138**, 688–690.

Thorson, T. B. (1967). Osmoregulation in fresh water elasmobranchs. In "Sharks, Skates and Rays" (P. W. Gilbert, R. F. Mathewson, and D. P. Rall, eds.), pp. 265–270. Johns Hopkins Press, Baltimore, Maryland.

Urist, M. R. (1961). Calcium and phosphorus in the blood and skeleton of the elasmobranchii. Endocrinology 69, 778–801.

Urist, M. R. (1962a). Calcium and other ions in blood and skeleton of Nicaraguan freshwater Shark. Science 137, 984–986.

Urist, M. R. (1962b). The bone-body fluid continuum: Calcium and phosphorous in the skeleton of extinct and living vertebrates. Perspectives Biol. Med. 6, 75–115.

Urist, M. R. (1963). The regulation of calcium and other ions in the serums of hagfish and lamprey. Ann. N.Y. Acad. Sci. 109, 294–311.

Urist, M. R. (1964). Further observations bearing on the bone-body fluid continuum: Composition of the skeleton and serums of cyclostomes, elasmobranchs and bony vertebrates. In "Monographs on Bone Biodynamics" (H. M. Frost, ed.), pp. 151–179. Little, Brown, Boston, Massachusetts.

Urist, M. R. (1966). Calcium and electrolyte control mechanisms in lower vertebrates. In "Phylogenetic Approach to Immunity" (R. T. Smith, R. A. Good, and P. A. Miescher, eds.), pp. 18–28. Univ. of Florida Press, Gainesville, Florida.

Urist, M. R., and Van de Putte, K. A. (1967). Comparative biochemistry of the blood of fishes. In "Sharks, Skates and Rays" (P. W. Gilbert, R. F. Mathewson, and D. P. Rall, eds.), pp. 271–292. Johns Hopkins Press, Baltimore, Maryland.

Vasington, F. D., and Murphy, J. V. (1962). Ca++ uptake by rat kidney mitochondria and its dependence on respiration and phosphorylation. J. Biol. Chem. 237, 2670.

Welcker, H. (1858). Bestimmungen der Menge des Korperblutes und der Blutfarbkraft, sowie Bestimmungen von Zahl, Mass, Oberfläche und Volumn des einzelnen Blutkörperchen bei Thieren und bei Menschen. Z. Rationelle Med. 4, 145.

Wikgren, B. (1953). Osmotic regulation in some aquatic animals with special reference to the influence of temperature. Acta Zool. Fennica 71, 1–102.

Winkler, A. W., Elkinton, J. R., and Eisman, A. J. (1943). Comparison of sulfocyanate with radioactive chloride and sodium in the measurement of extracellular fluid. Am. J. Physiol. 139, 239.

Woodhead, P. M. J., and Woodhead, A. D. (1958). Effects of low temperature on the osmoregulatory ability of the cod (Gadus callarias) in arctic waters. Proc. Linnean Soc. London 169, 63–64.

Woodhead, P. M. J., and Woodhead, A. D. (1959). Effects of low temperature on the physiology and distribution of the cod, Gadus morhua L. in the Barent Sea. Proc. Zool. Soc. London 133, 181–199.

Yamazaki, F., and Donaldson, E. M. (1968). Unpublished observations.

Young, J. Z. (1950). "The Life of Vertebrates." Oxford Univ. Press, London and New York.

Zaks, M. G., and Sokolova, M. M. (1961). O mekhanismakh adaptatsii k izmeneniian solenosti vody u nerki—Oncorhynchus nerka (Walb). Vopr. Ikhtiol. 1, 333–346 (Fisheries Res. Board Can. Transl. No. 372).

Zaks, M. G., and Sokolova, M. M. (1965). Ismenenie tipa osmoregulizatsii v raynye periody migratsionnogo tsikla u nerki Oncorhynchus nerka (Walb.). Vopr. Ikhtiol. 5, 331–337 (Fisheries Res. Board Can. Transl. No. 773).

2

THE KIDNEY

CLEVELAND P. HICKMAN, JR., and BENJAMIN F. TRUMP

I. INTRODUCTION

The significance of the kidney in the physiology of fish can be summarized as follows:

(1) In freshwater forms, the kidney functions largely as a water

91

excretory device. With rare exceptions, this is accomplished by filtration at the renal glomerulus, and implies the presence of suitable cellular components to conserve filtered ions and excrete dilute urine.

(2) In marine teleost forms, the kidney functions chiefly as an excretory device for magnesium and sulfate ions. In glomerular marine forms associated machinery must be present to conserve water, monovalent ions, and other filtered plasma constituents.

(3) The cartilaginous sharks, skates, and rays, because they are hyperosmotic in a marine environment, have combined the principal functions of the kidneys of both freshwater and marine bony fishes.

The marine teleost nephron probably reaches its highest degree of specialization in the aglomerular teleosts, which have only two tubular segments; together these are capable of accomplishing the efficient removal of magnesium and sulfate. Since many fish are euryhaline, this implies the presence of kidneys which are able to shift from an emphasis on water excretion to an emphasis on divalent ion excretion and water conservation. The renal contribution to nitrogenous end product excretion, though present, is probably of relatively little importance.

It is evident that increased understanding of the nephrons of lower vertebrates has been and will continue to be of great importance in the development of our ideas regarding the relationship between structure and function in the nephron of mammals, including man. This is based upon observations, such as those developed in this chapter, of homologies within the nephrons of lower vertebrates beginning with the hagfish. Studies of the functions of these homologous segments are often much more readily accomplished in fish than they are in mammals, and model systems employing fish nephrons have been of great usefulness in the understanding of both normal (Forster, 1961) and abnormal nephron physiology (Ginn et al., 1968; Trump and Bulger, 1967, 1968a,b; Trump and Ginn, 1968, 1969). This is particularly apparent in the case of the specialized aglomerular marine teleosts, which possess only two of the segments that have evidently originated in freshwater fish. The kidney's participation in osmotic, and specific ionic regulation and its adaptive limitations are also of importance in understanding the total biology of fish. This, in turn, is of undoubted significance economically with respect to the preservation and management of commercial and sports fisheries.

Knowledge of the complexity of the nephron as it exists in present-day freshwater teleosts and dipnoans, continues to be a useful tool in exploring hypotheses of evolution. The homologies within the nephron can be readily assessed in varieties of present-day fish, and therefore provide morphological and functional markers through which the evolution of fish species can be inferred.

This chapter represents an attempt to assemble all available information concerning the structure and function of the fish kidney and to provide a unifying synthesis for the understanding of this organ's role in body fluid regulation, of specific nephron function in fishes, and of the evolutionary significance of regions of the nephron in fishes and higher vertebrates.

We have adopted the classification of Berg (1940) throughout this chapter except that we have retained the use of the popular term Teleostei or "teleosts" to include the higher Teleostomi (Actinopterygii) from the Clupeiformes to Pegasiformes.

II. CLASS MYXINI (HAGFISHES)

The hagfishes are entirely marine and are nearly in osmotic equilibrium with their environment. The high body fluid osmolality results from high levels of sodium and chloride, as in marine invertebrates, rather than from a combined retention of nitrogenous end products and electrolytes, as in the elasmobranchs. The plasma is isosmotic or, at most, 2% hyperosmotic to seawater (Robertson, 1954; McFarland and Munz, 1958; Morris, 1960, 1965), but its composition differs from seawater in all of the major ions. The individual regulation of the plasma ions appears to be the principal function of the hagfish kidney, a task probably shared by other organs, especially the liver (Rall and Burger, 1967). There is no evidence of special ion-transporting cells in the gill epithelium (Morris, 1965).

A. Kidney Morphology

Adult hagfishes have both a paired anterior pronephros (Fig. 1) and a paired mesonephros (Fig. 2). The portion of the kidney between the pronephros and mesonephros degenerates during development and the two become independent. Atubular glomeruli and aglomerular tubules persist in this region. The small pronephric kidneys lie in the walls of the pericardial cavity, dorsal to the portal heart and along the dorsolateral surface of the gut. Each contains a single, very large, elongate Malpighian corpuscle, possibly formed by the fusion of two or three glomeruli and their capsules (Conel, 1917; Holmgren, 1950); there is also a separate tubular portion (Fig. 1). Bowman's capsule does not empty directly into the tubules and the tubules have no direct connection to any excretory canal, since the Wolffian duct, which in young animals connects the

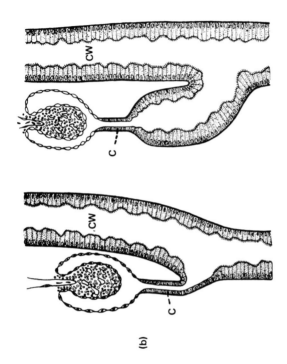

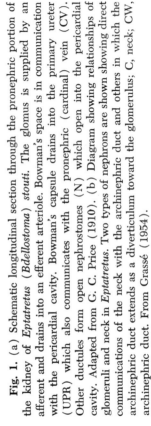

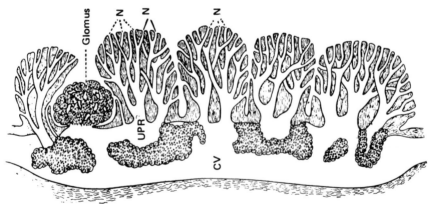

Fig. 1. (a) Schematic longitudinal section through the pronephric portion of the kidney of *Eptatretus* (*Bdellostoma*) *stouti*. The glomus is supplied by an afferent and drains into an efferent arteriole. Bowman's space is in communication with the pericardial cavity. Bowman's capsule drains into the primary ureter (UPR) which also communicates with the pronephric (cardinal) vein (CV). Other ductules form open nephrostomes (N) which open into the pericardial cavity. Adapted from G. C. Price (1910). (b) Diagram showing relationships of glomeruli and neck in *Eptatretus*. Two types of nephrons are shown showing direct communications of the neck with the archinephric duct and others in which the archinephric duct extends as a diverticulum toward the glomerulus; C, neck; CW, archinephric duct. From Grassé (1954).

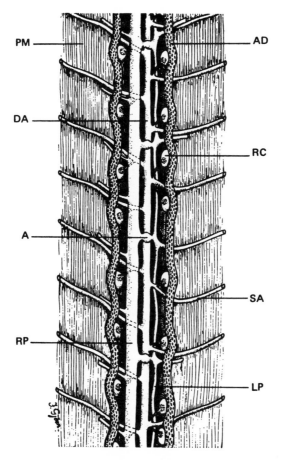

Fig. 2. Mesonephros of *Myxine glutinosa;* mesenteric vessels are not shown. Note the large renal corpuscles (RC) along the medial sides of the archinephric ducts (AD). The glomeruli are supplied by short branches from the segmental arteries (SA) which in turn arise from the dorsal aorta (DA). The right (RP) and left (LP) postcardinal veins are connected by bulblike anastomoses (A). PM, parietal muscle. From Fänge (1963).

pronephros and mesonephros, degenerates anteriorly to a thin lumenless thread in the adult (Muller, 1875). The function of the pronephros is not clear. It may serve in some ionic regulatory capacity, but critical experiments have never been performed to examine this possibility. Holmgren (1950) and Fänge (1963) review other possible functions (lymphoid, hemopoetic, and endocrine) that have been attributed to the pronephric kidney.

The mesonephros is considered the true functional kidney of the hagfishes (Fänge, 1963). Each side of the paired mesonephros consists of 30–35 very large, oval Malpighian corpuscles [averaging 0.68 mm in length with some as long as 1.02 mm (Nash, 1931)] arranged segmentally on the medial side of an archinephric duct (ureter) (Fig. 2). The glomeruli are composed of widely patent capillaries resembling those of glomerular freshwater teleosts. Each Bowman's capsule is connected to the archinephric duct by a short neck lined by flattened epithelium. Usually this neck is connected to a diverticulum of the archinephric duct. The diverticula and the duct are lined by identical epithelium.

The cells lining the duct have a well-developed brush border composed of closely packed microvilli. At the base of the microvilli, coated pits are abundant, and the apical cytoplasm contains numerous coated and uncoated vesicles and many large apical vacuoles, some of which contain a finely granular material (Ericsson, 1967; Ericsson and Trump, 1969). These are similar to the structures observed in the first proximal segment in the higher fishes and resemble those observed in the human proximal convolution (Trump and Bulger, 1968a). They presumably function in the uptake of materials from the lumen. In the midregion of the cell, there are abundant cytoplasmic bodies up to 3 μ in diameter which represent secondary lysosomes involved with processing of material taken in by endocytosis. The secondary lysosomes correspond to the numerous pigmented granules observed in these cells by earlier authors (Gerard, 1943; Fänge, 1963; Holmgren, 1950). This pigment appears similar to lipofuscin; hence, these structures, at least in part, represent residual bodies. Holmgren (1950) asserts that these yellow pigment bodies are excreted into the lumen of the tubule. If so, this process would be analogous to extrusion of lysosomal contents observed in mammalian liver (Bradford et al., 1969) and renal tubular cells (Ericsson et al., 1965; Griffith et al., 1967). The Golgi apparatus and endoplasmic reticulum are not prominent, and the mitochondria are comparatively small but moderately abundant. The lateral and basilar plasmalemmas are formed into small interlocking processes which usually do not contain mitochondria. At the base of the cell, the plasmalemma is deeply invaginated. The epithelium is deeply folded in the empty duct, becoming greatly stretched when the duct is filled with urine. Since true nephric tubules are virtually absent, the archinephric ducts of the hagfish assume whatever regulatory functions the kidney performs. The archinephric ducts open separately into a cloaca.

The morphology of the archinephric duct epithelium suggests a homology with the proximal convoluted tubule of mammals and with the first proximal segment of higher fishes (see Table XIII). It also fits into the hypothesis that this type of morphology is related to the presence of a glomerulus, and presumably to the presence of filtered materials within the tubular fluid.

The glomeruli are supplied by branches of segmental arteries arising from the dorsal aorta (Fig. 2). The capillary network of the ureters is served by the postglomerular circulation and by arteries directly from the aorta. There is no portal circulation. Juxtaglomerular cells, evidently present in all bony fishes, are reported absent from hagfishes (Sutherland, 1966; Capreol and Sutherland, 1968).

Small differences in the anatomy of the kidneys of the two species studied experimentally, the Atlantic (*Myxine glutinosa*) and Pacific [*Eptatretus* (= *Polistotrema, Bdellostoma*) *stouti*] hagfishes, are probably physiologically unimportant.

B. Kidney Function

In spite of its relatively simple morphology, the hagfish mesonephric kidney forms urine by the familiar sequence of ultrafiltration of the plasma, with subsequent modification of the filtrate in the tubules and ducts. The total filtering surface of the glomeruli is larger than in most marine teleosts and as large as that of many elasmobranchs and freshwater teleosts of equivalent body size (Nash, 1931). However, the glomerular filtration rate (GFR) of *Eptatretus*, varying between 0.16 and 0.41 ml/hr/kg (Munz and McFarland, 1964) is much lower than in freshwater fish, probably because of the very low blood pressure (3–9 mm Hg) in the dorsal aorta (Johansen, 1960). The large filtering surface may be of survival value, since these animals have been reported from waters of varying salinity (Fänge, 1963). The urine is isosmotic or slightly hypoosmotic to the blood (Munz and McFarland, 1964; Morris, 1965). Urine/plasma (U/P) ratios for inulin are near unity for both the Pacific and Atlantic hagfishes, indicating no net water movement across the duct epithelium (Munz and McFarland, 1964; Rall and Burger, 1967). While hagfishes are very permeable to water, having only a limited ability to regulate their body weight when subjected to osmotic variations in their environment (McFarland and Munz, 1965; Robertson, 1963), little water enters from their normal environment and urine flow is low (0.1–0.4 ml/hr/kg).

Since the urine remains isosmotic to the blood, there is no net change in total concentration of osmolytes in the urine. Yet the kidney does modify the filtrate by the reabsorption and secretion of materials, pri-

Table I

Renal Excretion of Plasma Solutes by the Pacific Hagfish, *Eptatretus stoutii*[a]

GFR = 0.307 ml/hr/kg; urine flow = 0.293 ml/hr/kg

Plasma solute	Seawater conc. (mmole/kg)	Serum conc. (mmole/kg)	Urine conc. (mmole/kg)	Urine/serum ratio	Quantity filtered[b] (μmole/hr)	Quantity excreted (μmole/hr)	Net quantity reabsorbed (μmole/hr)	Net quantity secreted (μmole/hr)	Reabsorbed or secreted (%)	Clearance (ml/hr)	Solute/inulin clearance ratio
Sodium	496	570	553	0.97	175	162	13	—	7.4	0.28	0.91
Potassium	10.3	7.0	11.0	1.57	2.1	3.2	—	1.1	34	0.46	1.5
Calcium	10.9	4.5	3.6	0.80	1.4	1.1	0.3	—	21	0.24	0.78
Magnesium	51.5	12.0	14.7	1.23	3.7	4.3	—	0.6	14	0.36	1.2
Chloride	543	547	548	1.00	168	161	7	—	4.2	0.29	0.94
Sulfate	25.7	0.9	7.3	8.59	0.3	2.1	—	1.8	86	2.3	7.5
Phosphate	0.0	2.3	8.9	3.91	0.7	2.6	—	1.9	73	1.1	3.6
Urea[c]	—	5.0	8.8	1.76	1.53	2.93	—	1.4	47.7	0.58	1.9
Glucose	—	1.75	0.35	0.2	0.54	0.102	0.44	—	81.4	0.058	0.19
pH	7.56	7.80	7.55	—	—	—	—	—	—	—	—

[a] From Munz and McFarland (1964) except as noted.
[b] Uncorrected for plasma protein binding of electrolytes or Donnan effect.
[c] From Fänge (1963).

marily in the archinephric ducts, although the neck may contribute. As shown in Table I, all of the divalent ions, with the possible exception of calcium, are concentrated in the urine (Fig. 47a). Potassium is secreted and sodium is probably reabsorbed. Chloride is about the same concentration in plasma as in urine. Urea appears to be actively secreted. Munz and McFarland (1964) argue that magnesium, sulfate, phosphate, and potassium are transported by independent mechanisms because the ratios of their concentrations vary in an apparent random fashion between individual urine samples. However, tubular function in the hagfish is difficult to evaluate owing to the great variation in time that the formative urine remains in the ureters. Micturition is associated with vigorous body movements; when active the animal may excrete urine virtually unmodified by ion transport activity in the archinephric duct. This may account for the large variations in urine ionic composition observed (Munz and McFarland, 1964; Morris, 1965; Rall and Burger, 1967).

An average of 80% of the filtered glucose is reabsorbed and the urine is occasionally glucose free (Munz and McFarland, 1964). Glucose excretion is greatly increased in animals made hyperglycemic (Falkmer and Matty, 1966). Unlike the gnathostomous fishes, the hagfish kidney is incapable of secreting phenol red and fluorescein (Fänge and Krog, 1963; Rall and Burger, 1967). This inability for organic acid transport may be related to the morphology at the base of the cell where the plasma membrane invaginations are less complex than those of the mammal or of proximal segments of higher fishes; furthermore, the compartments formed by the invaginations do not contain mitochondria.

The ultrastructural characteristics of the cell apex with its membrane invaginations and vacuoles implies a role in the uptake of filtered macromolecules as suggested earlier by Gerard (1943). It has been shown in the marine flounder, *Parophrys vetulus* (Bulger and Trump, 1969b), as well as in the rat (Ericsson, 1964), that macromolecules introduced into the lumen enter via these apical tubules and vesicles. They are then presumably transported to the large secondary lysosomes, where digestion occurs (Ericsson, 1964). This similarity to the apical morphology in the proximal tubules of mammals is also compatible with the demonstrated resorptive capacity of the hagfish duct for glucose.

Thus, while the epithelium of the archinephric duct actively secretes certain ions such as magnesium, sulfate, and phosphate and reabsorbs glucose, the hagfish certainly has the most ineffective of all vertebrate kidneys (Fig. 47a). However, its regulatory task is correspondingly small, and other excretory pathways assist the kidney. In *Myxine*, magnesium, calcium, and phenol red are all excreted by the liver. Rall and Burger (1967) found that the magnesium concentration in the bile was nearly twice its concentration in seawater and more than four times its concentration in plasma. In this primitive fish, the liver may be the locus of several specific excretory functions later assumed by the kidney tubules of more advanced vertebrates. The well-known slime secretion of hagfishes may also be a primitive ion secretory mechanism, especially for potassium (Munz and McFarland, 1964).

III. CLASS PETROMYZONES (LAMPREYS)

The patterns of body fluid regulation in hagfishes and lampreys have little in common. Lampreys are believed to employ regulatory mechanisms similar to those of teleost fishes and the ion composition of the blood plasma of lampreys living in freshwater is nearly identical to that of freshwater teleosts. However, only the migratory European river lamprey, *Lampetra* (= *Petromyzon*) *fluviatilis,* has been studied in any detail, and only during its spawning migration into freshwater.

A. Kidney Morphology

In the lamprey the opisthonephroi are paired, elongate structures which lie on either side of the midline, suspended from the dorsal body wall by mesenterylike membranes. The archinephric duct runs along the free, lateral margin of the kidney. In cross section each kidney is triangular with rather distinct regions formed by the glomus and tubules.

The nephrons of adult lampreys are composed of the following regions: renal corpuscle, ciliated neck segment, a first proximal segment, and a convoluted segment without brush border which ends in the archinephric duct (Wheeler, 1899; Regaud and Policard, 1902) (Fig. 3).

In at least some species, including *L. fluviatilis,* the glomeruli are fused into a single large glomus measuring 9 cm in length and 0.25 mm in width (Fig. 3). The fusion apparently involves both the capillaries and the urinary space. Efferent arterioles drain a sinus at the hilus which is supplied by branches of the aorta. There is no renal portal system. Individual nephrons leave the common urinary space and a single cross section of the glomus may show 2–6 such neck origins (Fig. 3). The glomerulus (or glomus) contains capillaries that are widely open as in freshwater fish (von Möllendorff, 1930). Capreol and Sutherland (1968) were unable to find juxtaglomerular cells in the lamprey.

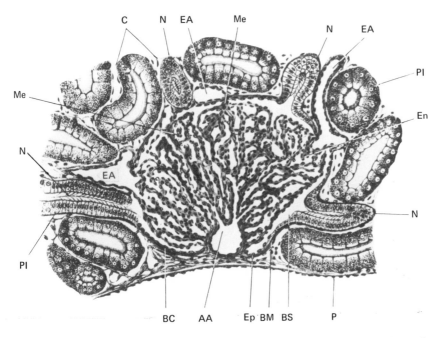

Fig. 3. Drawing of a glomus and tubules from the kidney of *Lampetra fluviatilis*. Note the large glomus with its afferent arteriole (AA) and several efferent arterioles (EA). Three necks (N) can be seen to arise from the glomus in this picture. Another section of the neck (N) appears toward the top. The other tubules are first proximal segments (PI). At the left a transition between the ciliated neck and the first proximal segment can be seen. Various features of the glomerulus such as the mesangial region (Me), the capillary endothelium (En), Bowman's capsule (BC) with epithelium (Ep), Bowman's space (BS), and basement membrane (BM) are shown. P, peritoneal covering of kidney; C, peritubular capillary. From von Möllendorff (1930), after Krause (1923).

The neck segment is short (200 μ) and straight; in cross section it is lined on two sides by tall columnar ciliated cells and on the other two sides by cuboidal nonciliated epithelium. This produces a slitlike lumen. The cilia originate in groups from the apex; each cilium has its own basal body as described below for the flounder, *Parophrys vetulus*.

The first proximal segment, which has numerous diverticula, resembles cytologically the archinephric duct of *Myxine glutinosa* and the proximal tubule of the mammalian kidney (Fig. 3). The cells contain numerous mitochondria that are concentrated in the basilar region and prominent secondary lysosomes in the supranuclear region. Schneider (1903) has demonstrated uptake and storage of intraperitoneally administered colloidal materials in these lysosomes.

The convoluted segment without brush border is smaller in diameter than the preceding segment and is lined by a cuboidal epithelium possessing few mitochondria (Fig. 3).

B. Kidney Function

The fragmentary information on renal function in lampreys in seawater has come from the river lamprey, *Lampetra fluviatilis*, studied after the beginning of its spawning migration into freshwater, but before it completely loses its ability to osmotically and ionically regulate in seawater (Morris, 1956, 1958, 1960). No studies have been published of kidney function in the sea lamprey, *Petromyzon marinus*, in either the marine or the freshwater environments, although the composition of the serum of this species during the four phases of its life cycle in the Great Lakes has been measured (Urist and Van de Putte, 1967). These authors also reported that the sea lamprey can maintain body fluid hypoosmolality when held in artificial seawater.

Prespawning adult river lampreys spend their early life history in brackish water estuaries and migrate into rivers in the autumn. After a 4–6 month residence in freshwater, they spawn and die. The Ammocoete larvae develop in freshwater and migrate to brackish water estuaries at the time of metamorphosis. At the beginning of the upstream spawning migration, adult river lampreys are capable of hypoosmotic regulation in 50% sea water (seawater 523 mOsm/liter, plasma 307 mOsm/liter) but not in full strength seawater (Bahr, 1952; Morris, 1958, 1960). Morris found that lampreys drank seawater, but that unlike marine teleosts, the gut fluid remained hyperosmotic to the plasma and contained a higher chloride content than the plasma. Four specimens drank an average 7.04 ml/hr/kg, of which 76% was absorbed.[*] Although divalent ions were not measured by Morris, the high chloride content of the gut fluid indicated that it could not be a divalent ion residue as in marine teleost fishes. Magnesium and sulfate, as well as monovalent ions, must have been absorbed into the plasma, yet were not excreted by the kidney since urine flow was absent or immeasurably low. These experiments clearly indicate that river lampreys do not hypoosmotically regulate exactly like marine

[*] The absorption of hyperosmotic gut fluid appears to indicate absorption of water against an osmotic gradient. However, studies are needed to determine whether this may indeed occur, as recently demonstrated in the isolated intestine of a euryhaline teleost (Skadhauge and Maetz, 1967) or whether the gut fluid is hyperosmotic only in the anterior gut, becoming isosmotic or hypoosmotic posteriorly.

teleosts, leaving unanswered the question of where the divalent ions are removed from the body.

Bentley and Follett (1963) found that the injection of blood hyperosmotic NaCl into prespawning river lampreys increased the body sodium by nearly 40% but had very little effect on renal excretion of sodium; the kidneys continued to reabsorb most of the filtered sodium. When lampreys were placed in hypoosmotic (50–100 mmoles/liter) NaCl solutions, the renal excretion of sodium actually decreased because of a significant drop in the GFR and urine flow. The animals maintained blood sodium homeostasis by readjusting sodium exchange across extrarenal pathways. These experiments show that after the river lamprey enters freshwater, the kidney functions as a volume regulator and loses whatever ability it may have had in the sea to regulate the ionic composition of the blood. Until experiments are carried out on lampreys fully adapted to seawater, it is not possible to decide whether the kidney of marine lampreys functions like that of a marine teleost or whether it is essentially nonfunctional in sea water, acquiring its definitive role of volume regulation once the animal has entered freshwater.

Lampreys in freshwater are considerably more water permeable than are teleosts (Bentley, 1962) and accordingly urine flow is very high. Reported estimates for the river lamprey range from 6–17 ml/hr/kg (Morris, 1956; Hardisty, 1956; Wikgren, 1953). Morris' (1956) value of 6.5 ml/hr/kg (at 16°–18°C) measured by catheterization of the urinary duct is lower and probably closer to normal than those collected by the "membrane" method (posterior end of body enclosed within a tube or bag sealed to the skin with a tight-fitting membrane). By an indirect method (initial weight loss when transferred to seawater from freshwater), Bull and Morris (1967) obtained a urine flow of 8.25 ml/hr/kg for the Ammocoete larva of the brook lamprey, Lampetra planeri. Sawyer (from Black, 1957) reported a urine flow rate of 6.6 ml/hr/kg for Petromyzon marinus in freshwater. As in teleost fishes, both the permeability of the body surface to water and the urine flow rise with increasing temperature (Wikgren, 1953).

Bentley and Follett (1963) found the tubular water reabsorption to be relatively constant at 36% over a wide range of filtration rates, induced by exposing the lampreys to different concentrations of medium. This linear relationship between GFR and urine flow is indirect evidence, but certainly not proof, that urine production is controlled by glomerular intermittency rather than by graded filtration. Although lampreys possess relatively few large glomeruli, constriction of portions of the segmental blood supply or of individual neck segments could control flow in each nephron independently.

Table II

Renal Excretion of Sodium, Potassium and Chloride by the River Lamprey, *Lampetra fluviatilis*, in Freshwater

GFR = 11.2 ml/hr/kg; urine flow = 6.5 ± 0.4 ml/hr/kg[a]

Electrolyte	Plasma conc.[b] (mmole/kg)	Urine conc.[c] (mmole/kg)	U/P ratio	Quantity filtered[d] (μmole/hr)	Net quantity reabsorbed (μmole/hr)	Reabsorbed (%)	Quantity excreted (μmole/hr)	Clearance	Clearance ratio
Sodium	119.6	18	0.15	1339	1222	91.2	117	0.98	0.088
Potassium	3.21	3.7	1.15	35.9	11.9	33.1	24	7.5	0.67
Chloride	95.9	0.7	0.0073	1074	1069	99.5	4.6	0.048	0.004

[a] Glomerular filtration rate calculated from urine flow reported by Morris (1956) and average creatinine U/P ratio of 1.73, relatively constant over a wide range of urine flows, reported by Bentley and Follett (1963).

[b] Values from Robertson (1954, p. 430).

[c] Na and K values from Bentley and Follett (1963), Table I; Cl value from Wikgren (1953).

[d] Uncorrected for Donnan effect or possible binding of electrolytes by plasma proteins.

Table II is a composite of data from several sources from which values for renal filtration, reabsorption, and excretion of Na, K, and Cl have been calculated. While the quantities filtered and excreted are considerably greater than in freshwater teleosts (see Table V) owing to the much higher GFR and urine flow in lampreys, the percent reabsorption of each electrolyte is not greatly different. Thus the tubular ion reabsorptive mechanism is about as effective in lampreys as in teleosts, although more ions are excreted by lampreys because of their greater urine production (Fig. 47a). Read (1968) found that as in most freshwater teleosts, Pacific lampreys (*Entosphenus tridentatus*) in freshwater excrete nearly all nitrogenous wastes as ammonia, predominantly across the gills. The small amount of urea formed in lampreys (less than 1% of the total nitrogen excreted) appears to be removed in the urine.

IV. CLASS ELASMOBRANCHII (SHARKS AND RAYS)

All of the sharks, skates, and rays are substantially (50–100 mOsm/ liter) hyperosmotic to seawater. The high osmolality is achieved by combining a total blood electrolyte concentration about the same as, or a little greater than, that of the marine teleost fishes, with the retention of urea and trimethylamine oxide (TMAO) in the blood at concentrations far above those in any other vertebrate group (H. W. Smith, 1929a, 1931b; Burger, 1967; Forster, 1967b). Consequently, water enters the body by osmosis as it does in freshwater teleosts. Electrolytes also tend to enter by diffusion because the concentration of salts in the blood, especially Na and Cl, is less than in seawater. In this respect, the elasmobranchs resemble marine teleosts. Since water enters osmotically, it is freely available for the formation of urine, and the cartilaginous fishes, in theory at least, need not drink seawater as must marine teleosts. Supplementing the kidney as a pathway for salt removal is the rectal gland, which forms a variable and intermittent secretion of colorless fluid, isosmotic to plasma, lacking any significant amount of urea, and containing NaCl at nearly twice its plasma concentration (Burger and Hess, 1960; Burger, 1962, 1965). In addition, it has recently been shown that sodium is extruded from the head, presumably the gills, of the spiny dogfish, and that the sum of sodium efflux through the kidney, rectal gland, and head of the dogfish is nearly equal to the unidirectional sodium influx through the head (Horowicz and Burger, 1968). This again suggests absence of seawater ingestion.

The freshwater and euryhaline elasmobranchs are much more commonly distributed than is generally recognized. Gunter (1942) notes that so many elasmobranchs penetrate into freshwater that they have been distinguished in the German literature as a group, the *"Süsswasser Elasmobranchier."* H. W. Smith (1936) lists 52 species of sharks and rays that have been collected from freshwater, and the list could probably be extended today. Most of the species of elasmobranchs that enter or are restricted to freshwater have been recorded only from equatorial rivers, perhaps, as Smith suggests, because in these regions the salinity gradient between the sea and completely freshwater is very gradual or because of the small annual temperature variation. Most of the species studied thus far have a blood urea and TMAO content approximately 25–45% of the level in marine elasmobranchs. This, together with a somewhat lower electrolyte content, reduces the total osmolality to about 480–540 mOsm/liter (Goldstein *et al.*, 1968; H. W. Smith, 1931b; Thorson, 1967; Urist, 1962). However, these species were euryhaline and not permanent residents of freshwater. Totally landlocked elasmobranchs, such as the stingray, *Potamotrygon* sp., may almost completely lack blood urea (Thorson *et al.*, 1967).

A. Kidney Morphology

1. GROSS STRUCTURE

Shark kidneys are long, retroperitoneal strap-shaped bodies which lie on either side of the aorta against the dorsal body wall. In the female skate, *Raja* sp., the kidneys consist of thick rounded bodies lying along the body wall on each side of the cloaca, and anterior degenerate portions composed of brownish tissue lying ventral to the dorsal aorta. In the male skate the anterior portions are firm cylindrical bodies lying on either side of the mid-dorsal line (Hyman, 1942).

2. ORGANIZATION OF THE NEPHRON

The nephrons of the elasmobranchs are typically very long and have large glomeruli exceeding 10,000 mm³ of glomerular volume per square meter of body surface (Nash, 1931). Kempton (1943, 1956) studied the kidney of the Atlantic dogfish, *Squalus acanthias;* we have confirmed his findings on *Squalus suckleyi*, the Pacific dogfish. In *Squalus suckleyi*, the glomeruli are composed of tufts of widely patent capillaries with thin mesangial areas. Granulated arteriolar cells (juxtaglomerular cells) are apparently absent (Bohle and Walvig, 1964). Some open nephrostomes reportedly exist (Hyman, 1942) although they communicate with short blind tubes. The nephron consists of the neck segment, first and second proximal segments, distal segment, and the collecting duct (Fig. 4). The

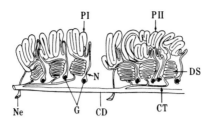

Fig. 4. Scheme of the architecture of the nephron of *Squalus acanthias:* glomerulus (G), neck (N), first proximal segment (PI), second proximal segment (PII), distal segment (DS), collecting tubule (CT), collecting duct (CD), and open nephrostome (Ne). From von Möllendorff (1930), after Haller (1902).

neck segment is long and composed of a small thin-walled tubule, lined by low cuboidal epithelium, which is ciliated at the proximal and distal ends (Figs. 5a and d). Three regions can be identified within the proximal tubule of *Squalus suckleyi.* The initial two represent the first proximal segment (Figs. 5a and b). The first region of the first proximal segment is the same diameter as the neck, and is lined with a tall cuboidal epithelium which shows a periodic acid-Schiff (PAS) positive brush border (Fig. 5a). It contains numerous, closely packed mitochondria. The second region shows a greater diameter, a higher epithelium, and is characterized by large numbers of lysosomes in the apical cell region. The second proximal segment, which constitutes most of the nephron, is much larger in diameter, with tall columnar cells showing a striated appearance resulting from the numerous, closely packed, elongate mitochondria (Figs. 5a and b).

The distal tubule, which appears to wrap around the neck region, is lined by cuboidal, basophilic cells, without brush border (Figs. 5a–d). The larger collecting ducts, lined by tall columnar cells, have a prominent zone of apical mucus granules (Trump, 1968). The nephron of the lesser electric ray, *Narcine brasiliensis,* is similar in all respects to that of *Squalus acanthias* (Kempton, 1962).

B. Kidney Function in Marine Elasmobranchs

1. GLOMERULAR FILTRATION AND THE REABSORPTION AND EXCRETION OF WATER

The elasmobranch kidney has been the object of a series of excellent investigations beginning at the close of the nineteenth century (H. W. Smith, 1936). Several representatives of the sharks, skates, and rays have been studied physiologically, but the spiny dogfish, *Squalus acanthias,* has received by far the most attention. This species holds the distinction

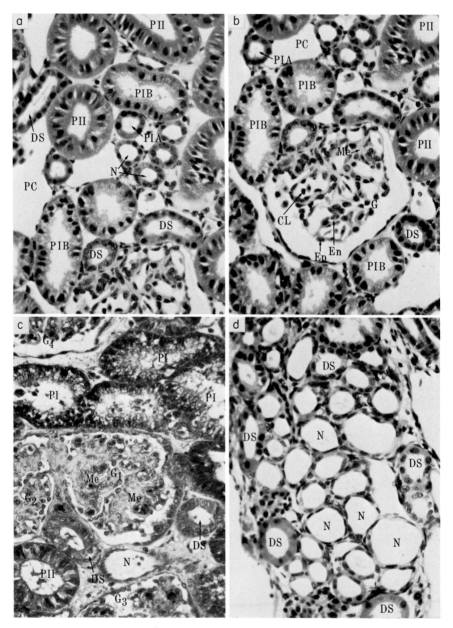

Fig. 5. (a) Photomicrograph of kidney of the Pacific dogfish, *Squalus suckleyi,* showing first and second parts of first proximal segment (PIA and PIB), distal segment (DS), and second proximal segment (PII). Note the numerous sections of neck (N). PC, peritubular capillary. ×280. (b) A view adjacent to (a) showing glomer-

of being the first glomerular fish, and very nearly the first vertebrate, to have its glomerular filtration rate measured with inulin (Shannon, 1934b), the polysaccharide that was to become universally accepted for this purpose, replacing xylose, sucrose, and creatinine.

The rates of glomerular filtration in marine elasmobranchs generally, as measured with inulin, range from 0.2 to 12 ml/hr/kg, and average about 4.0 ml/hr/kg (Shannon, 1934b, 1940; W. W. Smith, 1939b; Schmidt-Nielsen and Rabinowitz, 1964; Kempton, 1966; Burger, 1967). The GFR is much higher than in marine teleosts and approaches that of typical fresh-water teleosts. The open glomeruli with their very large filtering surface suggest that even these observed rates of filtration in marine elasmo-branchs are much lower than the potentially maximal rates. The GFR varies considerably and may cease altogether for several hours after the fish is handled or catheterized (Clarke and Smith, 1932; Clarke, 1935), in notable contrast to marine teleosts which typically respond to handling with a diuresis.

The filtration rate is thought to be controlled by intermittent glomeru-lar activity, that is, by the number of glomeruli active at any moment. The evidence for intermittent filtration is: (1) inulin clearance is linearly related to urine flow (Kempton, 1953, 1966); and (2) inulin clearance is directly associated with the maximal rates of phenol red secretion (W. W. Smith, 1939b; Shannon, 1940; Kempton, 1966). It is not known where the regulation of glomerular filtration is located (afferent arterioles, glomer-ular capillaries, or neck segment) or what endocrine factors, if any, are responsible.

Most of the filtered water is reabsorbed. Reported inulin U/P ratios range from as low as 1.1 (9% water reabsorption) to as high as 15.0 (93% water reabsorption), but typically only 15–30% of the filtrate escapes reabsorption (Kempton, 1953; Schmidt-Nielsen and Rabinowitz, 1964). The tubular reabsorption of water is closely associated with the reabsorp-

---

ulus (G) with well-vascularized tuft and thin capillar walls. Note the visceral epithe-lial cells (Ep), the endothelial cells (En), and the mesangial cells (Me). The capillary lumen (CL) contain erythrocytes. Also shown in this picture are the two divisions of the first proximal segment PIA and PIB, the second proximal segment (PII), and sec-tion of distal segment (DS). PC, peritubular capillary. ×280. (c) Photomicrograph of kidney of ratfish, *Hydrolagus colliei*, showing two adjacent glomeruli and surrounding tubules. Note the poorly vascularized glomeruli (G_1 and G_2) with very prominent mesangial regions (Me). The thin-walled neck (N), first (PI) and second (PII) proximal segments are shown. Portions of other glomeruli (G_3 and G_4) are shown. DS, distal segment. ×280. (d) *Squalus suckleyi* kidney showing highly convoluted neck segments (N) lined by flattened epithelium intermingled with the distal segment (DS) which appears to wrap around the neck region. ×280.

tion of urea, the most abundant single solute in the filtrate (Kempton, 1953). Although water reabsorption appears to passively follow the active reabsorption of urea especially, there is considerable resistance to water movement across the tubular epithelium. The urine of marine elasmobranchs is always blood hypoosmotic, generally varying between 50 and 250 mOsm/liter less concentrated than the blood (H. W. Smith, 1931b; Burger, 1967). In fact the transtubular osmotic gradient established in the marine elasmobranch kidney is nearly as great as, and may even exceed, the gradient developed across the tubules of the freshwater teleost kidney. Marine elasmobranchs exposed briefly to dilute seawater (H. W. Smith, 1931b) or to the stress of handling (Burger, 1965) may reduce the urine osmolality even further until the transtubular osmotic gradient reaches 500 mOsm/liter. Thus the permeability of the tubules can be adjusted, and this capability in combination with adjustments in filtration rate, make it possible for the marine elasmobranch to compensate for variations in water influx across the body surface.

The forces responsible for keeping the formative urine moving through the exceedingly long elasmobranch tubule are unknown. Except for the neck segment, most of the nephric tubule lacks numerous cilia. It is possible that peristaltic-like contractions of the tubular wall, as has been observed in isolated teleost nephrons, is the main propelling force.

2. Electrolyte Excretion

The evolution of the elasmobranch kidney as a water excretory device has created the problem of conserving valuable blood solutes, especially salts, filtered by the glomeruli. However, unlike the freshwater teleosts which are losing salts across the body surface as well as by renal filtration, marine elasmobranchs tend to gain salts from their seawater environment, because nearly all of the major electrolytes are more concentrated in the sea than in their body fluids (Table III). Consequently the requirements of the kidney are better expressed as the critical evaluation and regulation of blood electrolytes rather than just efficient retention. Predictably then, the elasmobranch kidney can regulate each ion species independently. Sodium, chloride, potassium, and calcium are reabsorbed from the filtrate (Fig. 47a). Sodium and chloride are usually reabsorbed at nearly equivalent rates with U/P ratios close to one. Burger (1967) suggests that the tubular reabsorption of sodium is active, and of chloride, passive. He noted that the urine chloride drops to as low as 30 mmoles/liter during infusion of magnesium sulfate or phosphate, and may rise to 372 mmoles/liter during infusion of magnesium chloride. This suggests that chloride is obligatorily paired with magnesium when the latter is present in high concentration.

Table III

Mean Renal Excretion of Ions and Other Solutes by the Spiny Dogfish, *Squalus acanthias*, in Seawater[a]

GFR = 3.50 ml/hr/kg[b]; V = 1.15 ml/hr/kg[b]

Solute	Seawater conc.[c] (mmole/kg)	Plasma conc.[c] (mmole/kg)	Urine conc.[c] (mmole/kg)	U/P ratio	Quantity filtered (µmole/hr)	Net quantity reabsorbed (µmole/hr)	Net quantity secreted (µmole/hr)	Reabsorbed or secreted (%)	Quantity excreted (µmole/hr)	Solute clearance	Solute clearance ratio
Sodium	440	250	240	0.96	875	599	—	68	276	1.10	0.31
Potassium	9	4	2	0.5	14	11.7	—	83.5	2.3	0.58	0.17
Calcium	10	3.5	3	0.86	12.3	8.8	—	71.5	3.5	1.0	0.29
Magnesium	50	1.2	40	33.3	4.2	—	41.8	91	46	38.3	10.94
Chloride	490	240	240	1.0	840	564	—	67	276	1.15	0.33
Sulfate	25	0.5[e]	70	140	1.8	—	78.7	98	80.5	161	46.0
Phosphate[f]	0	0.97	33	34	3.4	—	34.6	91	38	39.2	11.20
Glucose[d]	0	14	3	0.21	49	45.5	—	93	3.5	0.25	0.07
Urea	0	350	100	0.29	1225	1110	—	90.6	115	0.33	0.09
TMAO	0	70	10	0.14	245	233.5	—	95.3	11.5	0.16	0.05
Protein	0	8	—	—	—	—	—	—	—	—	—
pH	8.0	7.48	5.8	—	—	—	—	—	—	—	—
Osmolality	930	1000	800	—	—	—	—	—	—	—	—

[a] The plasma and urine values are taken mostly from Burger (1967) who compiled the averages for this species from a number of sources. Average concentrations of inorganic solutes in the plasma of other elasmobranchs can be found in Bernard et al. (1966).

[b] Average of all values from Table I in Shannon (1940).

[c] From Burger (1967) except as noted.

[d] From Kempton (1953).

[e] From Maren (1967).

[f] From Clarke and Smith (1932).

The discovery that the rectal gland functions efficiently to lower the plasma NaCl concentration (Burger and Hess, 1960) raises the question of why sodium and chloride also appear as the dominant solutes in the urine of sharks. NaCl loading experiments showed that only the rectal gland responded by increasing NaCl excretion; the renal excretion of chloride and the urine osmolality were unaffected (Burger, 1965). One possible explanation for the presence of sodium and chloride in the urine is that they serve together simply as osmotic ballast, which, when added to the divalent ions, urea, and the other minor urinary osmolytes, spare the kidney of the need to form an exceedingly dilute urine. Were sodium and chloride absent from urine, the urine would have an osmolality of only 200–250 mOsm/liter.

Magnesium, sulfate, and phosphate are all actively transported from the peritubular blood to the tubular lumen by powerful secretory mechanisms located in the tubular epithelium (Fig. 47a). In the dogfish, for example (Table III), the clearance ratios exceed 10, indicating that 10 times more of each of the divalent electrolytes is secreted than filtered. The magnesium secretory mechanism is normally operating at much below its potential since it is capable of transporting far greater amounts of magnesium if the dogfish is magnesium loaded (Burger, 1967). Similarly, phosphate excretion can be approximately doubled by phosphate loading (W. W. Smith, 1939a).

It is especially interesting that the renal excretion of magnesium and sulfate by dogfish may exceed that of marine teleosts. In the marine teleosts, these ions are believed to be derived entirely from ingested seawater (Hickman, 1968c). The elasmobranchs, however, reportedly do not drink seawater (H. W. Smith, 1931b; Burger, 1962, 1967) although the experimental evidence is not conclusive. Burger (1967) has calculated that most, or even all, of the magnesium appearing in the urine of unfed dogfish could come from internal sources such as tissue reserves mobilized during starvation. The presence of a high concentration of phosphate in the urine of these animals, which is entirely of endogenous origin during starvation, lends support to this possibility. The alternatives are that dogfish drink seawater in amounts sufficient to account for urinary and enteric excretion of magnesium and sulfate or that the body surface, in particular the gills, is highly permeable to magnesium and sulfate. Further work is clearly needed in this area.

In the dogfish, *Squalus acanthias*, urine is acidified in the proximal region of the tubule (Kempton, 1940). It has been estimated that the sum of titratable acid and total amine excretion is about 1.4 mEq/day, which is close to the maximal rate of renal excretion for these metabolites in this species (Murdaugh and Robin, 1967). The injection of acidic or basic solutions into the blood of dogfish produces acute disturbance of the arterial pH but almost no change in the urine pH, which remains fixed at about 5.7 (5.4–6.0) (W. W. Smith, 1939a; Hodler et al., 1955). However, Cohen (1959) showed that the urine pH rose as high as 6.3 following administration of trimethylamine which was excreted by net tubular secretion. Cohen suggests that the fixed pH of dogfish urine results from a limiting electrochemical gradient of about 1.7 pH units against which the tubular cell can secrete H^+. Despite a nearly inalterable H^+ concentration, the titratable acidity of the urine varies in proportion to the excretion of phosphate (W. W. Smith, 1939a; Hodler et al., 1955), which buffers the H^+ and allows more to be secreted within the limitations of the maximal transtubular pH gradient. Hodler et al. (1955) found that the intravenous administration of the carbonic anhydrase inhibitor acetazolamide (Diamox) did not alter the urine pH, indicating that the renal excretion of H^+ is not dependent on carbonic anhydrase. Compensation for acid–base disturbances is accomplished primarily by the gills, which can probably excrete H^+ and HCO_3^- directly in addition to CO_2 (Robin and Murdaugh, 1967).

3. EXCRETION OF UREA AND TMAO

The single most unique aspect of elasmobranch kidney function is the large net tubular reabsorption of urea and TMAO (Fig. 47a). The blood urea concentration in sharks averages about 350 mmoles/liter (Schmidt-Nielsen and Rabinowitz, 1964; Burger, 1967), but may vary over a surprisingly wide range [e.g., 260–764 mmoles/liter in the smooth dogfish, Mustelus canis (Kempton, 1953)]. Most marine elasmobranchs seem relatively indifferent to variations in both the urea content and the total osmolality of the blood, and this wide tolerance may explain why so many elasmobranchs are euryhaline.

Both urea and TMAO are always reabsorbed against their concentration gradient, but neither is completely removed from the urine. Urine/plasma ratios for urea average about 0.3 but may range from 0.07 to 0.89 in different species or in the same species at different times (H. W. Smith, 1931; Clarke and Smith, 1932; Kempton, 1953; Schmidt-Nielsen and Rabinowitz, 1964). In S. acanthias, 90–95% of the filtered urea and 95–98% of the filtered TMAO are reabsorbed by the tubules (Forster, 1967b).

Kempton (1953) studied the relationship between the filtered load and the tubular reabsorption of urea in the smooth dogfish, Mustelus canis. Both the total urea reabsorption and urea reabsorption per milliliter of filtrate rose as a linear function of plasma urea concentration. Re-

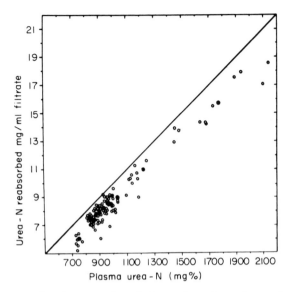

Fig. 6. Relationship between the tubular reabsorption of urea from the filtrate (ordinate) and the plasma urea concentration (abscissa) in normal smooth dogfish, *Mustelus canis*. Note that a relatively constant residuum of urea was not reabsorbed at all levels of plasma urea concentration observed. From Kempton (1953).

absorption of urea was somewhat more complete at low plasma levels (up to 99% of filtered urea reabsorbed) than at the highest plasma levels (as little as 70% reabsorbed) but, as a rule, a relatively constant residuum of about 1–2 mg (0.35–0.7 mmoles) urea per milliliter of filtrate was not reabsorbed (Fig. 6). There was no evidence of a transport maximum (Tm), even at the highest filtered loads. The tubular reabsorption of filtered water was more closely associated with the amount (concentration) of urea remaining in the urine than with either the absolute or percentage reabsorption of urea.

Thus the reabsorptive mechanism is unlike the transport-maximum-limited system for glucose and many other organic constituents; instead it appears to be limited by the gradient that can be established across the tubular epithelium (Pitts, 1963). The carrier mechanism for urea is not inhibited by probenecid given at a concentration sufficient to depress the secretion of p-amino hippuric acid (PAH) and its acetylated derivative PAAH (Forster and Berglund, 1957; Forster, 1967b). The transport specificity and inhibitor sensitivity characteristics of the elasmobranch urea reabsorptive system differ from the urea secretory system of the bullfrog (Forster, 1954) indicating that the two mechanisms do not utilize identical carrier systems rotated 180° (Forster, 1967b).

Squalus acanthias exposed to dilute seawater shows a progressive increase in GFR and urine flow (Clarke, 1935; Burger, 1965) which leads to decreased concentrations of urea, sodium, and chloride in plasma and urine (Smith, 1936). Similar decreases in plasma urea and chloride in clearnose skates (*Raja eglanteria*) in reduced salinities have been reported (Price and Creaser, 1967; Price, 1967). The plasma urea and TMAO concentrations of the lemon shark, *Negaprion brevirostris*, successfully adapted to half-strength seawater were reduced to 45 and 40%, respectively, of their levels in full-strength seawater (Goldstein *et al.*, 1968). The decline in plasma urea was caused by an approximate three-fold increase in total urea production which remained largely unaffected by the salinity change. The increased urea clearance could be the result of a physiological diuresis, since the gradient-time limited tubular reabsorptive mechanism for urea is compatible with this hypothesis. It is also possible that gill permeability to urea (and TMAO) was increased in the dilute environment. The data of Goldstein *et al.* did not permit them to choose between these possibilities.

4. Excretion of Organic Acids

The characteristics of the transport mechanisms for phenol red and certain other organic compounds appears to be much the same as those described for teleost fishes and higher vertebrates (see p. 189). At low plasma levels of phenol red (below 1 mg %) the phenol red/inulin clearance ratios were as much as 22.5 in the spiny dogfish (W. W. Smith, 1939b) and 25 in the smooth dogfish, *Mustelus canis* (Kempton, 1966). Phenol red secretion increased in proportion to the increase in plasma phenol red concentration and became nearly constant and independent of the plasma level at concentration above 4 mg %. The maximum rate of secretion was about 0.20 mg of phenol red per milliliter of filtrate. As the plasma level of phenol red is elevated, secretion accounts for less, and filtration for more, of the total phenol red excreted. Predictably, at very high plasma levels (ca. 100 mg %) the phenol red/inulin ratio drops below one, since about 15% of the phenol red is bound to the plasma proteins and not filtered (W. W. Smith, 1939b). Hippuran and PAH compete with phenol red for a common carrier (W. W. Smith, 1939b; Forster *et al.*, 1954).

Creatinine, a natural excretory product derived from muscle creatine and phosphocreatine, is secreted by the renal tubules of *Squalus acanthias*. The normal plasma level in the dogfish is 0.1–0.2 mg %, and the urine concentration is 1.0–10.0 mg % (Shannon, 1934a). If exogenous creatinine is injected into the musculature, the characteristics of the transport mechanism can be studied as the plasma level declines over a period of several days. Shannon (1934a,b, 1940) found the Tm for creatinine is 84 mg/100 ml glomerular filtrate, a value reached at plasma creatinine levels of about 7 mg %. The creatinine/inulin clearance ratio decline at plasma levels above 7 mg %, because, as with phenol red, the filtered moiety of creatinine makes up an increasingly larger proportion of the total excreted as the plasma level rises. Thus, both the phenol red and creatinine secretory mechanisms are Tm-limited systems.

C. Kidney Function in Freshwater Elasmobranchs

The investigation of Homer Smith and his wife (H. W. Smith, 1931a) on freshwater elasmobranchs in the Perak River of what is now Malaysia, is today nearly as complete a summarization of our meager knowledge of renal function in these forms as it was in 1931. Although the paper has been reviewed repeatedly (e.g., Krogh, 1939; Black, 1957; Potts and Parry, 1964; Thorson, 1967) the main points are again summarized to provide a convenient comparison with renal function in the marine elasmobranchs. H. W. Smith (1931a) studied four freshwater elasmo-branch species in all, but urine was collected and analyzed from a single species, the freshwater sawfish, *Pristis microdon*. The urine flows reported by Smith averaged 10.4 (6.3–19.2) ml/hr/kg (see, however, discussion of high urine flows on p. 143). These values are much higher than urine flows of normal freshwater teleosts of similar size (Table V). Glomerular filtration rate was not measured but is certain to have been greater than the urine flow.

Urine osmolality was 55 mOsm/liter, 10% of the blood osmolality of 550 mOsm/liter. Urea was the predominant urine osmolyte (average 14 mmoles/liter); average concentrations of the urine electrolytes measured by Smith were, in mmoles/liter: Cl$^-$, 6.3; PO$_4^{3-}$, 6.9; SO$_4^{2-}$, 0.3; K$^+$, 2.2; Ca^{2+}, 1.7; and Mg^{2+}, 1.3. Sodium was not measured, but we should expect at least as much sodium as chloride in the urine to balance the larger anion total. Sodium, chloride, and urea were reabsorbed against large concentration gradients. Urine/plasma ratios for the divalent ions cannot be calculated because plasma analyses were incomplete. However, Smith observed that the injection of Na$_2$SO$_4$ into the circulation produced a large increase in sulfate excretion, the urine concentration rising from 0.3 to 87 mmoles/liter. This indicates that the freshwater sawfish retains the tubular capacity to strongly secrete sulfate (and quite probably other divalent ions) in the manner of marine elasmobranchs.

Except that urea is present in plasma and urine, the overall pattern of kidney function, judged from these limited data, is not very different from that of freshwater teleosts. Even more striking is the fundamental similarity of kidney function in marine and freshwater elasmobranchs. Both are hyperosmotic to their environments. In both habitats, the glomerular filtration rate is high, and urea is largely reabsorbed from the urine. Both marine elasmobranchs and their freshwater counterparts reabsorb Na^+ and Cl^- against concentration gradients, and both form blood hypoosmotic urines. Furthermore, the freshwater elasmobranch kidney can apparently secrete divalent ions just as effectively as can the marine kidney, when the need arises. Differences in kidney function between the freshwater and marine habitats seem to be primarily quantitative rather than qualitative, and the invasion of freshwater by elasmobranchs has probably not necessitated any renal innovations. As would be expected, the rectal glands of freshwater elasmobranchs (Lake Nicaragua sharks) are nonfunctional, as evidenced by extensive regressive changes in structure (Oguri, 1964).

V. CLASS HOLOCEPHALI (CHIMAEROIDS)

Although the chimaeroids are frequently included with the sharks and rays in the class Chondrichthyes, palaentological evidence suggests that they may have evolved separately from placoderm ancestors and should be recognized as a separate vertebrate class (Romer, 1966). The chimaeroids are isosmotic or slightly hyperosmotic to seawater. The total ion concentration of the blood is about 10% higher and the urea concentration about 10% lower in the ratfish, *Hydrolagus colliei*, than in elasmobranchs (Urist and Van de Putte, 1967). The presence of a rectal gland in the chimaeroids, although structurally "more primitive" than in elasmobranchs (Fänge and Fugelli, 1962), suggests that the general pattern of salt and water balance in Holocephali and Elasmobranchii is fundamentally the same. There are no records of chimaeroids collected from habitats of low or variable salinity. We have included a description of the morphology of the chimaeroid kidney, although no studies of kidney function in this group have been reported.

Structure of the Kidney and Organization of the Nephron

In the ratfish, *Hydrolagus colliei*, the kidney extends from the anterior pole of the gonad to a point somewhat caudal to the anus. Six archinephric ducts leave the dorsal aspect of the kidney, coursing ventrally along the lateral surface between kidney substance and body wall, and in the male, usually emptying separately into the urogenital sinus. The absence of hematopoietic tissue in the urogenital system distinguishes the kidney of the Holocephali from the Elasmobranchii (H. P. Stanley, 1963).

The glomeruli are large but have few open capillary lumens. Most of their volume is occupied by an extensive mesangial region (Fig. 5c). The neck is lined by a very low cuboidal epithelium (Fig. 5c). Two segments of the proximal tubule are present (Fig. 5c; Table XIII). The first of these regions is lined by cuboidal epithelial cells with pale cytoplasm and relatively few mitochondria but with numerous very large lysosomes in the cell apices. The second segment, which is much longer, appearing to constitute most of the tubular length, is greater in diameter and lined by tall columnar epithelium. The cells contain numerous, elongate, closely packed mitochondria, oriented perpendicular to the basement membrane. The distal segment is much smaller in diameter and is lined by cuboidal epithelium with numerous mitochondria which reach from the cell apex to its base (Trump, 1968; Fig. 5c), thus the kidney of the ratfish resembles that of *Squalus*. The glomeruli, however are less vascular, the neck shorter, and the initial proximal segment is less complex.

VI. CLASS TELEOSTOMI (RAY-FINNED BONY FISHES)

Although the habitat of origin of the earliest vertebrates is controversial, evidence indicates that it was probably in the sea. Thus the jawless hagfish and lamprey are regarded as modern and highly specialized descendents of the early Ostracoderms. While the theater of early vertebrate evolution is debated, it is generally accepted that the Teleostomi evolved in freshwater. Subsequently many returned to the sea, where continued evolution of the more highly specialized marine teleosts such as the aglomerular forms presumably occurred. Thus the most complex nephrons (those of the dipnoans and freshwater spiny-rayed teleosts) originated in freshwater. The potentials of these nephrons were apparently retained or discarded according to the demands of continued existence in freshwater or saltwater, respectively.

In the kidney of bony fishes are seen both the highest development of basic structural and functional patterns introduced during the evolution of the early fishes and the greatest degree of secondary degenerative specialization. The freshwater habitat of the early Teleostomi necessitated hyperosmotic regulation. This was associated with the development of a structurally advanced filtration-reabsorption device, having no less than six cytologically distinct tubular regions in addition to the renal corpuscle, thus enabling the kidney of the advanced Teleostomi to produce the most monovalent ion-free urine of any vertebrate. The marine teleosts, which represent the vast majority of the 20,000–40,000 living Teleostomi species, have modified this versatile filtration–reabsorption system into a tubular divalent ion secretory device. With the possible exception of the marine lampreys, for which positive evidence is lacking, the marine bony fishes were the first vertebrates to evolve hypoosmotic body fluid regulation in the sea by drinking seawater and physiologically distilling it. The sea salts thus separated were removed by the gills and the kidney. The task of excreting the divalent ion component of these unwanted sea salts by tubular secretion logically devolved upon the kidney rather than the gills because the necessary carrier systems were already present and operative in the proximal segments of the kidney nephrons of all the more primitive fishes. Because the marine teleosts do not form a dilute urine, nearly all have lost the distal and intermediate segments of the nephron present in the freshwater teleost kidney. To eliminate the energy requirement of reabsorbing filtered monovalent ions and metabolically valuable organic solutes of the plasma, many marine teleosts have reduced the filtration area of the glomeruli and some have disposed of renal corpuscles altogether. These parsimonious modifications would appear to have headed the marine bony fishes into "a blind alley which does not admit of any great evolutionary advance" (H. W. Smith, 1953). Paradoxically, however, most of the truly euryhaline teleosts belong to marine teleost families, suggesting that the degenerative changes evident in the marine teleostean kidney may not be as limiting to evolutionary exploration as they appear to be.

In this section, we shall deal separately with the structure and function of the kidney of the present-day freshwater, marine, and true euryhaline bony fishes. Since there is no sharp dividing line between stenohalinity and euryhalinity, it is frequently difficult to decide into which of these three groups certain of the primitive teleosts, especially the anadromous forms, should be placed. Our decisions have been largely influenced by the publications of Gunter (1942, 1956) except that we have assigned the two *Salmo* species for which studies of kidney structure and function exist to the freshwater euryhaline rather than to the anadromous group.

In Berg's classification (1940), the class Teleostomi ($=$ most of the Osteichthyes) is divided into the subclasses Crossopterygii and Actinopterigii. No information on kidney structure or function in the extant Coelacanthi is available (see, however, comments in chapter by Forster and Goldstein, this volume, regarding high plasma urea concentrations in *Latimeria*). Contrary to Berg's scheme, the Actinopterygii are usually divided into three groups, the Chondrostei (paddlefishes and sturgeons),

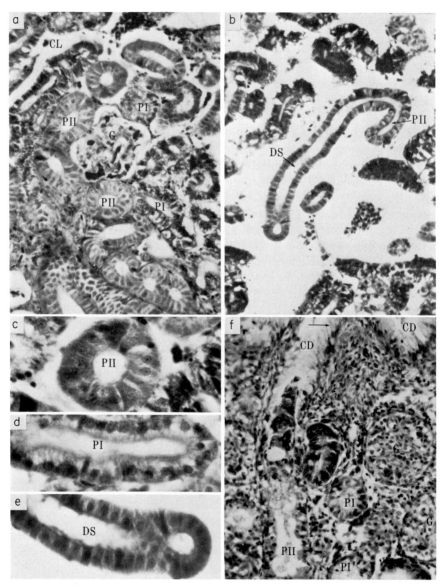

Fig. 7. (a) Photomicrograph of bluegill, *Lepomis macrochirus*, kidney showing glomerulus (G), first proximal segment (PI), second proximal segment (PII), and peritubular capillary lumen (CL). ×280. (b) Bluegill kidney showing transition from second proximal segment (PII) to distal segment (DS). ×280. (c) Second proximal segment from bluegill. ×560. (d) First proximal segment from bluegill. ×560. (e) Distal segment from bluegill. ×560. (f) Portion of kidney from the alligator gar, *Lepidosteus spatula*. Note the glomeruli (G), first proximal segment (PI), second proximal segment (PII), and collecting duct (CD) containing abundant apical mucus (free arrow). ×280.

the Holostei (bowfins and gars), and the Teleostei (higher bony fishes). We have retained this older, more familiar classification for the orders and families of the Teleostei. Unfortunately, no data on the structure or function of the chondrostean kidney are available.

A. Kidney Structure in the Holostei

In the alligator gar, *Lepidosteus spatula*, the kidney is tightly adherent to the dorsal body wall, dorsal to the air bladder. The kidney is segmentally arranged. Glomeruli are relatively large and well-vascularized (Fig. 7f). The neck region is short and composed of tightly packed low to the dorsal body wall, dorsal to the air bladder. The kidney is segment is lined by tall columnar cells with centrally or apically placed nuclei and a tall, PAS-positive brush border (Fig. 7f). The second proximal segment has a less well developed brush border, tall columnar cells with basal nuclei, and numerous mitochondria (Fig. 7f). The terminal portion of the second segment is smaller, but the cells are similar in appearance. The collecting tubule cells and collecting ducts are lined by a tall cuboidal cell with very prominent, PAS-positive apical mucus granules (Trump, 1968) (Fig. 7f). Some open nephrostomes are said to be present (Lagler *et al.*, 1962).

B. Kidney Structure in the Teleostei

1. Gross Structure of the Kidney

The marine teleost kidney is generally divided into two portions, the head kidney and the trunk kidney, although in many cases these regions cannot be distinguished by external examination. Generally there are no conspicuous differences in shape between the two sexes. Ogawa (1961a) has classified the marine teleostean kidney into five configurational classes (Fig. 8).

Type I: The two sides of the kidney are completely fused throughout; no clear distinction between trunk and head kidney. An example is the Clupeidae (herrings).

Type II: The middle and posterior portions only are fused; clear distinction between head and trunk kidney. Examples are the Plotosidae (marine catfishes) and Anguillidae (eels).

Type III: Posterior portion only is fused; anterior portion represented by two slender branches; clear distinction between head and trunk. Most marine fishes have this type of kidney. Examples are the Belonidae (bill

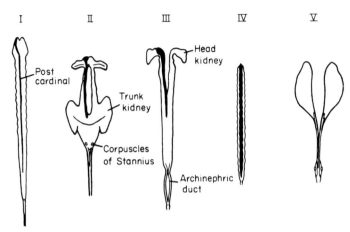

Fig. 8. Five configurational types of marine teleostean kidneys. See text for explanation. From Ogawa (1962).

fishes), Scopelidae (lantern fishes), Mugilidae (mullets), Scombridae (mackerels), Carangidae (pompanos and sacks), Cottidae (sculpins and sea ravens), and Pleuronectidae (flounders).

Type IV: Extreme posterior portion only is fused; the head kidney is not recognizable. An example is the Synganthidae (sea horses and pipe fish).

Type V: The two kidneys are completely separate. An example is the Lophiidae (anglers).

Ogawa (1961a) has observed that all of the freshwater teleosts species species that he examined can be grouped into the first three of the five groups described above (Fig. 8). Examples are as follows: Type I, the Salmonidae (salmon and trout); Type II, Cyprinidae (carps and minnows); and Type III, Cyprinodontidae (killifishes), Gasterosteidae (sticklebacks), and Cottidae (sculpins).

Kerr (1919) asserted that since the posterior definitive kidney of all gnathostomous fishes is derived from the entire caudal portion of the nephrotomic plate, it should be called an opisthonephros, representing both the embryonic mesonephros plus the caudal nephrogenic material which in higher vertebrates forms the metanephros.

Generally, the head kidney consists of lymphoid, hematopoietic, interrenal, and chromaffin (suprarenal) tissue. Variable amounts of hematopoetic and pigment cells are distributed among the tubules and vascular spaces in the trunk kidney (Figs. 14 and 28a). The corpuscles of Stannius are usually located on the dorsal side of the middle to posterior part of the kidney. Scattered throughout the renal parenchyma are many nerves

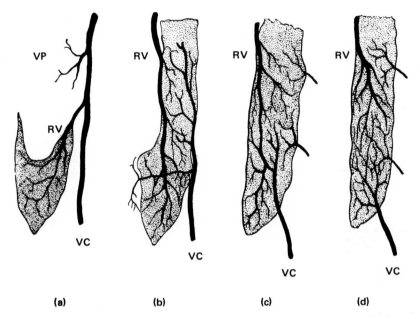

VP

RV   RV   RV   RV

RV

VC

VC

VC   VC

(a)   (b)   (c)   (d)

Fig. 9. Various patterns of venous supply in the trunk kidney of teleosts: (a) *Barbus fluviatilis* (minnow), (b) *Squalus cephalus* (shark), (c) *Anguilla anguilla* (*vulgaris*) (European eel), and (d) *Cepola rubescens* (marine band fish). Note that in the freshwater from *Barbus fluviatilis* there is no renal portal system. VP, parietal vein; RV, renal vein; and VC, caudal vein forming renal portal system in (b), (c), and (d). From Audigé (1910).

which are largely nonmyelinated; large nerve cell bodies are generally scattered alongside the nerve bundles (Bulger and Trump, 1968).

The two archinephric ducts always fuse; this fusion may occur at the posterior end of the kidney or at some point between the kidney and the urinary papilla. Dilations of the archinephric duct may form a bladderlike enlargement ("urinary bladder") where storage and modification of the urine can occur.

Arterial blood to the kidney is supplied by direct renal arteries arising from the dorsal aorta or by renal branches from segmental arteries. In glomerular forms these give rise to afferent arterioles which supply the glomerular capillaries which in turn drain into efferent arterioles; these break up into a network of sinusoids and peritubular capillaries. In marine, and probably all euryhaline teleosts, the peritubular capillaries also receive blood from various combinations of branches from the caudal and/or segmental veins, constituting a renal portal system. There are also various arrangements of drainage of renal blood into tributaries of the posterior cardinal vein (Fig. 9).

Table IV
The Nephron of Teleost Fishes

Habitat	Glomerulus	Neck	Proximal segment I	Proximal segment II	Intermediate segment	Distal segment	Collecting tubule	Family and species[30]
Freshwater	+	+	+	+	+	+	+	Cyprinidae: *Carassius auratus* (goldfish),[5,14,15,20] *Danio malabaricus* (great danio),[5,21,22] Centrarchidae: *Lepomis macrochirus* (bluegill)[19]
	+	+	+	+	0	+	+	Siluridae: *Parasiluris asotus* (mudfish),[15] Poeciliidae: *Platypoecilus maculatus* (red moon platy),[4,22] *Xiphophorus helleri* (swordtail)[4,22]
	+	+	+	+	?	+	+	Cyprinidae: *Cyprinus carpio* (carp),[19] *Hemibarbus barbus* (skincarp),[15] *Pseudorasbora parva* (ishimoroko),[15] *Misgurnus anguillicaudatus* (loach),[15] *Ptychocheilus oregonense* (squawfish),[19] Ictaluridae: *Ictalurus* (*Ameiurus*) *natalis* (bullhead),[19] Ophiocephalidae: *Ophiocephalus* (*Channus*) *argus*,[15] Centrarchidae: *Micropterus salmoides* (largemouth bass)[19]
Marine	+	+	+	+	+	+	+	Plotosidae: *Plotosus anguillaris* (ocean catfish),[15] Congridae: *Anago anago* (marine eel)[17]
	+	+	+	+	+	0	+	Ophichthyidae: *Ophisurus macrorynchus* (snake eel),[17] Muraenidae: *Gymnothorax kidalo* (moray),[15] *Muraena helena* (moray),[3,4] Gadidae: *Merluccius vulgaris* (hake),[3,4] Atherinidae: *Atherina presbyter* (silverside),[2,3] *Sphyraena pinguis* (red barracuda),[2,3] Serranidae: *Theropon oxyrhynchus* (grunt),[15] Sillaginidae: *Sillago shihama*,[15] Labracoglossidae: *Labracoglossa argentiventris*,[17] Carangidae[24]: *Caranz delicatissimu* (mackeral),[15] Menidae: *Coryphaena hippurus* (dolphin),[17] Pomadasydae: *Parapristipoma trilineatum*,[17] Lethrimidae: *Gymnocranius griseus* (sea bream),[15] Sparidae: *Chrysophrys major* (snapper),[15] Mullidae: *Upeneus bensasi* (red mullet),[15] Pempheridae: *Pempheris macrolepidotus*,[17] Scorpidae: *Ditrema temminchii* (surffish),[15] Embiotocidae: *Cymatogaster*

								aggregata (shiner seaperch),[17] Labridae: *Semicossyphus reticulatus* (cunner or wrass),[17] Blennidae[25]: *Salarias enosimae* (pickleback),[15] Brotulidae: *Brotula multibarbata* (brotula),[17] Scombridae: *Thynnus orientalis* (tuna),[17] Histiophoridae: *Makaira marlina* (marlin),[17] Gobiidae: *Acanthogobius flavimonus* (goby),[15] Cottidae: *Myoxocephalus octodecimspinosus* (longhorn sculpin),[4,6] Trachinidae: *Trachinus vipera* (weever fish),[2,3] Pleuronectidae: *Parophrys vetulus* (English sole),[1,29] Ostraciidae: *Ostracion tuberculatus* (trunk fish),[15] Tetradontidae: *Canthigaster rivulatus* (puffer)[15]
	0	0	0	+	0	0	+	Saccopharyngidae: *Gastrostomus bairdi*,[11] Syngnathidae[23,36]: *Hippocampus coronatus* (sea horse),[15] *Syngnathus schlegeli* (pipefish),[15] Batrachoididae: *Opsanus tau* (toadfish),[11] *Porichthys notatus* (midshipman),[18] Lophiidae: *Lophius americanus* (*piscatorius*) (goosefish),[3,12,27] Antennariidae: *Histrio* (*Pterophryne*) *histrio* (mouse fish)[7]
	0[35,36]							Callyonymidae[25]: *Callyonymous* (dragonet),[2] Cottidae[25]: *Cottus* sp. (sculpin),[2] Agonidae[25]: *Aspidophoroides* sp. (sea poacher),[2] Scorpaenidae: *Scorpaena scrofa*,[3,4] Tetradontidae[28]: *Tetrodon* (puffer),[10] Diodontidae[28]: *Diodon* (porcupine fish),[10] Gobiesocidae[26]: *Lepadogaster* sp. (clingfish),[3,9] Oncocephalidae: *Oncocephalus* (batfish)[18]
Catadromous	+	+	+	+	+	+	+	Anguillidae: *Anguilla rostrata* (American eel)[8,31,32] *Anguilla vulgaris* (European eel)[8,32]
Anadromous	+	+	+	+	0	+	+	Salmonidae: *Oncorhynchus gorbuscha* (Pacific pink salmon)[13,33]
True euryhaline, marine	+	+	+	+	0	+	+	Pleuronectidae: *Paralichthys lethostigma* (southern flounder),[19] *Platichthys stellatus* (starry flounder),[19,34] Cottidae: *Leptocottus armatus* (armored sculpin)[19,34]
	+	+	+	+	0	0	+	Cyprinodontidae: *Fundulus heteroclitus* (killifish),[4,5] Gasterosteidae: *Gasterosteus aculeatus* (threespine stickleback)[16,19]
True euryhaline, freshwater	+	+	+	+	+	+	+	Salmonidae: *Salmo gairdnerii* (rainbow trout)[19,32] *Salmo trutta* (brown trout)[19,32]
	+	+	+	+	0	+	+	Cyprinodontidae: *Oryzias latipes* (medaka)[15,33]

Table IV (*Continued*)

References and footnotes.

1. Bulger and Trump (1968). 2. Edwards (1928). 3. Edwards (1930). 4. Edwards (1935). 5. Edwards and Schnitter (1933). 6. Graffin (1937a). 7. Graffin (1937b). 8. Graffin (1937c). 9. Guitel (1906). 10. Hyrtl (1850) cited by Edwards (1928). 11. Marshall (1929). 12. Marshall and Graffin (1928). 13. Newstead and Ford (1960). 14. Ogawa (1961b). 15. Ogawa (1962). 16. Ogawa (1968b). 17. Ogawa (1968a). 18. H. W. Smith (1953). 19. Trump (1968). 20. Ogawa states that an intermediate segment is absent in the Crucian carp, *Carassius auratus*, and rare in the goldfish variety of this species. 21. Proximal segment II contains mucus in cell apex. 22. Neck segment lacks cilia. 23. Nine genera and 18 species of this family have been examined by Marshall and Smith (1930). Ogawa (1962), and Ogawa (1968a); all are aglomerular. 24. Five other species of this family studied by Ogawa (1962, 1968a) have similar kidney structures. 25. Although Edwards lists species of Cottidae, Gobiidae, Blenniidae, Agonidae, and Callyonymidae as aglomerular and cites Guitel (1906), this has not been confirmed by Marshall (1929). 26. Four other species of this family listed as aglomerular by Guitel (1906). 27. Occasional glomeruli found. 28. Marshall (1929) states that the aglomerular nature of Diodon and Tetradon cannot be documented. 29. In this nephron, the most extensively studied of any marine teleost, a third proximal segment which appears to represent a specialization of the terminal second segment was described (Bulger and Trump, 1968). 30. Ogawa (1962) lists additional species from several of the families listed in this table. 31. Double glomeruli sharing common capsular space often seen. 32. The terminal portion of the proximal segment is ciliated. 33. Presence or absence of intermediate segment not stated by author. 34. Positive proof of presence of distal tubule can be determined only by electron microscope examination; this has been done only for *Paralichthys lethostigma*. 35. Fish in this group are reported as aglomerular; no information on the structure of the nephric tubule is given. 36. Two freshwater Syngnathidae have been reported: *Microphis boaja* (Graffin, 1937b) and *Syngnathus nigrolineatus* (Lozovik, 1963).

Most freshwater teleosts appear to lack a true renal portal system (Moore, 1933) (Fig. 9). In many, the large caudal vein is continued through or above the kidney to become the right posterior cardinal vein. This vessel may receive veins from the kidney as it passes through but does not itself break up into venous sinusoids. The left posterior cardinal is usually much smaller than the right and drains only the left cephalic portion of the kidney (von Möllendorff, 1930; Moore, 1933; Gerard, 1954). In certain freshwater groups, for example, the perches (Percidae) and the pikes (Esocidae), some of the segmental veins from the body wall drain into the kidney capillary network, giving anatomical evidence of the existence of at least a partial renal portal system (Moore, 1933).

2. Structure of the Freshwater Glomerular Nephron

a. *Organization of the Nephron.* The typical freshwater teleost nephron is composed of the following regions: (1) a renal corpuscle containing a well-vascularized glomerulus with an inconspicuous mesangium, (2) a ciliated neck region of variable length, (3) an initial proximal segment with prominent brush border and numerous prominent lysosomes, (4) a second proximal segment with numerous mitochondria but a less well developed brush border, (5) a narrow ciliated intermediate segment which is variably present, (6) a distal segment with relatively clear cells and elongate mitochondria, and (7) a collecting duct system (Table IV).

b. *The Nephron of the Bluegill, Lepomis macrochirus.* Our original observations on the structure of the bluegill kidney are summarized in the figure legends: renal corpuscle and neck, Fig. 10; first proximal segment, Fig. 11; second proximal segment, Fig. 12; intermediate and distal segments, Fig. 13; and collecting tubule, Fig 13.

c. *Original Observations on Other Freshwater Species.* The plan of the nephron of the carp, *Cyprinus carpio* (Figs. 14a and b), is very similar to that of the bluegill. The relatively avascular glomeruli are smaller and the neck segment is thin and long; fewer lysosomes are present in the tubular cells. The intermediate segment is ciliated and well developed. Goblet cells are numerous in the collecting ducts.

In the yellow bullhead, *Ictalurus natalis* (Figs. 14c and d), the nephrons are characterized by large vascular glomeruli which are at least twice the diameter of the largest tubules. The neck is extremely long, provided with long cilia, and is lined by columnar epithelium which is only faintly eosinophilic. The first proximal segment is characterized by extremely large lysosomes consisting of eosinophilic droplets which extend to the apical cytoplasm. These were the largest such lysosomes

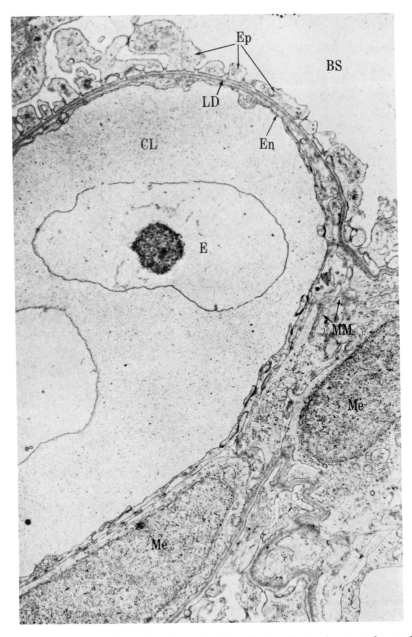

Fig. 10. Part of glomerulus from the bluegill. In this freshwater teleost, the glomerular capillary walls are very thin as can be seen in the upper left. It consists only of the epithelial cell processes (Ep), lamina densa (LD), and the capillary endothelium (En), which is quite thin. Mesangial cells (Me) are seen mainly near

observed in any of the species examined. The distal segment is also un-usual in that it is lined by tall, dome-shaped cells with numerous mito-chondria and extremely basophilic intercalated cells of uncertain signifi-cance. The second proximal segment and the collecting duct system are similar to those described for the bluegill.

In the largemouth black bass, *Micropterus salmoides,* the nephrons are basically similar to that described for the bluegill. Glomeruli are large and vascular, with widely patent capillary lumens. The neck is short, and the first part of the proximal tubule has occasional supranuclear lyso-somes in a well-developed brush border. The terminal part of the second proximal segment has a smaller diameter with a very tiny lumen, and probably corresponds to the intermediate segment. There is only a small amount of interstitial hematopoietic tissue.

In the northern squawfish, *Ptychocheilus oregonense,* the neck re-gion is composed of tightly packed basophilic cells and is short with long cilia. The first proximal segment has elongate mitochondria with a prominent brush border, and the second segment is similar to those described in other freshwater teleosts above. The apical region of the neck segment contains numerous PAS-positive granules. The main por-tions of the proximal and distal segments appear similar to those described in typical freshwater fish. A definite intermedial segment was not ob-served. There was a large amount of interstitial hematopoietic tissue.

Reported studies of kidney structure in freshwater teleosts include those of Edwards and Schnitter (1933), Edwards (1935), Moore (1933), and Ogawa (1961b, 1962).

d. Comparison of Nephrons of Freshwater Fish. The various segments that have been reported in the nephrons of freshwater teleosts are sum-marized in Table IV. It is apparent from this table that the basic archi-tecture of the nephron is similar in all species studied. Measurements of glomerular size and total filtration surface (Marshall and Smith, 1930; Nash, 1931; Ogawa, 1962) strongly indicate that the filtration surface expressed per square meter of body surface is significantly greater in freshwater forms (compare Figs. 14c and 26b). There are, however, ex-ceptions to this general rule. Some overlap exists at the lower limit of freshwater fish and the upper limit of marine teleosts. Examples of

the hilar region. Erythrocytes (E) are found within the capillary lumen (CL). At the hilus of the glomerulus, the afferent arteriole exhibits prominent PAS-positive granulated cells. In the neck region the thin parietal epithelium is continuous with that of the neck region which is composed of low, ciliated, closely packed cuboidal cells (not shown in this micrograph). The neck segment cells are quite basophilic and cilia extend well into the initial proximal segment. BS, Bowman's space; MM, mesangial matrix. ×6750.

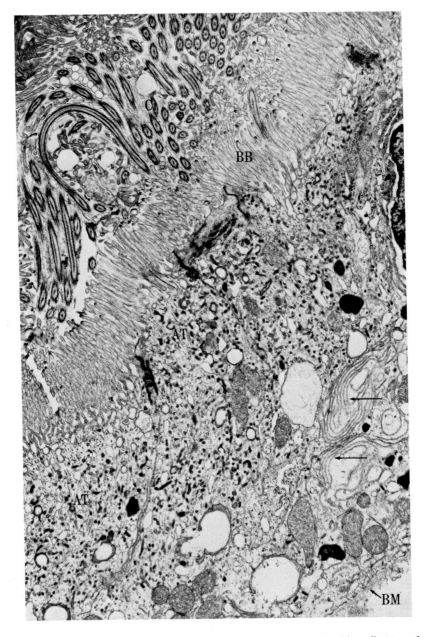

Fig. 11. First proximal segment near the neck region in the bluegill. Note the
numerous cilia in the tubular lumen. Most of them probably extend from the adjacent
neck region. In this transitional region from the neck, the first segment is low; however, the cell apex contains an abundance of apical tubules (AT). Note also the

these include goldfish, *Carassius auratus*, and *Copeina guttata*. Moore
(1933) has made the interesting observation that in the smallmouth bass,
Micropterus dolomieu, the glomeruli may exist as giant compound
glomeruli supplying up to 20 nephrons. This situation does not exist in
the closely related largemouth bass, *Micropterus salmoides* (Trump,
1968). Nash (1931) observed that in both marine and freshwater teleosts
the size of the glomeruli and the filtration area tend to increase with the
size of the fish, but not proportionately.

Because of the variability of the refinement of studies reported in the
literature, it is difficult to be certain about the precise morphology of
these various nephrons. The chief points of difference that are reported
in the literature and that are apparent from our own observations involve the neck segment, the intermediate segment, and the distal segment. Variation in the length of the neck segment and variation in the
number of cilia in this region are apparent. Some freshwater teleosts have
very short, inconspicuous necks; in others they are extremely long. Even
more variation appears to involve the intermediate segment which has
variously been reported as being present or absent, and in some instances
no mention of this segment has been made. It seems, however, that the
intermediate segment can best be regarded in some species as a specialization of the terminal end of the second part of the proximal tubule
(Fig. 14b). In some species the caliber of the lumen and the diameter of
the tubule appear to be reduced, whereas in others this is not the case.
Further detailed studies will be necessary in order to fully assess the
cytological characteristics and the presence or absence of this segment
in freshwater teleosts. It should also be noted that even in some species
such as the goldfish, in which the intermediate segment is reported as
present by some investigators, Ogawa (1962) has stated that it varies in
different nephrons and that in fact the presence of this segment is rare.

Other differences among species involve the presence in some species
such as the minnow, *Danio malabaricus*, of prominent mucus secretion
in the cell apices of the second part of the proximal tubule. The kidney
tubules of sexually mature male sticklebacks, *Gasterosteus aculeatus*, secrete a mucus which is used to glue together algal filaments in the construction of nests. The mucus is probably produced by the cells of the

complex images formed by the parallel plasmalemma infoldings from the cell base
(free arrows). Note the well-developed brush border (BB) with numerous cilia.
The transition from the neck segment is abrupt, and the cells of the first proximal
segment are moderately eosinophilic with numerous large supranuclear lysosomes.
The cell apex, with its numerous apical tubules, is similar to that of the proximal
segments of marine and euryhaline teleosts and the archnephric duct of the hagfish.
BM, basement membrane. ×11,250.

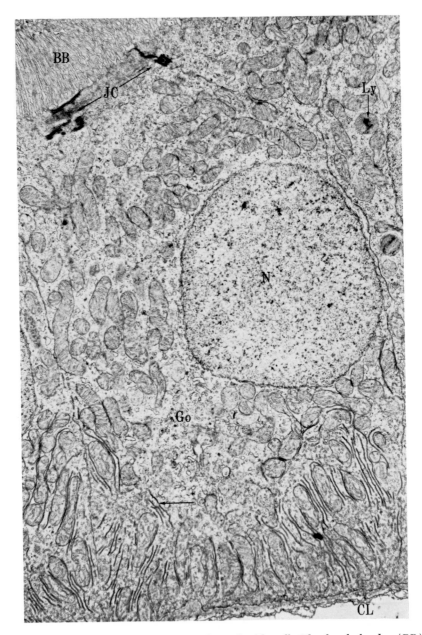

Fig. 12. Second proximal segment from the bluegill. The brush border (BB) appears at the top of the picture, the capillary lumen (CL) at the bottom. Note the junctional complexes (JC), the Golgi apparatus (Go), lysosomes (Ly), nucleus (N), and infoldings from basilar plasmalemma (free arrow). This segment is

second proximal segment which increase to three to four times their former height and accumulate secretory granules. Distinctive, clear, mucus-secreting cells also appear at this time. Female sticklebacks, which take no part in the nest building, do not show these changes in kidney structure (Courrier, 1922; Wai and Hoar, 1963; Ogawa, 1968b).

3. STRUCTURE OF THE FRESHWATER AGLOMERULAR NEPHRON

Information concerning aglomerular freshwater teleosts is scanty. Grafflin (1937b) and H. W. Smith (1932) have studied two specimens of *Microphis boaja* Bleeker. The nephron appears to consist of only two regions, one which appears to represent the second proximal segment in freshwater teleosts, and a collecting duct system. Grafflin (1937b) also studied a specimen of an unidentified *Microphis* species and observed at least two segments in addition to the collecting duct. Unfortunately, however, his study was incomplete and the significance of this finding is not clear. Although *Microphis boaja* is found in Malaysia, the Syngnathidae also have representatives in Panama which occasionally may be found in brackish or freshwater streams. Thus the nephron of *Microphis boaja* appears entirely comparable to the marine aglomerular teleost nephron described below. Lozovik (1963) states that representatives of the Syngnathidae live in low salinities in the Black Sea; one species (*Syngnathus nigrolineatus*) reportedly enters fresh water.

4. STRUCTURE OF THE MARINE GLOMERULAR NEPHRON

a. Organization of the Nephron. A typical nephron of a marine glomerular teleost consists of the following regions: (1) a renal corpuscle containing the glomerulus; (2) a neck segment of variable length; (3) two or three proximal segments which constitute the major portion of the nephron (the first of these is similar at the ultrastructural level to the proximal tubules of mammals); (4) a variably present intermediate segment between the first and second proximal segments; (5) collecting tubule; and (6) collecting duct system (Table IV). In the older literature, the entire brush border region is often referred to as the proximal tubule and compared in its entirety with the proximal convolution of mammals. We wish to emphasize that this may lead to erroneous

---

greater in diameter than the first proximal segment and is intensely eosinophilic as a result of the numerous large mitochondria. The apex of the cell is markedly different from that of the first segment since it does not contain the typical apical tubules; instead the apex contains smooth- and rough-surfaced profiles of endoplasmic reticulum. ×11,250.

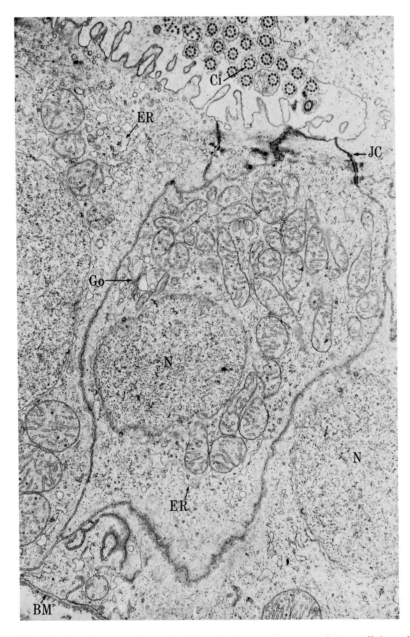

Fig. 13. Distal segment from the bluegill showing the general intracellular relationships. The tubular lumen containing a cluster of cilia (Ci) possibly derived from
adjacent intermediate segment appear at the top of the picture and the basement
membrane (BM) toward the bottom. Note the disposition of mitochondria, Golgi
apparatus (Go), endoplasmic reticulum (ER), and nuclei (N). Lateral cell mem-

conclusions since only the first proximal segment is similar to the mammalian proximal convolution.

As described above under freshwater teleosts the presence or absence of a distal segment is difficult to establish with certainty by light microscopy. Of approximately 100 species reported in the literature and studied by us, however, only two, the marine catfish, *Plotosus anguillaris* (Ogawa, 1962), and the marine eel, *Anago anago* (Ogawa, 1968a), have been reported to show a distal segment (Table IV). *Plotosus* sp., however, are reported to enter brackish water (Ong, 1968). Also of interest is the description by Edwards (1935) of a ciliated narrow segment between the first and second proximal segments in the Moray eel, *Muraena helena*. This, however, appears to represent a specialization of the terminal end of the first proximal segment and accordingly differs from the so-called intermediate segment described in the freshwater teleosts. The frequency of occurrence of such a region among marine teleosts cannot be stated presently.

b. The Nephron of the English Sole, Parophrys vetulus. A detailed analysis has been made of the nephron of *Parophrys vetulus*, the most extensively studied of any marine teleost nephron (Trump and Bulger, 1967; Bulger and Trump, 1968, 1969a). Each segment in the nephron of *Parophrys vetulus* is shown in Figs. 16–22, and the characteristics of each segment are summarized in the figure legends. A schematic representation of a teleost glomerulus is shown in Fig. 15.

5. STRUCTURE OF THE MARINE AGLOMERULAR NEPHRON

A review of the literature indicates that aglomerular fish occur in 6 families, 13 genera, and 23 species of marine teleosts (Table IV). Several

---

branes of the central cell can be seen and are somewhat irregular in this plane of section. A junctional complex (JC) occurs adjacent to the tubular lumen. This segment thus resembles the distal tubule of the amphibian. By light microscopy cells of the bluegill distal segment are less eosinophilic than cells of the proximal segment, and contain centrally placed nuclei, elongate, large, closely packed mitochondria, and very few lysosomes. The cilia seen within the lumen probably are derived in part from the preceding intermediate segment which occurs between the second proximal segment and the distal segment. The intermediate segment is lined with a low cuboidal epithelium and has an inconspicuous brush border. The morphology is otherwise similar to that of the second portion of the proximal segment. It presumably corresponds to the intermediate segments described by Edwards and Schnitter (1933) and Ogawa (1961b) in goldfish. The initial collecting duct, which follows the distal segment, is larger and lined by tall columnar cells with basilar nuclei. In the larger secondary and tertiary collecting ducts, the cells become taller and eventually, in the larger ducts, become pseudostratified. These cells often show prominent apical mucous granules and PAS-positive goblet cells. The largest ducts are surrounded by one or more complete layers of smooth muscle and show occasional wandering or intercalated cells. ×8550.

136   CLEVELAND P. HICKMAN, JR., AND BENJAMIN F. TRUMP

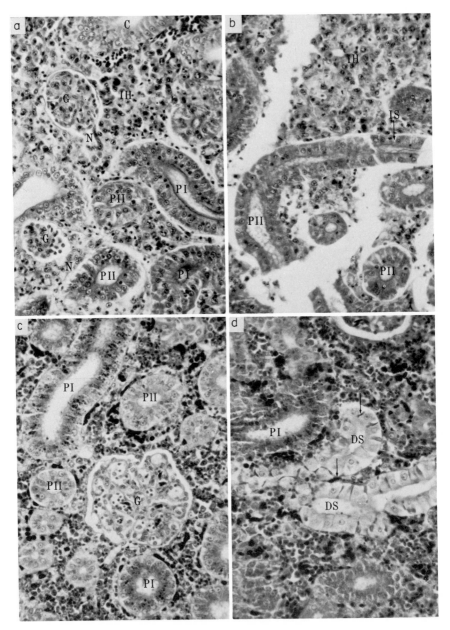

Fig. 14. (a) Portion of kidney from carp, *Cyprinus carpio*, showing glomeruli (G), beginning of neck segments (N), first proximal segments (PI), second proximal segments (PII), and collecting tubules (C). IH, interstitial hematopoietic tissue. ×280. (b) Kidney of carp showing transition between ciliated intermediate segment

of these have been studied by electron microscopy. The nephrons of the midshipman, *Porichthys notatus*, and the toadfish, *Opsanus tau*, consist of two regions: an initial segment with brush border and a terminal collecting duct system (Grafflin, 1931a, 1937b; Bulger, 1965; Bulger and Trump, 1965).

In the toadfish and midshipman the proximal portion of the nephron is essentially similar to the second proximal segment of glomerular marine teleosts (Figs. 23 and 24).

The collecting duct system in the aglomerular teleost is very similar to that described for glomerular marine teleosts.

Aglomerular Syngnathidae such as *Nerophis ophidion*, *Hippocampus* sp., and *Syngnathus* sp. show relatively similar morphology in the brush border segment with extensive interdigitating basilar processes containing mitochondria. In *Hippocampus* the nephrons show extensive arborization; *Lophius piscatorius*, on the other hand, is somewhat different with the absence of basal processes, and shallow interdigitations which are not related to mitochondria. It should be noted however that *Lophius piscatorius* is only partly aglomerular, and since complete studies on this species have not been performed, it is difficult to assess the significance of this difference.

6. STRUCTURE OF THE EURYHALINE GLOMERULAR KIDNEY

a. The Nephron of the Southern Flounder, Paralichthys lethostigma (Figs. 25–32). The most extensively studied euryhaline fish is the marine pleuronectid, *Paralichthys lethostigma*. In this species the nephron consists of the following regions: a small, relatively poorly vascularized glomerulus, a ciliated neck segment of moderate length, a first proximal segment, a second proximal segment, a distal tubule, and the collecting duct system. The cytological features of the light and electron microscope levels of this nephron are essentially identical with those described above for *Parophrys vetulus*, with the exception of the distal segment. The distal segments resembles that observed in freshwater teleosts or in higher forms, including amphibians, and is composed of cuboidal or low columnar cells having elongate mitochondria arranged in palisade fashion

---

(IS) and second proximal segment (PII). Note also the interstitial hematopoietic tissue (IH). ×280. (c) Kidney from yellow bullhead, *Ictalurus natalis*, showing extremely large, well-vascularized glomeruli (G), first proximal segments (PI) with numerous large droplets representing secondary lysosomes, and second proximal segments (PII). Note also interstitial hematopoietic tissue. (d) Yellow bullhead showing first proximal segment (PI) loaded with cytoplasmic droplets representing lysosomes and large clear cells lining distal segments (DS) with dark intercalated cells (arrows).

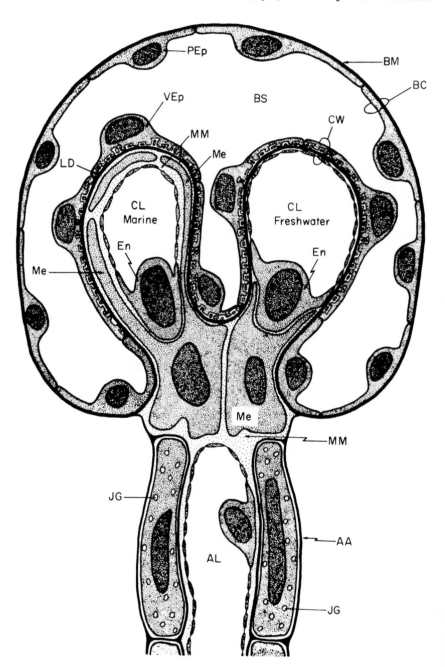

Fig. 15. Schematic representation of relationships in marine and freshwater glomeruli. The parietal epithelium (PEp) and basement membrane (BM) form

perpendicular to the basement membrane and closely paralleling complex invaginations of the basilar membrane (Trump *et al.*, 1968).

b. Comparison of Nephrons in Euryhaline Fish. The architecture of the nephrons in a variety of euryhaline fish that have been studied to date is given in Table IV. It is apparent from perusal of Table IV that some variation exists in nephron constitution. This variation is most apparent in the intermediate segments and in the distal segment. Thus far, terminal portions of the second proximal segment that might be equated with intermediate segments have been described with certainty only in the two species of *Anguilla* (Grafflin, 1937c) and two species of *Salmo* (Fig. 33a and b) (Trump, 1968). Although these regions might be classified as intermediate segments, they appear to consist of ciliated terminal portions of the second portion of the proximal segment. In the other species, intermediate segments are either absent, or their presence or absence cannot be established from the publications cited. Accordingly, intermediate segments have been established with certainty only in catadromous or freshwater euryhaline forms, both of which spend most of their adult lives living in freshwater.

Similar variation exists with respect to the distal segment. Although most of the species reported possess distal segments, two of them, *Fundulus heteroclitus* and *Gasterosteus aculeatus* (Fig. 33c), lack this segment (Edwards and Schnitter, 1933; Trump, 1968; Ogawa, 1968). The identification of the distal segment is subject to the limitations listed above under freshwater fish. However, an examination of these two species appears to show nephrons which are typical of marine teleosts. It is interesting that distal segments are lacking only in euryhaline

---

Bowman's capsule (BC). The parietal epithelium is continuous with the visceral epithelium (VEp) at the hilus. The visceral epithelium adjacent to Bowman's space (BS) has complex processes (foot processes or pedicels) which abut on the lamina densa (LD). The lamina densa is continuous with the parietal basement membrane, with the basement membrane surrounding the smooth muscle cells of the afferent arteriole (AA), and with the mesangial matrix (MM). The centrilobular areas of the tuft contain mesangial cells (Me) which have processes of variable length. In marine teleosts these processes often completely embrace the capillary lumen, whereas in freshwater teleosts they generally do not. These relationships result in a much thicker capillary wall (CW) in marine as compared with freshwater teleosts. The capillary wall is presumed to form the filtration barrier. The endothelial cells (En) are thick in the nuclear region, which occurs toward the center of the lobule, and attenuated peripherally where they are penetrated by numerous fenestrations. The afferent arteriole (AA) has modified smooth muscle cells forming the media. These cells contain numerous granules (JG) the so-called juxtaglomerular granules which are presumed to contain renin. AL, lumen of afferent arteriole. The efferent arteriole is not shown for simplicity.

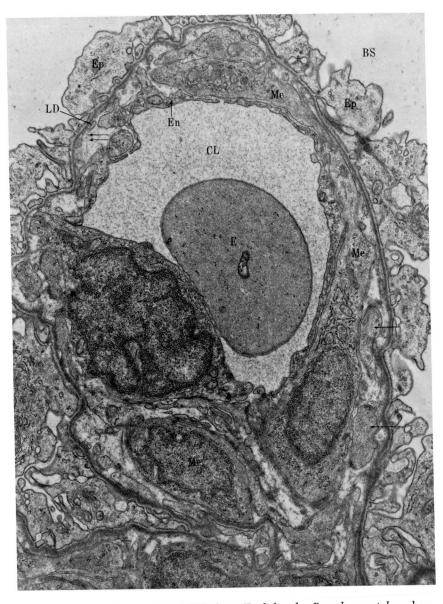

Fig. 16. Part of glomerular lobule from English sole, *Parophrys vetulus*, showing Bowman's space (BS), visceral epithelial cells (Ep), mesangial cell processes
and mesangial cells (Me), and capillary lumen (CL) containing an erythrocyte (E).
Note the lamina densa (LD) and the flocculent basement membrane material or
mesangial matrix in the space between the lamina densa and the endothelial cell
(double arrow). Note also numerous mesangial cell processes (free arrows) in this

species which spend the predominant portion of their lives in marine environments, since typical marine stenohaline teleosts also do not possess distal segments.

C. Kidney Function in Freshwater Teleostei

The freshwater teleosts, in common with all freshwater animals, are hyperosmotic regulators. The sensing and regulation of the ionic composition of the blood is carried out by extrarenal systems, presumably located within the gills (see chapter by Conte, this volume); the kidney primarily conserves filtered electrolytes. The urine is dilute, often nearly free of sodium and chloride, and its volume must balance the quantity of water entering the body from the animal's dilute environment. Two tubular characteristics are particularly essential to the efficient performance of this function: (1) a powerful monovalent ion reabsorptive mechanism which operates in conjunction with (2) a low tubular permeability to filtered plasma water. The addition of the distal and intermediate segments to the nephron (Table XIII) is doubtless responsible for the improved efficiency of the freshwater teleostean kidney as compared to that of the most primitive freshwater vertebrates, the lampreys (Petromyzones). In this section we will consider normal kidney function in freshwater teleosts and the functional adjustments possible when "stenohaline" freshwater teleosts are subjected to saline environments.

1. GLOMERULAR FILTRATION AND URINE FLOW

Reported measurements of inulin clearances and simultaneous urine flows in freshwater fish are given in Table V. Urine flow is, of course, greater in freshwater teleosts than in marine, but in general, more recent measurements have not confirmed the extremely high, "copius" flow rates reported by earlier workers.

---

space. Note the well-developed foot processes of the visceral epithelium and the thin filtration slit membranes which span them. Adjacent visceral epithelial cells show desmosomes. Note that the endothelial cell is thick near the mesangial region and attenuated peripherally. It is sometimes difficult to appreciate the relationships of cells within the glomerulus, and for this purpose the diagram shown in Fig. 15 is included. Note that the visceral epithelial cells have an extremely complex shape as originally described by Zimmermann (1911) for mammalian visceral epithelium. Note the relationships of the mesangial cells which have long processes, occasionally embracing the capillary lumen between the epithelial basement membrane and the capillary endothelium. The mesangial cells have numerous intracytoplasmic filaments and are often regarded as modified smooth muscle cells. ×17,500. From Bulger and Trump (1968).

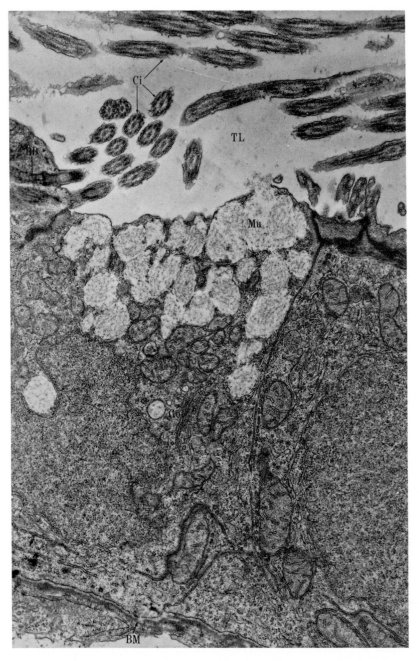

Fig. 17. Neck region from English sole, *Parophrys vetulus*. Note the numerous cilia (Ci) in the tubular lumen (TL). Basal bodies (BB) appear in the cell apex.

The urine flow has often been considered equal to the total permeability of the body surface to water. However, recent work showing that euryhaline fishes in freshwater drink their medium (Potts and Evans, 1967; Maetz and Skadhauge, 1968) suggests that this is no longer a safe assumption. Nevertheless, factors that are expected to modify surface water permeability also modify the urine flow. The larger weight-specific urine flows for small fish are probably at least partly associated with their relatively larger permeable surface. Surface permeability to water is also increased by injury to the integument and rising temperature. Mackay and Beatty (1968) report that both the urine flow and GFR of the white sucker, *Catostomus commersonii*, increased with rising temperature over the range 2–18°C. The Q_{10} value for urine flow was about 2.2 over the entire temperature range. No significant difference in the inulin U/P ratio (tubular water reabsorption) was observed over the temperature range 2–14°C. The authors conclude that the effects of temperature is on the permeability of the body surface to water; it has no direct effect on GFR or tubular water permeability. Glomerular filtration rate and urine flow respond secondarily to the changes in water influx. The increased water influx with increasing environmental temperature is probably the result of greater branchial irrigation and blood flow rather than an increase in membrane permeability per se (Hickman, 1965; Mackay and Beatty, 1968). Handling, anesthetization, surgery, and exposure to acute hypoxic stress frequently create a diuresis that can last several hours (Fleming and Stanley, 1965; J. G. Stanley and Fleming, 1966, 1967a; Hunn, 1969; Hunn and Willford, 1968). This probably results from a combination of increased surface permeability (stressing procedures create a heightened metabolic rate which subsides gradually, the influx of water across the gills rising during the period of increased respiration) and alterations in kidney tubular reabsorption of electrolytes, since the urine composition as well as urine flow may be radically altered.

It is characteristic of freshwater teleosts for both the GFR and urine production to vary markedly over extended collection periods (Fig. 34). The urine flow and GFR are directly, and usually linearly, related (Fig. 35), because a nearly constant proportion of the water filtered by the glomeruli is reabsorbed by the tubules.

Cells in the neck region have numerous mucous granules (Mu), a few mitochondria, and a small Golgi apparatus (Go). Basement membrane, BM. The neck cells are continuous with the parietal layer of Bowman's epithelium at the urinary pole and the neck region is usually short. The numerous granules in the apex are PAS positive. Cilia emerge in groups of approximately 12 from the cell apex and, in living preparations, the cilia are long and show continuous undulating movement. Smooth muscle cells surround the basement membrane. ×12,950. Courtesy of R. E. Bulger.

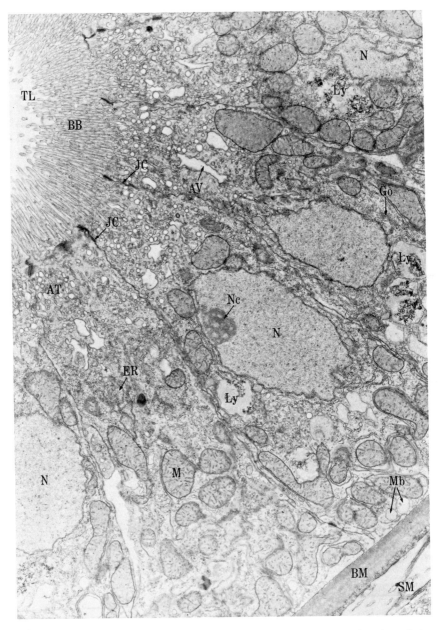

Fig. 18. First proximal segment of the nephron of the English sole. The tubular lumen (TL) and well-developed brush border (BB) are at the upper left; the basement membrane (BM) surrounded by smooth muscle cells (SM) is at the lower right. Numerous junctional complexes (JC) mark sites of intercellular attachment at

Unlike higher vertebrates tubular activities of salt and water transfer in freshwater teleosts appear to be mostly inflexible mechanisms that greatly modify the composition of the filtrate but do not control the final urine volume. In these forms, variation in urine flow is achieved by variations in GFR. The close association between urine flow and GFR argues for glomerular intermittency, since it has been difficult to explain in any other way the large changes in GFR and urine flow with little or no change in urine salt composition and fractional water reabsorption. Additional evidence is provided by an indirect method developed by Forster (1942). The Tm for glucose (maximal amount of tubular reabsorption of glucose during glucosuria) will vary in direct proportion to the inulin clearance if filtration is intermittent (variation in number of glomeruli operating). Such a proportionality has been observed in the white sucker, *C. commersonii* (Fig. 36), which shows large natural variations in GFR with time (Mackay and Beatty, 1968) and in the euryhaline flounder, *Platichthys flesus*, in which the GFR varies in response to salinity change (Lahlou, 1966). However, glomerular intermittency, considered in its original context as an all-or-none phenomenon, may be a hypothetical condition. Hickman (1965) suggested that some gradation of filtration was superimposed on intermittent control of filtration, and Hammond (1969) has questioned the validity of the entire concept of glomerular intermittency. Indeed, it should be recognized that the evidence presently interpreted to favor glomerular intermittency as an explanation for variations in GFR does not exclude the possibility that changes in filtration are uniform throughout the nephron population. That is, rather than the glomeruli operating in an all-or-none manner, filtration may be graded evenly between zero and some maximum. Graded changes in filtration may be coupled to proportional changes in effective tubular absorptive surface such that fractional fluid reabsorption and the ratio $Tm_{glucose}/C_{inulin}$ remain unchanged over a wide range in filtration rates. Direct observations with isolated flounder tubules

the apex. The nuclei (N) containing nucleoli (Nc) are in the cell center. Note the numerous apical tubules (AT) and large apical vacuoles (AV) in the cell apex. Large secondary lysosomes (Ly) flank the nucleus and are in the vicinity of the Golgi apparatus (Go). The cell also contains numerous mitochondrial profiles (M) and portions of endoplasmic reticulum, both smooth and rough surfaced (ER). The large lysosomes are PAS positive and acid phosphatase positive. These are correlated with the uptake of materials through the apical tubular system which occurs in this region. The junctional complexes are composed of tight junctions, intermediate junctions, and desmosomes and focal tight junctions also are observed along the lateral cell borders. Fine filaments from the cell apex extend into the microvilli and may be related to movements of these villi. Microbodies (Mb) are seen near the basement membrane. ×7600. From Trump and Bulger (1967).

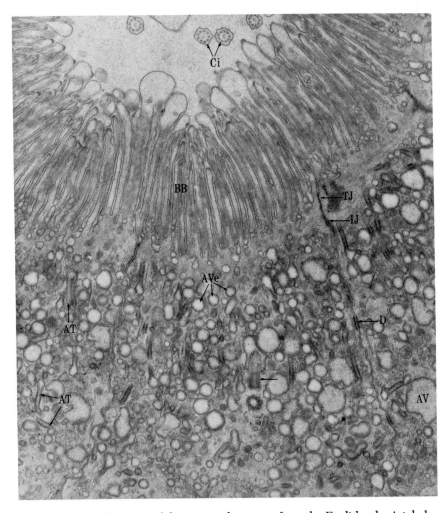

Fig. 19. Apical portion of first proximal segment from the English sole. A tubular lumen containing cross section of cilia (Ci) is seen at the top of the picture. Note the closely packed microvilli which compose the brush border (BB). A junctional complex composed of tight junction (TJ), intermediate junction (IJ) and desmosome (D) is seen. The cell apex contains numerous profiles including apical vacuoles (AV), apical vesicles (AVe), and apical tubules (AT). All of these represent sections cut in various planes through tubular invaginations of the apical plasmalemma between the bases of the microvilli. The basal body of a cilium is seen (free arrow). ×16,200. Courtesy of R. E. Bulger.

(Trump, 1968) confirm that all segments of the nephron are highly distensible. While the effect on reabsorption of changing cross-sectional area of the lumen with different flow rates has not been studied in the fish kidney, numerous studies with tetrapods have established that tubular size effects absolute reabsorption (see Windhager, 1969, and references cited therein).

Although control of GFR in fish remains unexplained, some information on renal effects of blood pressure has appeared. Hammond (1969) has shown that in lake trout fluctuations in dorsal aortic blood pressure, produced by swimming activity, fright, or intravenous injections of catecholamines, are correlated positively with changes in GFR. However, GFR may vary over wide limits without corresponding changes in blood pressure. Chester Jones et al. (1969) present similar data for the European eel but do not comment on this observation. The important conclusion is that while GFR is responsive to changes in systemic blood pressure (autoregulation of the kind present in the mammalian kidney is absent), the large fluctuations in GFR and urine flow commonly seen in normal, nonstressed fish are determined by mechanisms other than systemic blood pressure. This does not exclude the possibility that GFR is governed by the afferent arterioles serving the glomeruli.

2. BLOOD FLOW THROUGH THE KIDNEY

The glomeruli of freshwater teleosts are supplied by branches of segmental arteries arising from the dorsal aorta. The efferent glomerular arteriole breaks up into a network of sinusoids and capillaries surrounding the tubules. This postglomerular circulation may constitute the sole blood supply to the tubule since, according to Moore (1933), most freshwater teleosts lack a true renal portal system (Fig. 9).

Attempts have been made to estimate blood flow through the freshwater teleostean kidney with PAH and Hippuran. Lavoie et al. (1959) obtained a maximum PAH clearance of 193.2 ml/hr/kg in a bullhead, *Ameiurus nebulosus*, having an inulin clearance of 9.7 ml/hr/kg. The average PAH and inulin clearance for three bullheads were 139.9 and 12.9 ml/hr/kg, respectively, providing a C_{PAH}/C_{in} ratio of 11.8. Hickman (1965) noted the absence of any consistent correlation between Hippuran and inulin clearances and very low Hippuran/inulin ratios (below 2) in the white sucker, *C. commersonii*. Hippuran extraction was far from complete and probably bore no relation to renal plasma flow. Chester Jones et al. (1969) arrived at similar conclusions about experiments with European eels given PAH. The poor extraction of PAH and Hippuran by the kidneys of freshwater as compared to marine teleosts (see p. 173) is probably related to the small fraction of total tubular mass, estimated at 10% in the goldfish, *Carassius auratus*, showing luminal uptake of dyes such as phenol red by secretion (Kinter and Cline, 1961).

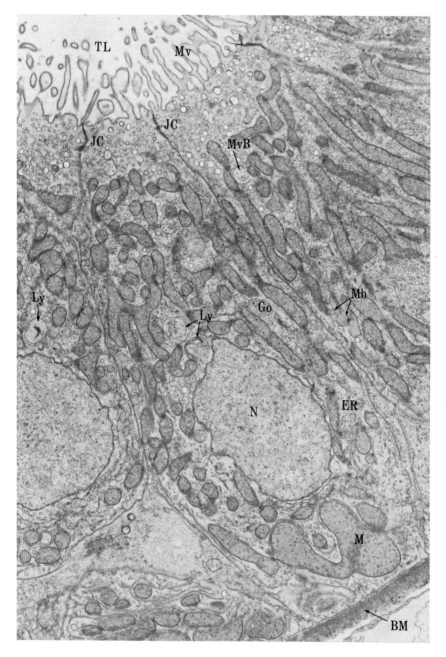

Fig. 20. Second proximal segment from English sole. The tubular lumen (TL) is at the upper left and the basement membrane (BM) at the lower right. Note the widely separated microvilli (Mv) which are in contrast to those in the first proximal

3. TUBULAR FUNCTION

a. Electrolyte Reabsorption. Once the plasma ultrafiltrate enters the capsular portion of the nephron, it begins to move down the tubule, driven by the net filtration pressure, the long cilia of the neck segment, the single cilia of cells of other nephron segments, and, in those species having it, the ciliated intermediate segment. At least some, and perhaps all, the Salmonidae lack a ciliated neck segment (Ogawa, 1962). In these, the residual filtration pressure is presumably the primary propulsive force. The tubules are invested with smooth muscle cells which, by peristaltic-like contractions, may provide some additional propulsion.

Na^+ and Cl^- are almost completely reabsorbed from the ultrafiltrate. As shown in Tables VI and VII the urine Na^+ and Cl^- concentrations of freshwater teleosts are usually below 20 mmoles/liter; concentrations as low as 0.05 mmole/liter have been measured in northern pike, *Esox lucius,* and suckers, *Catostomus commersonii* (Hickman, 1965). At these extremely low concentrations, over 99.9% of the filtered sodium and chloride has been reabsorbed.

No experiments have been reported that deal specifically with the characteristics of the active ion pumps involved. However, there are many morphological and functional similarities between the kidneys of freshwater teleosts and amphibians and it is not improbable that the characteristics of their ion carrier systems are identical. Should this be true, Na^+ is actively extracted from the tubular lumen and is passively accompanied by Cl^- (Pitts, 1963). The fact that urine concentration of Cl^- is frequently lower than Na^+ does not argue against passive Cl^- efflux,

---

segment. Note also that the apex contains a variety of vesicular profiles but no typical apical tubules characteristic of the first proximal segment. The mitochondria (M) are rather well-developed and extend throughout the cytoplasm and tend to be larger at the base of the cell. Profiles of endoplasmic reticulum (ER) can be seen throughout the cell. The Golgi apparatus (Go) is found near the nucleus and also in this region are several secondary lysosomes (Ly). Numerous multivesicular bodies (MvB) and junctional complexes are found in the cell apex. The second segment has cells that are taller and more eosinophilic than those of the first segment; however, the PAS-positive brush border is much less well developed as are the acid phosphatase-positive lysosomes. As shown in this picture, the mitochondria are large and numerous and related to the basilar infoldings. The apical region does not possess the apical tubules characteristic of the first segment and instead has numerous profiles of endoplasmic reticulum. Numerous microbodies (Mb) are also found in this region. The cells of this region undergo a transition to the cells of the third segment. The cells of the third segment are lower than those of the second, and the brush border is less conspicuous but otherwise is similar to the second region. It has a slightly more prominent rough-surfaced endoplasmic reticulum, probably related to the increased basophilia as seen by light microscopy. Nucleus (N) and junctional complexes (JC). ×7885. From Trump and Bulger (1967).

Table V

Glomerular Filtration and Urine Flow Rates in Freshwater Fishes

Species	Temp. (°C)	GFR (ml/hr/kg)	Urine flow (ml/hr/kg)	Average inulin U/P ratio	Average body weight	No. of fish	Notes	Reference
Salmo gairdnerii[a]								
Pre-smolt	"Seasonal"	7.3 ± 0.4	3.8	1.92	182	10	GFR and urine flow determined with separate groups of fish	W. N. Holmes and Stainer (1966)
Smolt	"Seasonal"	3.8 ± 0.3	1.6	2.38	169	9	—	—
Post-smolt	"Seasonal"	7.6 ± 0.9	4.7	1.62	169	9	—	—
Salmo irideus	10	—	3.4 (3.1-3.7)	—	215	—	1½-2-year-old fish	R. M. Holmes (1961)
Salvelinus namaycush	6	3.73 (2.22-5.91)	2.44 (1.51-3.79)	1.53	2700	1	—	Hammond (1969)
Ameiurus nebulosus	—	9.4 (3.5-24.2)	6.4 (1.3-14.2)	1.47	—	34	—	Lavoie et al. (1959)
	—	—	2.9 (1.7-3.9)	—	354	5	—	Haywood and Clapp (1942)
	18-23	17.1 (12.3-21.9)	10.0 (6.4-13.6)	(1.7)	—	—	—	Marshall (1934)
Carassius auratus	18-23	14.4 (12.6-17.2)	10.7 (8.2-15.0)	1.34	155	19	Xylose clearance, corrected	Bourquet et al. (1964)
		20.4 ± 0.65	13.7 ± 0.4	1.49	127	63	—	Maetz (1963)
Cyprinus carpio	6	—	1.1	—	540	1	—	Pora and Prekup (1960)
	12	—	2.0	—	560	1	—	—
	19	—	7.1	—	245	1	—	—
	28	—	10.8	—	380	1	—	—
Esox lucius	—	4.0	2.8	1.42	—	1	—	Ginetsinkii et al. (1961)
	10	2.45 (0.2-7.2)	0.67 (0.07-1.7)	3.65	1454	2	—	Hickman (1965)
Catostomus commersonii	4	2.20 (1.03-3.24)	1.15 (0.76-1.58)	1.91	1395	6	—	Hickman (1965)
	2	1.99	1.21 ± 0.19	1.64	1300	7	—	Mackay (1967)
	6	2.56	1.83 ± 0.20	1.40	1300	12	—	—
	10	3.67	2.45 ± 0.24	1.50	1300	11	—	—
	14	4.02	3.12 ± 0.47	1.29	1300	4	—	—
	18	5.96	4.39 ± 0.45	1.36	1300	2	—	—
Anguilla anguilla[a]	18	4.6 (2.3-7.8)	3.5 (1.9-6.7)	1.31	178	11	Silver eel stage	Sharratt et al. (1964)
Anguilla anguilla[a]	12	1.51 ± 0.18	1.1 ± 0.23	1.37	625	8	—	Chester Jones et al. (1969)
Platichthys flesus[a]	—	4.91 ± 0.52	2.87 ± 0.27	—	—	—	—	Motais (1967)
Fundulus kansae[a]	20	25	8.33	3.0	1.58	3	After 7 days' adaption to freshwater	Fleming and Stanley (1965)

[a] Euryhaline fishes adapted to freshwater.

Table VI

Composition of Plasma and Urine in Freshwater Teleost Fishes

Species and plasma solute	Plasma	Urine	U/P ratio	Reference
Salmo gairdneri				
Osmolality (mOsm/liter)	—	32 ± 6	—	Fromm (1963); 14–21 fish, av. weight
Sodium (mmole/liter)	144	9.3 ± 1.3	0.065	138 ± 5 g, 12–14°C; V = 4.2 ± 0.33
Potassium (mmole/liter)	6.0	1.3 ± 0.13	0.22	ml/hr/kg
Calcium (mmole/liter)	2.65	0.65 ± 0.8	0.24	
Chloride (mmole/liter)	151	10.2 ± 0.9	0.067	
pH	—	7.23 ± 0.05	—	
Total N (mg %)	—	11 ± 1.2	—	
Ammonia N (mg %)	—	6.6	—	
Salvelinus namaycush				
Osmolality (mOsm/liter)	328	36.2 (28–41)	0.11	Hammond (1969); 1 fish, 2700 g, 6°C, GFR =
Sodium (mmole/liter)	161	17.4 (16–28)	0.11	3.73 (2.22–5.91) ml/hr/kg; V = 2.44
Potassium (mmole/liter)	2.8	2.5 (1.9–2.9)	0.93	(1.51–3.79) ml/hr/kg
Calcium (mmole/liter)	2.05	0.95 (0.8–1.1)	0.46	
Magnesium (mmole/liter)	0.75	0.55 (0.2–0.8)	0.73	
Chloride (mmole/liter)	140.6	8.1 (5.4–10.5)	0.58	
Esox lucius				
Osmolality (mOsm/liter)	264	37.7 (37–38.4)	—	Hickman (1965); 2 fish, av. weight 1716 g,
Sodium (mmole/liter)	114	0.26 (0.13–0.40)	0.002	4°C; GFR = 2.9 ml/hr/kg, V = 1.6
Potassium (mmole/liter)	2.65	0.41	0.15	ml/hr/kg
Chloride (mmole/liter)	102	5.58 (5.0–6.17)	0.055	
Esox lucius				
Osmolality (mOsm/liter)	290	26	—	Hunn (1967); 1 fish, 350 g, 12°C,
Sodium (mmole/liter)	139	3.6	0.025	V = 1.9 (1.45–2.4) ml/hr/kg
Potassium (mmole/liter)	4.3	5.8	1.35	

Table VI (Continued)

Species and plasma solute	Plasma	Urine	U/P ratio	Reference
Calcium (mmole/liter)	2.55	0.14	0.054	
Magnesium (mmole/liter)	—	0.66	—	
Chloride (mmole/liter)	117	5.8	0.049	
Phosphate (mmole/liter)	2.87	2.26	0.79	
Channa argus				
Osmolality (mOsm/liter)	282 ± 15.7	49.7 ± 22.0	—	Oguri (1968); Oguri and Takada (1965); 11–41 fish, av. weight 410 g
Sodium (mmole/liter)	—	16.9 ± 3.16	—	
Potassium (mmole/liter)	—	2.75 ± 0.28	—	
Calcium (mmole/liter)	3.06 ± 0.63	0.79 ± 0.10	0.26	
Magnesium (mmole/liter)	—	0.52 ± 0.07	—	
Glucose (mmole/liter)	4.2 ± 0.25	0.73 (0.12–2.7)	0.17	
pH	7.53 ± 0.19	6.28 ± 0.28	—	
Fundulus kansae[a]				
Osmolality (mOsm/liter)	305 ± 4.4	82.3 ± 8.4	—	J. G. Stanley and Fleming (1967b); 4–11 fish, body wt. 0.7–2.5 g, 19°C; V = 12.67 ml/hr/kg
Sodium (mmole/liter)	157.8 ± 2.5	29.2 ± 5.3	0.18	
Potassium (mmole/liter)	3.8 ± 0.42	0.7 ± 0.14	0.18	
Calcium (mmole/liter)	2.65 ± 0.71	1.48 ± 0.2	0.56	
Anguilla anguilla[a]				
Osmolality (mOsm/liter)	323 ± 1.2	—	—	Sharratt et al. (1964); 6–8 fish, body wt. 145 ± 20 g, 18°C; GFR = 4.6 (2.3–7.8) ml/hr/kg; V = 3.5 (1.9–6.7) ml/hr/kg
Sodium (mmole/liter)	155 ± 3.2	18.9 ± 2.9	0.12	
Potassium (mmole/liter)	2.7 ± 0.12	0.65 ± 0.16	0.24	
Chloride (mmole/liter)	106 ± 3.1	Nil	—	

Anguilla anguilla[a]				Chester Jones et al. (1969); 8 fish, body wt. 250–1000 g, 12°C, GFR = 1.51 ± 0.18 ml/hr/kg; V = 1.1 ± 0.23 ml/hr/kg
Sodium (mmole/liter)	150 ± 1.4	13.1 ± 1.2	0.087	
Potassium (mmole/liter)	1.81 ± 0.3	1.14 ± 0.32	0.63	
Calcium (mmole/liter)	2.37 ± 0.04	0.63 ± 0.11	0.27	
Magnesium (mmole/liter)	2.07 ± 0.10	0.02 ± 0.01	0.097	
Chloride (mmole/liter)	88 ± 3.9	3.30 ± 0.47	0.038	
Phosphate (mmole/liter)	1.75 ± 0.12	4.47 ± 0.55	2.55	
Catastomus commersonii				Hickman (1965); 4 fish, body wt. 1005–1747 g, 4°C, GFR = 2.2 (1.03–3.24) ml/hr/kg; V = 1.15 (0.76–1.58) ml/hr/kg
Osmolality (mOsm/liter)	233 (208–248)	33.5 (21.2–54.8)	0.15	
Sodium (mmole/liter)	97.5 (80–133)	4.9 (0.6–8.6)	0.05	
Potassium (mmole/liter)	2.15 (1.8–3.8)	5.4 (1.4–9.4)	2.51	
Chloride (mmole/liter)	84.7 (64–98)	2.1 (0.66–3.9)	0.024	
Amerurus nebulosus				Hodler et al. (1955); 10 fish, wt. not given; V = 0.99 ml/hr (kg?)
Sodium (mmole/liter)	122	12.2	0.10	
Potassium (mmole/liter)	2.7	1.61	0.60	
Chloride (mmole/liter)	110	18.0	0.16	
Bicarbonate (mmole/liter)	3.4	0.4	0.12	
pH	7.54	6.37	—	
Carassius auratus				Maetz (1963); 63 fish, 56–197 g; GFR = 20.4 ± 0.65 ml/hr/kg; V = 13.7 ± 0.4 ml/hr/kg
Osmolality (mOsm/liter)	258.8 ± 2.65	35.9 ± 1.85	0.14	
Sodium (mole/liter)	115 ± 1.15	11.5 ± 0.85	0.10	
Potassium (mmole/liter)	3.6 ± 0.15	1.4 ± 0.085	0.39	

[a] Euryhaline teleosts adapted to freshwater.

Table VII

Average Renal Excretion of Sodium, Chloride and Potassium by the White Sucker, *Catostomus commersonii*[a]

GFR = 2.2 (1.03–3.24) ml/hr/kg; urine flow = 1.15 (0.76–1.58) ml/hr/kg

Plasma solute	Plasma conc. (mmole/liter)	Urine conc. (mmole/liter)	U/P ratio	Filtered (μmole/hr)	Net reabsorbed (μmole/hr)	Net secreted (μmole/hr)	Excreted (μmole/hr)	Reabsorbed or secreted (%)	Solute clearance	Solute/inulin clearance ratio
Sodium	97.5	4.9	0.05	214	208.4	—	5.6	97.3	0.043	0.019
Potassium	2.15	5.4	2.5	4.73	—	1.5	6.2	24.1	2.9	1.32
Chloride	84.7	2.1	0.025	186	183.6	—	2.4	98.7	0.028	0.013
Osmolality	223	33.5	0.29	—	—	—	—	—	—	—

[a] From Hickman (1965).

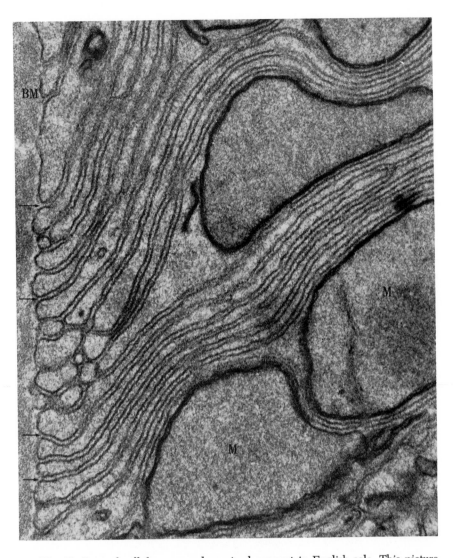

Fig. 21. Base of cell from second proximal segment in English sole. This picture shows the basement membrane (BM) and the numerous parallel membranes formed by infoldings of the basilar plasma membrane. When fixed with potassium permanganate, the continuity of these membranes with the surface can be readily appreciated (free arrows). Note also the unit membrane appearance of these membranes. Mitochondria (M). ×56,700. Courtesy of R. E. Bulger.

since Na$^+$ may leak back into distal regions of the tubule more rapidly than Cl$^-$.

Potassium may undergo either net secretion or net reabsorption against

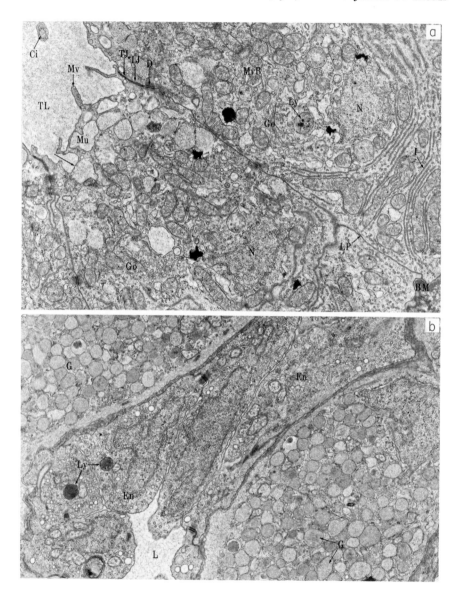

Fig. 22. (a) Collecting tubule from English sole. Tubular lumen (TL) is at the upper left and the basement membrane (BM) at the lower right. At the apex of the cell there is an occasional small microvillus (Mv) and an occasional cilium (Ci) is seen cut in cross section. Cell junctional complexes consist of tight junctions (TJ), intermediate junctions (IJ), and desmosomes (D). Mucous granules (Mu) are seen in the apex and some of them are in the process of discharging to the lumen (free arrow). They are PAS-positive. The Golgi apparatus (Go) is well developed, and

its concentration gradient. Both of these conditions have been observed in northern pike and white suckers (Hickman, 1965). The urine concentrations of K^+ and Na^+ vary inversely in the white sucker over prolonged periods of consecutive urine sample collection (Fig. 37). This relationship suggests, but does not prove, the existence of a Na^+ for K^+ exchange pump. An alternate explanation for the interdependence of Na^+ and K^+ is that if the permeability of the tubular epithelium were higher to K^+ than to certain anions present in the formative urine, K^+ might replace Na^+ to balance the total anionic charge when Na^+ reabsorption is especially high.

The few reported measurements of Mg^{2+} and Ca^{2+} in freshwater teleost urine (Table VI) indicate that their concentrations are low (Fromm, 1963; Oguri, 1968; Hunn, 1967; Hammond, 1969), and that Ca^{2+}, at least, is strongly reabsorbed against its concentration gradient (Reid et al., 1959; Oguri, 1968). There is evidence that phosphate may undergo active tubular secretion in the freshwater-adapted European eel (Chester Jones et al., 1969), but the presence or absence of tubular secretory mechanisms for divalent ions has not been determined in true stenohaline freshwater teleosts.

The active abstraction of electrolytes from the filtrate occurs, at least in part, without the osmotic accompaniment of water. During its travel through the tubule, the concentration of the filtrate is reduced from 220–320 mOsm/liter, to 20–80 mOsm/liter (Table VI). Usually less than half, and sometimes as little as 5%, of the filtered water is reabsorbed

---

numerous secondary lysosomes (Ly) are seen in the vicinity of the apparatus. Many of the mitochondrial profiles are above the nucleus (N). The lateral plasmalemma (LP) can be seen as well as numerous infoldings (I) of the basilar plasmalemma. Collecting tubules vary in length depending on the point of junction with collecting tubules from other nephrons and occasionally are very long. In some instances, two or more third proximal segments empty into a common collecting tubule. Wandering cells are especially numerous in this region. The collecting tubules empty into a collecting duct system in which the ducts converge to form larger channels which finally empty into the archinephric duct. As the ducts become larger, their cells become much higher. The zone of PAS-positive mucous granules is wider, and wandering cells are extremely prominent. In addition to a thick basement membrane, these larger collecting ducts are surrounded by multiple layers of smooth muscle cells. ×11,655. Modified from Bulger and Trump (1968). (b) Afferent arteriole from the English sole. The arteriolar lumen (L) and endothelium (En) can be identified. Note the desmosomes and other parts of the junctional complex joining the endothelial cells. The endothelial cells also contain moderate numbers of lysosomes (Ly). The smooth muscle cells are highly modified and filled with numerous granules (G) thought to contain renin. Basement membrane material surrounds the periphery of the smooth muscle cells and is seen between the endothelium and the modified smooth muscle cells. Multivesicular bodies (MvB). ×7875. Modified from Bulger and Trump (1969a).

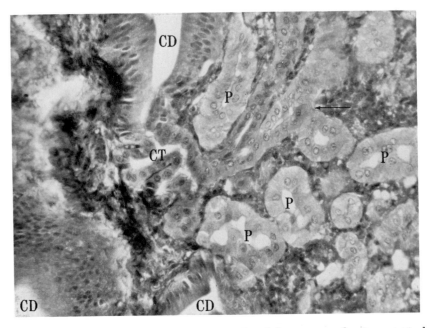

Fig. 23. Section of kidney of midshipman, *Porichthys notatus*, showing principal tubules (P), collecting tubules (CT), and collecting ducts (CD). A transition from principal to collecting tubules is indicated by the free arrow. ×380.

(Hickman, 1965; Mackay, 1967). The distal segment and collecting tubule and ducts must be nearly impermeable to water (Fig. 47a). How much, if any, of the proximal tubule is also impermeable is not known. The first segment of the proximal tubule is probably concerned with the reabsorption of filtered plasma organic constituents, such as glucose, amino acids, and macromolecules (Fig. 47a). The evidence for this is that this segment has been discarded by marine aglomerular teleosts which do not filter these materials and consequently have no need to reabsorb them. Futhermore, this anterior portion degenerates in conjunction with the progressive atrophy of the glomerulus in goldfish held in one-third seawater (Ogawa, 1961b). If tubular segments are functionally homologous in freshwater teleosts and amphibians, the reabsorption of electrolytes and organic solutes in both portions of the proximal tubule is accompanied by an osmotically equivalent quantity of water (Fig. 47a) (Whittembury *et al.*, 1959; Windhager *et al.*, 1959). However, the fact that as little as 5% of the filtered water is reabsorbed places a restriction on the amount of isosmotic NaCl reabsorption that can take place in the proximal segment. NaCl constitutes about 90% of the filtered osmolytes. Since nearly all of this is eventually reabsorbed, most must be actively transported across the water impermeable distal regions of the nephron. It is to be noted, however, that three species of Salmonidae (*Oncorhynchus masou* var *ishikawae, Salvelinus pluvius,* and *Plecoglossa altivelis*) reportedly lack the distal segment (Ogawa, 1962). However, this assessment is subject to the limitations described earlier. It would be of particular interest to know if the absence of the distal segment in these three species reduces the capacity of the kidney to form a dilute urine.

b. Urine pH. The urine of freshwater teleosts is characteristically acid as in marine teleosts, although the rainbow trout, *S. gairdneri,* reportedly excretes a slightly alkaline urine (Fromm, 1963; Hunn, 1969) (Table VI). Unlike marine teleosts, urine acidification involves a Diamox sensitive carbonic anhydrase system which is important in the reabsorption of bicarbonate. This circumstance suggests that as in mammals, the Diamox sensitive system resides in the distal segment. If carbonic anhydrase is inhibited with Diamox, the reabsorption of HCO_3^-, Na^+, K^+, and Cl^- is depressed (their urine concentrations increase) and the urine is alkalinized (Hodler *et al.,* 1955; Oguri and Takada, 1965). Hodler *et al.* (1955) noted a 37-fold increase in urine HCO_3^-, and an 8-fold increase in K^+ excretion following the administration of Diamox to the freshwater catfish, *A. nebulosus.* Urine flow doubled and urine pH rose from 6.37 to 7.43. These changes are very similar to those observed in mammals following Diamox administration (Pitts, 1963).

Hunn (1969) found that the normally neutral or slightly alkaline urine of rainbow trout became acid (to pH 6.55) following acute hypoxic stress. The drop in pH was associated with a massive release into the circulation of lactic acid and its appearance in the urine, and a rise in urinary phosphate, which, if secreted in acid form ($H_2PO_4^-$) would also tend to increase urine acidity.

c. Glucose Reabsorption. Oguri (1968) reported that glucose is largely, but never completely, removed from the filtrate of the freshwater teleost, *Channa argus.* This is possibly related to the short length of the first proximal segment in most freshwater teleosts (Fig. 47a). Urine and plasma glucose vary over wide ranges in this species (urine glucose 0.11–3.03 mmoles/liter; plasma glucose 1.9–7.1 mmoles/liter). Usually, the highest urine sugar levels were found in fish having the highest blood sugar levels (Oguri, 1968).

It has already been pointed out that the Tm for glucose in fish varies linearly with the GFR (Fig. 36), a fact that provides the strongest indirect evidence that glomerular activity is intermittent. Consequently, the Tm for glucose must be expressed as a ratio of the inulin clearance (Tm/C_{in}). By infusing glucose into the circulation of white suckers,

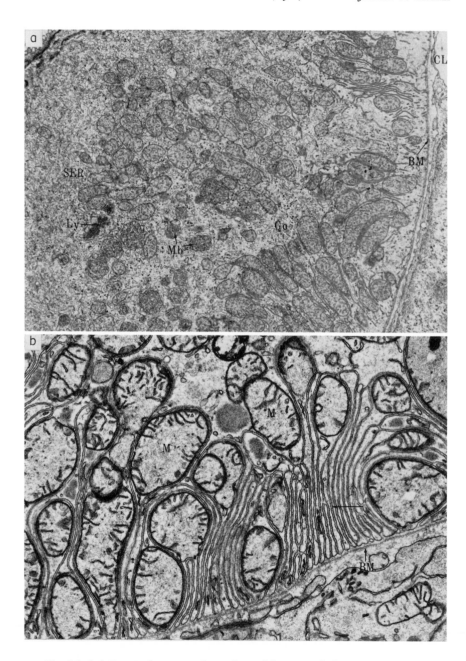

Fig. 24. (a) Principal segment from the midshipman tubules. Note the similarity between this tubule and the second and the third segments in glomerular teleosts. Note the mitochondria at the base of the cell. These are in proximity to infoldings

C. commersonii, to saturate the glucose reabsorptive mechanism, Mackay and Beatty (1968) reported that the Tm/C_{in} ratio increased gradually with temperature from 2.31 at 2°C to 3.11 at 19°C ($Q_{10} = 1.21$). Thus, contrary to the expected response of an energy-dependent metabolic system to temperature change, the glucose reabsorptive mechanism of this species became only slightly more efficient at high than at low temperatures.

High urine sugar levels are associated with increased urine flows; this is presumably an osmotic diuresis. Enomoto (1964) noted increased urine flows and proteinuria in rainbow trout, *S. gairdneri,* made glucosuretic with growth hormone.

d. Excretion of Nitrogenous Compounds. Of the total organic nitrogen excreted by freshwater fish, only a small proportion, estimated at 2.5–24.5% for different species, is excreted in the urine (H. W. Smith, 1929b; Pora and Prekup, 1960; Fromm, 1963) (Fig. 47a). Feeding fish excrete far more organic N than fasted fish; in rainbow trout, *S. gairdneri,* the total N and urinary N drop precipitously and proportionately during the first 6 days of starvation (Fromm, 1963). In this species ammonia comprised about 60% of the urinary N. Smith (1929b) gave the distribution of several different N-containing components in the urine of 8 species of freshwater fish. These were, in order of decreasing N concentration: creatine, urea, ammonia, amino acid, uric acid, and creatinine. In five fasted species, these identified compounds made up only about 50% of the total urinary N. Three actively feeding freshwater species had more urea and amino acid N, and less creatine, in the urine than the five fasted species studied by H. W. Smith (1929b). In the fed species only about 25% of the urinary N could not be accounted for. Since Smith (1929b) did not compare fed and fasted individuals of the same species, it is impossible to know whether the differences are nutritional or species specific. Trimethylamine oxide, present in urine of most marine teleosts, was reported absent from the urine of the carp, *Cyprinus carpio* (Hoppe-Seyler, 1928, 1930, summarized by Grafflin and Gould, 1936).

The kidney may be an important excretory pathway for the removal

---

of the basilar plasmalemma. In the cell apex are clusters of well-developed smooth endoplasmic reticulum (SER) and secondary lysosomes (Ly). Numerous microbodies (Mb) are found throughout the cell. Go, Golgi apparatus; CL, capillary lumen; BM, basement membrane. ×6935. Courtesy of R. E. Bulger. (b) Part of a principal segment of the midshipman tubule showing the basement membrane (BM), mitochondria (M), and numerous infoldings of the basilar plasma membrane indicated by the free arrows. Note that in this tissue fixed in potassium permanganate, the continuity of these infolded membranes with the base of the cell can be readily demonstrated. ×18,250. Courtesy of R. E. Bulger.

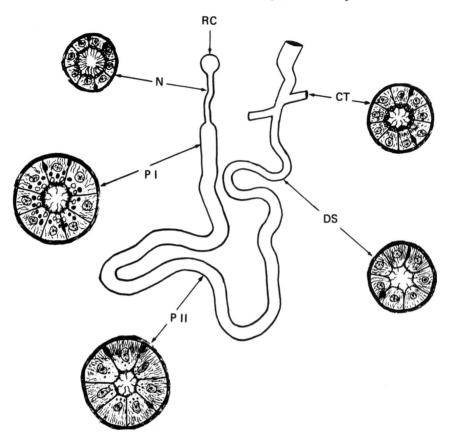

Fig. 25. Schematic representation of the nephron segments in the southern flounder, *Paralichthys lethostigma*. This drawing was made to scale from a three-dimensional reconstruction such as that shown in Fig. 26a. Note the characteristic morphology of each region. Renal corpuscle (RC), neck (N), first proximal segment (PI), second proximal segment (PII), distal segment (DS), and collecting tubule (CT).

of certain minor nitrogenous compounds, such as creatine and uric acid, even though the gills are clearly the principal route by which ammonia and urea, the major nitrogenous wastes, are removed from the body (see chapter by Forster and Goldstein, this volume).

4. RENAL ADJUSTMENTS IN SALT AND WATER
   EXCRETION IN SALINE MEDIA

Rarely in nature is the kidney of freshwater fish required to adjust to osmotic changes in the environment. Yet many freshwater fish can survive

Table VIII
Glomerular Filtration and Urine Flow Rates in Glomerular Teleost Fishes in Seawater

Species	Temp. (°C)	GFR (ml/hr/kg)	Urine flow (ml/hr/kg)	Average inulin U/P ratio	Average body weight (g)	No. of fish	Salinity (%)	Notes	Reference
Myozocephalus octodecimspinosus (longhorn sculpin)	—	0.70 (0.46–1.16)	0.13 (0.1–0.18)	5.4	171	9	Atlantic seawater	Xylose clearance corrected for xylose/inulin ratio (0.81, Clarke, 1936)	Clarke (1934)
Myozocephalus octodecimspinosus (longhorn sculpin)	—	0.68 (0.16–1.18)	0.55 (0.11–0.98)	1.2	—	—	Atlantic seawater	Xylose clearance corrected for xylose/inulin ratio (0.81, Clarke, 1936)	Marshall and Graffin (1932)
Myozocephalus octodecimspinosus (longhorn sculpin)	17(?)	2.88 (0.5–5.86)	1.48 (0.39–3.57)	1.95	—	24	Atlantic seawater	Many fish diuretic	Forster (1953)
Myozocephalus scorpius (daddy sculpin)	17(?)	0.13 (0–0.7)	0.88 (0.08–1.67)	0.15	588	34	Atlantic seawater	Mostly aglomerular	Forster (1953)
Paralichthys flesus (flounder)[a]	16–18	2.4 ± 0.27	0.60 ± 0.05	4.0	200	39	40	—	Lahlou (1967)
Paralichthys lethostigma[a] (southern flounder)	10	0.5 (0.005–1.34)	0.36 (0.11–0.85)	1.4	1452	5	28–36	Seasonal change in GFR: high in summer, low in winter	Hickman (1968a)
	15.5	1.38 (1.06–1.69)	0.24 (0.21–0.28)	5.8	1775	2	30–32	—	
	21	1.69 (1.41–2.10)	0.22 (0.11–0.31)	7.7	1167	4	30–34	—	
Anguilla anguilla[a] (European eel)	18	1.03 (0.56–2.16)	0.63 (0.3–0.9)	1.6	181	7	Seawater	Adapted > 10 days in seawater	Sharratt *et al.* (1964)
	12	0.43 ± 0.06	0.25 ± 0.04	1.72	625	13	Seawater	Adapted > 2 weeks in seawater	Chester Jones *et al.* (1969)
Salmo gairdnerii[a] (Rainbow trout)	6	10.1 ± 2.6	—	—	163.5	10	80% seawater	Adapted 10 days to 80% seawater	W. N. Holmes and McBean (1963)
Fundulus kansae[a] (Plains killifish)	20 ± 1	1.35 (0.83–1.88)	0.52 (0.21–0.83)	2.6	1.6	3	Seawater	Adapted 7 days to seawater	Fleming and Stanley (1965)

[a] Euryhaline.

163

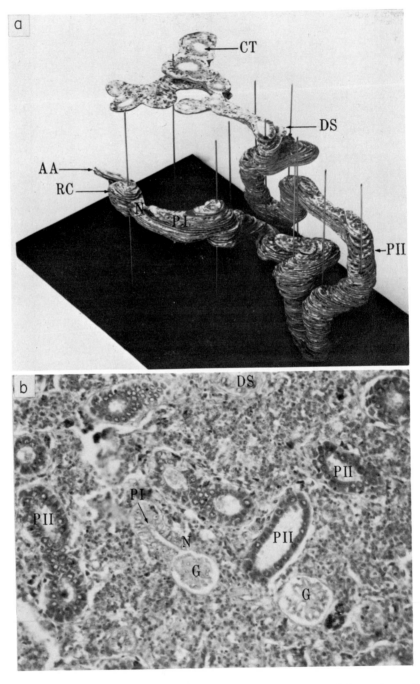

Fig. 26. (a) Photograph of a reconstruction from serial sections of the nephron of

prolonged periods in dilute saline media (Schwartz, 1964). Goldfish, for example, can live in half-strength seawater for several months (Ogawa, 1961b). These environmental conditions are abnormal and might always be avoided, were the fish offered a choice of environments. Nevertheless, it comes as no great surprise that even stenohaline species can survive in saline conditions, provided the osmotic pressure of the environment does not equal or exceed that of its body fluids, and provided the medium solute is predominantly NaCl. The fish needs only to reduce urine production by the kidney and NaCl influx across the gills. It can certainly do this, since it must have the capability of adjusting both of these functions to meet changing environmental conditions in the normal freshwater environment.

Beyond these primary adjustments, which can be achieved within a matter of minutes, or at most an hour or two, most freshwater teleosts can improve their pattern of regulation in saline media by allowing the urine osmolality to rise. This renal tubular adjustment requires several hours to several days, and is not observed in short-term experiments.

For example, when perch, *Perca fluviatilus*, and northern pike, *Esox lucius*, were subjected to a slightly hyperosmotic medium for 3 or 4 hr, the blood osmolality increased (becoming once more hyperosmotic to the new environment), and the GFR and urine flow dropped markedly (Ginetsinkii *et al.*, 1961). If the artificial medium exceeded about 430 mOsm/liter, the urine flow stopped. Yet neither the urine sodium concentration nor the total urine osmolality changed appreciably, indicating that in these species tubular electrolyte reabsorption was undiminished even while the blood NaCl concentration rose to lethal levels. Bourguet *et al.* (1964) reported that the intraperitoneal injection of dilute (20–35 mM) NaCl solutions into goldfish increased the GFR and urine flow 75–85% after a delay of an hour or so, but it had no effect on sodium excretion or tubular water reabsorption. The injection of a hyperosmotic (700 mM) NaCl solution effected a parallel drop in GFR and urine flow after a 2-hr delay and a relatively small (40%), paradoxical, decrease in renal sodium excretion. This drop resulted from the combined effects of decreased urine flow and increased sodium retention.

If freshwater teleosts are allowed several hours or days to adapt to a saline medium, the tubular reabsorption of monovalent ions will gradually

the southern flounder. Afferent arteriole (AA), renal corpuscles (RC), neck (N), first proximal segment (PI), second proximal segment (PII), distal segment (DS), and collecting tubule (CT). (b) General view of kidney of the southern flounder. Note the small poorly vascularized glomeruli (G), neck (N), and transition to proximal segment (PI). Note also numerous profiles of second proximal segment (PII) and small portions of the distal segment (DS). ×380.

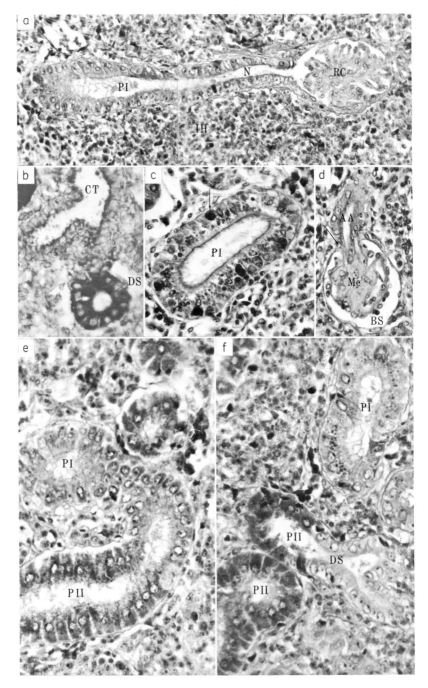

Fig. 27. (a) Light micrograph of kidney from southern flounder showing renal

decrease. Martret (1939) found that when carp, *Cyprinus carpio,* were placed in NaCl solutions more concentrated than the urine but less concentrated than the blood, urine osmolality rose until it nearly equaled that of the external environment. Favre and Hourday, quoted in Bourguet *et al.* (1964), found that when goldfish were held for a month in isosmotic NaCl and then returned to freshwater, the urine remained almost isosmotic to plasma until the "excess salt" was eliminated.

From these different studies, it seems clear that the primary and most effective renal adjustment to increased external salinity (or hyperosmotic saline injections) is reduction of the GFR, achieved by progressively shutting down individual glomeruli or portions thereof. If the external salinity approaches body fluid osmolality, glomeruli filtration ceases altogether and the glomeruli begin to atrophy. Ogawa (1961b) has provided an excellent description of the progressive degeneration of the glomeruli and second portion of the proximal tubules and considerable fluid accumulation within the distal tubule of goldfish surviving in one-third strength seawater. Since the goldfish cannot survive in an environment more concentrated than its blood, it seems unlikely that divalent electrolytes are being actively transported into the tubular lumen, as occurs in true hypoosmotic regulation by marine teleosts.

D. Kidney Function in Marine Teleostei

The kidney of marine teleosts has become a functional specialist. The relatively simple structure of the nephron, lacking the distal segment characteristic of many euryhaline and nearly all freshwater species, and often with the glomeruli degenerative or absent (Figs. 26b and 27d; Table XIII), is largely a consequence of the limited, though essential, part that the kidney performs in the animal's complex pattern of hypo-

corpuscle (RC), neck region (N), first proximal segment (PI) and interstitial hematopoietic tissue (IH). ×420. (b) Distal segments and collecting tubules from southern flounder showing the high activity of DPN diaphorase in the distal segment (DS) and the small amount of activity confined to the cell apices in the collecting tubule (CT). ×140. (c) First proximal segment (PI) from the southern flounder. Note the brush border and the numerous large secondary lysosomes (arrow) which fill most of the cytoplasm. ×700. (d) Renal corpuscle from southern flounder showing the afferent arteriole (AA) and the glomerulus lying within Bowman's space (BS). At the free arrow the continuity between the parietal epithelium, afferent arteriole, and visceral epithelium can be seen. Note the prominent mesangial region (Me) with few open capillary lumens. ×560. (e) Transition between first (PI) and second (PII) proximal segments in the southern flounder. ×420. (f) Portion of kidney of southern flounder showing first proximal segments (PI) with large supranuclear lysosomes, second proximal segments (PII) with transition to the distal segment (DS). ×420.

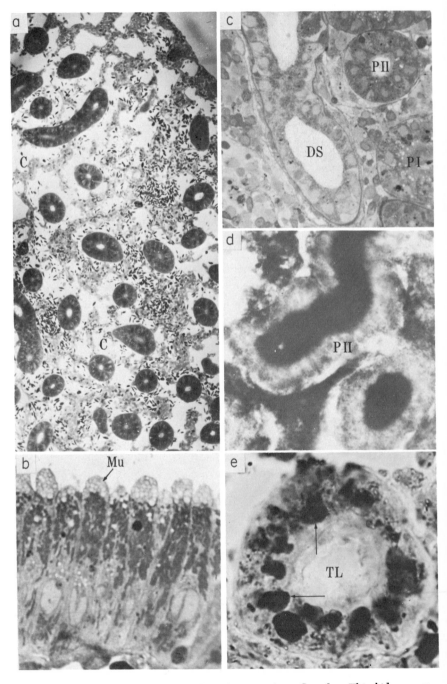

Fig. 28. (a) Survey picture of kidney from southern flounder. This kidney was

osmotic regulation. This specific function is the excretion of magnesium and sulfate, which are osmoregulatory by-products rather than metabolic wastes. They enter the intestine in seawater deliberately swallowed by the fish, as the first step in the replacement of water lost exosmotically across the body surface. A small fraction of the divalent ions swallowed, normally less than 20%, penetrates the intestinal mucosa to enter the bloodstream. The available evidence indicates that all of the absorbed magnesium and sulfate is excreted exclusively by the kidney (Hickman, 1968c). Other ions, notably sodium, chloride, potassium, and calcium are also absorbed from the intestine and appear in the urine, but the kidney does not serve as the exclusive pathway for the excretion of any of these.

It would scarcely do the marine teleostean kidney justice to consider it purely a magnesium sulfate pump, for it does much more than this, yet most of the morphological and functional specializations of the organ can be understood by recognizing this to be its dominant role. The following discussion deals primarily with kidney function in stenohaline marine fishes, of which only two species, the glomerular longhorn sculpin, *Myoxocephalus octodecimspinosus*, and the goosefish, *Lophius americanus* (= *piscatorius*), have been at all well studied. We have additionally included information from renal function studies of the southern flounder, *Paralichthys lethostigma*, in seawater because in this environment the kidney of this euryhaline species appears to function in the manner of stenohaline forms (Fig. 47b).

1. GLOMERULAR FILTRATION

a. Glomerular Development. Because the dominant activity of the marine teleostean kidney is MgSO$_4$ secretion, the glomeruli have become superfluous structures, at least to those species which never venture into reduced salinities. Consequently, all degrees of glomerular degeneration are evident among stenohaline marine teleosts, from those having well-vascularized glomeruli to those lacking glomeruli altogether (Marshall and Smith, 1930; Nash, 1931). The completely aglomerular teleosts are

fixed by intravascular perfusion and shows the widely patent peritubular capillaries (C) surrounding the tubules. ×225. (b) Collecting tubule from southern flounder showing apical mucous granules (Mu). ×900. (c) Epon section showing appearance of mitochondria and organelles in distal segment (DS), first proximal segment (PI), and second proximal segment (PII) from the southern flounder. ×540. (d) Second proximal segment (PII) from southern flounder showing localization of alkaline phosphatase (black precipitate) along brush border. ×720. (e) First proximal segment from southern flounder showing tubular lumen (TL) and large acid phosphatase positive lysosomes (free arrows) in the cytoplasm. ×900.

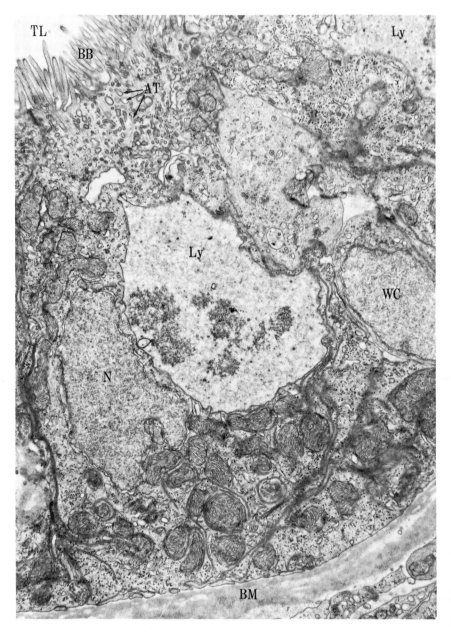

Fig. 29. First proximal segment from southern flounder. Tubular lumen (TL) is at the upper left and the basement membrane (BM) is near the lower right. Note the closely packed microvilli forming the brush border (BB). In the apical zone beneath the brush border are numerous images of circular or elongate profiles which

scattered rather randomly among many unrelated phyletic groups, although all the members of certain families of marine teleosts, for example, the Syngnathidae, Gobiesocidae, and Saccopharyngidae (Smith, 1931c), may turn out to be entirely aglomerular when all representatives have been examined (Table IV). About the only generalization possible is that many aglomerular species are among the most exotic, highly specialized and least active fishes known (Marshall and Smith, 1930; Marshall, 1934; H. W. Smith, 1931c, 1953; Ogawa, 1962).

Among the glomerular teleosts, estimates of total filtering surface (Nash, 1931) or number and size of glomeruli (Marshall and Smith, 1930; Ogawa, 1962; Lozovik, 1963) have shown that marine fishes generally have less well developed glomeruli than freshwater fishes. It has been assumed that glomerular development is correlated with the filtration rate, but so few measurements of GFR have been made (Table VIII) that, excepting the congenital aglomerular species and those showing obvious glomerular degeneration during growth (pauciglomerular), it is impossible to know if GFR and apparent filtration surface bear any consistent relationship to each other. Furthermore the size and number of glomeruli and degree of capillary branching may be less important in determining filtration rate than several factors which cannot be estimated from morphological appearance alone. These factors include patency of the capillaries, vasoconstrictive ability of the afferent arterioles, capillary membrane pore size, filtration pressure, contractility of the mesangium, tubular back pressure, and possible contractility of neck region or other tubular segments. The best index of glomerular development is the inulin clearance, and it is to be hoped that comparative renal physiologists soon will examine some of the many neglected groups of glomerular oceanic teleosts as techniques for collection, holding, and measurement are improved and simplified.

b. Relationship between GFR and Urine Flow. Unlike freshwater teleosts, urine flow of marine teleosts is usually determined far more by tubular activities than by the volume of glomerular filtrate. In aglomerular teleosts the urine volume is most probably determined solely by the amount of water that diffuses into the tubule secondary to the secretion of divalent ions. In such animals, Bieter (1931c) found a

---

represent sections in various planes through the apical invaginations of the plasma membrane or apical tubules (AT). Note the numerous irregular mitochondrial profiles throughout the cell. In the midregion of the cell near the nuclei are extremely large secondary lysosomes (Ly) which typify this region. Nucleus (N) is seen to the left adjacent to these large secondary lysosomes. A wandering cell (WC), probably a lymphocyte, is seen between two epithelial cells. ×11,000.

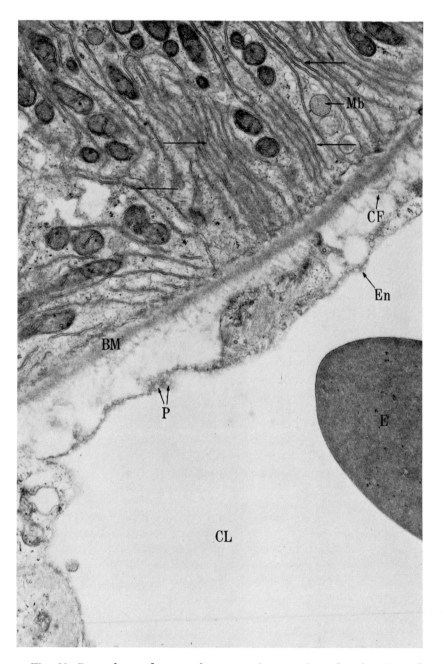

Fig. 30. Base of second proximal segment from southern flounder. Note the numerous infoldings of the basilar plasma membrane in this region indicated by the free arrows. These are in proximity to the mitochondrial profiles. One microbody

secretion pressure greater than dorsal aortic blood pressure. In glomerular species, the contribution that the filtrate makes to urine volume depends on glomerular vascular development and glomerular activity. Urine flow is independent of, and always exceeds, the GFR in the daddy sculpin, *M. scorpius*, which shows progressive glomerular degeneration during the life of the fish (Grafflin, 1933; Forster, 1953). But in marine teleosts having relatively high filtration rates [for example, the longhorn sculpin, *M. octodecimspinosus* (Forster, 1953), and the southern flounder *P. lethostigma* (Hickman, 1968a)] the GFR and urine flow are proportional, just as in freshwater teleosts. This relationship is not easily explained because even at high filtration rates, nearly all of the filtered sodium and most of the chloride is reabsorbed, leaving a urine composed predominantly of secreted divalent ions. Since the tubular epithelium is permeable to water, the urine volume is simply a function of solute load. Consequently the positive relationship between the separate activities of magnesium sulfate secretion and glomerular filtration is unexpected and not yet understood. In this connection, it may be significant that PAH secretion is also positively related to the GFR, suggesting that tubular secretory activities may be governed, or at least affected, by variations in renal blood flow (see below).

2. Blood Flow to the Kidney

The blood supply to the kidney of wholly marine teleosts is predominantly venous (Fig. 9). The main afferent venous supply is usually the caudal vein which enters the posterior end of the kidney, but other veins such as the segmental, subcardinal, and epibranchial may contribute. In at least some aglomerular fishes, the caudal accessory vein is larger than the caudal and presumably supplies more blood (Marshall and Grafflin, 1928; von Möllendorff, 1930; Gerard, 1954). The kidney is drained entirely by a large posterior cardinal vein.

In juvenile *Lophius* the arterial supply serves glomeruli distributed in the periphery of the kidney but these nearly all degenerate, and in adults, arterial blood reaches only the capsules of the few glomeruli remaining (Nizet and Wilsens, 1955). Even in marine teleosts with relatively well vascularized glomeruli, the arterial supply is meager. The anatomical picture is supported by PAH clearance estimates of renal plasma flow. Forster (1953) reported that the average PAH clearance of

(Mb) is seen. The tubule is separated from the interstitium by basement membrane (BM). The interstitium contains scattered collagen fibrils (CF). A peritubular capillary is seen with a very thin endothelial cytoplasm (En), perforated by thin fenestrations or pores (P). An erythrocyte (E) is in the capillary lumen (CL). ×17,575.

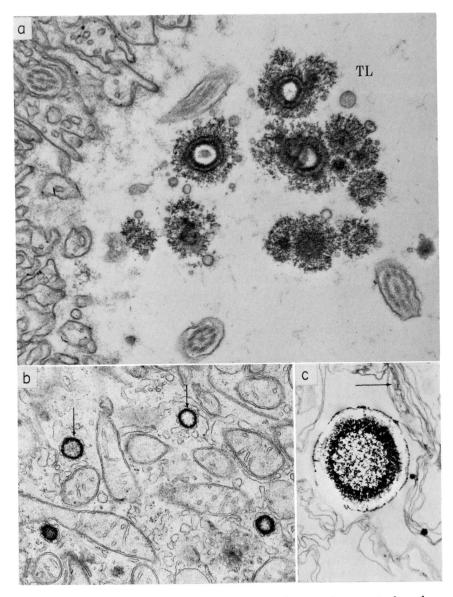

Fig. 31. (a) Tubular lumen (TL) from second proximal segment of southern flounder, showing cluster of microcrystalline aggregates probably a form of calcium phosphate. ×27,000. (b) Portion of cytoplasm from the same cell showing micro-crystalline aggregates within the endoplasmic reticulum (free arrows). ×16,650. (c) Section of urine sediment from southern flounder showing identical-appearing micro-crystalline aggregates in urine sediment. Free arrow indicates membranous debris, probably resulting from cellular disorganization. ×22,500.

24 glomerular longhorn sculpins, *M. octodecimspinosus*, was 108 ml/hr/ kg as compared to an average inulin clearance of 2.9 ml/hr/kg, giving a C_{PAH}/C_{in} ratio of 37.2. In the nearly aglomerular shorthorn sculpin, *M. scorpius*, the PAH clearance of two specimens averaged 133 ml/hr/kg as compared to the average inulin clearance 0.102 mg/hr/kg (C_{PAH}/C_{in} ratio = 1304). These are minimal estimates of the renal plasma flow since it is unlikely that PAH extraction by the tubules was complete. Thus in both species the fraction of the renal plasma flow supplying the glomeruli is very low, although it is not possible to estimate this fraction since PAH from both postglomerular and renal portal blood is secreted by the tubules. Forster found that the PAH clearance of the longhorn sculpin increased with the GFR and urine flow, indicating that either the extraction of PAH or the renal plasma flow varied in proportion to the number of nephrons functioning. The arterial supply to the kidney of the European eel, *Anguilla anguilla*, is said to comprise "some 30%" of the total vascular supply (Chester Jones *et al.*, 1965), suggesting that both the arterial and renal portal components of the euryhaline teleostean kidney may be well developed.

As one would expect, significant increases in dorsal aortic blood pressure produced by injections of arginine vasotocin had no detectable effect on urine flow in the aglomerular toadfish (Lahlou *et al.*, 1969). However, aglomerular kidney function may be influenced by changes in blood flow through the venous bed surrounding the nephric tubules. Brull *et al.* (1953) found that urine flow from the blood-perfused kidney of *Lophius americanus* rose exponentially with increasing perfusion pressure and flow (Brull *et al.*, 1953; Brull and Cuypers, 1954b). Although the authors interpret this as variable "water secretion," a more acceptable explanation might be that increased blood flow enables greater tubular secretion of divalent ions which secondarily determine the output of near-isosmotic urine. The exponential leveling of urine flow observed at high rates of renal blood flow implies a progressive saturation of the divalent ion transport mechanism.

3. TUBULAR FUNCTION

a. Urine Composition. In Tables IX and X are given the urine composition and renal excretion values for the southern flounder, *Paralichthys lethostigma*, in seawater, and the goosefish, *Lophius americanus*. It is unfortunate that the pauciglomerular goosefish, which has received more attention by renal physiologists than any other marine teleost, is especially prone to the well-known and troublesome phenomenon known as osmotic or "laboratory" diuresis after capture, when the renal excretion

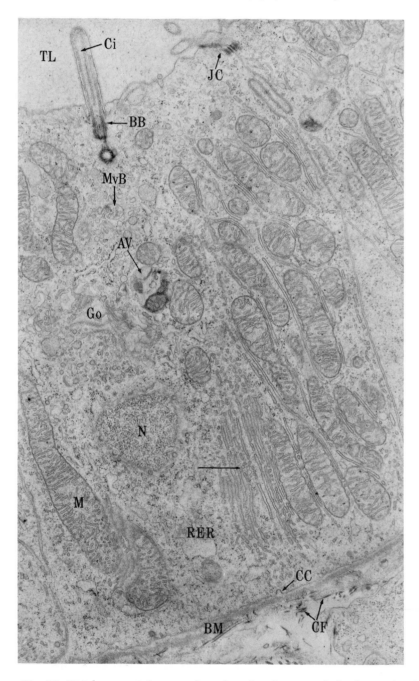

Fig. 32. Distal segment from southern flounder showing tubular lumen (TL), cilium (Ci) with basal body (BB), and basement membrane (BM). Note the elongate

of electrolytes and water increases rapidly and the urine composition is radically altered. There is so much reported variation in urine composition that it is nearly impossible to be certain of the normal pattern for this species. Even the composition of residual bladder urine, collected immediately upon capture and thought to represent normal urine (Forster and Berglund, 1956), is highly variable. Consequently, the values given in Table X were selected from the literature subjectively and represent what we think come closest to being "normal." Most are from Brull and Nizet (1953). One criterion used in this selection was similarity to the urine composition of the relatively unexcitable southern flounder, which, unless injured, never experiences more than a transient diuresis after capture or operative procedures (Hickman, 1968a,b). To assist those wanting information on urine composition for a particular marine teleost species, we have also included a source list to references (Table XI) that give the analyses of two or more urine constituents. With two exceptions (Denis, 1914; Sulze, 1922), we have omitted papers earlier than 1928. Reference to the earlier literature is found in H. W. Smith (1932). Despite the great range of concentrations that virtually any of the urine evident (Tables IX and X and Figs. 38 and 47b). Magnesium is the electrolytes can have in different species, a typical pattern is clearly dominant cation and chloride the dominant anion in the urine. The sulfate concentration is usually less than half that of magnesium (Marshall and Grafflin, 1928; H. W. Smith, 1930b; Brull and Nizet, 1953; Hickman, 1968b) but may occasionally exceed that of magnesium (Pitts, 1934; Forster and Berglund, 1956). In the southern flounder magnesium evidently penetrates the intestinal mucosa more rapidly than sulfate because the concentration ratio of magnesium to sulfate in urine (2.4:1) is usually greater than their concentration ratio in the ingested seawater (1.9:1) (Hickman, 1968b). Except during the diuretic state, or during periods of very high phosphate excretion, the sum of the equivalent concentrations of sulfate and chloride is usually about equal to the equivalent concentration of magnesium (Fig. 38). Chloride is a normal constituent of urine, and its presence does not indicate abnormality as was once believed (Clarke, 1934; Pitts, 1934; Grafflin and Ennis, 1934; Grafflin, 1935; Forster and Berglund, 1956). The abundance and high mobility of chloride suggests that it is obligatorily paired with magne-

---

mitochondria (M) which are perpendicular to the basement membrane and parallel to the plasmalemmal infoldings (free arrow). In the upper part of the cell is a multivesicular body (MvB), a Golgi apparatus (Go), and an autophagic vacuole (AV). Near the basement membrane a coated caveolus (CC) is seen. Numerous collagen fibrils (CF) appear in the interstitium. RER, rough-surfaced endoplasmic reticulum; JC, junctional complex; N, nucleus. $\times 11{,}250$.

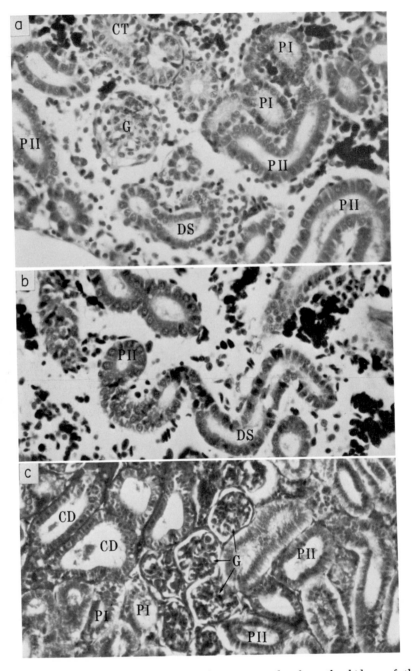

Fig. 33. (a) and (b) These two photomicrographs show the kidney of the rainbow trout, *Salmo gairdneri*. Glomeruli (G), first proximal segment (PI), second

sium, calcium, and sodium when their cumulative concentration exceeds that of sulfate, phosphate, and bicarbonate. Pitts (1934) noted that the urine phosphate diminished as the urine chloride rose in concentration. The concentrations of sodium and potassium in urine are normally low and independent of the GFR.

Urinary phosphate derives entirely from food since it is virtually absent from seawater. Its concentration is highly variable, ranging from practically zero to nearly 140 mmoles/liter (Edwards and Condorelli, 1928; Marshall and Grafflin, 1928, 1933; H. W. Smith, 1930b; Pitts, 1934; Grafflin and Ennis, 1934; Grafflin, 1936a; Forster and Berglund, 1956; Hickman, 1968b; Chester Jones et al., 1969). These great differences presumably relate to the amount of dietary phosphorus and overall nutritional state at the time urine was collected, although this question has not been directly studied.

The U/P ratio for calcium is always greater than one, but unlike magnesium and sulfate, calcium is excreted by other, unknown, routes as well as by the kidney. In the southern flounder, an average 70% of the calcium swallowed with seawater was absorbed from the intestine, as compared to 15.5% of the magnesium and 11.3% of the sulfate, but only 11.4% of the absorbed calcium was excreted via the kidney (Hickman, 1968c). The remainder may be removed across the gills or elsewhere across the body surface, especially the fins, as suggested by the experiments of Mashiko and Jozuka (1964) with the marine teleost *Duymaeria flagellifera* (see chapter by Conte, this volume).

b. Precipitates in the Urine. An expected consequence of the high concentration of divalent ions in the urine of marine teleosts at normal ranges of urine pH is the appearance of sediments in the urine and urinary passages (Guitel, 1906; Grafflin and Ennis, 1934; Grafflin, 1936a; Pitts, 1934; Lahlou, 1967; Hickman, 1968b). Many combinations of the electrolytes present in urine will yield salts with low solubility products, but only two, $MgHPO_4 \cdot 3H_2O$ in the longhorn sculpin (Pitts, 1934) and $CaHPO_4 \cdot 2H_2O$ in the southern flounder (Hickman, 1968b) have been positively identified. It is probable that calcium and magnesium carbonates and oxalates also appear as precipitates in marine fish urine. Guitel (1906) made detailed observations of phosphate-containing calculi in the urinary passages (tubules and archinephric ducts) of clingfishes

---

proximal segment (PII), distal segment (DS), and collecting tubule (CT). In (b) a transition between a second proximal segment (PII) and a distal segment (DS) is shown. ×360. (c) General view of kidney of the marine stickleback, *Gasterosteus aculeatus*, showing cluster of glomeruli (G), first proximal segments (PI), second proximal segments (PII), and collecting ducts (CD). ×360.

Table IX

Plasma solute	Seawater conc. (mmole/liter)	Plasma conc.	% Ultra-filterable	Urine conc. (mmole/liter)	U/P ratio[b]	Clearance (ml/hr/kg)	High GFR 2 ml/hr/kg Filtered[c] (μmole/hr)	Net reabsorbed (μmole/hr)	Net secreted (μmole/hr)	Excreted (μmole/hr)
Sodium	460.7	174.2	78.4	17.1	0.098	0.029	273.1	268.0	—	5.13
Potassium	9.8	2.86	88.0	1.42	0.49	0.15	5.03	4.61	—	0.42
Calcium	10.0	2.73	18.5	19.3	7.07	2.12	1.01	—	4.78	5.79
Magnesium	52.5	1.07	54.6	133.4	124.7	3.74	1.17	—	38.8	40.0
Chloride	537.3	145.9	91.1	120.6	0.83	0.25	265.8	229.6	—	36.2
Sulfate	27.7	0.19	—	68.5	360	108.4	0.38	—	20.2	20.6
Phosphate	0.00025	2.73	48.7	9.6	3.52	1.05	2.66	—	0.22	2.88
H₂O	—	937.9	—	937.6	—	—	1.88	1.59	—	0.29
pH	8.10	7.82	—	7.40	—	—	—	—	—	—
Osmolality	972	318	—	304	—	—	—	—	—	—
Totals (μmole/hr)	—	—	—	—	—	—	549.2	502.2	64.0	111.0

[a] Values from Hickman (1968b).
[b] Ratio uncorrected for protein binding and Donnan effect.
[c] Quantity filtered = plasma concentration (mmole/liter) × % ultrafilterable/100 × inulin clearance (ml/hr/kg).

(family Gobiesocidae), which as adults, have both pronephric and mesonephric kidneys. Calculi were always present, were highly variable in size, shape, color, and texture, and were occasionally sufficiently large to partly block the archinephric duct, causing dilation of the lumen upstream from the occlusion. Trump et al. (1968) have noted dense masses of microcrystalline aggregates of precipitated calcium in the endoplasmic reticulum of the second proximal segment of the southern flounder kidney (Figs. 31a–c). Their presence suggests that at least some of the urine sediments are formed intracellularly and are secondarily moved into the tubular lumen. The normally acid pH of the urine of marine teleosts does not completely prevent the formation of calcium and magnesium precipitates but undoubtedly does help to reduce their concentration in the urine.

c. Tubular Secretion of Divalent Ions. The urine concentration of magnesium sulfate and often, phosphate, are greater than their plasma concentrations. The U/P ratios for magnesium and sulfate may reach 100–300, leaving no doubt that these ion species are actively secreted into the tubular urine (Fig. 47b). Most magnesium and sulfate appearing in the urine are secreted because the quantity of the diffusable form of these ions in the filtrate of glomerular marine fishes is very low. In the southern flounder, for example, usually less than 3% of the excreted mag-

Average Renal Excretion of Ions and Other Solutes by the Southern Flounder,
Paralichthys lethostigma, at Three Rates of Glomerular Filtration[a]
Urine flow 0.3 ml/hr

Low GFR 0.3 ml/hr/kg				Zero GFR 0.0 ml/hr/kg			
Filtered[c] (μmole/hr)	Net reabsorbed (μmole/hr)	Net secreted (μmole/hr)	Excreted (μmole/hr)	Filtered[c] (μmole/hr)	Net reabsorbed (μmole/hr)	Net secreted (μmole/hr)	Excreted (μmole/hr)
40.97	35.8	—	5.13	0	0	5.13	5.13
0.75	0.33	—	0.42	0	0	0.42	0.42
0.15	—	5.64	5.79	0	0	5.79	5.79
0.18	—	39.9	40.0	0	0	40.0	40.0
39.9	3.7	—	36.2	0	0	36.2	36.2
0.057	—	20.5	20.6	0	0	20.6	20.6
0.40	—	2.48	2.88	0	0	2.88	2.88
0.281	—	0.011	0.292	0	0	0.292	0.292
—	—	—	—	—	—	—	—
82.4	39.8	68.4	111.0	0	0	111.0	111.0

nesium and sulfate is filtered (Hickman, 1968b), although this fraction may rise during an unusual combination of high GFR and very low urine flow.

Experiments by Bieter (1931a, 1933, 1935), Berglund and Forster (1958), and Hickman (1968b) have shown that the tubular electrolyte secretory mechanism responds directly and specifically to increases in the blood concentration of magnesium, calcium, and sulfate. The injection of a salt containing one of these divalent electrolytes into the circulation of either the aglomerular toadfish, *Opsanus tau* (Bieter, 1931a, 1933), or the aglomerular goosefish, *Lophius americanus* (Berglund and Forster, 1958), caused a significant diuresis. The infusion of $MgCl_2$ into the circulation of the glomerular southern flounder, *Paralichthys lethostigma*, caused an increase in both the urine magnesium concentration and the urine flow (Fig. 39). The activity of the divalent ion transport system appears to be governed directly by the quantity of Mg^{2+} and SO_4^{2-} in the peritubular blood and presumably sensed at the vascular surfaces of the tubular cells. Bieter (1935) reported that the injection of $MgSO_4$ into the toadfish caused an ipsilateral diuresis. The urine flow of the opposite kidney increased significantly only if the injected dose was so large that some of the salt reached it by recycling through the systemic circulation. This experiment appears to rule out the par-

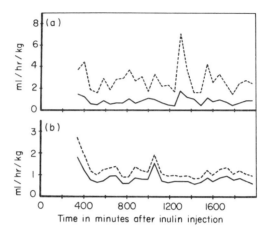

Fig. 34. Normal variations in GFR (dashed line) and urine flow (solid line) in (a) northern pike, *Esox lucius*, and (b) white sucker, *Catostomus commersonii*. The fish were held in a flow-through chamber and urine collected by catheter placed in the urinary bladder; GFR was determined from the clearance of C-14 inulin. From Hickman (1965).

ticipation of any systemic hormone which would be expected to affect both kidneys alike.

Efforts to characterize the specificity and capacity limitations of the divalent ion secretory mechanism have not been entirely conclusive. The

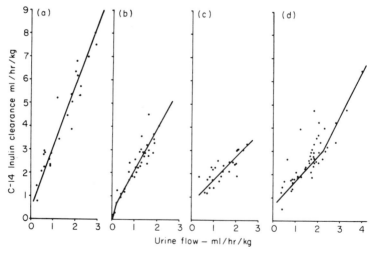

Fig. 35. Relationship between inulin clearance and urine flow in white sucker and northern pike. The near-linear relationship indicates that the percentage tubular reabsorption of filtered water is nearly constant in individual fish despite wide variations in filtration rate. (a)–(c), Sucker; (d) pike. From Hickman (1965).

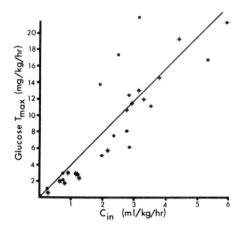

Fig. 36. Relationship between inulin clearance (GFR) and glucose transport maximum in the white sucker. The data were obtained at 4°C. The four different symbols represent separate experiments on four different fish. From Mackay (1967).

kidney does not appear to actively transport divalent ions, such as strontium and mercury, which are not normally excreted by the kidney (Bieter, 1933). From their study of tubular secretion in the goosefish, Berglund and Forster (1958) concluded that there were at least two separate divalent ion transport systems, one for cations and another for anions. In this species, the injection of $MgCl_2$ depressed the excretion of calcium and sodium thiosulfate depressed the excretion of sulfate.

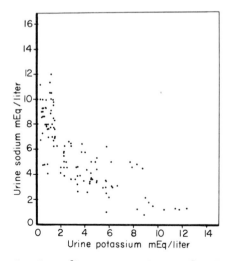

Fig. 37. Variation in urine sodium concentration as a function of urine potassium concentration in the white sucker. From Hickman (1965).

184 CLEVELAND P. HICKMAN, JR., AND BENJAMIN F. TRUMP

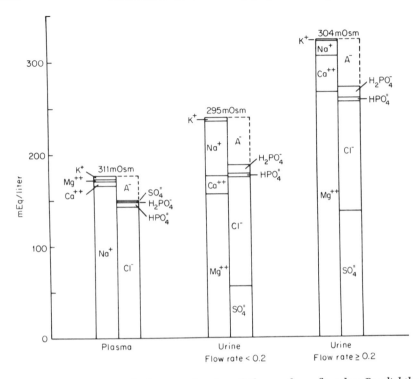

Fig. 38. Ionograms of plasma and urine of the southern flounder, *Paralichthys lethostigma.* Magnesium, sulfate, and chloride are the major electrolytes in the urine of marine fishes. In the southern flounder, sodium was more concentrated in the urine of fish having the lowest urine flows (flow rate <0.2 ml/hr/kg). The proportions of monobasic and dibasic orthophosphate in the urine are determined by the pH which is usually acidic. The unknown anion component of urine is probably mostly bicarbonate; organic anions, including protein, and trace quantities of inorganic anions comprise the remainder. Nonelectrolytes such as urea are in relatively low concentration. The two urine ionograms differ greatly in total equivalent concentration but not greatly in total osmotic concentration, because the high flow rate urine sample to the right contains proportionately more divalent ions. Because of the abundance of divalent ions having low activity coefficients, the sum of the molar concentrations of urine electrolytes exceeds the total urine osmolality. From Hickman (1968b).

However, the infusion of $MgCl_2$ into the southern flounder, while causing a drop in the urine concentrations of calcium and sulfate, did not reduce their excretion because of a compensatory increase in urine flow (Fig. 39). The capacity limitations of the divalent ion transport system, although not yet defined, are far in excess of the loads normally transported by the kidney. During osmotic diuresis or exposure to concentrated seawater the excretion of magnesium sulfate and calcium may increase several fold.

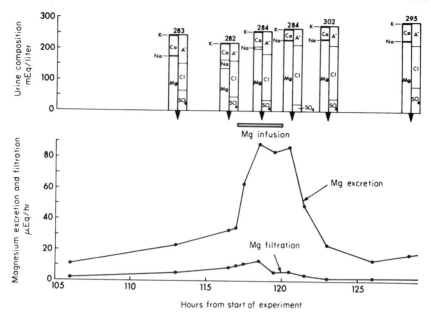

Fig. 39. Response of the kidney of the southern flounder to intravascular infusion of magnesium (as MgCl₂). Urine composition is shown in the ionograms; samples shown were collected at the times indicated by the pointer beneath each ionogram; the numbers above indicate urine osmolality. Magnesium infusion, which began at hour 117 and ended at hour 120, promoted a large rise in the excretion of magnesium. Since the magnesium filtration declined coincident to a drop in the GFR during infusion, increased tubular secretion of magnesium entirely accounted for the increased magnesium discharge. Note that the urine concentrations of calcium and sulfate were depressed during the magnesium diuresis. From Hickman (1968b).

The U/P ratio for phosphate is usually greater than unity and may even exceed 50, indicating that it undergoes net secretion into the tubular lumen (Fig. 47b). Phosphate excretion is probably associated with the phosphate content of the diet and rate of release from tissue phosphate stores, but it is not known whether urinary phosphate is derived directly from plasma inorganic phosphate or from some organic phosphate precursor. Phosphate excretion in the glomerular longhorn sculpin may spontaneously increase several fold following the disturbance of handling; this may represent phosphate flushed into the circulation from muscle when this usually slugglish species suddenly becomes active (Grafflin, 1936a).

In the aglomerular toadfish, *Opsanus tau,* and the pauciglomerular goosefish, *Lophius americanus,* phosphate excretion appears to be independent of the blood inorganic phosphate level. Marshall and Grafflin (1933) found that large intravenous or intramuscular injections of phosphate did not increase renal excretion of phosphate, even though the

Table X

Average Renal Excretion of Ions and Other Solutes by
Lophius americanus (= *piscatorius*)
Urine flow[a] = 0.44 (0.35–0.52) ml/hr/kg

Plasma solute	Seawater conc.[b] (mmole/liter)	Plasma conc.[c] (mmole/liter)	Urine conc.[c] (mmole/liter)	U/P ratio	Net secreted and excreted (μmole/hr)
Sodium	470.2	185	11	0.059	4.8
Potassium	9.96	5.1	1.8	0.35	0.79
Calcium	10.24	3.2	7.2	2.25	3.2
Magnesium	53.6	2.5	137	54.5	60.3
Chloride	548.3	153	132	0.86	58.1
Sulfate	28.2	1.2[d]	42	35.0	18.5
Phosphate	0.003	5.3[e]	1.5	0.28	0.66
Glucose	—	—	—	—	—
Urea	—	0.30	0.57	1.9	0.25
Ammonia	—	0.23	13.7[f]	59.5	6.0
TMAO	—	8.2	12.5[f]	1.5	5.5
Creatinine	—	0.07	0.33[f]	4.7	0.15
Creatine	—	0.05	4.7[f]	93	2.1
Uric acid	—	0.03	0.005	0.16	0.002
pH	—	7.07[d]	6.56[d]	—	—
Osmolality	991	452	406	—	—

[a] From Shannon (1938b).
[b] From Table 36, Sverdrup et al. (1942).
[c] From Potts and Parry (1964), Table VII. 5, after Brull and Nizet (1953), except as noted otherwise.
[d] From Forster and Berglund (1956).
[e] From H. W. Smith (1929c).
[f] From Brull and Cuypers (1954a).

blood inorganic phosphate level might be nearly tripled. Yet the tubules of aglomerular teleosts unquestionably have the capacity to secrete phosphate because the phosphate U/P ratios may frequently exceed unity, although never reaching the ratios observed in glomerular fishes.

Sodium and potassium appear in the urine of aglomerular and pauciglomerular forms but only at U/P ratios less than one (Edwards and Condorelli, 1928; Marshall and Grafflin, 1928; Forster and Berglund, 1956). These ions probably passively leak into the tubular urine from the peritubular blood (Fig. 47b). The U/P ratio for chloride frequently exceeds one when it is obligatorily paired with magnesium and other cations in the urine. There is no evidence of active tubular secretory mechanisms for any of the monovalent ions except H⁺.

d. Secretion of Hydrogen Ion. The urine of marine teleosts is usually, although not invariably, acid (W. W. Smith, 1939a; Brull *et al.*, 1953;

Table XI

Source List to Articles Giving Original Analyses of the
Composition of Urine of Marine Teleost Fishes

Species	References
Family Anguillidae	
Anguilla rostrata (American eel)	H. W. Smith (1930b)
Anguilla anguilla (European eel)	Sharratt *et al.* (1964); Chester Jones *et al.* (1969)
Family Muraenidae	
Muraena sp. (moray)	Edwards and Condorelli (1928)
Family Gadidae	
Merluccius bilinearis (silver hake)	Pitts (1934)
Gadus callarias (cod)	Marshall and Grafflin (1928); H. W. Smith (1929b)
Family Syngnathidae	
Syngnathus sp. (pipefish)	Edwards and Condorelli (1928)
Hippocampus sp. (seahorse)	Edwards and Condorelli (1928)
Family Blenniidae	
Xiphister atropurpureus (blenny)	Evans (1967)
Family Stichaeidae	
Cryptacanthodes maculatus (wrymouth)	Grafflin (1936b)
Family Cottidae	
Hemitripterus americanus (sea raven)	Grafflin (1936b)
Myoxocephalus octodecimspinosus (longhorn sculpin)	H. W. Smith (1930b); Marshall (1930); Grafflin (1931b, 1935, 1936a,b); Grafflin and Gould (1936); Clarke (1934); Forster (1953); Hodler *et al.* (1955)
Myoxocephalus scorpius (shorthorn or daddy sculpin)	Pitts (1934); Grafflin (1936b); Forster (1953)
Family Pleuronectidae	
Platichthys flesus (flounder)	Lahlou (1967)
Pseudopleuronectes americanus (winter flounder)	Marshall (1930); Grafflin (1936a,b)
Paralichthys lethostigma (southern flounder)	Hickman (1968b,c)
Glyptocephalus cynoglossus (witch flounder)	Pitts (1934)
Limanda ferruginea (yellow tail)	Pitts (1934)
Family Tetrodontidae	
Sphaeroides maculatus (puffer)	H. W. Smith (1929b)

Table XI (*Continued*)

Species	References
Family Batrachoididae	
Opsanus tau (toadfish)	Marshall (1930); Marshall and Grafflin (1933); Grafflin (1931b); Shannon (1938a);Lahlou *et al.* (1969)
Family Lophiidae	
Lophius americanus (= *piscatorius*) (goosefish)	Denis (1914); Sulze (1922); Edwards and Condorelli (1928); Marshall and Grafflin (1928); Grollman (1929); H. W. Smith (1929b, 1930b); Marshall (1930); Pitts (1934); Grafflin (1936b); Brull *et al.* (1953); Forster (1953); Berglund and Forster (1958); Forster and Berglund (1956)

Hodler *et al.*, 1955; Forster and Berglund, 1956; Fanelli and Nigrelli, 1963; Hickman, 1968b). The urine pH of marine sculpins is almost completely unaffected by injections of inorganic phosphate or bicarbonate (H. W. Smith and J. H. Jarofsky, reported by W. W. Smith, 1939a) or the carbonic anhydrase inhibitor acetozolamide (Hodler *et al.*, 1955), suggesting that similar to the elasmobranchs, but unlike freshwater teleosts, the urine pH of marine teleosts is fixed on the acid side. However, Hickman (1968) observed that the urine pH of the euryhaline southern flounder in seawater varied considerably, ranging from 5.68 to 8.24, tending to be more alkaline at higher rates of urine flow. In individual fish showing substantial variation in urine phosphate excretion, urinary pH was lowest during those periods when the urine concentration of phosphate was highest (Hickman, 1969). This suggests that phosphate was secreted in the acid form ($H_2PO_4^-$). This may be a necessary adaptation effected to avoid excessive precipitation of insoluble Ca^{2+} or Mg^{2+} aggregates of phosphate or carbonate in the tubules. Hydrogen ions and/or acid phosphate ions are probably injected in the brush border regions of the tubule since studies of phenol red secretion indicate a low intraluminal pH in these regions.

    e. Tubular Transport of Organic Anions. Studies of organic acid transport systems of the fish nephron have been of great importance in the development of our present notions regarding renal function as a whole. They have also provided several important methods for evaluation of renal function in intact animals and man.

The prototype of organic acid transport systems is that which secretes phenol red. The observation that *Lophius americanus* and other aglomerular fish could excrete phenol red provided the first firm evidence that tubular secretion did, in fact, exist as an excretory mechanism (Marshall, 1930). Our present knowledge of these transport processes represents the direct development of these early observations of Marshall and others. Much of the early literature concerning the transport of organic compounds appears in the review by Forster (1961).

The characteristics of transport systems have been studied for organic anions, organic cations, and certain other organic compounds. It has been suggested, moreover, that close relationships exist between some of these transport systems and those for the transport of certain inorganic molecules.

The precise functional significance to the animal of these transport systems remains essentially unknown. It does, however, seem unlikely that they were evolved to secrete foreign compounds such as phenol red, Diodrast, or PAH. There are data which suggest a role for these systems in the excretion of certain organic nitrogenous compounds, for transcellular substrate transport and a relationship to the divalent ion pumps.

Of all the transport mechanisms, that for organic anions has been most widely studied both *in vivo* and *in vitro*. Many of the studies of organic anion transport are based upon the observation originally made by Forster (1948) that isolated nephrons from the flounder actively transport and accumulate intraluminally various organic anions (Fig. 40). The most studied ones are chlorphenol red, because of its convenience for light microscopic observation, and Diodrast. More recently these studies have been extended to include ultrastructural observations of the relationship between cellular fine structure and transport mechanisms and to study the relationship between cell injury and kidney tubular function (Trump and Bulger, 1967, 1968a,b). It has now been demonstrated that isolated nephrons from the flounder are capable of concentrating dye in the lumen several thousand-fold as compared with the incubation medium (Kinter, 1966) (Figs. 40a and b). The temperature coefficient for PAH transport is 2.0 in fishes and the kinetics are characteristic of a first-order reaction.

A variety of studies have suggested that the organic acid transport system is energy dependent and probably related to intracellular production or concentration of high energy compounds possibly ATP. A series of investigations in our laboratory (B.F.T.), involving a wide spectrum of metabolic inhibitors, indicates that the transport system in

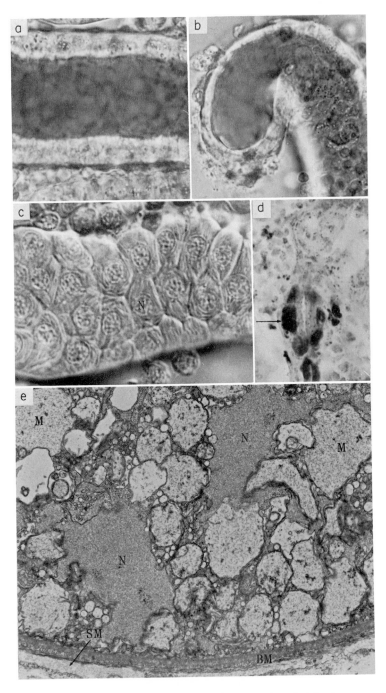

Fig. 40. (a) Isolated southern flounder nephron incubated 4 hr in oxygenated

the flounder, *Paralichthys lethostigma*, is dependent upon respiration and oxidative phosphorylation but is not inhibited by inhibitors of protein synthesis, inhibitors of nucleic acid systhesis, or inhibitors or activators of lysosomal enzymes. Furthermore, agents such as the polyene antibiotics, that directly modify the structure and function of the plasmalemma, are associated with marked inhibition of dye accumulation. Agents that affect dye transport are also associated with marked alterations in cellular ultrastructure (Fig. 40e).

It was noted early by Forster and colleagues (Forster, 1948; Forster and Hong, 1958; Forster *et al.*, 1954) and by Wasserman *et al.* (1953) that two stages in dye transport could be defined. Step I is from the peritubular fluid across the base of the cell and Step II is from the intracellular compartment to the tubular lumen. It was observed that Step I was dependent upon potassium and Step II on calcium. It appears that in isolated tubules, Step I is limiting. When tubules are incubated in media devoid of calcium, dye accumulates intracellularly. Under normal conditions this was not observed. When potassium is removed from tubules in media devoid of calcium, intracellular dye accumulation is also inhibited, indicating the dependence of Step I on potassium. Recently, these steps have been related to changes in cell ultrastructure (Bulger and Trump, 1969c). In tubules incubated in calcium-free media, marked alterations of the cell apex occur. These include disorganization and disappearance of junctional complexes and marked distortions in the contour of the microvillous border (Figs. 41 and 42a). On the other hand, tubules incubated in potassium-free media show simplification and disappearance of the complex basilar cell infoldings (Fig. 42b). It thus

Forster's buffer containing chlorphenol red. Observe the intraluminal dye concentration. This is the black area occupying the tubular lumen and was magenta in the original preparation. Note the appearance of the epithelial cells lining the tubule. Cell junctions and traces of the brush border can be observed. ×750. (b) Same preparation showing that the ends of the dissected nephrons seal off. ×750. (c) Goldfish, *Carassius auratus*, tubule functioning *in vitro* and photographed using Nomarski phase-interference optics. Note the appearance of the nuclei (N) and the striations which represent mitochondria and infolded plasma membrane along the cell base which is viewed in optical tangential section. ×750. (d) Afferent arteriole from southern flounder showing acid phosphatase activity in the juxtaglomerular granules (free arrow). This activity is found within the granules and thus they may be interpreted as forms of primary lysosomes. ×225. (e) Isolated tubule from English sole incubated for 4 hr in Forster's medium containing 3×10^{-3} M potassium cyanide. Note the marked changes in ultrastructure. The nuclei (N) are very irregular and have lost most of their DNA. The mitochondria (M) are extremely swollen and show disorganization of internal membranes and flocculent intramatrical deposits probably representing denatured protein. The rest of the cytoplasm is filled with vesicular profiles that derive from altered endoplasmic reticulum and lysosomes. Basement membrane (BM) and surrounding smooth muscle cells (SM) can be seen. ×7125.

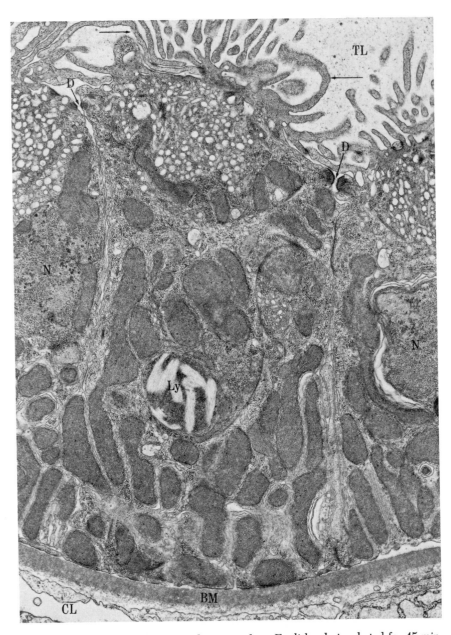

Fig. 41. Isolated second proximal segment from English sole incubated for 45 min in calcium-free Forster's medium showing marked changes. The tubular lumen (TL) is toward the top of the picture. The basement membrane (BM) is toward the bottom. Note the disposition of the junctional complex and observe the desmosomes (D) which are undergoing wide separation. All of the structures in the cell apex are distorted and the microvilli are curved (free arrows) forming flowerlike arrangements. (N) nucleus, (Ly) lysosome, (CL) capillary lumen. ×10,250. Courtesy of R. E. Bulger.

appears that the two steps, postulated on the basis of transport data, may have as their basis ultrastructural alterations in the configuration of these two respective regions.

It was observed very early that various organic anions act as competitive inhibitors for the transport system, as evidenced by experiments in which the sum of two or more maximum transfer rates is less than the sum of each when calculated singly. Furthermore, the tubular maxima for different compounds are widely different. In the aglomerular goosefish, the Tm_{PAH} is 25 μmoles/kg/hr and the $Tm_{DIODRAST}$ is 5 μmoles/kg/hr. Diodrast, however, apparently has the greater affinity for the transport system and acts as an effective competitive inhibitor. In general, effective competitors have high affinities for the carrier which they easily saturate but are only slowly transported themselves. In a series of phenosulfonphthaleins, the more effective competitors had slow rates of transfer, were characterized by intracellular accumulation, and had a great affinity for the carrier. Forster has pointed out that these characteristics are similar to those of carrier-mediated transport systems in erythrocytes under conditions approximating saturation.

Kinter (1966) has made detailed studies of the kinetics of chlorphenol red transport in isolated flounder tubules and has concluded that the influx data are consistent with Michaelis–Menten saturation kinetics. He also evaluated separately the influx and efflux processes. Two important characteristics of chlorphenol red efflux, which are also observed in other vertebrate renal tissue preparations, are (1) movement down the concentration gradient proportional to the intraluminal concentration of chlorphenol red with no evidence of competition from the luminal side, and (2) biphasic action of competitor anions present on the medium side. The latter is based on the observation of the enhancement of chlorphenol red efflux by low concentrations, and depression by high concentration, of competitors in the medium.

These efflux data are of great interest, particularly with respect to countertransport phenomena which may provide a partial explanation of the significance of these organic anion transport systems in terms of coupled transport processes with ions such as sodium or particularly with other anions which may eventually provide the basis for greater understanding regarding the control mechanisms involved in secretion vs. reabsorption. Unfortunately, however, the presently available data have not permitted a satisfactory explanation of competitor enhancement of anion efflux. Three explanations have been proposed: (1) run-out from lumens as leakage by simple back diffusion; (2) displacement from postulated intracellular binding or trapping mechanism by competitors with greater affinities for the latter; and (3) exchange diffusion. Kinter's

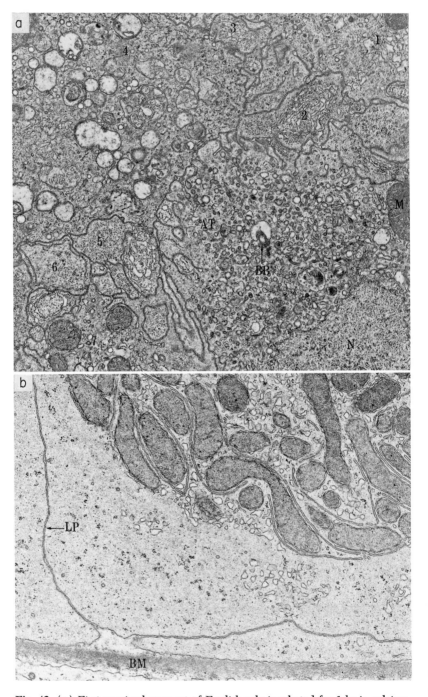

Fig. 42. (a) First proximal segment of English sole incubated for 1 hr in calcium-

data (1966) on isolated flounder tubules could not support any of these proposed explanations.

Kinter's data do, however, establish that efflux behaves as a first-order reaction and that competitors do act from the medium side to enhance or depress dye efflux. The work with flounder tubules also clearly shows that organic anions almost certainly move back across the cells rather than escaping from the ends [which seal off in these tubule preparations (Fig. 40b)].

The nature of the influx process has not been defined although it is known that it follows the pattern of classical Michaelis–Menten kinetics. Although, as Kinter states, modern membrane carrier theory has evolved as the most satisfactory explanation for such transport, conclusive identification of the underlying process, e.g., isolation of carrier, has not been achieved. Furthermore, the precise mechanism by which metabolic energy can be coupled to many specific carriers for transport is not clear. Although the suggestion by Crane (1965), of a series of mobile double-site exchange carriers which bind both the sodium ion and the second molecule, is very attractive and, although data suggesting such a proportionality in frog have been obtained, this has not been confirmed in mammal or flounder. Comparisons of *in vivo* and *in vitro* data suggest that in the flounder influx is the primary determinant of tubular secretion.

Recent ultrastructural evidence on the isolated flounder nephron indicates that organic anions enter the cell base, possibly through the complex infolded plasma membranes. Work with protein synthesis inhibitors indicates that a carrier protein, if present, must be relatively stable, and not synthesized on demand, since the process proceeds unabated in the presence of protein synthesis inhibition. The normal state of the basilar plasma membranes evidently is dependent on the presence of potassium. The interference with transport induced by modification of plasmalemmal composition, as with amphotericin, suggests that the binding site may reside within the membrane itself. It is of interest that amphotericin does not exert a primary effect on mitochondrial mem-

free Forster's buffer. After this time the relationships of the cell apex have completely changed. The junctional complexes have completely dissociated and disappeared and the lumen is filled with interlocking apical cell processes such as those numbered 1–7 in the micrograph. Other apical components such as the apical tubules (AT) have moved toward the base of the cell. This type of cell shows marked intracellular but no intraluminal dye accumulation. Nucleus (N), mitochondria (M), and basal body (BB). ×10,000. (b) Isolated tubule from English sole incubated for 1 hr *in vitro* in Forster's buffer free of potassium. Note the basilar part of the cytoplasm adjacent to the basement membrane (BM). In this region, only the lateral plasmalemma (LP) can be seen. In contrast to control preparations, this zone is largely clear, and the mitochondria and infolded membranes appear to be greatly displaced toward the middle of the cell. Rough- and smooth-surfaced endoplasmic reticulum are visible. ×14,800. Courtesy of R. E. Bulger.

branes and, in the toad bladder, appears to facilitate sodium transport by increasing sodium influx. The morphology of the base of the tubule suggests that mitochondria, which are in proximity to the membrane infoldings, are probably strategically located to supply ATP to the basilar anion "pump." The presumed relationship between mitochondria, plasmalemmel infoldings, and Na^+ transport in the opposite direction brings to mind Crane's hypothesis mentioned above.

After entering the cell, the transported anion is evidently transferred to the cell apex, either within the cell sap or in channels such as the endoplasmic reticulum. At the cell apex the organic anions are transferred to the lumen, across the membrane in a calcium-dependent step. This step is clearly related to the morphological specializations of the apex since incubation in calcium-free medium is associated with pronounced changes in the morphology of that region. The lysosomes do not appear to be involved in active transport of organic anions. They may, however, represent the intracellular sites of accumulation of slowly transported materials such as neutral red.

f. *Excretion of Nitrogenous End Products.* Several nitrogenous wastes appear in the urine of both glomerular and aglomerular marine teleosts, but these account for only a small percentage of the total nitrogen excreted by fish. Ammonia, urea, and TMAO, the major nitrogenous end products of the teleosts, are excreted largely across the branchial surfaces (see chapter by Forster and Goldstein, this volume). It is probable, nevertheless, that the kidney is essential for the excretion of less diffusable materials such as creatine, creatinine, and uric acid. In *Lophius* the U/P ratios are characteristically above one for all of these materials except uric acid (Table X). However, it is unlikely that all are actively secreted by the tubular epithelium (Grollman, 1929; H. W. Smith, 1929b; Grafflin and Gould, 1936; Marshall and Grafflin, 1932; Shannon, 1938a; Brull and Nizet, 1953; Brull and Cuypers, 1954a). The small concentration gradients for urea and creatinine (U/P ratios less than five) could result from passive inward diffusion in the proximal tubule and subsequent concentration by removal of salts and water in the distal regions of the nephron and in the urinary bladder. Malvin and Fritz (1962) found no evidences of active urea secretion by the kidney of goosefish. Urine/plasma ratios for urea in goosefish injected intramuscularly with urea averaged about unity, and they attributed the high U/P ratios reported by Grafflin (1936b) to erroneous sampling procedure.

Creatine and ammonia are so strongly concentrated in the urine that they must be either secreted by some carrier-mediated process or formed *de novo* in the tubular epithelium (Fig. 48). There are indications that deamination systems exist in the kidneys of teleosts (see Chapter by Forster and Goldstein, this volume) suggesting that ammonia may be formed in the tubular cells from amino and amide groups of amino acids. Ammonia could then diffuse into the tubular urine and combine with H^+ to be carried away as ammonium ion (NH_4^+). Preformed free ammonia in the peritubular blood may also enter the urine, its diffusion being favored by the continuous combination with H^+ and removed in the urine as NH_4^+. Ammonium ion could also be secreted into the lumen in exchange for Na^+ by a process similar, if not identical, to that demonstrated in the gill (Maetz and Romeu, 1964; see also chapters by Conte and Forster and Goldstein, this volume), but this possibility has not been examined experimentally.

g. Tubular Reabsorption of Electrolytes. Tubular reabsorption of sodium, potassium, and chloride which leak or are filtered into the tubule occurs in all marine teleosts. In glomerular forms, the quantity of NaCl filtered and reabsorbed may greatly exceed the quantity of $MgSO_4$ secreted. A comparison of the amounts of electrolytes filtered and reabsorbed by southern flounder at three different filtration rates, are shown in Table IX. When the GFR is nearly maximal (2 ml/hr), 98% of the sodium, 86% of the chloride, and 92% of the potassium are reabsorbed. At a lower GFR, the filtration of monovalent ions drops, and their reabsorption is decreased accordingly. These figures clearly illustrate why a high rate of glomerular filtration is of no evident benefit to marine fish. The greater the GFR, the greater must be the energy-dependent tubular reabsorption of NaCl while Mg^{2+} and SO_4^{2-} secretion proceeds mostly unabated. These facts notwithstanding, a relatively small proportion of marine teleosts have disposed of glomeruli (Table IV). Consequently, the majority must have the tubular capability to actively reabsorb filtered NaCl to a greater or lesser degree. To this must be added the unknown amount of sodium that leaks into the tubule (Fig. 47b). The urine of aglomerular fish always contains sodium, not infrequently at concentrations in excess of 100 mmoles/liter (Marshall and Grafflin, 1928; Edwards and Condorelli, 1928; Forster and Berglund, 1956). Thus, while the urine is largely a product of tubular divalent ion secretion, the reabsorption of monovalent ions originating from the filtrate and/or tubular leak may be quantitatively much greater and require the expenditure of substantially more energy than the secretion of divalent ions. This certainly is true if the GFR exceeds the urine flow, and it may also be true in completely aglomerular forms. This fact becomes important when one attempts to assign specific functions to the different regions of the nephron.

It is especially interesting that the urine of both glomerular and aglo-merular marine teleosts is nearly always a few mOsm/liter hypoosmotic to plasma, although exceptions have been reported in euryhaline teleosts (see below). In aglomerular fish the primary secretion of divalent ions, together with the inward leakage of NaCl, should produce a hyperos-motic urine, because water is drawn into the tubular lumen secondary to the establishment of a transtubular osmotic gradient. If there were absolutely no resistance to water movement, the formative urine under these conditions would at best be blood isosmotic. Since in fact, the urine is hypoosmotic, the reversal of the osmotic pressure gradient must occur in the distal regions of the nephron. Here, the final abstraction of sodium, chloride, and potassium would create an outwardly directed net ionic flux favorable to the withdrawal of water. The magnitude of the final drop in urine osmolality would be determined by the transepithelial re-sistance to water movement.

If divalent ion secretion and monovalent ion reabsorption are region-ally separated as suggested, any condition that severely mitigates or abolishes distal ion reabsorption should permit the excretion of a hyper-osmotic urine. Such instances have been reported. Hickman (1968b) noted that the urine of the southern flounder was often 5–40 mOsm/liter hyperosmotic to the blood for several hours after handling and catheter-ization. Urine flow and urine sodium concentration increased markedly during this interval, suggesting that distal sodium reabsorption was in-terrupted. Especially noteworthy is the report that the plains killifish, Fundulus kansae, produces a urine more than 100 mOsm/liter more concentrated than the plasma during the first few days following transfer to seawater from freshwater (J. G. Stanley and Fleming, 1964). Con-sistent with the hypothesis above, urinary sodium and potassium were especially elevated during this interval. Other changes are probably as-sociated with the production of a hyperosmotic urine in this species and discussed in the section on euryhaline teleosts.

It must be admitted, however, that while these explanations can explain observations of both blood hyperosmotic and blood hypoosmotic urine in marine fish without the need to postulate active transport of water, one can scarcely remain satisfied until the structural and func-tional relationships within the cell are understood. In particular, the func-tion of the complex infoldings of the basilar plasmalemma of both proxi-mal and distal regions of the tubule (e.g., Figs. 21 and 22a) has yet to be satisfactorily established. Their compartmentalized kind of organiza-tion into deep basilar plates in proximity to mitochondria strongly sug-gests that the active transport of electrolytes occurs across their surfaces. Furthermore, it is possible that a hyperosmotic urine could be formed by the passive movement of water from luminal fluid to the peritubular blood in a sequence as postulated for certain vertebrates and invertebrates (Diamond, 1965; Berridge and Gupta, 1967). In this system, separated as it is from immediate contact with the extracellular fluid, extrusion of sodium and chloride into the folded plasmalemma compartments would create local osmotic gradients down which water would obligatorily follow.

Finally, to these proposed explanations of hyperosmotic urine formation must be added the presently unpopular possibility that water is actively transported in one or both directions across the tubular epithelium, since there are indications that pure water pumps exist in the animal kingdom (Beament, 1965; Potts and Parry, 1964).

h. Tubular Reabsorption of Glucose. Glucose normally is absent from the urine of both glomerular and aglomerular marine teleosts, or appears in trace amounts only (Marshall, 1930; Marshall and Grafflin, 1928, 1932; Lahlou, 1966; Malvin and Fritz, 1962; Malvin *et al.*, 1965). Lahlou (1966) reported that the blood glucose concentration of fed flounder, *Platichthys flesus*, was 44 ± 3.7 mg % (2.44 mmoles/liter) and that the urine was free of glucose. The Tm for glucose was found to vary with the GFR as a proportional relationship which could be expressed by the equation: $Tm_{glucose} = 2.34\ C_{in} + 0.04$. The proportionality was offered as evidence that the GFR of this species was determined by intermittent glomerular activity.

Contrary to the reports of earlier workers (Marshall and Grafflin, 1928; Marshall, 1930), glucose does appear in small amounts as a normal constituent of the urine of the aglomerular goosefish, *Lophius americanus* (Table X). Malvin *et al.* (1965) found that both glucose loading and phlorizin administration increased the renal excretion of glucose, indicating that glucose normally enters the tubule by diffusion and is actively reabsorbed. In this species, the total reducing substance greatly exceeds the glucose concentration in plasma by a factor of 3.5 and in urine by a factor of about 16 (Malvin *et al.*, 1965).

i. Uptake of Macromolecular Materials. The luminal surfaces of the tubular epithelial cells are constantly exposed to plasma proteins and other macromolecular materials that pass through the glomerular filter in glomerular forms (Bieter, 1931b). Recent evidence indicates that different mechanisms for handling this material have evolved in the various regions of the nephron.

In the first proximal segment, present only in glomerular forms, the macromolecules in the tubule lumen pass into the apical tubular invaginations at the base of the microvilli (Bulger and Trump, 1969b)

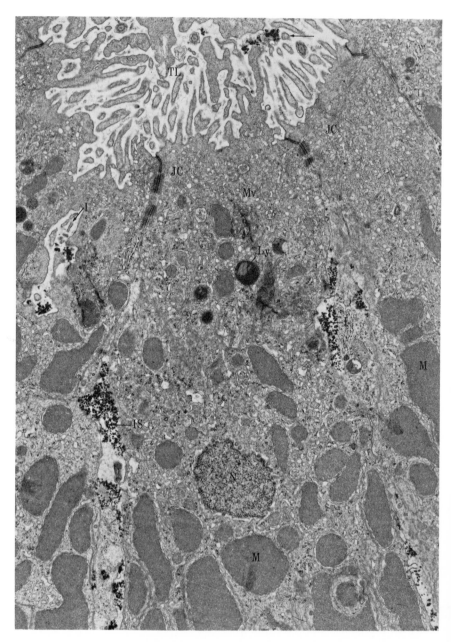

Fig. 43. Second proximal segment from English sole, taken a few minutes after retrograde injection of India ink into the archinephric duct. In this segment such particles (free arrow) as well as protein molecules are taken up from the tubular lumen (TL) in massive invaginations of the plasma membrane (I) which rapidly

(Figs. 11, 18, 19, and 29). From there the particles gain access to the large apical vacuoles, either through direct continuity of vacuoles and tubules or via membrane fusions. The apical vacuoles probably move into the cell center where they fuse with primary or preexisting secondary lysosomes (Figs. 11, 18, 28e, and 29). This fusion brings the contents of the vacuoles into contact with acid hydrolases such as acid phosphatase (Fig. 28e), cathepsin c, DNase, RNase, and polysaccharidases such as β-glucuronidase or glucosidase. These enzymes are capable of rapidly splitting most biologically occurring substances to their subunits such as amino acids, monosaccharides, and nucleotides. These can presumably enter the cellular pool for reutilization. This mechanism therefore potentially protects the animal against excessive loss of amino acids from filtered proteins. Essentially similar mechanisms exist in the mammalian proximal tubule (Ericsson, 1964).

It is of great interest that fish have evolved a second type of mechanism for handling macromolecular materials that is thus far not known in mammals. This mechanism, well developed in the second proximal segments in marine forms, but also present in other segments, constitutes a rapid, bulk transport system which permits rapid translocation of quanta of particles from the tubular lumen to the lateral intercellular spaces (Bulger and Trump, 1969b). Within minutes after the introduction of particles to the tubular lumen, enormous invaginations of the apical plasmalemma engulf the particles and transfer them, via fusion, to the lateral intercellular spaces where they accumulate (Fig. 43). In this system the particles are not brought into contact with the lysosomal enzymes, and thus digestion of particles such as proteins does not occur. Such a mechanism might serve to conserve biologically useful molecules such as antibodies or enzymes that find their way, through glomerular filtration or transtubular leak, into the filtrate. Indeed, Maack and Kinter (1969) observed rapid transport of active lysozymes from flounder tubule masses to incubation medium *in vitro*.

j. Urine Modification in the Urinary "Bladder." The terminal expansion of the archinephric duct (urinary "bladder") of fish may function to modify the kidney urine by moving electrolytes and water across the

---

fuse with the lateral membrane below the junctional complex (JC) releasing the particles into the lateral intercellular spaces (IS). This appears to constitute a rapid transport system for bulk transfer of macromolecular materials from the lumen to the extracellular space (Bulger and Trump, 1969b). Note also that the multivesicular bodies (Mv) and the lysosomes (Ly) are free of particles. In the first proximal segment particles are taken into the cell by pinocytosis and moved to the secondary lysosomes. First segment cells also show the rapid transport system. N, nucleus; M, mitochondria. ×13,505. Courtesy R. E. Bulger.

bladder wall. Inulin and bicarbonate have been shown to move bidirectionally across the bladder of the goosefish, *Lophius americanus* (Murdaugh *et al.*, 1963). In the euryhaline European flounder, *Platichthys flesus*, the urine concentration of sodium and potassium, and the urine osmolality, are significantly reduced in the urinary bladder (Lahlou, 1967). Urine chloride may be reduced to nondetectable levels in the bladder of the toadfish, *Opsanus tau* (Lahlou *et al.*, 1969). These reports emphasize the importance of understanding the extent of bladder modification of urine composition in renal function studies with marine teleosts. The bladder certainly serves more than the storage function previously assigned it. It may be responsible for concentrating certain materials such as urea and creatinine (Table X) which may not be actively secreted by the tubular epithelium.

E. Kidney Function in Euryhaline Teleostei

Euryhaline teleosts can survive a wide range of salinities by maintaining steady state body fluid regulation when adapted to either hyperosmotic or hypoosmotic environments. Gunter, who has compiled a list of euryhaline fishes of North and Middle America (Gunter, 1942, 1956), has defined a euryhaline fish as "one which had been recorded from both fresh-water and sea water by competent observers" (Gunter, 1956). This definition includes anadromous and catadromous forms but excludes those species that enter bays of low salinities. Several points of interest emerge from Gunter's lists. A much greater percentage of primitive teleosts (Clupeiformes =Isospondyli) than higher spiny-rayed teleosts are euryhaline. On the North American continent all but two of the anadromous species belong to the primitive Clupeiformes (e.g., shad, alewives, salmon, trout, and smelt). Most of the northern euryhaline species belong to the Clupeiformes, while in more tropical waters the majority belong to more advanced teleost orders. According to Gunter (1956), there are more than 10 times as many marine species that enter freshwater as there are freshwater species that enter seawater. This suggests that the kidney of marine fishes has wider adaptive capabilities than the kidney of freshwater fishes. Unfortunately, kidney function has been studied in only a very few of the 150 euryhaline species listed by Gunter. Many of the marine species recorded from freshwater are infrequent, and perhaps even accidental, visitors to the freshwater habitat. To what extent these species survive the dilute environment by tolerating internal change or by actively regulating against internal change cannot, in most cases, be answered at the present time. The many factors that determine or modify tolerance of fish to salinity change are reviewed by Parry (1966) (see also chapter by Conte, this volume).

There is a tendency for the nephron of the euryhaline fish to resemble that of stenohaline fish of the primary environment. Since most euryhaline fish are members of marine families, most would be expected to have kidneys with typical marine teleost glomerular nephrons, consisting of a proximal segment subdivided into two or three brush border segments, a collecting tubule and duct, but to lack the type of distal segment found in nearly all freshwater species (Table XIII). It must be emphasized, however, that kidney structure in only a few euryhaline fishes has been examined in detail (Table IV). One of these, the pleuronectid southern flounder, *Paralichthys lethostigma*, possesses a cytologically distinct distal segment which is absent from the nephron of the stenohaline pleuronectid, *Parophrys vetulus*. Since the Pleuronectidae are unquestionably a primarily marine family, these differences in tubular structure suggest that the distal segment may have arisen as a secondary adaptation consequent to the animals' tendency to invade freshwater and functions specifically to form a dilute urine. Nevertheless, the distal segment clearly is not essential for this function since it is absent from at least two euryhaline species (Table IV). In this section we will consider some of the functional attributes of the kidneys of euryhaline forms associated with the transition from the seawater to the freshwater habitat and vice versa.

1. GLOMERULAR DEVELOPMENT

Euryhaline teleosts in general have well-developed glomeruli, yet there are several significant exceptions. For example, the glomeruli appear somewhat better developed in the marine stenohaline English sole, *Parophrys vetulus*, than in the euryhaline southern flounder, *Paralichthys lethostigma* (Bulger and Trump, 1968; Trump *et al.*, 1968). The extreme exception is seen in the aglomerular teleosts that have either permanently reinvaded freshwater [e.g., the Thai freshwater pipefish, *Microphis boaja*, and other representatives of the Syngnathidae in Panama (Marshall and Smith, 1930; Grafflin, 1937b; Gunter, 1942, 1956; H. W. Smith, 1953)] or make occasional invasions into fresh or brackish water [the toadfish, *Opsanus tau* (Marshall and Smith, 1930; Lahlou *et al.*, 1969)]. Thus poor glomerular development, or even the absence of glomeruli, clearly does not preclude euryhalinity.

Despite these, and other, exceptions, there is a definite correlation between glomerular development and habitat. Lozovik (1963) found that euryhaline and stenohaline marine teleosts living in the Black Sea (salinity 17–18‰) had, on the average, larger and more numerous glomeruli than did marine teleosts living in the more saline Mediterranean Sea, Barents Sea, and Khadzhibeiskii Bay (salinity 33–38‰). Certain species of the families Gobiidae and Pleuronectidae living in both the Black Sea and Khadzhibeiskii Bay showed signs of "glomerular degeneration" in the latter, more saline, environment. This presumably occurred during the life of these fish and was caused by continued glomerular inactivity.

Euryhaline species typically have significantly higher filtration rates in freshwater than in seawater and continuous exposure to one habitat or the other may be morphologically imprinted on the glomeruli. For example, Ford (1958) reported that pink salmon fry, *Oncorhynchus gorbuscha,* raised in freshwater had significantly more glomeruli than those raised in seawater. Conversely, guppies, *Lebistes reticulatus,* reared for 2½ years in seawater had 40% fewer glomeruli than their controls in freshwater, and the renal corpuscles tended to be smaller in the seawater group (Daikoku, 1965). It is not unlikely that pronounced morphological variations in the nephron can be induced during the life of individual euryhaline fish by constant exposure to one salinity, since the kidney tissue of fish retains the ability to renew morphogenetic processes throughout life. It is noteworthy that Olivereau and Lemoine (1968) could induce pronounced structural changes in the kidney tubules of the European eel with injections of ovine prolactin. These included increased mitotic activity in all regions of the nephric tubule, marked nuclear changes and increased diameter of the collecting tubules, and the appearance of numerous buds which rapidly differentiated into new, functional kidney tubules. The morphogenetic responsiveness of the fish kidney to changed conditions of the environment may explain the variable presence of the distal segment among euryhaline species (Table IV).

2. Physiological Adjustments to Salinity Change

The extension in recent years of functional studies of the fish kidney to include several euryhaline species has been of great importance in developing our present understanding of adaptive capabilities of the nephron. Although observations have in general been restricted to the more readily quantified parameters of glomerular filtration, urine flow, and tubular reabsorption of water and monovalent ions, these have nonetheless served to establish guidelines defining adaptive limits and temporal responsiveness of the whole organ, and to some extent, of nephron components, to salinity change. Most, if not all, the structural components of the nephron appear to be active in the adaptive adjustments of the whole organ. Glomerular and tubular adjustments do not necessarily occur simultaneously or in the same sequence during bidirectional salinity changes, although the two almost certainly are coordinated by a presently undefined regulatory system. There is considerable pharmacological evidence indicating that renal water and electrolyte excretion is responsive to a variety of hormones and gland extracts, but at present no single hormone has been firmly established to regulate any specific aspect of kidney function (Bern, 1967; Maetz, 1968). The subject of endocrine control of renal function is discussed in other chapters.

   a. Glomerular Filtration. It has been demonstrated repeatedly that the GFR and urine flow decline when freshwater-adapted euryhaline fish are introduced into seawater, although the magnitude of these changes is highly variable between different species studied thus far (Table XII; Fig. 47b). After transfer to seawater, the GFR and urine flow may drop to 1–8% of their rates in freshwater in *Salmo gairdneri* (W. N. Holmes and McBean, 1963), *Salmo irideus* (R. M. Holmes, 1961), and *Fundulus kansae* (J. G. Stanley and Fleming, 1964; Fleming and Stanley, 1965). In other species, such as the European eel and the European flounder, the reduction in GFR and urine flow is less pronounced (Table XII). In the Japanese eel, *Anguilla japonica,* and in certain euryhaline flounders of the United States east coast, renal adaptation to seawater is a biphasic process. When freshwater Japanese eels are transferred into seawater, the GFR and urine flow drop within 6 hr to about 30% of their rates in freshwater. Later, as the eel becomes fully adapted to seawater, the GFR recovers and may equal or even exceed the average GFR of freshwater-adapted eels. At the same time, tubular water permeability increases. Consequently, the urine flow continues to fall while the GFR rises, since increased tubular reabsorption of filtered water more than offsets the increased filtration rate (Oide and Utida, 1968). Similar results have been obtained with the euryhaline southern flounder, *P. lethostigma,* and summer flounder, *P. dentatus* (Hickman, 1969). Thus, in all of these species, the rapid reduction in glomerular filtration appears to be a transient adaptation which serves to reduce water loss in the seawater environment, during the period that the permeability of the tubular epithelium to water gradually increases.
   The recent observation that the GFR of the euryhaline southern flounder varies seasonally, suggests that environmental salinity is not the only factor governing glomerular activity. Adult southern flounder of the southeastern United States migrate seaward in the fall, spawn during the winter on the continental shelf, and return in the spring to the dilute waters of sounds and estuaries. Experiments with female flounder which

Table XII

Species	Habitat[a]	GFR (ml/hr/kg)	Urine flow (ml/hr/kg)	Osmolality (mOsm/liter)	Na+ (mmole/liter)	Cl- (mmole/liter)
Salmo irideus	FW	—	3.44	—	—	5–12
	SW	—	0.031	—	—	200–220
Anguilla anguilla[d]	FW	4.60 ± 0.53	3.52 ± 0.41	—	18.9 ± 3	Nil
(European eel)	SW	1.03 ± 0.21	0.63 ± 0.09	—	6.5 ± 1	119 ± 5.7
Anguilla anguilla	FW	1.51 ± 0.18	1.1 ± 0.23	—	13.1 ± 1.2	3.3 ± 0.47
	SW	0.43 ± 0.06	0.25 ± 0.04	—	64.4 ± 5.4	123.0 ± 8.1
Fundulus kansae	FW	25	8.33	14.1 ± 0.7	12.8 ± 0.7	—
(plains killifish)	SW	1.35	0.52	184 ± 5	140 ± 16	—
Platichthys flesus	FW	4.16 ± 0.22	1.78 ± 0.09	48.6 ± 9.5	16.4 ± 3.6[b]	16.3 ± 1.1
(flounder)	SW	2.4 ± 0.27	0.60 ± 0.05	283.1 ± 8.3	35.8 ± 8.9[b]	103.0 ± 13.0
Paralichthys lethostigma	FW	3.88 (3.4–4.1)	2.9 (2.6–3.0)	56.7 (51–64)	27.9 (23–31)	6.4 (32–7.4)
(southern flounder)	SW	1.69 (1.4–2.1)[c]	0.22 (0.11–0.31)	304 (275–333)	17.1 (0.1–32)	120.6 (51–148)
Anguilla japonica	FW	2.80 ± 0.26	2.26 ± 0.17	—	7.46 ± 1.16	2.68 ± 0.55
(Japanese eel)	SW	3.13 ± 0.78	0.38 ± 0.04	—	20.6 ± 4.86	41.3 ± 1.47

[a] Here FW stands for freshwater and SW stands for seawater.

[b] Concentrations are for constituents of bladder urine; "free" urine, which had not remained in bladder, had higher concentration of Na and Cl.

[c] Glomerular filtration rate varies seasonally in this species in seawater. Average of summer GFR given.

[d] In a separate study with this species, Butler (1966) found similar urine sodium and potassium values but obtained much lower urine flows (FW = 2.03 ml/hr/kg; SW = 0.17 ml/hr/kg).

were captured and held throughout the year in an estuarine pond revealed that the GFR was typically high in the summer and very low, even zero, in the winter (Hickman, 1968a). This pattern would appear to be a useful adaptation to the flounder in its chosen summer and winter habitats because of the salinity differences between these habitats. However, the absence of any significant seasonal variation in salinity within the ponds containing the experimental flounder indicates that salinity was not the factor governing seasonal changes in GFR. Some other environmental factor, such as temperature or photoperiod, or endogenous factor such as gonad maturation, is implied; but a satisfactory explanation must await more explicit data on the effects of these variables on renal function and the pathways over which the effects are conveyed to the kidney. Of interest in this respect is the observation that the increase in size of renal corpuscles of the three-spine stickleback, *Gasterosteus aculeatus*, transferred to freshwater, is greater in late spring than in early winter fish (Ogawa, 1968). This species normally migrates into freshwater in the spring to spawn and is able to hyperosmotically regulate much more perfectly at this time than in the winter. Lam and Hoar (1967) and Lam and Leatherland (1969) have presented evidence showing that injection of prolactin, a pituitary hormone which promotes survival of euryhaline fishes in freshwater (Bern, 1967; Maetz, 1968), causes an increase in glomerular size and enables winter sticklebacks to hyperregulate in fresh-

Euryhaline Teleosts in Freshwater and Seawater, Comparing GFR, Urine Flow, and
Urine Composition after Adaptation to Freshwater or Seawater

K+ (mmole/liter)	Ca²⁺ (mmole/liter)	Mg²⁺ (mmole/liter)	SO₄²⁻ (mmole/liter)	PO₄³⁻ (mmole/liter)	pH	Reference
—	—	—	—	—	—	R. M. Holmes (1961)
—	—	—	—	—	—	—
65 ± 0.16	—	—	—	—	—	Sharratt $et\ al.$ (1964)
08 ± 0.9	—	—	—	—	—	—
14 ± 0.32	0.63 ± 0.11	0.02 ± 0.01	—	4.47 ± 0.55	—	Chester Jones $et\ al.$ (1969)
58 ± 0.33	8.20 ± 0.52	28.7 ± 2.8	—	0.62 ± 0.14	—	—
17 ± 0.01	0.4 ± 0.1	—	—	—	—	J. G. Stanley and Fleming (1964)
2.0 ± 0.1	19.9 ± 2.4	—	—	—	—	Fleming and Stanley (1965)
55 ± 0.88	6.5 ± 1.7	—	—	—	—	Lahlou (1967)
82 ± 0.47	18.5 ± 2.5	—	—	—	—	—
(1.9–2.2)	1.1 (1.0–1.2)	0.42 (0.35–0.55)	0.008	2.8 (2.6–3.1)	7.98	Hickman (1969)
2 (0.47–3.6)	19.3 (10.6–25)	133 (116–148)	68.5 (58–75)	9.6 (0.2–19.2)	7.40	Hickman (1968a,b)
—	—	—	—	—	—	Oide and Utida (1968)

water as effectively as the late spring fish. This again suggests that the
kidney of fish is responsive to seasonally changing patterns of hormonal
balance.

b. Tubular Function. Two tubular adjustments are enacted during
adaptation to freshwater: (1) nearly complete cessation of tubular secretion of magnesium and sulfate, and (2) reduction of tubular water permeability (Fig. 47b). The first of these changes happens almost as soon
as the fish stops drinking seawater. Much more time is required for the
development of tubular water impermeability; this varies with different
species, and probably with other, less well defined factors such as body
size and environmental temperature. Very small euryhaline species such
as *Fundulus heteroclitus*, *Fundulus kansae*, and *Periophthalmus* sp. seem
especially tolerant of abrupt salinity change. In small (100–300 g) specimens of the flounder, *Platichthys flesus*, transferred rapidly from seawater
to freshwater, the urine osmolality declines slowly at first, then drops
rapidly at about 6 hr after transfer. Complete adjustment requires 3–4
days in this species (Lahlou, 1967). During this period of adjustment
the urine sodium rises considerably. In large (ca. 1 kg) specimens of
the euryhaline southern flounder, *Paralichthys lethostigma*, 12–24 hr
are required for the kidney to begin forming a dilute urine. If the transfer to freshwater is abrupt, magnesium and sulfate virtually disappear

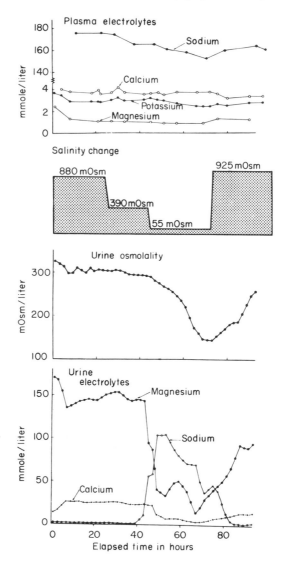

Fig. 44. Kidney function in the euryhaline southern flounder during exposure to salinity variation. Urine was collected by bladder catheter led to a fraction collector changing at hourly intervals. Blood samples were withdrawn at irregular intervals by caudal vein cannula. Note the sudden decrease in urine magnesium concentration coincident with the salinity drop from 390 to 55 mOsm/liter and note further the absence of any detectable change in the plasma magnesium concentration at this time. The activity of the magnesium pump must be governed by minute changes in the blood magnesium concentration by responding directly to its level in the peritubular blood or indirectly through an unknown hormonal intermediate. As the urine magnesium fell, it was replaced as the dominant cation by sodium. As the kidney

from the urine and are replaced by sodium and chloride (Fig. 44). The high concentrations of sodium and chloride gradually decline as the kidney tubules develop sufficient impermeability to water to enable the reabsorption of monovalent ions without the accompaniment of water (Hickman, 1969) (Fig. 47b).

Considerable interest has attached to the findings of Stanley and Fleming (1964; Fleming and Stanley, 1965) who studied the euryhaline plains killifish, *Fundulus kansae*. During the first few days of adaptation to seawater following direct transfer from freshwater, these animals produced a strongly blood hyperosmotic urine: Average urine osmolality was 362 ± 28 mOsm/liter compared to a serum value of 242 ± 16 mOsm/liter. Although urine sodium and potassium were much higher during this transient period than before transfer or after complete adaptation to seawater at 20 days, it does not seem possible to explain these results solely on the basis of a simple abrogation of distal sodium reabsorption because the fish were not diuretic during the period of adaptation. Leakage or osmosis through the papilla would also appear to be ruled out (J. G. Stanley, 1968). It is significant that hypophysectomized fish did not form a hyperosmotic urine even though urine flows were similar to those of intact fish (J. G. Stanley and Fleming, 1966). Furthermore, once the fish was fully adapted to seawater, blood hyperosmotic urine was never observed, even at low urine flows. Although decisive interpretation is not possible because of the absence of Mg^{2+} and SO_4^{2-} determinations, the explanation is probably related to the initiation of divalent ion secretion which is superimposed on a reduced volume of filtrate and high tubular impermeability to water. It is additionally possible that sodium is secreted into the tubules to assist in the removal of some of the sodium load which rapidly accumulates during the initial period of adaptation to seawater.

Students of comparative renal physiology have long speculated over how the freshwater aglomerular pipefish, *Microphis boaja*, and other freshwater Syngnathidae in Thailand, Panama, and the Black Sea maintain salt and water balance (Marshall and Smith, 1930; H. W. Smith, 1932, 1953; Grafflin, 1937b; Lozovik, 1963). Some insight into this question has been provided by recent observations of Lahlou *et al.* (1969) on the aglomerular toadfish, *Opsanus tau*, which has long been known to survive dilute salinities. Lahlou *et al.* found that the toadfish could

---

tubules gradually developed water impermeability, sodium (and chloride, not shown in the figure) were reabsorbed more completely, thus reducing the urine osmolality. Magnesium reappeared in quantity in the urine when the salinity was restored to that of full strength seawater. The changes during the first six hours occurred during the equilibration time in the chamber following anesthesia.

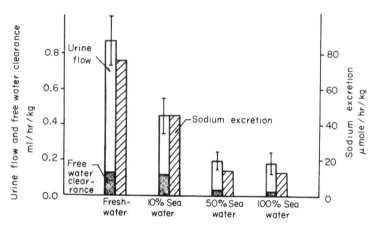

Fig. 45. Urine flow, free water clearance, and sodium excretion in the toadfish, *Opsanus tau*. Urine flow is read from the base of the bar. From Lahlou *et al.* (1969).

survive up to 3 weeks in freshwater if adapted gradually to this medium from seawater. Urine flow was low in seawater (about 0.2 ml/hr/kg) and increased to about 0.9 ml/hr/kg in freshwater (Fig. 45). In freshwater the kidney was unable to form a significantly dilute urine, urinary sodium was high, and the free water clearance was not much greater than in seawater. Thus, the kidney is evidently capable of excreting all of the water that enters osmotically from the animal's dilute environment, but it is unable to do so efficiently since it cannot conserve monovalent ions. The poor performance of the kidney is, however, offset by especially effective branchial adjustments to salinity change. Lahlou and Sawyer (1969) found that this species shows an immediate reduction in sodium outflux across the gills when transferred to freshwater. This change was identical in pattern to that observed in the euryhaline flounder, *Platichthys flesus* (Motais *et al.*, 1966), postulated to be an exchange diffusion effect.

The question of how water is made to enter the aglomerular nephron remains unstudied. Presumably it follows the secretion of electrolytes, perhaps sodium and chloride, which, in freshwater, might substitute for the primary tubular secretion of divalent ions in the marine habitat. The availability of sufficient Na^+ and Cl^- is provided by efficient branchial absorption. While the toadfish kidney lacks the capacity to withdraw monovalent ions from the tubule without the accompaniment of water, the freshwater Syngnathidae may well have improved upon this scheme by reducing the water permeability of the distal regions of the nephron. Grafflin (1937b) commented upon the presence of cytologically distinctive areas of the nephron of an unnamed species of *Microphis* that could be associated with this provision.

VII. CLASS DIPNOI (LUNGFISHES)

This class contains three extant species of lungfishes included in two families. A discussion of the habits of these animals as they relate to nitrogen metabolism is found in the chapter by Forster and Goldstein, this volume.

A. Structure of the Nephron

Rather extensive studies of the nephron of *Protopterus aethiopicus* and *Lepidosiren paradoxa* have been made by Edwards (1935), Grafflin (1937d), and Guyton (1935). The nephrons in both fish appear to be quite similar, consisting of the following regions:

(1) A renal corpuscle containing a glomerulus with capillaries and inconspicuous mesangial regions.

(2) A ciliated neck segment.

(3) A proximal tubule containing first and second segments, both of which resemble the corresponding portions in freshwater teleosts. The first segment, as in all teleosts that have been studied, contains conspicuous cytoplasmic droplets which correspond to lysosomes.

(4) A ciliated intermediate segment which is lined by low cuboidal ciliated cells.

(5) A distal segment with a striated cytoplasm formed by the elongate perpendicular mitochondria.

(6) A collecting duct system.

The cytological features of the nephron in this group are thus similar in all aspects to certain of the typical freshwater teleosts which have been described in more detail above (pp. 127–133).

B. Kidney Function

Kidney function has been studied only in the African lungfish, *Protopterus aethiopicus*, by H. W. Smith (1930a), who measured the concentration of nitrogenous end products in the urine from a single animal, and by Sawyer (1966), who studied the renal response to injections of neurohypophysial octapeptides. Lungfish are freshwater forms that are completely anuric during estivation. After awakening and breaking out of its cocoon, the lungfish begins to excrete a urine which is most respects differs little from that of freshwater teleosts. The urine is very dilute (16.5–18.8 mOsm/liter) and contains about 5 mEq/liter of sodium and about 3.5 meq/liter of ammonium, which together probably constitute most of the cation present. Inulin clearances and urine flows of five cannulated lungfish, weighing 98–250 g, averaged 11.0 and 5.9 ml/kg/hr, respectively (Sawyer, 1966). Sawyer reported that in this species arginine vasotocin in doses as low as 0.5 μg/kg produced significant increases in GFR, free water clearance, and sodium excretion. Components of these responses were similar to those reported by Maetz *et al.* (1964) for the goldfish, *Carassius auratus*, and by Bentley and Follett (1963) for the lamprey, *Lampetra fluviatilus*. This reinforces the indication of the fundamental similarity of renal function in all fishes living in freshwater.

VIII. SYNTHESIS: FUNCTIONAL AND EVOLUTIONARY SIGNIFICANCE OF HOMOLOGOUS REGIONS IN THE FISH NEPHRON

In Table XIII the structure of the fish nephron, phylogenetically arranged in representative species, is compared with habitat and type of osmotic regulation. It is apparent from the table that in the most primitive fish shown, the hagfish, the nephron consists of a renal corpuscle, a neck, and a first proximal segment which, in this species, forms the archinephric duct. This animal is an ionic regulator but is nearly in osmotic equilibrium with its environment.

In all the other species listed, beginning with the lamprey, the animals do regulate osmotically. It is apparent that the addition of another portion of the nephron, the collecting tubule, is essential for osmotic regulation. It is also evident that the nephron reaches its greatest complexity in the lungfishes and freshwater teleosts. Moreover, an essentially identical pattern exists in amphibians and, with modification, in mammals. In these freshwater forms three segments are added: a second portion of the proximal tubule, a variable present intermediate segment, and a distal tubule. It would appear that these additional specializations, although not essential for primitive osmotic regulation, may be of significance for the more sophisticated regulatory physiology of higher forms by improving the efficiency of specific excretory or regulatory activities. In the marine glomerular teleosts the distal and intermediate segments appear to have been discarded. In some pauciglomerular teleosts such as the shorthorn sculpin, *Myoxocephalus scorpius*, and the moray eel, *Muraena helena*, the glomeruli become degenerate and the neck segments begin to disappear. This parsimony is most extreme in the marine aglomerular teleosts, exemplified by the midshipman, *Porichthys notatus*, in

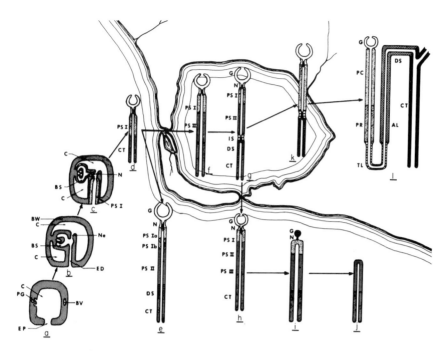

Fig. 46. Schematic representation of our interpretation of the evolutionary development of the vertebrate nephron. This diagram shows the early development in the sea (a → d), the invasion of freshwater (d → f), evolution in freshwater (f → g), evolution of land vertebrates including amphibians (k) and mammals (l), reinvasion of the sea (g → h), later evolution of marine teleosts (h → j), and the alternative pathway leading to the elasmobranchs (d → e). For complete description, see text. (a) and (b) Hypothetical early protovertebrates; (c) Myxini such as *Myxine glutinosa* (hagfish); (d) Petromyzones such as *Lampetra fluviatilus* (lamprey); (e), Elasmobranchii such as *Squalus suckleyi* (Pacific dogfish); (f) Holostei such as *Lepidosteus spatula* (alligator gar); (g) freshwater Teleostei such as *Lepomis macrochirus macrochirus* (bluegill); (h) marine Teleostei such as *Parophrys vetulus* (English sole); (i) pauciglomerular marine Teleostei such as *Lophius americanus* (goosefish); (j) aglomerular marine Teleostei such as *Porichthys notatus* (midshipman); (k) Amphibia such as *Rana esculenta* (European frog); and (l) Mammalia such as man. AL, ascending limb of Henle's loop; BS, Bowman's space; BV, blood vessel; BW, body wall; C, coelom; CT, collecting tubule; DS, distal segment (distal convoluted tubule in Amphibia and Mammalia); G, glomerulus; ED, excretory duct; EP, excretory pore; IS, intermediate segment; N, neck; Ne, nephrostome; PC, pars convoluta of proximal tubule; PG, primitive glomerulus; PR, pars recta of proximal tubule; PSI, first proximal segment; PSIa and PSIb, first and second regions of elasmobranch first proximal segment; PSII, second proximal segment; PSIII, third proximal segment; and TL, thin limb of Henle's loop.

Table XIII

Phylogenetic Comparisons of Nephron Morphology in Fishes

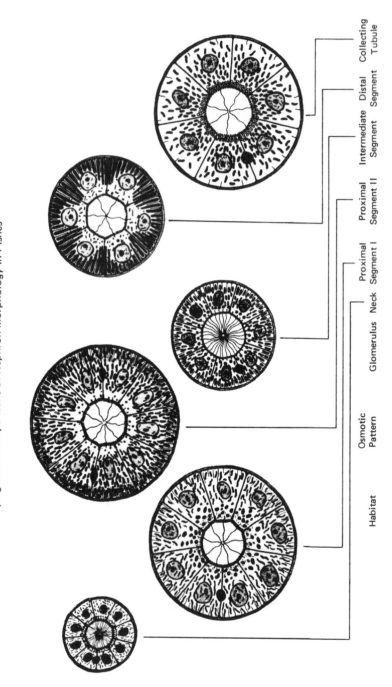

	Habitat	Osmotic state							
Myxini									
Myxine glutinosa (hagfish)	Marine	Isosmotic	+	+	+[a]	0	0	0	0
Petromyzones									
Lampetra fluviatilis (lamprey)	Anadromous	Hyperosmotic (F.W.)	+	+	+	0	0	0	+
Elasmobranchii									
Squalus acanthias (dogfish)	Euryhaline	Hyperosmotic (F.W. and S.W.)	+	+	+[b]	+	0	+	+
Holocephali									
Hydrolagus colliei (ratfish)	Marine	Hyperosmotic	+	+	+	+	0	+	+
Teleostomi									
Holostei									
Lepidosteus spatula (alligator gar)	Euryhaline	Hyperosmotic (F.W.)	+	+	+	+	0	0	+
Teleostei									
Lepomis macrochirus (bluegill)	Freshwater	Hyperosmotic	+	+	+	+	+	+	+
Salmo gairdneri (rainbow trout)	Euryhaline	Hyperosmotic (F.W.) / Hypoosmotic (S.W.)	+ / +	+ / +	+ / +	+ / +	+ / +	+ / +	+ / +
Paralichthys lethostigma (southern flounder)	Euryhaline	Hypoosmotic (S.W.)	+	+	+	+	+	+	+
Parophrys vetulus (English sole)	Marine	Hyperosmotic (F.W.) / Hypoosmotic	+ / +	+ / +	+ / +[c]	+ / +	0 / +	+ / 0	+ / +
Porichthys notatus (midshipman)	Marine	Hypoosmotic	0	0	+	+	0	0	+
Dipnoi									
Lepidosiren paradoxa (lungfish)	Freshwater	Hyperosmotic	+	+	+	+	+	+	+

[a] Archinephric duct in this species.

[b] Kempton (1943) has described a "fine" segment in the proximal tubule but its position and cytology have not been worked out.

[c] The terminal portion of the proximal tubule is different in *Parophrys*; this may eventually be classified as a unique segment in marine glomerular teleosts.

which only two segments have been retained, the second proximal segment and the collecting duct system (Fig. 23). With only these two segments, such animals are able to carry out the essential excretory activities of marine teleosts.

In Fig. 46, this evidence is summarized in an interpretation of nephron evolution based on comparative morphology and embryology. The most primitive form of kidney (Fig. 46a right) is envisioned as a blood vessel (BV) separated from the coelomic cavity (C) by a layer of mesenchyme and the mesothelial lining. In this form, excretion would occur by low pressure filtration across the coelomic surface to its cavity with excretion through a simple pore (EP) to the outside. In the next stage (Fig. 46a left), the efficiency of filtration is improved by making the glomerulus (PG) a simple recess of the coelomic cavity in proximity to blood vessels. The next stage (Fig. 46b) would involve partial separation of the duct portion (ED) of the coelomic cavity with retention of the connection to the general cavity in the form of so-called open nephrostomes (Ne). Bowman's space (BS) is thus a diverticulum of the coelom. It should be noted that invertebrates appear to have evolved analogous types of excretory systems such as the antennal gland in crayfish, which are based upon filtration to provide a medium containing materials which can then be modified by other portions of the nephron.

The next stage (Fig. 46c) in the proposed evolutionary sequence is represented by the hagfish, in which separation of the nephron from the coelomic cavity is virtually complete. In this form, the glomerulus is composed of a capillary tuft, followed by a long ciliated neck segment (N) which presumably evolved to improve the efficiency of flow in these low pressure systems. The neck in turn drains into a specialized portion which appears for the first time in this form. This specialized terminal portion, often termed the "archinephric duct" (PSI), is, in fact, homologous with the first proximal segment of higher fish species as well as that of higher vertebrates, including man. In the first vertebrates to enter freshwater, as exemplified by the present-day anadromous lamprey (Fig. 46d), a collecting duct portion (CT) was added which, we propose, was essential to permit the formation of a dilute urine. It appears that the first permanent residents of freshwater represented by the present-day alligator gar (Fig. 46f) added a second proximal segment (PSII) which goes on to become the longest part of the nephron in later forms. The dominant role assumed by this segment attests to its metabolic versatility. As discussed below it is evidently involved in divalent cation excretion, sodium reabsorption, organic anion and cation secretion, and hydrogen ion ejection. Since the hagfish does not possess the organic anion secretory system, it is tempting to speculate that the addition of this segment is related to the acquisition of this function which, as discussed below, is almost certainly related to more important functions. These might include Na^+-K^+ exchange or substrate transport, rather than the mere excretion of compounds such as phenol red which are foreign to the animal. Moreover, with the appearance of this segment, the capability for the extensive excretion of divalent cations, so essential to hypoosmotic regulation in the marine environment, was manifest.

The highest degree of nephron development is found in the freshwater Teleostomi and Dipnoi (Fig. 46g) which added a variably present ciliated intermediate segment (IS), and a distal tubule (DT) between the proximal segment and the collecting tubule. We suspect that these segments improved hyperosmotic regulation and served as useful preadaptations which presaged the movement to land as shown by the essential similarity of these nephrons to those of modern amphibians (Fig. 46k). The apparent homologies between the amphibian and the mammalian nephron are diagrammed in Fig. 46l.

Another evolutionary pattern evidently occurred when the freshwater teleosts reinvaded the sea. These later marine teleosts exhibit a regression of nephron complexity presumably associated with greater specialization of function. In marine glomerular teleosts (Fig. 46h) the intermediate segments and distal tubule are lost; however, there is evidence of further development of the terminus of the second proximal segment forming a third such segment. Further regression occurs in pauciglomerular forms such as the goosefish (Fig. 46i) in which glomeruli and neck segments are degenerate. This disappearance apparently lessens the need for the first proximal segment which is specialized for uptake of macromolecular materials that pass the glomerular filter. The greatest parsimony in these marine forms is exemplified by modern aglomerular marine teleosts (Fig. 46j) in which the nephrons consist only of second proximal segments and collecting tubules.

Present-day elasmobranchs (Fig. 46e), which are generally considered as representatives of a parallel evolutionary development to the bony fishes from an early vertebrate ancestor, have evolved a nephron which vies with that of the freshwater teleosts for complexity. Though *Squalus* is primarily a marine form, its kidney has many of the functional and morphological attributes associated with freshwater fish. These include a high glomerular filtration rate, high urine flow, and reabsorption of monovalent ions. This necessitates an extensive first proximal segment which is composed of two subdivisions (PSIa and PSIb). The great length of the nephron, mainly in the second proximal segment (PSII), is probably related to the unique requirement of elasmobranchs for conservation of urea and TMAO. The euryhaline propensities of these fish are pre-

HAGFISH

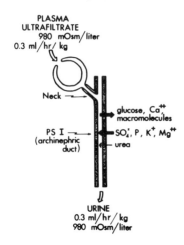

PLASMA
ULTRAFILTRATE
980 mOsm/liter
0.3 ml/hr/kg

Neck

glucose, Ca⁺⁺
macromolecules

PS I
(archinephric
duct)

SO_4^-, P, K⁺, Mg⁺⁺

urea

URINE
0.3 ml/hr/kg
980 mOsm/liter

FRESHWATER LAMPREY

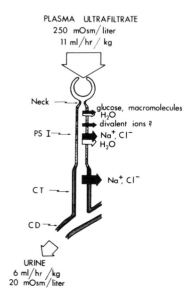

PLASMA ULTRAFILTRATE
250 mOsm/liter
11 ml/hr/kg

Neck

glucose, macromolecules
H_2O
divalent ions ?

PS I

Na⁺, Cl⁻
H_2O

CT

Na⁺, Cl⁻

CD

URINE
6 ml/hr/kg
20 mOsm/liter

MARINE ELASMOBRANCH

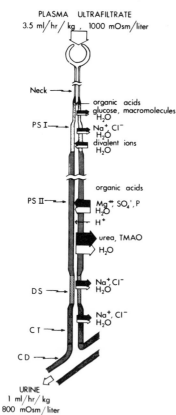

PLASMA   ULTRAFILTRATE
3.5 ml/hr/kg , 1000 mOsm/liter

Neck

organic acids
glucose, macromolecules
H_2O

PS I

Na⁺, Cl⁻
H_2O

divalent ions
H_2O

organic acids

PS II

Mg⁺⁺, SO_4^-, P
H_2O

H⁺

urea, TMAO
H_2O

Na⁺, Cl⁻
H_2O

DS

Na⁺, Cl⁻
H_2O

CT

CD

URINE
1 ml/hr/kg
800 mOsm/liter

FRESHWATER TELEOST

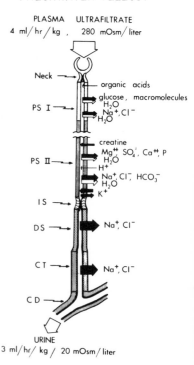

PLASMA   ULTRAFILTRATE
4 ml/hr/kg ,   280 mOsm/liter

Neck

organic acids

PS I

glucose, macromolecules
H_2O
Na⁺, Cl⁻
H_2O

creatine

PS II

Mg⁺⁺, SO_4^-, Ca⁺⁺, P
H_2O
H⁺
Na⁺, Cl⁻, HCO_3^-
H_2O
K⁺

IS

DS

Na⁺, Cl⁻

CT

Na⁺, Cl⁻

CD

URINE
3 ml/hr/kg / 20 mOsm/liter

Fig. 47a.

MARINE GLOMERULAR TELEOST

PLASMA
ULTRAFILTRATE
0.5 ml/hr/kg, 450 mOsm/liter

Neck

PS I — organic acids
glucose, macromolecules
H_2O
Na^+, Cl^-
H_2O
Mg^{++}, $SO_4^=$, Ca^{++}, P
H_2O

PS II — organic acids
Mg^{++}, $SO_4^=$, Ca^{++}, P
H_2O
H^+
NH_3, TMAO, creatine
urea, creatinine, uric acid

PS III — Na^+, Cl^-

CT — Na^+, Cl^-, K^+
H_2O

CD

URINE
0.3 ml/hr/kg
410 mOsm/liter

MARINE AGLOMERULAR TELEOST

PLASMA
450 mOsm/liter

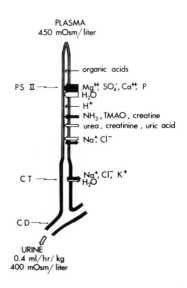

organic acids

PS II — Mg^{++}, $SO_4^=$, Ca^{++}, P
H_2O
H^+
NH_3, TMAO, creatine
urea, creatinine, uric acid
Na^+, Cl^-

CT — Na^+, Cl^-, K^+
H_2O

CD

URINE
0.4 ml/hr/kg
400 mOsm/liter

EURYHALINE TELEOST

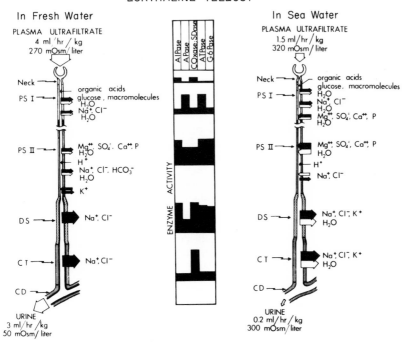

In Fresh Water

PLASMA ULTRAFILTRATE
4 ml/hr/kg
270 mOsm/liter

Neck

PS I — organic acids
glucose, macromolecules
H_2O
Na^+, Cl^-
H_2O

PS II — Mg^{++}, $SO_4^=$, Ca^{++}, P
H_2O
H^+
Na^+, Cl^-, HCO_3^-
H_2O
K^+

DS — Na^+, Cl^-

CT — Na^+, Cl^-

CD

URINE
3 ml/hr/kg
50 mOsm/liter

In Sea Water

PLASMA ULTRAFILTRATE
1.5 ml/hr/kg
320 mOsm/liter

Neck

PS I — organic acids
glucose, macromolecules
H_2O
Na^+, Cl^-
H_2O
Mg^{++}, $SO_4^=$, Ca^{++}, P
H_2O

PS II — Mg^{++}, $SO_4^=$, Ca^{++}, P
H_2O
H^+
Na^+, Cl^-

DS — Na^+, Cl^-, K^+
H_2O

CT — Na^+, Cl^-, K^+
H_2O

CD

URINE
0.2 ml/hr/kg
300 mOsm/liter

ENZYME ACTIVITY

AlPase
APase
COxase, SDase
AlPase
G6Pase

Fig. 47b.
219

sumably reflected by the distal tubule (DT) which probably functions to conserve sodium in freshwater. This function, though redundant in a marine environment, is compensated in that situation by the rectal gland.

In Figs. 47a and b, schematic representations of the nephron of the major groups of fishes are given, showing the probable movements of materials across the different tubular regions. These speculations are based on: (1) the calculations given earlier of the amounts of solutes and water added to or subtracted from the plasma filtrate, (2) comparisons with amphibian micropuncture data, (3) changes in urine composition following alterations in plasma electrolyte composition, and (4) electron microscopic and cytochemical observations. The functions of each region of the nephron are discussed below and are related to ultrastructure in Fig. 48 and its legend.

Glomerulus (Figs. 10, 15, and 16). It has been asserted that the glomerulus evolved as a device to rid the animal of excess water (Marshall and Smith, 1930; Marshall, 1934; H. W. Smith, 1932), and that this was related to the proposed evolution of protovertebrates in freshwater. This idea has been challenged by Robertson (1957) who argues that the glomerulus must have originated for some other reason because it is found in myxinoids which are nearly isosmotic with their marine environment, and because marine invertebrates, especially the decapod Crustacea, often have filtration devices of a similar type. Furthermore, careful studies of the paleontological evidence supports the theory that the earliest vertebrates evolved in a marine environment (Denison, 1956; Robertson, 1957) although the matter is still debated (Romer, 1968). Based on Robertson's evidence and on our own comparative studies of fish nephrons, it would appear much more likely that the glomerulus evolved not as a device to rid the animal of excess water, but as a device to

Fig. 47. Schematic representation of nephron structure and function: (a) Hagfish, freshwater lamprey, marine elasmobranch, and freshwater teleost; (b) marine glomerular teleost, marine aglomerular teleost, and euryhaline teleost. These diagrams show the activities of morphological regions identifiable within the nephron of fishes. The shading of each segment is the same in all pictures. Solid arrows indicate active, and open arrows passive, movements. The width of the arrows is proportional to the rate of movement of substances across the tubular epithelium. The presence of a solid line along the luminal surface of distal segments and collecting tubules indicates relative impermeability to water. The enzyme profiles shown for the euryhaline teleost consist of histograms showing relative activity as estimated from histochemical preparations. AlPase, alkaline phosphatase; APase, acid phosphatase; ATPase, adenosinetriphosphatase; CD, collecting duct; COxase, cytochrome oxidase; CT, collecting tubule; DS, distal segment; G6Pase, glucose-6 phosphatase; IS, intermediate segment; P, all oxidation states of orthophosphate ($H_2PO_4^-$, HPO_4^{2-}, PO_4^{3-}); PSI, first proximal segment; PSII, second proximal segment; PSIII, third proximal segment; SDase, succinic dehydrogenase.

initiate regulatory functions by providing a filtrate which could then be modified selectively. This would be consistent with the paleontological evidence reviewed by Denison (1956). While the glomerulus can indeed assist materially in the removal of water from the body, we view it rather as a useful preadaptation which was retained when the early Teleostomi moved into freshwater. Significantly, it is not an essential device for life in freshwater, evidenced by the presence of several freshwater aglomerular Syngnathids and by the demonstrated survival of aglomerular toadfish held in freshwater. It is also apparent that the presence of glomeruli is compatible with life in the marine environment. Reduced glomerular filtration rates, however, are more efficient for marine forms; in this regard it is interesting that in general the glomeruli of marine teleosts are relatively avascular as a result of the marked development of mesangial cells which produce additional basement membrane material (Fig. 15). Certain marine forms thrive with no glomeruli at all, e.g., the toadfish, *Opsanus tau*, and midshipman, *Porichthys notatus;* moreover, there are intermediate, pauciglomerular fish, such as the daddy sculpin, *Myoxocephalas scorpius*, which have degenerating glomeruli and neck segments (Grafflin, 1935). The glomeruli of freshwater fish and elasmobranchs on the other hand have widely patent capillary lumens with very thin capillary walls and high rates of filtration. It is of interest that the pattern of degeneration in the daddy sculpin, obliteration of capillary lumens, and thickening of filtration barrier by extension of the mesangium is a sequence that is retained in the mammal where it becomes apparent in the type of glomerular degeneration that commonly accompanies human kidney disease (Faith and Trump, 1966). Changes in glomerular filtration rate in euryhaline forms also occur on a seasonal basis and provide an adaptation related to freshwater or marine migrations.

From the standpoint of comparative morphology it is extremely interesting to note that the basic pattern of glomerular ultrastructure with complex foot processes of the visceral epithelial cells, the appearance of the lamina densa, and the relationships of the endothelium and mesangial cells are retained throughout the animal kingdom.

Neck Region (Fig. 17). The neck region is a very primitive portion of the nephron, being present in all forms except marine aglomerular teleosts. Although the function of the neck has not been established, it seems most likely that the ciliary activity in this region is of importance in the movement of materials from Bowman's space into the tubular lumen. It would appear that this is especially important in low pressure filtration systems such as exist in all fish. The neck region disappears in higher vertebrates, but is retained in amphibians.

First Proximal Segment (Figs. 11, 18, 19, and 29). This is another

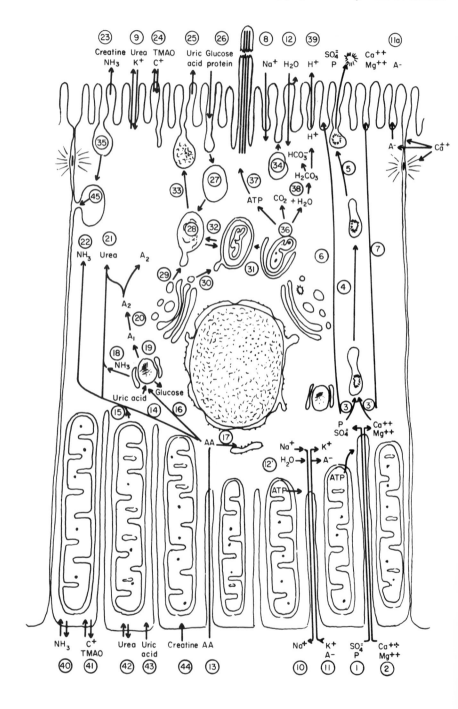

Fig. 48. Composite representation of tubular epithelial cell of fish nephron showing relationship between ultrastructure and functional activities. Only certain of the numerous activities of the tubular epithelial cells are depicted in this diagram. The cell shown is a composite of all segments in virtually all types of fish with differences as noted below.

A. Divalent ion transport system. Ca^{2+} and Mg^{2+} enter the cell base (2), probably in closely related, if not identical, ATP-requiring systems. Sulfate and orthophosphate appear to enter by a related but different system (1). The cations may enter the endoplasmic reticulum (RER) (3,3') where precipitation of complex minerals such as apatites occurs, probably dependent on other factors such as relative amounts of anion and cation and pH. These microconcretions are excreted by reverse pinocytosis (4 and 5) into the tubular lumen. An alternative pathway would involve passage of cation and anion into the cell sap (6 and 7) with again active extrusion into the lumen at the cell apex.

B. Na^+ and K^+ transport system. Sodium probably enters the cell apex from the lumen by passive diffusion (8), passes through the cell and is actively extruded at the base (10) into narrow compartments formed by the infolded basilar plasmalemma. This active sodium extrusion (net reabsorption) is believed to be coupled to inward movement of potassium (11), is ATP-requiring and involves a Na-K dependent, ouabain-sensitive ATPase which is probably located in these complex basilar infoldings. Potassium must have some active step at the cell apex as well since net secretion can occur in some regions (9). The Na–K pump at the cell base appears to be related in some way to the uptake and transport of organic anions (11). The extrusion of sodium into the long, narrow compartments formed between the complex basilar plasmalemmal infoldings may create local osmotic gradients which promote the passive efflux of water from the cell compartment into the extracellular space (12'). The relationship between the plasmalemmal infoldings and elongate mitochondria are especially evident in the distal segments of fish and higher forms.

C. Organic anion transport system. The organic anions (A^-) enter the base of the cell in a potassium dependent, ATP-requiring step (11) which may interdigitate with that of the Na^+ extrusion system. Organic anions pass through the cell in unknown compartment(s) to the apex where active, ATP- and calcium-dependent, extrusion occurs (11a).

D. Organic bases. Systems somewhat analogous to those for organic anions exist for certain organic bases (C^+). The most important of these are probably TMAO and creatine (41 and 24). Creatine, however, seems to possess a slightly different system from that of the other organic bases (44 and 23).

E. Nitrogen metabolism. Amino acids, ammonia, urea, and uric acid enter the cell base via poorly understood transport mechanisms (13, 40, 42, and 43). Some ammonia, urea, and uric acid probably pass through the tubule to the lumen unaltered (21, 22, 23, and 25); other portions take part in the following pathways. Amino acids may be deaminated in several ways including mitochondrial reactions (15) (glutamic dehydrogenase) and oxidation via amino acid oxidases in microbodies (16 and 18). The resultant ammonia enters the cellular pool and some of it can be utilized to form urea (21) via the ornithine cycle. Amino acids in the cell can also result from autophagic or heterophagic digestion within the lysosome system. Ammonia can combine with H^+ to form NH_4^+.

Purines, produced from cellular turnover, at least some of which is lysosomal, can also result from synthesis involving glutamate metabolism. Purine catabolism results in the formation of uric acid which is oxidized by urate oxidase, found in the microbody, to allantoin (19) which in turn is converted to allantoic acid (20). In some forms allantoic acid is excreted intact (A_2); in others, further conversion to urea, which enters the cellular urea pool (21), occurs. It is also evident that some of the entering amino acids or amino acids derived from intra- or extralysosomal protein turnover participate in polypeptide synthesis (17) on free- or membrane-associated polysomes.

These considerations emphasize the complex interrelationships of nitrogen metabolism in the tubular cell and also focus on the microbody (found also in the liver) in purine metabolism, urea formation and gluconeogenesis.

F. Glucose reabsorption. Glucose is actively reabsorbed from the tubular lumen by a phlorizin-sensitive carrier possibly related to the brush-border-associated alkaline phosphatase. Some glucose probably enters via pinocytosis (26).

G. Lysosome system. In the first proximal segment the macromolecules such as proteins are taken into the apical tubules (26) then find their way into the apical vacuoles (27) which initially are free of acid hydrolases. These apical vacuoles then move into the cell and acquire acid hydrolase activity by fusion with primary lyso-somes (Golgi vesicles) (29) or with preexisting secondary auto- or heterolysosomes (32). After acquisition of hydrolase activity the phagosome (apical vacuole) is called a secondary heterolysosome (28). Digestion of pinocytosed material then occurs with release of intermediates to the cell sap. In other regions rapid transport systems appear to exist in which large apical invaginations (35) discharge directly to the lateral intercellular space below the junctional complex (45).

The primary lysosomes (29) are believed to be elaborated as vesicles from the maturing surface of the Golgi apparatus. The hydrolases are assumed to be syn-thesized on polysomes of the rough ER under the influence of mRNA from the nucleus and injected into the lumen of the ER in which they are transported to the forming face of the Golgi apparatus.

The lysosome also participates in the digestion or turnover of the cell's own organelles such as ER, glycogen, and microbodies which are enclosed in a double-walled sac formed in an ATP-requiring step from ER cisternae (31) (Arstila and Trump, 1968). The inner membrane disintegrates and the outer membrane thickens. These autophagosomes (devoid of acid hydrolase activity) acquire the latter by fusion with primary (30) or secondary (28) lysosomes. Digestion of material within secondary auto- or heterolysosomes results in the intralysosomal accumulation of indigestible debris, much of it lipid in nature. The debris-filled residual bodies are autofluorescent, contain lipid, and are usually yellow-brown in color (lipofuscin or aging pigment). At least some release their content to the tubular lumen by migra-tion to the cell apex followed by reverse pinocytosis (33).

H. Water permeability. There are indications that the distal regions of the nephron in euryhaline and freshwater fishes have the capacity to control or limit the water permeability of the apical surface (12). We speculate that this is related in some way to the mucopolysaccharide coating of the surface plasmalemma, the so-called glycocalyx. The apical mucous granules found in these latter regions may contribute to the elaboration, possibly under hormonal control, of a water-tight external coating for the apical plasmalemma (34). We do not presently know the relationships of the junctional complexes to water impermeability. The tight junc-tions, in which the plasmalemmas are fused, seem to be relatively water tight; in the intermediate junction the cells are closely apposed and acid mucopolysaccharide fills the interstices. It is possible that hormonal factors may influence the water perme-ability of this junction.

primitive region of the nephron, constituting, as it does, the principal portion of the kidney in hagfish. Its morphology even at the ultrastructural level is retained throughout all of the more advanced forms, with the exception of the marine aglomerular teleosts. It should also be noted that the proximal convoluted tubule in higher vertebrates, including mammals, is of essentially identical morphology with this first proximal segment.

From the evidence given above, it appears that this region has several functions. Its well-developed lysosome system, the high content of acid phosphatase, and the apical tubular system which has been shown to be involved in pinocytic uptake of particles are probably related to the resorption of filtered proteins and other macromolecules (Bieter, 1931b). This function would evidently be unnecessary in aglomerular fish in which this segment is absent. Such considerations lead to the notion that some of the specialization within the fish nephron may be induced by characteristics of the environment or characteristics of the tubular fluid. Other probable functions of this region include reabsorption of glucose and amino acids, isosmotic reabsorption of sodium and chloride, and secretion of organic molecules such as phenol red. These functional assignments are made largely on the basis of the similarity with the proximal convolution of amphibians and mammals, but in some, such as macromolecular reabsorption and organic acid transport, direct evidence has been obtained in fish. It is likely that at least some divalent ion secretion occurs in this segment, since it is the only functional segment in the hagfish. On the other hand, its presence is not essential for divalent ion excretion since this segment is absent in aglomerular marine teleosts.

Second Proximal Segment (Figs. 12, 20, 21, and 30). This constitutes the largest region of the nephron of both marine and freshwater Teleostomi. It probably is a metabolically active region being filled with numerous mitochondria. Its lysosome and pinocytic vesicle systems are poorly developed, suggesting that its role in macromolecular uptake is minimal. On the other hand, it has been demonstrated to participate in organic acid secretion. Since it constitutes the longest portion of the nephron, and is the only portion of the proximal tubule in aglomerular

---

I. Acidification. One mechanism for acidification is the combination of CO_2 and water to form carbonic acid (38) in a reaction catalyzed by carbonic anhydrase. This ionizes to form HCO_3^- and H^+; excretion of the proton occurs at the plasmalemma through the action of proton pumps (39) or exchange mechanisms of unknown type. The CO_2 is in part derived from mitochondrial oxidations (36) which also form ATP, some of which is required for ciliary movements (37). The latter are evidently of importance in propulsion of urine in the tubule in these low pressure filtration systems.

marine teleosts, it seems that this region is responsible for the principal part of the divalent ion secretion. The participation of the second proximal segment in divalent ion excretion is strengthened by the observation of similar-appearing microcrystalline aggregates within the urine sediment and within the endoplasmic reticulum of this region. It may also participate in isosmotic sodium reabsorption and hydrogen ion secretion, although its role in these processes has not as yet been defined.

Intermediate Segment (Fig. 14b). This is a somewhat controversial segment that has been described in certain freshwater teleosts. Typically, it is a highly ciliated region and may constitute a specialized portion of the second proximal segment. The functional significance of this region, which is absent in many freshwater teleosts, cannot be assessed. It may be a secondary pump to assist in the propulsion of fluid along the nephron. This would be advantageous in freshwater forms because it would minimize water reabsorption.

Distal Segment (Figs. 13 and 32). The distal segment appears in freshwater and some euryhaline teleosts, elasmobranchs, and dipnoans. Its morphology with the elongate mitochondria and numerous infoldings of the basilar plasmalemma is strikingly similar to the distal segment of the amphibian and to the ascending limb of Henle and parts of the distal tubule in the mammal (Trump and Bulger, 1968a) (Fig. 46). In these areas it has fairly clearly been shown to participate in active sodium resorption. This sodium resorption is not isosmotic, and it appears that in euryhaline forms the water permeability of this region can be varied. It would appear, then, that the distal segment may be important in improving the efficiency of monovalent ion retention and urinary dilution in freshwater and euryhaline forms.

Collecting Tubule and Ducts (Fig. 22a). The collecting tubule and duct system appear to be essential for the formation of a dilute urine by reabsorbing monovalent ions from the filtrate. The hagfish, which does not possess this region, is incapable of hyperosmotic regulation. It is present, however, from the lamprey to the higher forms. The collecting tubule has been retained in all marine forms, in which it probably functions to remove sodium that filters or leaks into the tubular fluid. The large number of mitochondria and the high activity of mitochondrial enzymes in this region, coupled with structural similarities to known water-impermeable ion-reabsorption systems such as the amphibian bladder, suggest that it is in this terminal region that controlled water permeability may occur. It is also possible that the archinephric duct system and the so-called urinary bladder may function in this regard in higher fish.

ACKNOWLEDGMENTS

We thank Mr. William C. Ober for making the drawings reproduced in Figs. 15, 25, 44, 46, and 47, and Table XIII, Mrs. Jessie Calder for technical assistance, and Mr. Bernard Bell for photographic assistance. We wish especially to thank Dr. Ruth Ellen Bulger who supplied several electron micrographs as noted in the text.

REFERENCES

Arstila, A. U., and Trump, B. F. (1968). Studies on cellular autophagocytosis. The formation of autophagic vacuoles in the liver after glucagon administration. *Am. J. Pathol.* 53, 687–733.

Audigé, J. (1910). Contribution a l'étude des reins des poissons téléostéens. *Arch. Zool. Exptl. Gen.* [5] 4, 275–624.

Bahr, K. (1952). Beitrage zur Biologie des Flussneunauges, *Petromyzon fluviatilis* L. (Lebensraum und Ernahrung). *Zool. Jahrb.* (*Syst.*) 81, 408–436.

Beament, J. W. L. (1965). The active transport of water: Evidence, models and mechanism. *Symp. Soc. Exptl. Biol.* 19, 273–298.

Bentley, P. J. (1962). Permeability of the skin of the cyclostome *Lampetra fluviatilis* to water and electrolytes. *Comp. Biochem. Physiol.* 6, 95–97.

Bentley, P. J., and Follett, B. K. (1963). Kidney function in a primitive vertebrate, the cyclostome *Lampetra fluviatilis. J. Physiol.* (*London*) 169, 902–918.

Berg, L. S. (1940). "Classification of Fishes, Both Recent and Fossil," Vol. 5, No. 2. Trav. Inst. Zool. Acad. Sci., U.S.S.R. (Russian and English texts; also reprint, Ann Arbor, Michigan, 1947).

Berglund, F., and Forster, R. P. (1958). Renal tubular transport of inorganic divalent ions by the aglomerular marine teleost, *Lophius americanus. J. Gen. Physiol.* 41, 429–440.

Bern, H. A. (1967). Hormones and endocrine glands of fishes. *Science* 158, 455–462.

Bernard, G. R., Wynn, R. A., and Wynn, G. G. (1966). Chemical anatomy of the pericardial and perivisceral fluids of the stingray, *Dasyatis americana. Biol. Bull.* 130, 18–27.

Berridge, M. J., and Gupta, B. L. (1967). Fine-structural changes in relation to ion and water transport in the rectal papillae of the blowfly, *Calliphora. J. Cell Sci.* 2, 89–112.

Bieter, R. N. (1931a). The action of some diuretics upon the aglomerular kidney. *J. Pharmacol. Exptl. Therap.* 43, 399–406.

Bieter, R. N. (1931b). Albuminuria in glomerular and aglomerular fish. *J. Pharmacol. Exptl. Therap.* 43, 407–412.

Bieter, R. N. (1931c). The secretion pressure of the aglomerular kidney. *Am. J. Physiol.* 97, 66–68.

Bieter, R. N. (1933). Further studies concerning the action of diuretics upon the aglomerular kidney. *J. Pharmacol. Exptl. Therap.* 49, 250–256.

Bieter, R. N. (1935). The action of diuretics injected into one kidney of the aglomerular toadfish. *J. Pharmacol. Exptl. Therap.* 53, 347–349.

Black, V. S. (1957). Excretion and osmoregulation. In "The Physiology of Fishes" (M. E. Brown, ed.), Vol. 1, p. 163–205. Academic Press, New York.

Bohle, A., and Walvig, F. (1964). Beitrag zur vergleichenden Morphologie der epitheloiden Zellen der Nierenarteriolen unter besonderer Berücksichtigung der epitheloiden Zellen in den Nieren von Seewasserfischen. *Klin. Wochschr.* **42,** 415–421.

Bourguet, J., Lahlou, B., and Maetz, J. (1964). Modifications expérimentales de l'équilibre hydrominéral et osmorégulation chez *Carassius auratus. Gen. Comp. Endocrinol.* **4,** 563–576.

Bradford, W. D., Elchlepp, J. G., Arstila, A. U., Trump, B. F., and Kinney, T. D. (1969). Iron metabolism and cell membranes. I. Relation between ferritin and hemosiderin in bile and biliary excretion of lysosome contents. *Am. J. Pathol.* (in press).

Brull, L., and Cuypers, Y. (1954a). Quelques charactéristics biologiques de *Lophius piscatorius* L. *Arch. Intern. Physiol.* **62,** 70–75.

Brull, L., and Cuypers, Y. (1954b). Blood perfusion of the kidney of *Lophius piscatorius* L. II. Influence of perfusion pressure on urine volume. *J. Marine Biol. Assoc. U.K.* **33,** 733–738.

Brull, L., and Nizet, E. (1953). Blood and urine constituents of *Lophius piscatorius* L. *J. Marine Biol. Assoc. U.K.* **32,** 321–328.

Brull, L., Nizet, E., and Verney, E. B. (1953). Blood perfusion of the kidney of *Lophius piscatorius* L. *J. Marine Biol. Assoc. U.K.* **32,** 329–336.

Bulger, R. E. (1965). The fine structure of the aglomerular nephron of the toadfish, *Opsanus tau. Am. J. Anat.* **117,** 171–192.

Bulger, R. E., and Trump, B. F. (1965). Effects of fixatives on tubular ultrastructure of the aglomerular midshipman, *Porichthys notatus,* and the glomerular flounder, *Parophrys vetulus. J. Histochem. Cytochem.* **13,** 719.

Bulger, R. E., and Trump, B. F. (1968). Renal morphology of the English sole, (*Parophrys vetulus*). *Am. J. Anat.* **123,** 195–225.

Bulger, R. E., and Trump, B. F. (1969a). Ultrastructure of granulated arteriolar cells (juxtaglomerular cells) in kidney of a fresh and a salt water teleost. *Am. J. Anat.* **124,** 77–88.

Bulger, R. E., and Trump, B. F. (1969b). A mechanism for rapid transport of colloidal particles by flounder renal epithelium. *J. Morphol.* **127,** 205–224.

Bulger, R. E., and Trump, B. F. (1969c). Ca^{++} and K^+ effects on ultrastructure of isolated flounder kidney tubules. *J. Ultrastruct. Res.* (in press).

Bull, J. M., and Morris, R. (1967). Studies on freshwater osmoregulation in Ammocoete larva of *Lampetra planeri* (Bloch). I. Ionic constituents, fluid compartments, ionic compartments and water balance. *J. Exptl. Biol.* **47,** 485–494.

Burg, M., and Orloff, J. (1962). Effect of strophanthidin on electrolyte content and PAH accumulation of rabbit kidney slices. *Am. J. Physiol.* **202,** 565–571.

Burger, J. W. (1962). Further studies on the function of the rectal gland in the spiny dogfish. *Physiol. Zool.* **35,** 205–217.

Burger, J. W. (1965). Roles of the rectal gland and the kidneys in salt and water excretion in the spiny dogfish. *Physiol. Zool.* **38,** 191–196.

Burger, J. W. (1967). Problems in the electrolyte economy of the spiny dogfish, *Squalus acanthias. In* "Sharks, Skates and Rays" (P. W. Gilbert, R. F. Mathewson, and D. P. Rall, eds.), pp. 177–185. Johns Hopkins Press, Baltimore, Maryland.

Burger, J. W., and Hess, W. N. (1960). Function of the rectal gland in the spiny dogfish. *Science* **131,** 670–671.

Butler, D. G. (1966). Effect of hypophysectomy on osmoregulation in the European eel (*Anguilla anguilla* L.). *Comp. Biochem. Physiol.* 18, 773–781.
Capreol, S. V., and Sutherland, L. E. (1968). Comparative morphology of juxtaglomerular cells. I. Juxtaglomerular cells in fish. *Can. J. Zool.* 46, 249–256.
Chester Jones, I., Chan, D. K. O., and Rankin, J. C. (1969). Renal function in the European eel (*Anguilla anguilla* L.): Changes in blood pressure and renal function of the freshwater eel transferred to sea-water. *J. Endocrinol.* 43, 9–19.
Chester Jones, I., Henderson, I. W., and Butler, D. G. (1965). Water and electrolyte flux in the European eel (*Anguilla anguilla* L.). *Arch. Anat. Microscop.Morphol. Exptl.* 54, 453–468.
Clarke, R. W. (1934). The xylose clearance of *Myoxocephalus octodecimspinosus* under normal and diuretic conditions. *J. Cellular Comp. Physiol.* 5, 73–82.
Clarke, R. W. (1935). Experimental production of diuresis in the dogfish. *Bull. Mt. Desert Isl. Biol. Lab.* pp. 25–26.
Clarke, R. W. (1936). Simultaneous xylose and inulin clearances in the sculpin. *Bull. Mt. Desert Isl. Biol. Lab.* p. 25.
Clarke, R. W., and Smith, H. W. (1932). Absorption and excretion of water and salts by the elasmobranch fishes III. The use of xylose as a measure of the glomerular filtrate in *Squalus acanthias. J. Cellular Comp. Physiol.* 1, 131–143.
Cohen, J. J. (1959). The capacity of the kidney of the marine dogfish, *Squalus acanthias*, to secrete hydrogen ion. *J. Cellular Comp. Physiol.* 53, 205–213.
Conel, J. (1917). The urogenital system of Myxinoids. *J. Morphol.* 29, 75–163.
Courrier, R. (1922). Etude préliminaire du déterminisme des caractères sexuels secondaires chez les poissons. *Arch. Anat.* (*Strasbourg*) 1, 118–144.
Crane, R. K. (1965). Na⁺-dependent transport in the intestine and other animal tissues. *Federation Proc.* 24, 1000–1006.
Daikoku, T. (1965). Renal corpuscles of the guppies, *Lebistes reticulatus*, reared in sea water. *Physiol. Ecol.* (*Kyoto*) 13, 49.
Denis, W. (1914). Metabolism studies on cold-blooded animals. II. The blood and urine of fish. *J. Biol. Chem.* 16, 389–393.
Denison, R. H. (1956). A review of the habitat of the earliest vertebrates. *Fieldiana, Geol.* 11, 359–457.
Diamond, J. M. (1965). The mechanism of isotonic water absorption and secretion. *Symp. Soc. Exptl. Biol.* 19, 329–347.
Edwards, J. G. (1928). Studies on aglomerular and glomerular kidneys. I. Anatomical. *Am. J. Anat.* 42, 75–107.
Edwards, J. G. (1930). Studies on aglomerular and glomerular kidneys. III. Cytological. *Anat. Record* 44, 15–27.
Edwards, J. G. (1935). The epithelium of the renal tubule in bony fish. *Anat. Record* 63, 263–279.
Edwards, J. G., and Condorelli, L. (1928). Studies on aglomerular and glomerular kidneys. II. Physiological. *Am. J. Physiol.* 86, 383–398.
Edwards, J. G., and Schnitter, C. (1933). The renal unit in the kidney of vertebrates. *Am. J. Anat.* 53, 55–87.
Enomoto, Y. (1964). A transient glucosuria (diabetes mellitus) of rainbow-trout (*Salmo irideus*) induced by bovine growth hormone injection. *Bull. Japan. Soc. Sci. Fisheries* 30, 533–536.
Ericsson, J. L. E. (1967). Unpublished data.
Ericsson, J. L. E. (1964). Absorption and decomposition of homologous hemoglobin

in renal proximal tubular cells. An experimental light and electron microscopic study. *Acta Pathol. Microbiol. Scand.* Suppl. 168, 1–121.

Ericsson, J. L. E., and Trump, B. F. (1969). Electron microscopy of the uriniferous tubules. *In* "The Kidney: Morphology, Biochemistry and Physiology" (C. Rouiller and A. Muller, eds.). Vol. I, pp. 351–440. Academic Press, New York.

Ericsson, J. L. E., Trump, B. F., and Weibel, J. (1965). Electron microscopy studies of the proximal tubule of the rat kidney. II. Cytosegresomes and cytosomes: Their relationship to each other and to the lysosome concept. *Lab. Invest.* 14, 1341–1365.

Evans, D. H. (1967). Sodium, chloride and water balance of the intertidal teleost, *Xiphister atropurpureus*. II. The role of the kidney and the gut. *J. Exptl. Biol.* 47, 519–534.

Faith, G. C., and Trump, B. F. (1966). Comparative electron microscopic observations on the glomerular capillary wall in human renal disease. Acute glomerulonephritis, systemic lupus erythematosus and preeclampsia-eclampsia. *Lab. Invest.* 15, B1682–1719.

Falkmer, S., and Matty, A. J. (1966). Blood sugar regulation in the hagfish, *Myxine glutinosa*. *Gen. Comp. Endocrinol.* 6, 334–346.

Fanelli, G. M., Jr., and Nigrelli, R. F. (1963). Renal excretion of tetracycline in the aglomerular toadfish, *Opsanus tau*. *Proc. Soc. Exptl. Biol. Med.* 114, 582.

Fänge, R. (1963). Structure and function of the excretory organs of myxinoids. *In* "Biology of Myxine" (A. Brodal and R. Fänge, eds.), pp. 516–529. Oslo Univ. Press, Oslo.

Fänge, R., and Fugelli, K. (1962). Osmoregulation in chimaeroid fishes. *Nature* 196, 689.

Fänge, R., and Krog, J. (1963). Inability of the kidney of the hagfish to secrete phenol red. *Nature* 199, 713.

Fleming, W. R., and Stanley, J. G. (1965). Effects of rapid changes in salinity on the renal function of a euryhaline teleost. *Am. J. Physiol.* 209, 1025–1030.

Ford, P. (1958). Studies on the development of the kidney of the Pacific pink salmon *Oncorhynchus gorbuscha* (Walbaum). II. Variation in glomerular count of the kidney of the Pacific pink salmon. *Can. J. Zool.* 36, 45–47.

Forster, R. P. (1942). The nature of glucose reabsorptive process in the frog renal tubule: Evidence for intermittency of glomerular function in the intact animal. *J. Cellular Comp. Physiol.* 20, 55–69.

Forster, R. P. (1948). Use of thin kidney slices and isolated renal tubules for direct study of cellular transport kinetics. *Science* 108, 65–67.

Forster, R. P. (1953). A comparative study of renal function in marine teleosts. *J. Cellular Comp. Physiol.* 42, 487–510.

Forster, R. P. (1954). Active cellular transport of urea by frog renal tubules. *Am. J. Physiol.* 179, 372–377.

Forster, R. P. (1961). Kidney cells. *In* "The Cell" (J. Brachet and A. E. Mirsky, eds.) Vol. 5, pp. 89–161. Academic Press, New York.

Forster, R. P. (1967a). Renal transport mechanisms. *Federation Proc.* 26, 1008–1019.

Forster, R. P. (1967b). Osmoregulatory role of the kidney in cartilaginous fishes (Chondrichthys). *In* "Sharks, Skates and Rays" (P. W. Gilbert, R. F. Mathewson, and D. P. Rall, eds.), pp. 187–195. Johns Hopkins Press, Baltimore, Maryland.

Forster, R. P., and Berglund, F. (1956). Osmotic diuresis and its effect on total electrolyte distribution in plasma and urine of the aglomerular teleost, *Lophius americanus*. *J. Gen. Physiol.* 39, 349–359.

Forster, R. P., and Berglund, F. (1957). Contrasting inhibitory effects of probenecid on the renal tubular excretion of PAH and on the active reabsorption of urea in the dogfish, *Squalus acanthias*. *J. Cellular Comp. Physiol.* **49**, 281–285.

Forster, R. P., and Hong, S. K. (1958). In vitro transport of dyes by isolated renal tubules of the flounder as disclosed by direct visualization. Intracellular accumulation and transcellular movement. *J. Cellular Comp. Physiol.* **51**, 259–272.

Forster, R. P., Sperber, I., and Taggart, J. V. (1954). Transport of phenolsulfonphthalein dyes in isolated tubules of the flounder and in kidney slices of the dogfish. Competitive phenomena. *J. Cellular Comp. Physiol.* **44**, 315–318.

Fromm, P. O. (1963). Studies on renal and extra-renal excretion in a freshwater teleost, *Salmo gairdneri*. *Comp. Biochem. Physiol.* **10**, 121–128.

Gerard, P. (1943). Sur le mésonéphros des Myxinides. *Arch. Zool. Exptl. Gen. Notes Rev.* **83**, 37–42.

Gerard, P. (1954). Organes uro-genitaux. In "Traité de zoologie" (P.-P. Grassé, ed.), Vol. 12, pp. 974–1043. Masson, Paris.

Ginetsinkii, A. G., Vasileva, V. F., and Natochin, Yu. V. (1961). Reaktsiya ryb na izmenenie solenosti sredy. (The reaction of fishes to a change in the salinity of the water.) In "Problemy Evolyutsii Funktsii i Enzimochimii Protsessov Vozbuzhdehiya" (Problems of the evolution of function and of enzyme chemistry of the processes of stimulation.), pp. 89–102. Akad. Nauk. S.S.S.R., Moscow.

Ginn, F. L., Shelburne, J. D., and Trump, B. F. (1968). Disorders of cell volume regulation. I. Effects of inhibition of plasma membrane adenosine triphosphatase with ouabain. *Am. J. Pathol.* **53**, 1041–1071.

Goldstein, L., Oppelt, W. W., and Maren, T. H. (1968). Osmotic regulation and urea metabolism in the lemon shark *Negaprion brevirostris*. *Am. J. Physiol.* **215**, 1493–1497.

Grafflin, A. L. (1931a). The structure of the renal tubule of the toadfish. *Bull. Johns Hopkins Hosp.* **48**, 269–271.

Grafflin, A. L. (1931b). Urine flow and diuresis in marine teleosts. *Am. J. Physiol.* **97**, 602–610.

Grafflin, A. L. (1933). Glomerular degeneration in the kidney of the daddy sculpin (*Myoxocephalus scorpius*). *Anat. Record* **57**, 59–79.

Grafflin, A. L. (1935). Renal function in marine teleosts. I. Urine flow and urinary chloride. *Biol. Bull.* **69**, 391–402.

Grafflin, A. L. (1936a). Renal function in marine teleosts. IV. The excretion of inorganic phosphate in the sculpin. *Biol. Bull.* **71**, 360–374.

Grafflin, A. L. (1936b). Renal function in marine teleosts. III. The excretion of urea. *Biol. Bull.* **70**, 228–235.

Grafflin, A. L. (1937a). The structure of the nephron in the sculpin, *Myoxocephalus octodecimspinosus*. *Anat. Record* **68**, 145–163.

Grafflin, A. L. (1937b). Observations upon the aglomerular nature of certain teleostean kidneys. *J. Morphol.* **61**, 165–169.

Grafflin, A. L. (1937c). Observations upon the structure of the nephron in the common eel. *Am. J. Anat.* **61**, 21–62.

Grafflin, A. L. (1937d). The structure of the nephron in fishes. Representative types of nephron encountered; the problem of homologies among the differentiated portions of the proximal convoluted segments. *Anat. Record* **68**, 287–303.

Grafflin, A. L., and Ennis, D. (1934). The effect of blockage of the gastrointestinal tract upon urine formation in a marine teleost, *Myoxocephalus octodecimspinosus*. *J. Cellular Comp. Physiol.* **4**, 283–296.

Grafflin, A. L., and Gould, R. G., Jr. (1936). Renal function in marine teleosts. II. The nitrogenous constituents of the urine of sculpin and flounder, with particular reference to trimethylamine oxide. *Biol. Bull.* **70,** 16–27.

Grassé, P.-P. (1954). Classe des cyclostomes. In "Traité de zoologie" (P.-P. Grassé, ed.), Vol. 13, p. 98. Masson, Paris.

Griffith, L. D., Bulger, R. E., and Trump, B. F. (1967). The ultrastructure of the functioning kidney. *Lab. Invest.* **16,** 220–246.

Grollman, A. (1929). The urine of the goosefish (*Lophius piscatorius*): Its nitrogenous constituents with special reference to the presence in it of trimethylamine oxide. *J. Biol. Chem.* **81,** 267–278.

Guitel, F. (1906). Recherches sur l'anatomie des reins de quelques Gobiésocidés (*Lepadogaster, Caularchus, Gobiesox, Syciases* et *Chorisochismus*). *Arch. Zool. Exptl. Gen.* [4] **5,** 505–698.

Gunter, G. (1942). A list of the fishes of the mainland of North and Middle America recorded from both fresh water and sea water. *Am. Midland Naturalist* **28,** 305–326.

Gunter, G. (1956). A revised list of euryhaline fishes of North and Middle America. *Am. Midland Naturalist* **56,** 345–354.

Guyton, J. S. (1935). The structure of the nephron in the South American lungfish, *Lepidosiren paradoxa. Anat. Record* **63,** 213–230.

Hammond, B. R. (1969). Renal function and the effects of arginine vasotocin in lake trout, *Salvelinus namaycush* (Walbaum). Doctoral thesis, Cornell Univ.

Hardisty, M. W. (1956). Some aspects of osmotic regulation in lampreys. *J. Exptl. Biol.* **33,** 431–447.

Haller, B. (1902). Über die Urniere von *Acanthias vulgaris. Morph. Jahrb.* **29,** 283–316.

Haywood, C., and Clapp, M. J. (1942). A note on the freezing points of the urines of two freshwater fishes: the catfish (*Ameiurus nebulosus*) and the sucker (*Catostomus commersonii*). *Biol. Bull.* **83,** 363–366.

Hickman, C. P., Jr. (1965). Studies on renal function in freshwater teleost fish. *Tran. Roy. Soc. Can. Sect. III* [4] **3,** 213–236.

Hickman, C. P., Jr. (1968a). Glomerular filtration and urine flow in the euryhaline southern flounder, *Paralichthys lethostigma,* in sea water. *Can. J. Zool.* **46,** 427–437.

Hickman, C. P., Jr. (1968b). Urine composition and kidney function in southern flounder, *Paralichthys lethostigma,* in sea water. *Can. J. Zool.* **46,** 439–455.

Hickman, C. P., Jr. (1968c). Ingestion, intestinal absorption and elimination of sea water and salts in the southern flounder, *Paralichthys lethostigma. Can. J. Zool.* **46,** 457–466.

Hickman, C. P., Jr. (1969). Unpublished observations.

Hodler, J., Heinemann, H. O., Fishman, A. P., and Smith, H. W. (1955). Urine pH and carbonic anhydrase activity in the marine dogfish. *Am. J. Physiol.* **183,** 155–162.

Holmes, R. M. (1961). Kidney function in migrating salmonids. *Rept. Challenger Soc. (Cambridge)* **3,** No. 13.

Holmes, W. N., and McBean, R. L. (1963). Studies on the glomerular filtration rate of rainbow trout (*Salmo gairdneri*). *J. Exptl. Biol.* **40,** 335–341.

Holmes, W. N., and Stainer, I. M. (1966). Studies on the renal excretion of electrolytes by the trout (*Salmo gairdneri*). *J. Exptl. Biol.* **44,** 33–46.

Holmgren, N. (1950). On the pronephros and the blood of *Myxine glutinosa. Acta Zool. (Stockholm)* **31,** 233–348.

Horowicz, P., and Burger, J. W. (1968). Unidirectional fluxes of sodium ions in the spiny dogfish, *Squalus acanthias. Am. J. Physiol.* **214,** 635–642.

Hunn, J. B. (1967). Personal communication.

Hunn, J. B. (1969). Chemical composition of rainbow trout urine following acute hypoxic stress. *Tran. Am. Fish. Soc.* **98**, 20–22.

Hunn, J. B., and Willford, W. A. (1968). Flow rates and chemical composition of urine from rainbow trout, *Salmo gairdneri*, after MS 222 or methylpentynol anesthesia. Manuscript.

Hyman, L. H. (1942). "Comparative Vertebrate Anatomy." Univ. of Chicago Press, Chicago, Illinois.

Johansen, K. (1960). Circulation in the hagfish, *Myxine glutinosa* L. *Biol. Bull.* **118**, 289–295.

Kempton, R. T. (1940). The site of acidification of urine within the renal tubule of the dogfish. *Bull. Mt. Desert Isl. Biol. Lab.* p. 34.

Kempton, R .T. (1943). Studies on the elasmobranch kidney. I. The structure of the renal tubule of the spiny dogfish (*Squalus acanthias*). *J. Morphol.* **73**, 247–263.

Kempton, R. T. (1953). Studies on the elasmobranch kidney. II. Reabsorption of urea by the smooth dogfish, *Mustelus canis. Biol. Bull.* **104**, 45–56.

Kempton, R. T. (1956). The problem of the "special segment" of the elasmobranch kidney tubule. *Year Book Am. Phil. Soc.* pp. 210–212.

Kempton, R. T. (1962). Studies on the elasmobranch kidney. III. The kidney of the lesser electric ray, *Narcine brasiliensis. J. Morphol.* **111**, 217–225.

Kempton, R. T. (1966). Studies on the elasmobranch kidney. IV. The secretion of phenol red by the smooth dogfish, *Mustelus canis. Biol. Bull.* **130**, 359–368.

Kerr, J. G. (1919). "Textbook of Embryology," Vol. II. Macmillan, New York.

Kinter, W. B. (1966). Chlorphenol red influx and efflux: Microspectrophotometry of flounder kidney tubules. *Am. J. Physiol.* **211**, 1152–1164.

Kinter, W. B., and Cline, A. L. (1961). Exchange diffusion and runout of Diodrast I^{131} from renal tissue in vitro. *Am. J. Physiol.* **201**, 309–317.

Krause, R. (1923). "Mikroskopische Anatomie der *Wirbeltiere* bei Einzeldarstellungen." Berlin and Leipzig.

Krogh, A. (1939). "Osmotic Regulation in Aquatic Animals." Cambridge Univ. Press, London and New York.

Lagler, K. F., Bardach, J. E., and Miller, R. R. (1962). "Ichthyology." Wiley, New York.

Lahlou, B. (1966). Mise en évidence d'un "recrutement glomérulaire" dans le rein des Téléostéens d'après la mesure du Tm glucose. *Compt. Rend.* **262**, 1356–1358.

Lahlou, B. (1967). Excretion renale chez un poisson euryhalin, le flet (*Platichthys flesus* L.): Caractéristiques de l'urine normale en eau douce et en eau de mer et effets des changements de milieu. *Comp. Biochem. Physiol.* **20**, 925–938.

Lahlou, B., and Sawyer, W. H. (1969). Sodium exchanges in the toadfish, *Opsanus tau*, a euryhaline aglomerular teleost. *Am. J. Physiol.* (in press).

Lahlou, B., Henderson, I. W., and Sawyer, W. H. (1969). Renal adaptations by *Opsanus tau*, a euryhaline aglomerular teleost, to dilute media. *Am. J. Physiol.* (in press).

Lam, T. J., and Hoar, W. S. (1967). Seasonal effects of prolactin on freshwater osmoregulation of the marine form (*trachurus*) of the stickleback *Gasterosteus aculeatus. Can. J. Zool.* **45**, 509–516.

Lam, T. J., and Leatherland, J. F. (1969). Effects of prolactin on the glomerulus of the marine threespine stickleback, *Gasterosteus aculeatus* L., form *trachurus*, after transfer from seawater to freshwater, during late autumn and early winter. *Can. J. Zool.* **47**, 245–250.

Lavoie, M. E., Chidsey, J. L., and Forster, R. P. (1959). Glomerular filtration rate and renal plasma flow in the freshwater brown bullhead or horned pout, *Ameiurus nebulosus. Bull. Mt. Desert Isl. Biol. Lab.* pp. 50–51.

234   CLEVELAND P. HICKMAN, JR., AND BENJAMIN F. TRUMP

Lozovik, V. I. (1963). Effect of water salinity on the development of the glomerular apparatus in the kidneys of marine Teleostei. *Dokl. Akad. Nauk. SSSR* 153, 225–228.

Maack, T., and Kinter, W. B. (1969). Transport of a small molecular weight protein, lysozyme, in dissected and incubated tubular masses of the flounder kidney. *Bull. Mt. Desert Isl. Biol. Lab.* 7 (in press).

McFarland, W. N., and Munz, F. W. (1958). A re-examination of the osmotic properties of the Pacific hagfish, *Polistotrema stouti. Biol. Bull.* 114, 348–356.

McFarland, W. N., and Munz, F. W. (1965). Regulation of body weight and serum composition by hagfish in various media. *Comp. Biochem. Physiol.* 14, 383–398.

Mackay, W. C. (1967). Plasma glucose levels in the northern pike, *Esox lucius* and the white sucker, *Catostomus commersonii*, and renal glucose reabsorption in the white sucker. Master's thesis, University of Alberta.

Mackay, W. C., and Beatty, D. D. (1968). The effect of temperature on renal function in the white sucker fish, *Catostomus commersonii. Comp. Physiol. Biochem.* 26, 235–245.

Maetz, J. (1963). Physiological aspects of neurohypophysial function in fishes with some reference to the amphibia. *Symp. Zool. Soc. London* 9, 107–140.

Maetz, J. (1968). Salt and water metabolism. *In* "Perspectives in Endocrinology. Hormones in the Lives of Lower Vertebrates" (E. J. W. Barrington and C. G. Jorgensen, eds.), pp. 47–162. Academic Press, New York.

Maetz, J., and Garcia Romeu, F. (1964). The mechanisms of sodium and chloride uptake by the gills of a freshwater fish, *Carassius auratus*. III. Evidence for NH_4^+/Na^+ and HCO_3^-/Cl^- exchanges. *J. Gen. Physiol.* 47, 1209–1227.

Maetz, J., and Skadhauge, E. (1968). Drinking rates and gill ionic turnover in relation to external salinities in the eel. *Nature* 217, 371–373.

Maetz, J., Bourguet, J., Lahlou, B., and Hourdry, J. (1964). Peptides neurohypophysaires et osmoregulation chez *Carassius auratus. Gen. Comp. Endocrinol.* 4, 508–522.

Malvin, R. L., and Fritz, I. B. (1962). Renal transport of urea and some carbohydrates in *Lophius piscatorius. J. Cellular Comp. Physiol.* 59, 111–116.

Malvin, R. L., Cafruni, E. J., and Kutchai, H. (1965). Renal transport of glucose by the aglomerular fish *Lophius americanus. J. Cellular Comp. Physiol.* 65, 381–384.

Maren, T. H. (1967). Special body fluids of the elasmobranch. *In* "Sharks, Skates and Rays" (P. W. Gilbert, R. F. Mathewson, and D. P. Rall, eds.), pp. 287–292. Johns Hopkins Press, Baltimore, Maryland.

Marshall, E. K., Jr. (1929). The aglomerular kidney of the toadfish (*Opsanus tau*). *Bull. Johns Hopkins Hosp.* 45, 95–100.

Marshall, E. K., Jr. (1930). A comparison of the function of the glomerular and aglomerular kidney. *Am. J. Physiol.* 94, 1–10.

Marshall, E. K., Jr. (1934). The comparative physiology of the kidney in relation to theories of renal secretion. *Physiol. Rev.* 14, 133–159.

Marshall, E. K., Jr., and Grafflin, A. L. (1928). The structure and function of the kidney of *Lophius piscatorius. Bull. Johns Hopkins Hosp.* 43, 205–230.

Marshall, E. K., Jr., and Grafflin, A. L. (1932). The function of the proximal convoluted segment of the renal tubule. *J. Cellular Comp. Physiol.* 1, 161–176.

Marshall, E. K., Jr., and Grafflin, A. L. (1933). Excretion of inorganic phosphate by the aglomerular kidney. *Proc. Soc. Exptl. Biol. Med.* 31, 44–46.

Marshall, E. K., Jr., and Smith, H. W. (1930). The glomerular development of the vertebrate kidney in relation to habitat. *Biol. Bull.* 59, 135–153.

Martret, G. (1939). Variations de la concentration moléculaire et de la concentration en chlorures de l'urine des Téléostéens stenohalins en fonction des variations de salinité du milieu extérieur. *Bull. Inst. Oceanog.* **774**, 1–38.

Mashiko, K., and Jozuka, K. (1964). Absorption and excretion of calcium by teleost fishes with special reference to routes followed. *Annotations Zool. Japon.* **37**, 41–50.

Moore, R. A. (1933). Morphology of the kidneys of Ohio fishes. *Contrib. Stone Lab. Ohio Univ.* **5**, 1–34.

Morris, R. (1956). The osmoregulatory ability of the lampern (*Lampetra fluviatilis* L.) in sea water during the course of its spawning migration. *J. Exptl. Biol.* **33**, 235–248.

Morris, R. (1958). The mechanism of marine osmoregulation in the lampern (*Lampetra fluviatilis* L.) and the causes of its breakdown during the spawning migration. *J. Exptl. Biol.* **35**, 649–665.

Morris, R. (1960). General problems of osmoregulation with special reference to cyclostomes. *Symp. Zool. Soc. London* **1**, 1–16.

Morris, R. (1965). Studies on salt and water balance in *Myxine glutinosa* (L). *J. Exptl. Biol.* **42**, 359–371.

Motais, R. (1967). Les mécanismes d'échanges ioniques branchiaux chez les téléostéens. *Ann. Inst. Oceanog.* (*Paris*) [N.S.] **45**, 1–83.

Motais, R., Garcia Romeu, F., and Maetz, J. (1966). Exchange diffusion effect and euryhalinity in teleosts. *J. Gen. Physiol.* **50**, 391–422.

Muller, W. (1875). Ueber das Urogenitalsystem des Amphioxus und der Cyclostomen. *Jena. Z. Naturw.* **9**, 94–129.

Munz, F. W., and McFarland, W. N. (1964). Regulatory function of a primitive vertebrate kidney. *Comp. Biochem. Physiol.* **13**, 381–400.

Murdaugh, H. V., Jr., and Robin, E. D. (1967). Acid-base metabolism in the dogfish shark. *In* "Sharks, Skates and Rays" (P. W. Gilbert, R. F. Mathewson, and D. P. Rall, eds.), pp. 249–264. Johns Hopkins Press, Baltimore, Maryland.

Murdaugh, H. V., Jr., Soteres, P., Pyron, W., and Weiss, E. (1963). Movement of inulin and bicarbonate ion across the bladder of an aglomerular teleost, *Lophius americanus*. *J. Clin. Invest.* **42**, 959.

Nash, J. (1931). The number and size of glomeruli in the kidneys of fishes, with observations on the morphology of the renal tubules of fishes. *Am. J. Anat.* **47**, 425–445.

Newstead, J. D., and Ford, P. (1960). Studies on the development of the kidney of the Pacific pink salmon *Oncorhynchus gorbuscha* (Walbaum). III. The development of the mesonephros with particular reference to the mesonephric tubule. *Can. J. Zool.* **38**, 1–7.

Nizet, E., and Wilsens, L. (1955). Quelques aspects de l'anatomie du rein de *Lophius piscatorius* L. *Pubbl. Staz. Zool. Napoli* **26**, 36–41.

Ogawa, M. (1961a). Comparative study of the external shape of the teleostean kidney with relation to phylogeny. *Sci. Rept. Tokyo Kyoiku Daigaku* **B10**, 61–68.

Ogawa, M. (1961b). Histological changes of the kidney in goldfish in sea water. *Sci. Rept. Saitama Univ.* **B4**, 1–20.

Ogawa, M. (1962). Comparative study on the internal structure of the teleostean kidney. *Sci. Rept. Saitama Univ.* **B4**, 107–129.

Ogawa, M. (1968a). Personal communication.

Ogawa, M. (1968b). Seasonal difference of glomerular change of the marine form of the stickleback, *Gasterosteus aculeatus* L., after transferred into fresh water. *Sci. Rept. Saitama Univ.* **B5**, 117–123.

Oguri, M. (1964). Rectal glands of marine and freshwater sharks: Comparative histology. *Science* **144**, 1151–1152.

Oguri, M. (1968). Urinary constituents of snake-head fish, with special reference to urine sugar. *Bull. Japon. Soc. Sci. Fisheries* **34**, 6–10.

Oguri, M., and Takada, N. (1965). pH, calcium level and osmotic concentration of the urine in the snake-head fish, *Channa argus*. *Bull. Japan. Soc. Sci. Fisheries* **31**, 293–296.

Oide, H., and Utida, S. (1968). Changes in intestinal absorption and renal excretion of water during adaptation to sea-water in the Japanese eel. *Marine Biol.* **1**, 172–177.

Olivereau, M., and Lemoine, A. (1968). Action de la prolactine chez l'anguille intacte. III. Effet sur la structure histologique du rein. *Z. Zellforsch.* **88**, 576–590.

Ong, K. S. (1968). Personal communication.

Parry, G. (1966). Osmotic adaptation in fishes. *Biol. Rev.* **41**, 392–444.

Pitts, R. F. (1934). Urinary composition in marine fish. *J. Cellular Comp. Physiol.* **4**, 389–395.

Pitts, R. F. (1963). "Physiology of the Kidney and Body Fluids." Year Book Publ., Chicago, Illinois.

Pora, A. E., and Prekup, O. (1960). A study of the excretory processes of freshwater fish. Part II. Influence of environmental temperature on the excretory processes of the carp and Cruscian carp. *Vopr. Ikhtiol.* **15**, 138–147; *Biol. Abstr.* **37**(4), 13391 (1962).

Potts, W. T. W., and Evans, D. H. (1967). Sodium and chloride balance in the killifish, *Fundulus heteroclitus*. *Biol. Bull.* **47**, 461–470.

Potts, W. T. W., and Parry, G. (1964). "Osmotic and Ionic Regulation in Animals." Macmillan, New York.

Price, G. C. (1910). The structure and function of the adult head kidney of *Bdellostoma stouti*. *J. Exptl. Zool.* **9**, 849–864.

Price, K. S., Jr. (1967). Fluctuations in two osmoregulatory components, urea and sodium chloride, of the clearnose skate. *Raja eglanteria* Bosc 1802. II. Upon natural variation of the salinity of the external medium. *Comp. Biochem. Physiol.* **23**, 77–82.

Price, K. S., Jr., and Creaser, E. P., Jr. (1967). Fluctuations in two osmoregulatory components, urea and sodium chloride, of the clearnose skate, *Raja eglanteria* Bosc 1802. I. Upon laboratory modification of external salinities. *Comp. Biochem. Physiol.* **23**, 65–76.

Puck, T. T., Wasserman, K., and Fishman, A. P. (1952). Some effects of inorganic ions on the active transport of phenol red by isolated kidney tubules of the flounder. *J. Cellular Comp. Physiol.* **40**, 73–88.

Rall, D. P., and Burger, J. W. (1967). Some aspects of hepatic and renal excretion in *Myxine*. *Am. J. Physiol.* **212**, 354–356.

Read, L. J. (1968). A study of ammonia and urea production and excretion in the fresh-water-adapted form of the Pacific lamprey, *Entosphenus tridentalus*. *Comp. Biochem. Physiol.* **26**, 455–466.

Regaud, C., and Policard, A. (1902). Etude sur le tube urinifère de la lamproie. *Compt. Rend. Assoc. Anat.* **4**, 245–261.

Reid, D. F., Ego, W. T., and Townsley, S. J. (1959). Ion exchange through epithelia of fresh and sea water adapted teleost studied with radioactive isotopes. *Anat. Record* **134**, 628.

Robertson, J. D. (1954). The chemical composition of the blood of some aquatic chordates, including members of the Tunicata, Cyclostomata and Osteichthyes. *J. Exptl. Biol.* 31, 424–442.

Robertson, J. D. (1957). The habitat of the early vertebrates. *Biol. Rev.* 32, 156–187.

Robertson, J. D. (1963). Osmoregulation and ionic composition of cells and tissues. In "Biology of Myxine" (A. Brodal and R. Fänge, eds.), pp. 503–515. Oslo Univ. Press, Oslo.

Robin, E. D., and Murdaugh, H. V., Jr. (1967). Gill gas exchange in the elasmobranch, *Squalus acanthias.* In "Sharks, Skates and Rays" (P. W. Gilbert, R. F. Mathewson, and D. P. Rall, eds.), pp. 221–247. Johns Hopkins Press, Baltimore, Maryland.

Romer, A. S. (1966). "Vertebrate Paleontology," 3rd ed. Univ. of Chicago Press, Chicago, Illinois.

Romer, A. S. (1968). "Notes and Comments on Vertebrate Paleontology." Univ. of Chicago Press, Chicago, Illinois.

Sawyer, W. H. (1966). Diuretic and natriuretic responses of lungfish (*Protopterus aethiopicus*) to arginine vasotocin. *Am. J. Physiol.* 210, 191–197.

Schmidt-Nielsen, B., and Rabinowitz, L. (1964). Methylurea and acetamide: Active reabsorption by elasmobranch renal tubules. *Science* 146, 1587–1588.

Schneider, S. (1903). Ein Beitrag zur Kenntre's der Physiologie der Niere niederer Wirbeltiere. *Skand. Arch. Physiol.* 14, 383–389.

Schwartz, F. J. (1964). Natural salinity tolerances of some freshwater fishes. *Underwater Naturalist* 2, 13–15.

Shannon, J. A. (1934a). Absorption and excretion of water and salts by the elasmobranch fishes. IV. The secretion of exogenous creatinine by the dogfish, *Squalus acanthias.* *J. Cellular Comp. Physiol.* 4, 211–220.

Shannon, J. A. (1934b). The excretion of inulin by the dogfish, *Squalus acanthias.* *J. Cellular Comp. Physiol.* 5, 301–310.

Shannon, J. A. (1938a). Renal excretion of exogenous creatinine in the aglomerular toadfish, *Opsanus tau. Proc. Soc. Exptl. Biol. Med.* 38, 245–248.

Shannon, J. A. (1938b). The renal excretion of phenol red by the aglomerular fishes, *Opsanus tau* and *Lophius piscatorius. J. Cellular Comp. Physiol.* 11, 315–323.

Shannon, J. A. (1940). On the mechanism of the renal tubular excretion of creatinine in the dogfish, *Squalus acanthias. J. Cellular Comp. Physiol.* 16, 285–291.

Sharratt, B. M., Chester Jones, I., and Bellamy, D. (1964). Water and electrolyte composition of the body and renal function of the eel (*Anguilla anguilla* L.). *Comp. Biochem. Physiol.* 11, 9–18.

Skadhauge, E., and Maetz, J. (1967). Etude in vivo de l'absorption intestinale d'eau et d'electrolytes chez *Anguilla anguilla* adapte a des milieux de salinites diverses. *Compt. Rend.* 265, 347–350.

Smith, H. W. (1929a). The composition of the body fluids of elasmobranchs. *J. Biol. Chem.* 81, 407–419.

Smith, H. W. (1929b). The excretion of ammonia and urea by the gills of fish. *J. Biol. Chem.* 81, 727–742.

Smith, H. W. (1929c). The composition of the body fluids of the goosefish (*Lophius piscatorius*). *J. Biol. Chem.* 82, 71–75.

Smith, H. W. (1930a). Metabolism of the lungfish, *Protopterus aethiopicus. J. Biol. Chem.* 88, 97–130.

Smith, H. W. (1930b). The absorption and excretion of water and salts by marine teleosts. *Am. J. Physiol.* 93, 480–505.

Smith, H. W. (1931a). The absorption and excretion of water and salts by the elasmo-branch fishes. I. Fresh water elasmobranchs. *Am. J. Physiol.* **98**, 279–295.

Smith, H. W. (1931b). The absorption and excretion of water and salts by the elasmobranch fishes. II. Marine elasmobranchs. *Am. J. Physiol.* **98**, 296–310.

Smith, H. W. (1931c). The regulation of the composition of the blood of teleost and elasmobranch fishes, and the evolution of the vertebrate kidney. *Copeia* **1931**, 147–152.

Smith, H. W. (1932). Water regulation and its evolution in the fishes. *Quart. Rev. Biol.* **7**, 1–26.

Smith, H. W. (1936). The retention and physiological role of urea in the Elasmo-branchii. *Biol. Rev.* **11**, 49–82.

Smith, H. W. (1953). "From Fish to Philosopher." Little, Brown, Boston, Massachusetts.

Smith, W. W. (1939a). The excretion of phosphate in the dogfish, *Squalus acanthias*. *J. Cellular Comp. Physiol.* **14**, 95–102.

Smith, W. W. (1939b). The excretion of phenol red in the dogfish, *Squalus acanthias*. *J. Cellular Comp. Physiol.* **14**, 357–363.

Stanley, H. P. (1963). Urogenital morphology in the chimaeroid fish *Hydrolagus colliei* (Lay and Bennett). *J. Morphol.* **112**, 99–128.

Stanley, J. G. (1968). Personal communication.

Stanley, J. G., and Fleming, W. R. (1964). Excretion of hypertonic urine by a teleost. *Science* **144**, 63–64.

Stanley, J. G., and Fleming, W. R. (1966). Effect of hypophysectomy on the function of the kidney of the euryhaline teleost, *Fundulus kansae*. *Biol. Bull.* **130**, 430–441.

Stanley, J. G., and Fleming, W. R. (1967a). Effect of prolactin and ACTH on the serum and urine sodium levels of *Fundulus kansae*. *Comp. Biochem. Physiol.* **20**, 199–208.

Stanley, J. G., and Fleming, W. R. (1967b). The effect of hypophysectomy on the electrolyte content of *Fundulus kansae* held in fresh water and in sea water. *Comp. Biochem. Physiol.* **20**, 489–497.

Sulze, W. (1922). Beitrage zur Physiologie der Aufsaugung und Absonderung I. Untersuchungen uber den Salzgehalt des Harnes mariner Knochenfische. *Zeit. Biol.* **75**, 221–238.

Sutherland, L. E. (1966). Immunological and functional aspects of juxtaglomerular cells. Doctoral thesis, University of Toronto.

Sverdrup, H. U., Johnson, M. W., and Fleming, R. H. (1942). "The Oceans." Prentice-Hall, Englewood Cliffs, New Jersey.

Thorson, T. B. (1967). Osmoregulation in freshwater elasmobranchs. In "Sharks, Skates and Rays" (P. W. Gilbert, R. F. Mathewson, and D. P. Rall, eds.), pp. 265–270. Johns Hopkins Press, Baltimore, Maryland.

Thorson, T. B., Cowan, C. M., and Watson, D. E. (1967). *Potamotrygon* spp.: Elasmobranchs with low urea content. *Science* **158**, 375–377.

Trump, B. F. (1968). Unpublished data.

Trump, B. F., and Bulger, R. E. (1967). Studies of cellular injury in isolated flounder tubules. I. Correlation between morphology and function of control tubules; observations of autophagocytosis and mechanical cell damage. *Lab. Invest.* **16**, 453–482.

Trump, B. F., and Bulger, R. E. (1968a). The morphology of the kidney. In "The

Structural Basis of Renal Disease" (E. L. Becker, ed.), pp. 1–92. Harper (Hoeber), New York.

Trump, B. F., and Bulger, R. E. (1968b). Studies of cellular injury in isolated flounder tubules. III. Light microscopic and functional observations of alterations in tubules treated with cyanide. *Lab. Invest.* **18**, 721–730.

Trump, B. F., and Bulger, R. E. (1968c). Studies of cellular injury in isolated flounder tubules. IV. Electron microscopic observations of changes during the phase of altered homeostasis in tubules treated with cyanide. *Lab. Invest.* **18**, 731–739.

Trump, B. F., and Ginn, F. L. (1968). Studies of cellular injury in isolated flounder tubules. II. Cellular swelling in high-potassium media. *Lab. Invest.* **18**, 341–351.

Trump, B. F., and Ginn, F. L. (1969). The pathogenesis of subcellular reaction to lethal injury. *In* "Methods and Achievements in Experimental Pathology" (E. Bajusz and G. Jasmin, eds.), Vol. IV, pp. 1–29. Karger, Basel.

Trump, B. F., Green, K., and Hickman, C. P., Jr. (1968). The morphology of the nephron of the southern flounder (*Paralichthys lethostigma*). Manuscript.

Urist, M. R. (1962). Calcium and other ions in blood and skeleton of Nicaraguan freshwater shark. *Science* **137**, 984–986.

Urist, M. R., and Van de Putte, K. A. (1967). Comparative biochemistry of the blood of fishes: Identification of fishes by the chemical composition of serum. *In* "Sharks, Skates and Rays" (P. W. Gilbert, R. F. Mathewson, and D. P. Rall, eds.), pp. 271–285. Johns Hopkins Press, Baltimore, Maryland.

von Möllendorff, W. (1930). Der Exkretionsapparat. *In* "Handbuch der mikroskopischen Anatomie der Menchen" (W. von Möllendorff, ed.), Vol. 7, Part 1, pp. 184–199. Springer, Berlin.

Wai, E. H., and Hoar, W. S. (1963). The secondary sex characters and reproductive behavior of gonadectomized sticklebacks treated with methyl testosterone. *Can. J. Zool.* **41**, 611–628.

Wasserman, K., Becker, E. L., and Fishman, A. P. (1953). Transport of phenol red in the flounder renal tubule. *J. Cellular Comp. Physiol.* **42**, 385–393.

Wheeler, M. W. (1899). The development of the urinogenital organs of the lamprey. *Zool. Jahrb.* **13**, 1–88.

Whittembury, G., Oken, D. E., Windhager, E. E., and Solomon, A. K. (1959). Single proximal tubules of *Necturus* kidney. IV. Dependence of H_2O movement on osmotic gradients. *Am. J. Physiol.* **197**, 1121–1127.

Wikgren, B. (1953). Osmotic regulation in some aquatic animals with special reference to the influence of temperature. *Acta Zool. Fennica* **71**, 1–102.

Windhager, E. E. (1969). Kidney, water, and electrolytes. *Ann. Rev. Physiol.* **31**, 117–172.

Windhager, E., Whittembury, G., Oken, D. E., Schatzmann, H. J., and Solomon, A. K. (1959). Single proximal tubules of the *Necturus* kidney. III. Dependence of H_2O movement on NaCl concentration. *Am. J. Physiol.* **197**, 313–318.

Zimmermann, K. W. (1911). Zur Morphologie der Epithelzellen der Saugetierniere. *Arch. Mikroskop. Anat. Entwicklungsmech.* **78**, 199.

3

SALT SECRETION

I. INTRODUCTION

Certain ancestral vertebrates, when faced with a diversity of environments, have selectively evolved a modulating system that can control both the volume of water and the concentration of electrolytes of the internal body fluids to within very narrow limits. This type of homeostasis

241

is termed "osmoregulation." Reviews by Potts (1968), Parry (1966), Prosser and Brown (1962), Potts and Parry (1964), and Shaw (1960) have summarized our present knowledge on the physiology of osmotic and ionic regulation in animals. For the sake of clarity of presentation, it will be difficult to avoid repetition of certain aspects. However, the present chapter will attempt to describe the role that salt-secreting organs play in those fishes which are living in a hyperosmotic medium. Hypoosmoregulation is evidenced by those animals which can maintain the salt concentration of their extracellular fluid below that of the environment in which they live. Many hypoosmoregulating species inhabit marine or brackish waters, whereas others are located inland in saline lakes. Hyperosmoregulators on the other hand occupy terrestrial and/or freshwater. habitats. The development of diverse anatomical structures within the various vertebrate classes that deal with salt and water balance suggest that at least two, and possibly more, independent solutions to the problem of osmoregulation have arisen. The evolution and development of the mammalian renal system appears to be an example of a system which couples the mechanism of regulating deviations of the extracellular solutes and excess solvent within the same structure. The development of extrarenal organs, such as the salt secretory glands of marine birds and reptiles, represents an entirely different approach to the problem of maintaining hydromineral balance. This system appears to have focused on the problem of eliminating the excess inorganic salts which accumulate in the extracellular fluids by developing an intercellular sequestration mechanism. Comparison of the histology of several types of salt-secreting glands discloses that the cellular organization of the secretory epithelium is quite similar. The discriminating evidence comes from ultrastructural investigations that show the trademark of cells which are involved in the active transport of salts to be an abundance of mitochondria and a highly developed labyrinth of agranular membranes. If differences in fine structure are noted between types of ion secretory cells, it appears to be restricted to the form and shape of the membranous labyrinth which serves as an extension of cellular surface. This distinctive cytological design is definitely evidenced in the secretory cells of the nasal gland of marine birds (Doyle, 1960b; Komnick, 1963), the lachrymal gland of the turtle (Ellis and Abel, 1964), the filamental cells of the crustacean gill epithelium (Copeland, 1964b, 1967), and the cells of the anal papillae of mosquito larvae (Copeland, 1964a). The epithelial cells of the gill filaments of fishes and the rectal gland of selachians offer no exception to this rule.

This chapter will consider recent studies on the cellular differentiation and its related biochemical changes in regard to the development of extrarenal osmoregulation.

II. AGNATHA

A. Physiological Studies of Hypoosmoregulation in Myxinoids

Recent studies on electrolyte and water balance in myxinoids (McFarland and Munz, 1958, 1965; Morris, 1965; Rall and Burger, 1967; review by Robertson, 1963) have shown that body fluids of hagfish, *Myxine glutinosa* and *Polistotrema stouti*, are isosmotic with the external environment (see Table I). Earlier work (Dekhuyzen, 1904; Greene, 1904; H. Smith, 1932) suggested that a slight degree of hyperosmotic regulation was achieved by myxinoids, but according to McFarland and Munz (1958) this is probably a result of handling effects that cause enhanced slime production which concomitantly elevates the ionic concentration of the blood. Rather complete analyses of the ionic composition of body fluids have been reported (Robertson, 1963). Interestingly, there is little difference between the environmental concentrations of sodium and chloride and the values found in blood serum and urine. Munz and McFarland (1964) have shown that water is not reabsorbed from the glomerular filtrate and support the concept that there is no apparent mechanism in the mesonephric duct cells which can actively reabsorb sodium, a lack which is unique among vertebrates. However, divalent ions such as Mg^{2+}, Ca^{2+}, SO_4^{2-}, and the monovalent ion, K^+, are maintained at significantly different concentrations between blood and urine, which suggests that there is a renal mechanism for regulation of these ions. Morris (1965), investigating the relationship between the external medium and the gut fluid, showed that the majority of the experimental animals did not swallow seawater. Therefore, swallowing is not a regular and integral part of the water balance mechanism. Thus, there is little evidence to indicate that myxinoids hypoosmoregulate.

B. Morphological Studies of Salt Secretion in Myxinoids

1. SKIN

The morphology of the skin and its derivatives has been reviewed by Blackstad (1963). The slime glands have attracted much attention because of their great capacity for slime production. Munz and McFarland (1964), investigating the cationic composition of slime, found the material is rich in Mg^{2+}, Ca^{2+}, and K^+ (see Table II). They postulate that the thin layer of slime which the skin is constantly producing may serve as a pathway for the elimination of these cations. Therefore, the slime gland may represent a very primitive mechanism for salt secretion.

Table I

Osmotic Concentration of Body Fluids of Myxinoids Relative
to the External Environment[a,b]

Genus–Specie	External conc. (°C)	Plasma conc. (°C)	Urine conc. (°C)	Reference
Myxine glutinosa	1.90	2.00	1.87	Morris (1965)
	1.84	1.88	1.84	
	1.89	1.81	1.82	
	1.88	1.85	—	H. Smith (1932)
	—	1.93	—	
	—	1.98	—	
	1.26	1.40	—	Schmidt-Nielsen and Schmidt-
	1.85	1.85	—	Nielsen (1923)
	2.32	2.32	—	
	1.73	1.74	—	Dekhuyzen (1904)
	—	1.83	—	
Eptatretus[c] stouti	1029 ± 4[d]	1031 ± 4[d]	1051[d]	Munz and McFarland (1964)
	1.86	1.89	—	McFarland and Munz (1958)
	—	1.94	—	
	—	1.89	—	
	1.88	1.89	—	
	1.88	1.89	—	
	1.92	1.97	—	Greene (1904)

[a] Modification from McFarland and Munz (1958).
[b] External environment was normal seawater.
[c] *Eptatretus* (= *Bdellostoma, Polistotrema*).
[d] Concentration in mOsm/kg (see review by Robertson, 1963).

Recently, Wilson and Arnheim (1968) have found a small organic molecule in hagfish tissues, possibly related to the cardiac agent, eptatretin, that could act as an osmotic agent. Permeability studies have not been made in hagfish skin.

Table II

Comparison of Slime to Intracellular Cation Concentrations[a]

	Na+	K+	Ca²⁺	Mg²⁺	TMAO[b]	Total cation
"Tongue" muscle cells[c]	132[d]	144	2.6	17.5	211	507
Liver cells[c]	110	161	1.3	9.8	234	516
Slime[e]	95	207	10.7	38.5	—	(351)

[a] Reproduced from Munz and McFarland (1964).
[b] TMAO stands for trimethylamine oxide.
[c] Data from Bellamy and Chester Jones (1961).
[d] Concentrations are in millimoles per kilogram of water.
[e] Data from Munz and McFarland (1964).

2. GILL

The histology of the gill epithelium of hagfish shows that it does not possess a population of mitochondria-rich cells. Morris (1965) observed that the only cells which contained mitochondria in quantity in hagfish gill were in an unsuitable position for contributing to salt transport since they lie within the gill epithelium and rarely have a free border. Most of these cells were destined to become mucous secreting cells.

C. Physiological Studies of Hypoosmoregulation in Petromyzonids

Regulation of water and inorganic salts in a freshwater environment primarily involves the kidney (see chapter by Hickman and Trump, this volume). Burian (1910) found that the blood serum osmotic concentration of *P. marianus* was hypoosmotic ($\Delta - 0.59°C$) to seawater ($\Delta - 2.3°C$), which certainly indicates that water and electrolyte balance in petromyzonids might be regulated in a manner similar to marine teleosts. Attempts to return migrating lampreys, which have been captured in freshwater, into full strength seawater have failed (Fontaine, 1930; Galloway, 1933; Wikgren, 1953). Morris (1956, 1958), utilizing diluted seawater ($\Delta - 0.66$ to $- 0.97°C$), found that a few fresh run lampreys, *L. fluviatilis*, could regulate hypoosmotically for only a short period of time. Salt-loading experiments by Bentley and Follett (1963) showed, following intraperitoneal injection of hypertonic salt solutions into animals being maintained in tap water, that they neither expressed an increase or decrease in volume of urine nor net gain in body weight. Measurements of the total excretion of electrolytes indicated that only 13% of injected salt was attributable to loss via the kidney. However, 39% was attributed to leakage by sites which were anatomically outside the renal system. Morris (1958), attempting to delineate the possible extrarenal mechanisms involved in hypoosmoregulation, assessed the swallowing capacities of lampreys. The results show that the lampreys which were hypoosmotically regulating had swallowed between 10 and 22 ml of seawater per 100 g of body weight per day. Calculations indicated that more than 75% of the swallowed water was absorbed through the intestine and that 85% or more of the intestinal chloride had entered into the extracellular fluid compartment. Since these animals showed very little net change in body weight and blood chloride concentration, plus a negligible urinary output of water or chloride, it was concluded that chloride must have been excreted via an extrarenal system. Since Na^+ was not assessed, it was assumed that this ion accompanied the chloride ion. The route of excretion of other ions, mainly the divalent ions (Ca^{2+},

Mg^{2+}, and So_4^{2-}) was uncertain but was thought to be similar to the pathway found in marine teleosts. Failure to osmoregulate in seawater was thought to result from (1) increased water permeability of external surface, (2) decrease in swallowing capacity, and (3) decrease in extrarenal excretion.

D. Morphological Studies of Salt Secretion in Petromyzonids

1. GILL

Morris (1957) has described the gross and microscopic anatomy of the lamprey gill. The gill epithelium contains six cytologically distinct types of epithelial cells which were classified as the platelet cell, the chloride excretory cell, the basal cell, the mucous cell, the chloride uptake cell, and the glandular cell. Morris (1958) attempted to quantify the numbers of chloride excretory cells by serial sectioning the gill pouches and correlating their concentration with plasma osmotic values. Individual animals were exposed to diluted seawater ($\Delta - 0.97 \rightarrow -1.00$) and following exposure were sacrificed. He showed that the number of chloride excretory cells appears to increase with salinity (Fig. 1). It can be seen that the correlation between the hypoosmoregulatory capabilities and the number of chloride excretory cells is variable. This is probably because of the inability to precisely quantify the cell population of a large surface area of the gills.

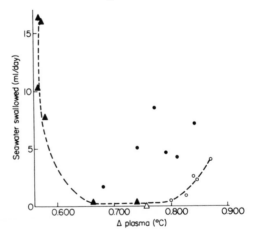

Fig. 1. The relationship between swallowing capacity, chloride excretory cell assessment, and plasma freezing-point depression (Δ) in fresh-run lampreys immersed in 50% seawater. Chloride excretory cell assessment: ($\blacktriangle$) many, ($\bullet$) few, ($\triangle$) very few, and ($\bigcirc$) nil. Reproduced from Morris (1958).

2. SKIN

Bentley (1962)—investigating the permeability of the skin of the freshwater petromyzonid, L. fluviatilis, to water and electrolytes in hypoosmotic solutions—found that the skin was highly permeable to water when compared to teleosts. The estimated time for 1 ml of water to pass 1 cm^2 of skin at a pressure difference of 1 atm was 91 days for Lampetra and 5 years for Anguilla. No measurements were made using hyperosmotic solutions.

3. INTESTINE

The intestinal wall offers little resistance to the diffusion of salts and water into the extracellular fluid compartment. However, during the spawning migration, the gut undergoes degeneration and this probably contributes a great deal to the inability of mature animals to restore hypoosmoregulation once they have entered freshwater (Morris, 1958).

E. Summary

(1) Myxinoids do not hypoosmoregulate. There is some evidence that the slime gland may represent a very primitive mechanism for salt secretion. They do not possess a mitochondria-rich cell in the gill epithelium.

(2) Petromyzonids are capable of hypoosmoregulation. The animals ingest seawater to offset the water loss resulting from the diffusion gradient of the external medium. The intestinal mucosa absorbs the monovalent ions and water. Urinary output is reduced. Excess salts are eliminated extrarenally, apparently through the gill or skin.

(3) Petromyzonids have a population of mitochondria-rich cells in the gill epithelium. They appear to increase in number following seawater immersion, but quantitation of the cell population remains inconclusive.

III. CHONDRICHTHYES

A. Physiological Studies of Hypoosmoregulation in Selachians

H. Smith (1931) was intrigued by the ratio of Mg^{2+}:Cl^- found in the urinary fluid of sharks when compared to the ratio of the concentrations of these ions in the plasma and seawater. From this observation, he suggested that elasmobranchs, in addition to the regulation of these ions carried out by the kidneys, can probably excrete either one or both of these ions extrarenally. Hodler *et al.* (1955) sought to explain this irregularity in chloride concentrations by investigating the mechanism of acid–base balance in the dogfish, *Squalus acanthias.* He found that the carbonic anhydrase enzyme system was not involved in the acidification of the urine by the kidney but that it played a major role in the branchial excretion of sodium bicarbonate. However, it was not shown if the HCO_3^- elimination was coupled with Cl^- secretion. Thus, these experiments did not yield any definitive information as to the existence of an extrarenal salt-secreting system.

Rather unexpectedly, Burger and Hess (1960) established the role of the rectal gland as being a salt-secreting organ. Their experiments showed that the fluid secreted by the gland was essentially a hypertonic sodium chloride solution with a salt concentration twice that of the plasma (Table III). The volume of fluid secreted by the gland was large enough to convince these investigators that it could play a significant role in the mineral economy of the elasmobranchs. Burger (1962, 1965) has evaluated several aspects of the problem. A comparison was made of the effects of salt-loading fish which had intact glands to those which were glandless. It was found that the rate of secretion increased under a salt load and was proportional to the amount of salt injected. Other substances such as sucrose, urea, KCl, $NaHCO_3$, Na_2SO_4, and antipyrine were effective in evoking a response, but less so than NaCl. Therefore, the response appears to be dependent upon an osmotic component, a volume component, and a sodium chloride component.

Piperocaine applied to the structure leading to the gland did not suppress the response to NaCl. Unlike the nasal gland of birds (Fänge *et al.*, 1958) secretion was not stimulated by acetylcholine, metacholine, pilocarpine, eserine, epinephrine, or pitressin. From this evidence Burger (1962) concluded that the control of the gland does not involve the nervous system. Secretory rates were not inhibited by atropine, tetraethylammonium bromide, dibenamine, L-8 vasopressin.

Comparison between the volume of rectal gland fluid secreted to urine produced over long periods of time (> 100 hr) showed that the two fluids were nearly equal in volume in the intact animal. However, Mg^{2+} was lost exclusively through the urine. Even so, there is a difference in urinary flow between glandless and control fish. Glandless fish had twice the rate as compared to the controls. Comparison of the effect varying external salinities upon glandless and intact fish showed that regardless of being in dilute or full strength seawater, the glandless fish have greater urine volumes than do intact fish. Injection of hypertonic salt solution into glandless fish did not invoke a renal response to concentrate chloride

Table III

Comparison of the Composition of Fluid from the Rectal Gland, Plasma, and Urine[a]

Fish (No.)	Wt (kg)	Fluid	pH	Osmolarity (mOsm)	Concentration (mmole/liter)							Flow (ml/kg/hr)
					Na	K	Cl	Ca	Mg	Urea	CO_2	
1	4.3	Rectal gland	7.0	1011	522	5.6	510	—	1	—	—	0.051
		Plasma	7.55	1011	—	—	—	—	—	—	—	—
2	3.1	Rectal gland	6.8	1018	535	6.8	549	1	1	20	—	0.320
		Plasma		1018	320	7.0	250	—	—	349	—	—
3	4.7	Rectal gland	6.8	1036	580	8.4	562	1	1	11	2.8	0.850
		Plasma		1036	300	4.4	247	—	—	347	—	—
		Urine		754	352	2.0	170	—	—	—	—	—
4	5.0	Rectal gland	6.9	1020	542	—	552	—	—	14	—	0.820
		Plasma		1020	—	—	—	—	—	—	—	0.012
5	4.8	Rectal gland		1001	502	—	490	—	—	28	—	0.012
		Plasma		1001	254	—	252	—	—	356	—	—
		Urine		806	339	—	286	10	50	—	—	—
8	6.4	Rectal gland	6.7	1020	560	7.5	549	1	1	13	—	1.300
		Plasma		1020	283	—	242	2.6	3.7	352	—	—
		Urine		780	327	—	174	4	25	—	—	0.810
9	3.1	Rectal gland		—	—	—	521	1	1	—	—	—
		Plasma		—	—	—	239	—	—	—	—	—
		Urine		—	—	—	182	—	—	—	—	—
Average of		Rectal gland		1018	540	7.1	533	1	1	14.5	—	—
		Plasma		1018	286	—	246	—	—	351	—	—
		Urine		780	337	—	203	—	—	—	—	—
		Seawater		925–935	440	9.1	492–500	10	51	—	—	—

[a] Reproduced from Burger and Hess (1960).

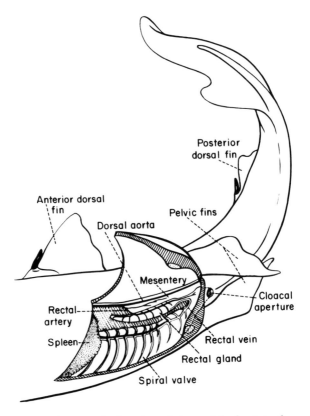

Fig. 2. Diagram showing the location of the rectal gland in the spiny dogfish, *Squalus acanthias*. Reproduced from Bulger (1963).

above plasma levels. Hence, it was concluded that the kidneys could not serve as a substitute rectal gland, but the absence of a rectal gland, however, does not prevent the elasmobranchs from maintaining salt balance. Presumably the response is to reduce the uptake of salt so that the influxing solution is nearly equal to or slightly below plasma concentration. The gill epithelium may be involved in the uptake of salt (Burger and Tosteson, 1966).

B. Morphological Studies of Salt Secretion

1. RECTAL GLAND

Several studies dealing with the macroscopic anatomy of the rectal gland have been published (Crofts, 1925; Hoskins, 1917; Sullivan, 1908). The rectal gland has been described as a compound tubular gland with a central canal that continues as a duct which emerges from the posterior ventral part of the gland and curves forward before it drains into the intestinal canal posterior to the spiral valve. Figure 2 is a schematic diagram showing the location of the rectal gland in the spiny dogfish, *Squalus acanthias*. If microscopic sections were made of the glandular portion, it would show three concentric layers: (1) an outer capsule containing small arteries plus a peripheral connective and muscle tissue layer (Fig. 3), (2) an inner layer consisting of veins and ducts arranged around a central canal that terminates in a central duct (Fig. 4), and (3) a middle glandular layer consisting of a zone of secretory tubules (Figs. 5 and 6).

Electron microscopic studies by Doyle (1960a, 1962) of the rectal gland of the round stingray, *Urolophus*, and by Bulger (1963, 1965) of the dogfish, *Squalus suckleyi*, and Komnick and Wohlfarth-Bottermann (1966) in several species have revealed some interesting facets with regard to cell boundaries of the tubular cell. Figure 7 shows the complex infoldings and interdigitations of the basal and lateral plasmalemma which separate groups of mitochondria. As can be seen, mitochondria are abundant and appear spherical or rod-shaped. The distribution of mitochondria in the cell shows a basal–apical orientation with the basal compartment filled so tightly that little cytoplasm appears. Small amounts of rough and smooth endoplasmic reticulum do occur as either small cisternal tubules, irregular profiles or as vesicular elements. Other structures such as surface microvilli, golgi apparatus, multivesicular bodies and dense composite bodies are frequently found. Nearly identical ultrastructural features are observed in the salt-secreting lachrymal gland of the turtle (Ellis and Abel, 1964).

2. GILL

Doyle and Gorecki (1961) have reported the existence of mitochondria-rich cells in the gill epithelium of the elasmobranch, *Urobatis*, which also have a tubular endoplasmic reticulum characteristic of salt secretory cells. Chan *et al.* (1967), studying the hormonally induced electrolyte changes in the lip-shark, *Hemiscyllium plagiosum*, shows that the gill does participate in the elimination of Na$^+$ from the animal, but it is less important than the contributions made by the rectal gland or the kidney. This confirms the earlier work on the elasmobranch gill reported by Maetz and Lahlou (1966).

3. SKIN

Information is lacking on the role that the skin epithelium plays in salt secretion in elasmobranchs.

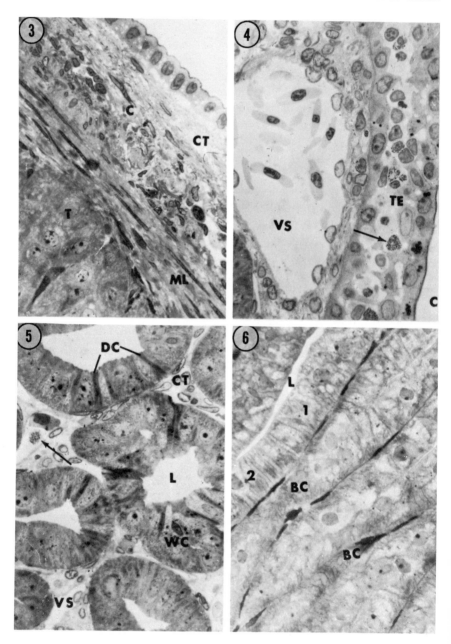

Fig. 3. A section (1 μ) showing the capsule (C) and the outer part of the tubular layer (T). The capsule shows the peripheral connective tissue layer (CT) and the inner muscle layer (ML). ×530. Reproduced from Bulger (1963).

Fig. 4. A section (1 μ) taken near the central lumen (C) to show the transitional

C. Biochemical Studies of Salt Secretion

The biochemical knowledge of the selachian rectal gland is very limited. Recent cytochemical studies of the rectal gland by Chan and Phillips (1967), Bonting (1966), and Bernard and Hartmann (1960) have found similar enzymic patterns for the glandular parenchyma in several species of sharks. Such enzymes as adenosinetriphosphatase, carbonic anhydrase, acid phosphatase, alkaline phosphatase, and succinic dehydrogenase were reported as being present in high concentrations. No evidence for the presence of cholinesterase was found, which supports the idea that the gland is not under direct nervous control. The tubular cells are reported to be rich in phospholipids and contain an acid-mucopolysaccharide substance which coats the cell membrane surfaces. Oguri (1964), comparing the histological appearance of rectal glands from marine and freshwater sharks, found that the freshwater environment produced a regressive change in the staining properties of the tubular cells. Bonting (1966) expanded the enzyme study of the rectal gland by investigating the physicochemical characteristics of the adenosinetriphosphatases. Table IV compares the mean enzymic activity for the Na–K activated adenosinetriphosphatase and Mg-activated adenosinetriphosphatase per kilogram body weight that was isolated from nine species of elasmobranchs. It appears that the spiny dogfish, *Squalus acanthias,* has the highest Na–K ATPase activity. Interestingly, it was shown that species with radially arranged tubules have higher Na–K ATPase than those with a lobular arrangement of the glandular parenchyma. The significance of this finding is not clear. Skou (1957, 1960) first implicated the Na–K ATPase activity with the monovalent ion transport system in salt-secreting tissues. Bonting (1966) found that

epithelium (TE) found lining the lumen. A venous sinus (VS) can also be seen. ×530. Reproduced from Bulger (1963).

Fig. 5. A section (1 μ) of the inner region of the tubular layer. The tubules are loosely packed in this region. The tubular lumina (L) are of variable but often large size. The connective tissue (CT) is more abundant. Venous sinuses (VS) often appear large and may contain red and white blood cells. The cytoplasm of some tubular cells appears darker (DC). Atypical nuclei are seen in some tubules and probably belong to wandering cells (WC). Arrow indicates wandering leucocytes with ovoid eosinophilic granules. ×530. Reproduced from Bulger (1963).

Fig. 6. A section (1 μ) showing the tubules in the outer region of the tubular layer. It can be seen that they are closely packed and the tubular lumina (L) are generally small in size. Basal striations are seen in region 1. Interdigitating lamellae are probably located in region 2. Basal compartments (BC) can be seen in several places. ×530. Reproduced from Bulger (1963).

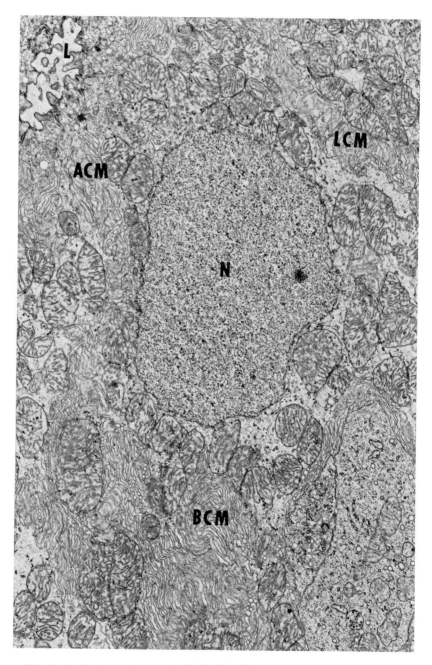

Fig. 7. An electron micrograph of tubular cells from the outer part of the tubular layer near the lumen (L) to demonstrate cellular form. The lateral cell membrane (LCM) demonstrates repeated infoldings. The basal cell membrane (BCM) under-

Table IV

ATPase Activities in Rectal Gland of Nine Elasmobranch Species

Species	No. of animals	Body wt (kg)	Gland wt (g)	Mg ATPase Wet[a]	Mg ATPase Dry[b]	Na–K ATPase Wet[a]	Na–K ATPase Dry[b]	Na–K ATPase mmoles hr/kg body wt[c]
Spiny dogfish	32	2.27–4.54	0.97–2.09	0.69	2.90	1.32	5.68	0.60
Sand shark	10	9.1–18.2	1.69–3.33	0.59	2.56	0.75	3.25	0.12
Sickle-shaped shark	1	6.80	1.29	0.79	3.08	0.54	2.13	0.10
Smooth dogfish	6	1.36–1.70	0.27–0.39	0.81	3.51	0.36	1.57	0.079
Clear-nosed skate	5	0.57–1.81	0.12–0.35	0.67	2.80	0.32	1.31	0.059
Eagle ray	6	0.79–6.80	0.11–1.28	0.67	2.62	0.33	1.31	0.051
Butterfly ray	1	36.3	3.13	0.67	2.61	0.43	1.65	0.037
Stingray	4	—	0.94–2.05	0.92	4.42	0.39	1.88	—
Monkfish	4	10.0–15.4	0.88–1.68	0.36	1.62	0.36	1.56	0.032

[a] Enzymic activities in moles per kilogram wet weight of gland per hour.
[b] Enzymic activities in moles per kilogram dry weight of gland per hour.
[c] Average of individual values, calculated with the appropriate values of body weight and gland weight. For stingray no individual body weights were available, hence the activity per kilogram body weight could not be calculated. Reproduced from Bonting (1966).

50–60% of the total ATPase activity is Na–K dependent, whereas the remainder is Mg-dependent. Also, the Na–K ATPase system was ouabain-sensitive. Ouabain is a cardiac glycoside which specifically inhibits sodium transport. The pH optima for the Na–K ATPase was much more acidic (Fig. 8) and is significantly lower than those found in other vertebrate species. Otherwise, in all of its characteristics, the Na–K ATPase system of the selachians is very similar to systems described for a variety of species and organs. Comparison between the rectal gland Na–K ATPase activity with the Na excretion rate gives a ratio of 2.2 mEq Na excreted per millimole of ATP hydrolyzed, which is well within the range of Na–K ATPase ratios previously determined ($1.8 \rightarrow 3.1$) Palmer (1966) reported to Bonting (1966) that perfusion of 10^{-4} ouabain inhibited sodium secretion in a functional, arterially perfused rectal gland. Apparently the Na–K ATPase system has the role of being the primary and rate-limiting system for regulating salt secretion in the rectal gland of elasmobranchs.

---

goes a process of infoldings and then secondary infolding and interdigitation. The apical cell membrane (ACM) is specialized to surround interdigitating compartment, alternate ones connecting to the cytoplasm of the cell beneath. Numerous large mitochondria and a large oval nucleus (N) are present. Fixed with osmium tetroxide, stained with Millonig's lead stain. ×7,220. Reproduced from Bulger (1963).

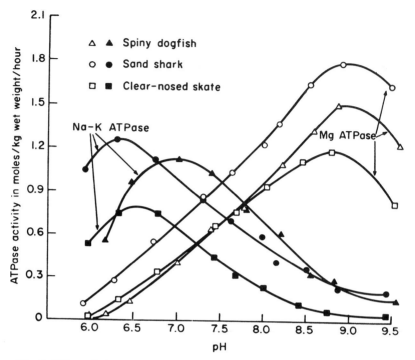

Fig. 8. Effect of pH on Mg ATPase and Na–K ATPase activities in elasmobranch rectal gland homogenates. Reproduced from Bonting (1966).

D. Summary

(1) The rectal gland of selachians functions as an auxillary excretory organ and is used to remove excess salt from the body fluids. Salt loads probably occur intermittently at times of feeding because of the seawater ingested with the food. Indirect evidence for this hypothesis is afforded by Chan and Phillips's study (1966) on the embryology of the rectal glands where at the time of the opening of the uterus, in which the body fluids are mixed with freshwater, there is a significant increased rate of growth for the rectal gland above the overall body growth rate of the developing fish.

(2) The anatomy and ultrastructure of the shark rectal gland is quite similar to other salt-secreting organs of birds and reptiles.

(3) The gill contains secretory cells but plays a much smaller role in mineral regulation than does the kidney or rectal gland.

(4) The contribution of the skin is unknown at this time.

IV. OSTEICHTHYES

A. Physiological Studies of Hypoosmoregulation in Teleosts

Most of the experimental studies of extrarenal osmoregulation in teleosts can be arbitrarily classified into one of the following categories: (a) phylogenetic relationship of euryhaline species in the evolution and development of hypoosmoregulation, (b) maintenance of the nutritional balance of water and/or electrolytes during steady and nonsteady state, and (c) kinetic relationships which exist between certain ionic constituents of the external environment and various internal fluid compartments.

1. PHYLOGENETIC RELATIONSHIP IN CERTAIN EURYHALINE SPECIES WITH REGARD TO THE DEVELOPMENT OF SALTWATER ADAPTATION

a. Development of Hypoosmoregulation in Isolated Subpopulations. Studies by Gordon (1959a) have been shown that different populations of brown trout and sea trout, *Salmo trutta*, although isolated genetically for many generations, are quite similar in osmotic and ionic regulation when stressed over a wide range of salinities $(0 \rightarrow 32\%_{0})$. Conte and Wagner (1965) in their investigation of steelhead and rainbow trout, *Salmo gairdneri*, found these two populations to be quite homoiosmotic. Wagner et al. (1969), investigating the onset of the osmoregulatory mechanism in the embryonic and postembryonic stages of two populations of chinook salmon, *Oncorhynchus tshawytscha*, have found the two groups to be identical. Thus, it appears that several species of salmonids may have undergone many generations of genetic isolation, but they have not altered the basic pattern in the development of hypoosmoregulation.

b. Species Differentiation in Regard to Seawater Tolerance. Investigations by Parry (1960) on the development of salinity tolerance in juvenile *Salmo salar, Salmo trutta,* and *Salmo gairdneri,* showed that young fish were not completely homoiosmotic in full strength seawater and that osmoregulation in different salinities was dependent upon the size and age of the fish as well as upon the species. Conte and Wagner (1965) separated these two factors (age and size) by observing the salinity tolerance of a slow-growing and a fast-growing population of steelhead trout. Experimental fish, because of the differential in growth rate, would reach certain selected size groups at different times of the year. Comparison of identical size groups revealed that salinity tolerance was the same for individuals within each size group and was independent of the time of year the assay was made. Therefore, the chronological age of the animals did not enhance the adaptive mechanism unless there was an alteration in growth. These results support the concept that growth regulating factors must control the maturation of the hypoosmoregulatory system (Elson, 1957; Evropietseva, 1957; Houston and Threadgold, 1963; Parry, 1960). In Table V are listed the survival times for several species

Table V

Comparative Survival for Certain Species of Salmonids
at Different Salinities for Various Sizes

Species	Stage in life history	Size range (cm)	30‰	28‰	26‰	24‰	22‰	20‰	Control (0–5‰)
O. nerka[a]	Fry	3–4	72	150	720	720	720	720	∞
O. tshawytscha[b]	Fry	3–4	60	720	720	720	720	720	∞
O. kisutch	Fry	3–4	44	48	720	720	720	720	∞
S. gairdneri[c]	Fry	3–4	24	48	48	144	576	720	∞
S. clarki	Fry	3–4	24	24	48	96	268	720	∞
O. nerka	Parr	5–6	720	720	720	720	720	720	∞
O. tshawytscha	Parr	5–6	720	720	720	720	720	720	∞
S. salar[d]	Parr	3–4	4	7	10	22	50		∞
O. kisutch	Parr	7–8	720	720	720	720	720	720	∞
S. gairdneri	Parr	7–8	36	72	336	720	720	720	∞
S. clarki	Parr	7–8	24	Data not available at this time					∞
S. salar	Parr	7–10	19	30	72	∞	∞	∞	∞
S. trutta	Parr	8–10	22	36	50	72	100	200	∞
O. nerka	Smolt	6–8	720	720	720	720	720	720	∞
O. tshawytscha	Smolt	7–15	720	720	720	720	720	720	∞
O. kisutch	Smolt	9–15	720	720	720	720	720	720	∞
S. gairdneri	Smolt	9–15	216	Data not available at this time					∞
S. salar	Smolt	12–15	100	∞	∞	∞	∞	∞	∞
S. trutta	Smolt	12–15	25	48	75	150	∞	∞	∞

[a] Conte and Bailey (1968).
[b] Wagner et al. (1969).
[c] Accumulated data for both sea-run and resident form.
[d] Data calculated from Parry (1960).

of salmonids at various sizes and salinities. Since several size groups may exist in any one juvenile stage of the life cycle, it is thought convenient to indicate what stage of the life history that each group was in with respect to the time of the assay. The data indicates that for fry (3–4 cm), parr (7–10 cm), and smolts (9–15 cm) the survival order of the species is as follows:

$$O.\ tshawytscha \geq O.\ kisutch \geq O.\ nerka > S.\ gairdneri >$$
$$S.\ salar \geq S.\ trutta$$

Generally speaking, these findings show that significant differences exist between species and is further amplified when comparisons are made between different genera.

c. *Intergeneric Differences Possibly Related to the Parr–Smolt Metamorphosis.* It has been reported many times that during the development of hypoosmoregulation in migratory fishes such as *Anguilla, Salmo,* and *Oncorhynchus* there is a temporal parameter which appears to influence the response of the osmoregulatory system to electrolyte stress, especially so in those juvenile fish which are undergoing a maturation of the tissues of the body termed the "parr–smolt" transformation (see reviews by Black, 1957, and Parry, 1966). The parr–smolt transformation has been correlated with changes in tissue organic compounds (Lovern, 1934) and tissue salts (Chartier-Baraduc, 1960; Houston and Threadgold, 1963), plasma organic concentrations (Fontaine, 1960; Zaks and Sokolova, 1962), and nonorganic composition (Conte *et al.*, 1966; Conte and Wagner, 1965; Fontaine, 1960; Houston, 1959, 1960; Houston and Threadgold, 1963; Koch *et al.*, 1959; Kubo, 1953, 1954, 1955; Parry, 1958, 1961; Wagner *et al.*, 1969), purine metabolism (Johnston and Eales, 1966), and protein and fat metabolism (Malikova, 1957).

However, it must be pointed out that many juvenile fish from species which exhibit the parr–smolt transformation can be transferred from freshwater to full strength seawater ($> 30\%_0$) several months prior to the transformation (Conte *et al.*, 1966; Conte and Wagner, 1965). These animals can hypoosmoregulate for long periods of time and do not need the parr–smolt transformation to trigger or maintain the regulatory mechanism. It has been observed that variations occur in the hypoosmoregulatory powers which seem to coincide with seasonal changes (Gordon, 1959a; D. Smith, 1956). The greatest degree of variability appears in the late spring and early summer months when the juvenile fish are at the postsmolt stage. The interesting point of the temporal pattern in migrating species of salmonids is not the fact that development of hypoosmoregulation precedes the parr–smolt transformation, it is the deactivation of the mechanism when it is at its peak.

The purpose for the regression must be of ecological importance to the species. One could surmise that its usefulness lies in preventing animals from pursuing seaward movement during unfavorable conditions. The recent findings of McInerney (1963), who described a temporal sequence in the development of salinity preference for five species of

Pacific salmon, support this concept. Similarly, Baggerman (1960), investigating certain species of *Oncorhynchus* for voluntary salinity preference, found that two of the four species which did not exhibit parr–smolt transformations (*O. keta* and *O. gorbuscha*) did not show temporal regression, whereas *O. kisutch* and *O. nerka*, like *S. trutta* and *S. gairdneri*, did effect a temporal regression. Conte *et al.* (1966) and Wagner *et al.* (1969) in their studies did not find this to be the case for *O. kisutch* or *O. tshawytscha*. The deactivation of hypoosmoregulation mechanism may be partly responsible for stream residualism or "land-locked" forms. If such were true, then comparison between *O. nerka* in which there are land-locked races (kokanee trout) to other species of *Salmo* would be most helpful in understanding this relationship. Zaks and Sokolova (1966), investigating *O. nerka*, found that regression does occur for this specie.

In summary, it appears that the parr–smolt transformation is associated with a type of biological clock which informs those smolting species of the precise time for them to seek a marine environment.

2. NUTRITIONAL BALANCE OF WATER AND ELECTROLYTES DURING TRANSITIONS FROM FRESHWATER TO SEAWATER

a. Water Loss Resulting from the Ionic Gradient. The marine environment causes an obligatory water loss from the body fluids owing to the differential in osmotic concentrations. This loss occurs through the body surfaces such as the skin and gill. The kidney, to a very minor extent, loses some water because of the need to produce sufficient urine to rid the body of metabolic wastes. The urine is usually scanty and hypotonic to blood, but an exception occurs in the euryhaline plains killifish, *Fundulus kansae*, where it has been reported that the kidney produces a hypertonic urine relative to the blood during adaptation to seawater (Stanley and Fleming, 1966; Fleming and Stanley, 1965). Further details are given in the chapter by Hickman and Trump, this volume. Skin permeability is of little importance in controlling hypoosmoregulation, except when damage allows an uncontrolled flux of water and ions to occur (Parry, 1966).

b. Replacement of Water Loss by Ingestion of Environmental Seawater. H. Smith (1930) first demonstrated that marine fish ingest seawater to replace the water loss resulting from dehydration. Since that time, the amount and rate of water ingested in several species of teleosts have been reported. Table VI summarizes these findings and shows that for most species the rate of drinking is between 0.3–1.5% body weight per hour.

Table VI

Measurement of the Ingestion Rate of Seawater
by Teleosts Living in a Marine Environment

Species	Ingestion rate (ml/hr/kg)	Reference
American eel *Anguilla rostrata*	2.77	H. Smith (1930)
European eel *Anguilla anguilla*	3.25	Maetz and Skadhauge (1968)
Rainbow trout *Salmo gairdneri*	5.37	Shehadeh and Gordon (1969)
Southern flounder *Paralichthys lethostigma*	4.57	Hickman (1968)
Euryhaline flounder *Platichthys flesus*	10.00	Motais and Maetz (1965)
Sea perch *Serranus scriba*	5.00	Motais and Maetz (1965)
Killifish *Fundulus heteroclitus*	15.4–23.0	Potts and Evans (1967)
Blenny *Blennius pholis*	60.0	Mullins (1950)
Stickleback *Gasterosteus aculeatus*	40.0	Mullins (1950)
Cichlid *Tilapia mossambica*	234.00	Evans (1968)
Blenny *Pholius gunnellus*	266.00	Potts et al. (1967)
Blenny *Xiphister atropurpureus*	8.14	Evans (1967)

c. *Composition of Intestinal Fluid following Drinking of Seawater.* After swallowing sea water, the fish needs to extract water from the internal brine solution in order to maintain its hemodynamic balance. Shehadeh and Gordon (1969) have measured the changes in ionic composition of the fluid in different segments of the intestinal tract of rainbow trout, *S. gairdneri*, by means of *in situ* catheters. Changes in concentrations of Na^+, Mg^{2+}, SO_4^{2-}, Ca^{2+} and K^+ in the ingested medium were followed during passage through the intestine. Fish was acclimated at various salinities, and the ionic composition of the intestinal fluids investigated are shown in Table VII. It can be seen that monovalent ions (Na^+, K^+, and Cl^-) were absorbed with the water. Calculations indicate that between 60 and 80% of the water in the ingested salt solution was absorbed and that 95% of the salt that was absorbed was in the form of NaCl. The residue which remained in the lumen was made up of free Mg^{2+} and SO_4^{2-} plus an accumulation of insoluble mixed carbonate salts.

Table VII

Inorganic Composition, pH, and Osmolarities of Intestinal Fluid Collected from Catheters, $\bar{X}$ ± S.E.[a]

State of acclimatization	Intestinal fluid (mmoles/liter)						Osmolarity (mOsm/liter)	pH
	Na	K	Ca	Mg	SO$_4$	Cl		
Freshwater	170 ± 3 (4)[b]	4 ± 0.3 (4)	2 ± 0 (4)	0	0	70 ± 5 (4)	300 ± 3 (4)	8.5 ± 1 (4)
$\frac{1}{3}$ Seawater	140 ± 6 (11)	4 ± 0.4 (11)	3 ± 0.3 (11)	50 ± 4 (11)	30 ± 4 (8)	80 ± 5 (9)	310 ± 6 (10)	8.4 ± 0.1 (6)
$\frac{1}{2}$ Seawater	80 ± 7 (13)	3 ± 0.2 (13)	3 ± 0.2 (13)	90 ± 5 (11)	60 ± 6 (11)	70 ± 3 (13)	320 ± 6 (13)	8.4 ± 0.1 (7)
Seawater	20 ± 3 (4)	1 ± 0.1 (4)	2 ± 0 (4)	120 ± 2 (4)	110 ± 1 (4)	50 ± 2 (4)	300 ± 9 (4)	8.1 ± 0.1 (4)
				Medium				
$\frac{1}{3}$ Seawater	146	3	3	17	10	170	330	7.8
$\frac{1}{2}$ Seawater	220	5	5	25	15	267	480	7.8
Seawater	450	10	9	50	30	530	960	8.1

[a] Reproduced from Shehadeh and Gordon (1969).
[b] ( ) = number of animals in experimental group.

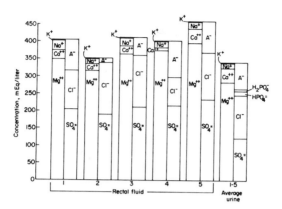

Fig. 9. Ionograms of rectal fluid and urine of five southern flounders, *Paralichthys lethostigma*. Courtesy of C. P. Hickman, Jr.

Hickman (1968) has found similar evidence for the southern flounder, *Paralichthys lethostigma* as shown in the inonograms for rectal fluid and urine (Fig. 9). These experiments are consistent with earlier work in other species.

d. Possible Pathways for the Elimination of Excess Divalent Salts. i. Storage reservoirs. The removal of monovalent ions by storing them in some innocuous manner within soft tissues such as muscle has been ruled out by Gordon (1959b). However, Houston (1964) maintains that attention must be given the alterations in "volume" resulting from the ionic shifts he observed between intracellular and extracellular compartments. Investigations which would measure body fluid compartments during freshwater to seawater transitions would help to clarify this point. L. Smith (1966) has made preliminary measurements of blood volume for steelhead trout, *S. gairdneri*, in both seawater (6.9%) and freshwater (6.2%) and suggests that blood volumes may be larger in seawater, rather than smaller as suggested by changes in osmolarity. Hard tissue (bone) has been shown to be a calcium reservoir, but its importance as a monovalent ion sink has not been determined (Boroughs *et al.*, 1957; Mashiko and Jozuka, 1964; Rosenthal, 1961).

ii. Intestinal mucous tubes. It has been shown that insoluble carbonates of Mg^{2+} and Ca^{2+} are trapped in mucous tubes and could lead to alterations of the bicarbonate–carbonate ratio (Shehadeh and Gordon, 1969). This alteration of the $HCO_3^-:CO_3^{2-}$ would affect the pH of the intestinal fluid and could markedly change the rate of entry of sodium and chloride. This aspect requires further investigation to clarify how it relates to the total acid–base balance of the body fluids (Busnel, 1943).

iii. Intestinal wall. Several *in vitro* studies on isolated segments of intestine have shown that a larger amount of water passes through the intestine of seawater (SW) adapted fish than freshwater (FW) fish (Utida *et al.*, 1967; Oide and Utida, 1967). This phenomenon can be affected by metabolic inhibitors (Oide, 1967), hormones (Hirano, 1967), and the stage of the life cycle of the animal (Utida *et al.*, 1967). Significance of these findings is difficult to correlate with the information on the gill, but it appears likely that regulation of salt uptake at the intestinal wall can be an important aspect of hypoosmoregulation.

iv. Kidney. See chapter by Hickman and Trump, this volume, for the role of the kidney in regulating divalent ions.

e. Gill–Head Region as Principal Site for Elimination of Monovalent Ions. Classic studies by Keys (1931a,b) and Krogh (1939) identified the gill–buccal region as the probable site for extrarenal secretion in euryhaline fishes. Tosteson (1962a,b) and Tosteson *et al.* (1962a–d), studying the net chloride transport out of the gill epithelium in an eel heart–gill preparation, found that the electrical potential difference between the fluid surrounding the gill epithelium and the heart ranged between 15 and 30 mV with heart positive to the mouth. Thus, the electrical potential difference is directed against the chloride transport, indicating that chloride transport is an active process. Preliminary attempts to characterize the electrical potential difference across the intracellular boundaries proved very interesting. When 0.15 M NaCl bathed the outside of the gills, the average cell potential was -15 mV to the outside solution and -17 mV to the blood. When the concentration of NaCl was reduced (0.005 M) the cell potential was -30 mV to the outside solution and -50 mV to the blood. Hence, the net negative potential was at the basal cell boundary. Increasing the salt concentration of the outside fluid (seawater) reversed the net negative potential, -10 mV to the outside and -5 mV to the blood, it now being at the apical cell surface. Marked variations in recorded cell potentials in successive punctures suggest that the potential is not the same in all cells of the gill epithelium.

3. Kinetic Studies on the Active Ion Transport across
    the Gill Epithelium

Major advances in the understanding of the dynamics of active ion transport across the gill epithelial surface in fishes have been made by Maetz and his colleagues (Maetz and Romeu, 1964; Motais, 1961a,b, 1967; Motais *et al.*, 1965, 1966; Motais and Maetz, 1965; Romeu and Maetz, 1964). Utilizing radioactive tracer techniques, these investigators have been able to determine the quantity and rate of individual ions which enter and exit the extracellular fluid compartment from the external medium for the stenohaline seawater perch, *Serranus scriba*, the European eel, *Anguilla anguilla*, and the euryhaline flounder, *Platichthys flesus*. Table VIII shows the sodium turnover rate, sodium outflux, sodium influx, and net flux for *Platichthys* adapted to various salinities. The net sodium loss through the gills is approximately equal to the net sodium uptake across the intestine (Tables VI, VII, and VIII). The sodium permeability of the gills is, however, 5–10 times that of the intestinal wall such that a much larger total flux occurs across the gill epithelium. The external permeability rates are dependent upon the concentration of each specific monovalent ion that exists in the external medium (Gordon, 1959b, 1963). Thus sodium transport is independent of chloride transport and precludes the need for a coupled mechanism.

It can be seen that when the steady state animal (SW → SW or FW → FW) is contrasted to the nonsteady state animal (SW → FW or FW → SW) two significant situations become apparent (Motais *et al.*, 1966):

(1) When FW-adapted fish are transferred to higher salinities, Na influxes are significantly higher than Na outfluxes (Table VIII). This relationship can now explain the immediate excursion of plasma sodium concentrations which appear in animals following rapid transfer between FW → SW (Conte, 1965; Parry 1966) and is responsible for what Houston (1959) terms the "adjustive" phase in osmoregulation. This

Table VIII

Sodium Turnover Rates, Outflux, Influx, and Net Flux
of *Platichthys* Adapted to Various Salinities[a]

Adaption media	Turnover rate at steady state[b]	Type of flux[c]	Media in which fluxes are measured			
			SW	$\frac{1}{2}$SW	$\frac{1}{4}$SW	FW
SW	$\lambda_1 = 46.9$	Outflux	2.600	2.08	1.20	0.310
		Influx	2.250	2.00	1.36	0.000
		Net flux	+0.35	+0.08	−0.16	+0.310
$\frac{1}{2}$SW	$\lambda_1 = 17.3$	Outflux	1.10	0.97	0.60	0.240
		Influx	0.93	0.69	0.31	0.00
		Net flux	+0.17	+0.28	0.29	0.240
FW	$\lambda_1 = 0.4$	Outflux	0.083	0.047	0.040	0.022
		Influx	0.430	0.475	0.310	0.014
		Net flux	−0.349	−0.428	0.270	+.008

[a] Reproduced from Motais *et al.* (1966).
[b] λ_1 is given in per cent per hour.
[c] Flux value in mEq/hr/100 g body weight.

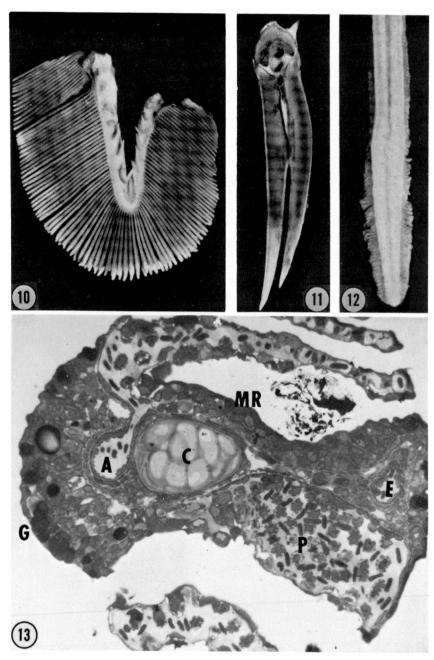

Fig. 10. Sagittal view of an isolated gill arch taken from an Atlantic salmon, *Salmo salar*. Reproduced from Van Dyck, 1967.

cellular component appears to be rapidly assembled and must be responsible for controlling the inward "leakiness" of the gill epithelium. It is dependent upon the sodium concentration in the external media and can account for nearly 85% of the total exchangeable sodium.

(2) When SW-adapted fish which have already been tightened to sodium entrance are transferred to lower salinities, about 15% of the turnover outflux remains unaffected. The net outward directed flux (0.35 mEq/hr) represents the active ion secretion by the gill epithelium. Recall that the amount of seawater swallowed ($\sim$ 5.0 ml/hr/kg) provides for water balance and the excess salt ($\sim$ 0.3 mEq/hr) accumulates in the extracellular fluid. This salt buildup is offset by the amount that is being secreted and hydromineral balance can be maintained. This second component is termed by Houston (1959) as the "regulatory" phase. It is not responsive to the sodium concentration in the external media but appears to respond to excursions of sodium in the extracellular fluids (plasma, etc). The cellular machinery of the gill epithelium appears to initiate *de novo* synthesis immediately, but it requires at least several days to complete.

B. Morphological Studies of Salt Secretion

1. ANATOMY OF THE GILL

The gross morphology of the gills from many species of fishes is quite similar. Lying beneath an opercular cover are found four branchial arches, which extend from either side of the pharynx. Each branchial arch bears two hemibranchs consisting of two rows of tapered and flattened gill filaments which lie parallel to one another and are perpendicular to the arch (Figs. 10, 11, and 12). A lamellar system, which is comprised of flattened, leaflike structures which extend above and below the axial plane of the gill filament exists and is referred to either as the respiratory lamellae or secondary lamellae (Hughes and Grimstone, 1965; Newstead, 1967). The upper and lower rows of lamellae are axially offset

Fig. 11. Dorsal surface of isolated gill hemibranchs showing openings of the blood vessels.

Fig. 12. Sagittal plane of an isolated gill filament showing numerous respiratory lamellae. Reproduced from Van Dyck (1966).

Fig. 13. Transverse section (1 μ) of an isolated gill filament showing an afferent artery (A) which lies close to the eccentrically placed cartilage (C). An efferent artery (E) is located at a distance from the support structure. Pillar cells (P), mucous goblet cell (G) and mitochondria-rich cell (MR) are showing. $\times$200. Reproduced from Newstead (1967).

from one another and curve toward the distal extremity of the filament. Therefore, histological sections often contain lamellae which are cut normal to their flattened aspect on one side of the filament (transversely sectioned) but parallel to their flattened aspect (frontally sectioned) on the other side (Fig. 13). An eccentrically placed cartilaginous rod, which lies near the afferent artery, extends from the main gill arch well into the body of the gill filament. This rod serves as the primary supporting structure of the filament.

The vascular anatomy of the filament has been recently investigated by Steen and Kruysse (1964) in several species of teleosts. They found that the filament has a double circulation. Blood can enter into the efferent filamental artery from the afferent vessels by several routes. Since the efferent blood vessel lies at the opposite side of the support rod, the blood from the afferent filamental artery can pass (1) through the flat lacunar blood channels within the respiratory lamellae; (2) through the peripheral vessels at the tip of the filaments, thus bypassing the surfaces of the respiratory lamellae; and (3) through a network of capillaries and sinuses that lie within the connective tissue compartment which is beneath the epithelial surface of the filament. This epithelial surface lies between the respiratory lamellae and should be termed "interlamellae filamental epithelium." Newstead (1967) could not find any structural specialization in the walls of the afferent and efferent vessels which would account for shunting of the blood. Østlund and Fänge (1962) have shown that blood flow within an isolated gill perfusion apparatus can be altered by the addition of adrenaline or acetylcholine to the perfusate. These two hormones have different effects on the filaments. Adrenalin causes an increase in blood flow via the respiratory lamellae, whereas acetylcholine causes blood flow to decrease, apparently by the shunting, of blood through the central compartment. Keys and Bateman (1932) showed that adrenalin causes an increase in the ionic exchange of the gills. However, a detailed study is lacking on the hemodynamics of this double circulation during periods of salt secretion by the gill.

2. Histology of the Gill Filament

a. *Respiratory Epithelium.* This is the structural barrier that lies between the oxygen-bearing water and the blood sinus connecting the afferent and efferent filamental arteries and consists of a layer of epithelial cells (one or two), a basal lamina, and a thin layer of connective tissue (see Figs. 13 and 14). In some species, the separation between basal surface of epithelial cells and the basal lamina is large enough to accom-

Table IX

Structure of the Blood–Water Barrier in Some Teleostean Species

Species	Reference	Thickness of barrier structures (μ)[a]				
		Basal lamina	Connective tissue	Epithelium	Cytoplasm flange	Total
O. maculosus	Newstead (1967)	0.19–1.3 (0.75)	0.1–1.5 (0.9)	0.3–2.6 (2.26)	0.04–0.75 (0.26)	3.6
G. meandricus	Newstead (1967) (one animal)	0.04–0.05	Absent	No data	0.04–0.08	—
A. purpurescens	Newstead (1967)	0.2–0.25 (0.21)	Absent	No data	0.2–0.4 (0.33)	—
P. notatus	Newstead (1967)	0.04–0.11 (0.07)	0.13–0.84 (0.25)	1.3–3.4 (118)	0.1–0.7 (0.21)	3.15
P. vetulus	Newstead (1967)	0.2–0.7 (0.38)	0.1–0.45 (0.28)	1.1–2.4 (1.6)	0.05–0.35 (0.19)	2.62
O. kisutch	Newstead (1967)	0.04–0.25 (0.1)	Absent	0.5–2.9 (1.7)	0.10–0.34 (0.19)	2.0
G. pollachius (pollock)	Hughes and Grimstone (1965)	0.05	0.3	0.4–2.5	0.1–0.3	—
C. carassius (goldfish)	Schulz (1960)	400–665 Å	No information	No information	800–1150 Å	0.5–0.6 min
T. trichopterus (labyrinthine organ)	Schulz (1960)	Endothelium only present in some sites 600 Å/min	No information	No information		
Lebistes (guppy)	Hughes and Shelton (1962)	—	Present	—	—	—
Callionymus	Hughes and Shelton (1962)	—	Present	—	—	—

[a] As far as possible, measurements were made at or near the central region of the blood channels and were made only on lamellae in which membranes of epithelial and endothelial cells were sharply defined.

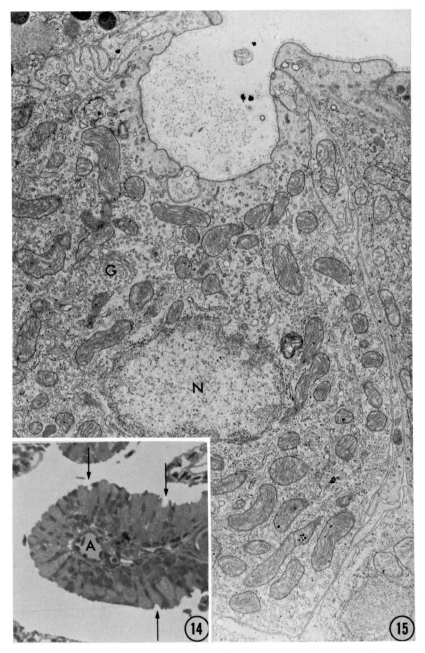

Fig. 14. A light micrograph of a gill filament cut in cross section. The central portion of a filament is composed of connective tissue, including a cartilaginous shaft,

modate wandering leukocytes. Newstead (1967) has reported on the
thickness of the barrier structure in those species in which it has been
measured (Table IX).

b. Salt Secretory Epithelium. This is the structural barrier which lies
between the salt-bearing water and the central compartment of the gill
filament and is comprised of from two to eight layers of cells (see Figs.
13 and 14). This cell population has been described in detail for many
species of fishes. In general, the four principal types of cells cited are:
(1) the squamous-type surface epithelial cell, (2) a nondifferentiated
cell, (3) the mucous goblet cell, and (4) the mitochondria-rich cell (referred to as "chloride" cell).

Each of these cell types has, at one time or another, been postulated
as being the site of ionic secretion (Bevelander, 1935; Datta-Munshi,
1964; Keys and Willmer, 1932; Vickers, 1961). Numerous histological investigations have been undertaken in various laboratories with the goal
of identifying either the intracellular ion secretory apparatus or the
changes of intracellular morphology that could be traced to the transitional manifestations caused by going from salt extrusion to salt uptake
activity. Cytological and cytochemical studies (Bergeron, 1956; Columbo,
1961; Copeland, 1947, 1950; Getman, 1950; Liu, 1942; Natochin and
Bocharov, 1962; Ogawa, 1962; Pettengill and Copeland, 1948; Van
Dyck, 1966) have shown a few specific changes in the cytoplasm, such as
a qualitative increase in certain intracellular enzymes (i.e., succinic dehydrogenase, alkaline phosphatase) but no definitive evidence for an ion
secretory apparatus. However, the name "chloride" cell is a physiological
misnomer because this cell may be involved in the transportation of more
than one species of ion. It has been found that the "chloride" cell does
have a unique ultrastructure.

*c. Ultrastructural Distinctiveness of Epithelial Cells during Ion Secre-
tion versus Ion Uptake.* Numerous electron microscopic investigations

and, at opposite edges, an afferent arteriole and an efferent arteriole. Only the edge
possessing the afferent arteriole (A) is shown here. Chloride cells predominate in the
epithelium in the region around this arteriole. Arrows point to apical crypts. Phosphate-buffered glutaraldehyde–osmium tetraoxide fixation. ×427.5. Micrograph by
Karl J. Karnacky, Jr., Rice University.

Fig. 15. From saltwater-adapted specimen of the euryhaline minnow, *Cyprinodon
variegatus.* The hallmark of the saltwater-adapted chloride cell is an apical crypt
which contains a polyanionic mucus. The subjacent region contains numerous droplets,
also filled with an anionic substance, and these are presumably derived from the Golgi
complex (G). The anastomosing tubular system is best seen in the lower right side
of the micrograph. The chloride cell is flanked by pavement epithelial cells, mucous
cells, and other chloride cells. Phosphate-buffered osmium tetroxide fixation. N =
nucleus. ×14,725. Micrograph by Karl J. Karnacky, Jr., Rice University.

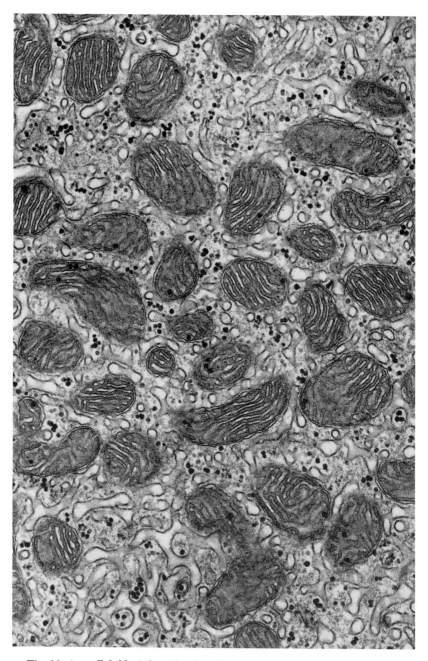

Fig. 16. A small field of the chloride cell cytoplasm showing the extensive smooth-surfaced tubular reticulum. This system is intimately associated with a rather uniform

(Doyle, 1960b; Doyle and Gorecki, 1961; Kessel and Beams, 1960, 1962; Newstead and Conte, 1969; Philpott and Copeland, 1963; Threadgold and Houston, 1964); have shown that the ultrastructural uniqueness lies in the fact that the cell possesses a very dense population of mitochondria and a well-developed agranular endoplasmic reticulum (Figs. 15 and 16). However, this information by itself constitutes only circumstantial evidence in support of its ion secretory function and some investigators have taken issue with this interpretation (Datta-Munshi, 1964; Doyle and Gorecki, 1961; Holliday and Parry, 1962; Parry et al. 1959; Straus, 1963). Comparisons have been made on the fine structure of the cytoplasm of mitochondria-rich cells in three species of Fundulus: F. similis, a marine form; F. heteroclitus, a euryhaline form; and F. chrysotus, a freshwater form (Philpott and Copeland, 1963). The change in ultrastructure which was observed and thought to be consistent with ion secretory activity was a swelling of the tubular elements of the smooth endoplasmic reticulum as they converged toward the region of the apical cavities. A polyanionic substance was found to be enclosed in both the tubular elements and the apical cavity. Philpott (1966) has found a large concentration of chloride in these apical cavities and suggests that this polyanionic material is a type of acid mucopolysaccharide which serves as the electrolyte carrier acting similar to an "ion exchanger." If the tubular labyrinth is the transporting system, then differences in electron density patterns observed would suggest that the polyanionic material and ions are combined in the basal region of the cytoplasm and then concentrated during transportation to the apical cavities where release to the environment occurs. Demonstration that physical continuities exist between the intracellular membranes of the tubular labyrinth and the lateral and basal cell membranes (Newstead and Conte, 1969; Philpott, 1966; Straus, 1963) reinforces the idea that the role of the tubular labyrinth is to allow electrolytes, which accumulate in the extracellular fluids trapped in the intercellular space, to gain entrance and be guided through the cytoplasmic matrix. Philpott (1966) has extended these findings to show that large molecules (MW = 40,000) are able to penetrate the mitochondria-rich cell from the blood side by way of the tubular system (Fig. 17).

---

population of mitochondria and is directly continuous with the plasmalemma. Dense glycogen granules are peppered throughout the continuous phase of the cytoplasmic ground substance. Mitochondria and glycogen reflect a high metabolic potential for this cell, and their close association with the extended cell surface as represented by the tubular reticulum is consistent with the presumed role of the chloride cell in high level electrolyte transport. Seawater adapted Fundulus. ×20,425. Courtesy of Dr. C. W. Philpott, Rice University.

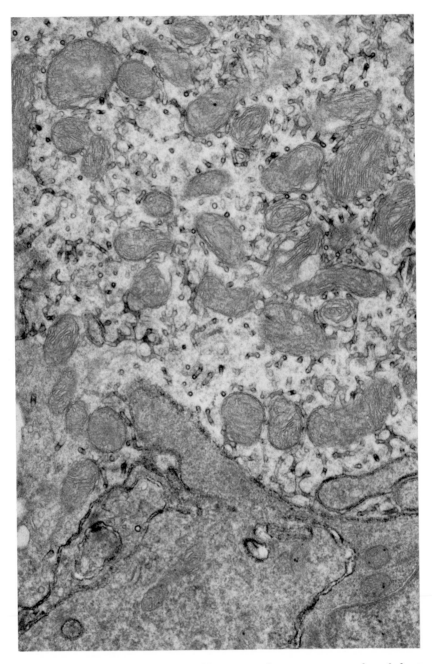

Fig. 17. Horseradish peroxidase, used here as an electron opaque marker of about

d. Cellular Dynamics of the Transcellular Epithelium during Ion Secretion. Doyle and Gorecki (1961), investigating the ultrastructure of gills from several species of fish, found that in *Fundulus* there appeared two types of mitochondria-rich cells which were distinguishable by their electron density. One type was referred to as a "dense" form and the other as a "light" form. It was found that in those specimens kept in seawater, the "dense" form appeared to be more abundant. Threadgold and Houston (1961, 1964), studying the fine structure of the mitochondria-rich cell in four classes of juvenile Atlantic salmon, *Salmo salar,* reported similar results. Newstead and Conte (1969) found similar effects in juvenile coho salmon, *O. kisutch,* but these changes did not appear to be a result of a particular stage of the life history but rather of the accompanying shifts of the internal environment. Apical crypts which are a unique feature of the mitochondria-rich cells of *Fundulus* are not apparent in either *O. kisutch* or *S. salar.* These observations suggested that the renewal rate of the mitochondria-rich cells in transcellular epithelium might be changing during transitions from one environment into another. Conte (1965) showed that inhibition of cell division by exposure to X radiation would result in a loss of hypoosmotic regulation. Pursuing the problem further, Conte and Lin (1967), utilizing radioautographic techniques, showed that the site of cellular renewal in the gill filament in *O. kisutch* is principally in the interlamellar region of the gill filament (Figs. 18 and 19). Attempts were made to measure the turnover rate of the population of cells in the gill epithelium. Following injection of tritium-labeled thymidine (^{3}H-TdR), the kinetic relationships of the *de novo* synthesis of DNA are shown in Figs. 20 and 21. The turnover rate of labeled DNA yields a very rough index of the total transit time from cell formation to the extrusion from the cell compartments. The $T_{1/2}$ of labeled DNA for the gill epithelium in the freshwater animal was 15.8 ± 1 day, whereas in the seawater animal, the cell population had a $T_{1/2}$ of 5.8 ± 1 day. Other tissues such as the heart and pseudobranch did not reflect similar rate changes in labeled DNA.

40,000 MW, was injected directly into the bulbus arteriosus of a living *Fundulus.* After 15 min, the animal was sacrificed and the branchial epithelium processed to show sites of peroxidase activity. Images such as this convincingly demonstrate that the extensive tubular system of the chloride cell is continuous with the plasmalemma at sites along the basal and lateral cell surfaces. These results support the view of an unrestricted passage of rather large molecules from the blood to the epithelial intercellular spaces, and from there into the extensive tubular system of the chloride cell. It is likely that electrolytes follow a similar path in saltwater-adapted animals such as this example. Unstained except for peroxidase activity. ×35,245. Courtesy of Dr. C. W. Philpott, Rice University.

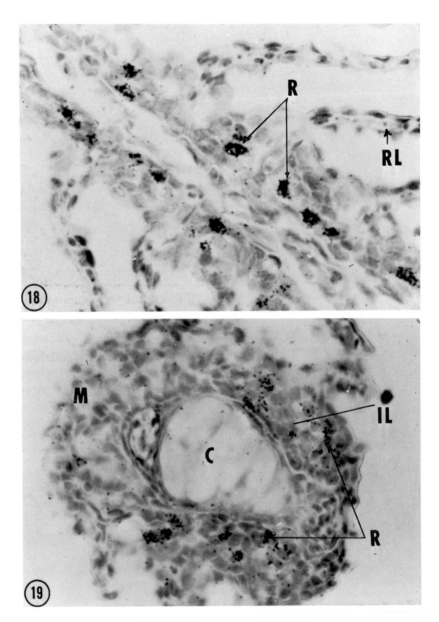

Fig. 18. Longitudinal section of gill filament 1 day postinjection showing the major site of the replacement cells (R) as being within the interlamellae epithelium. Note the absence of labeled cells occurring in the respiratory lamellae (RL). ×576.

Fig. 19. Transverse section of gill filament 1 day postinjection cut obliquely to the cartilaginous-supporting spine (C) showing many replacement cells in the interlamellae epithelium (IL). Very few labeled cells appear in the mucoid cell (M) region. R is replacement cells. ×576.

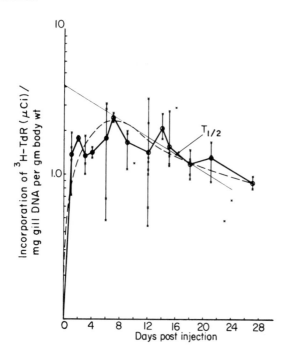

Fig. 20. Incorporation of tritiated thymidine into DNA of the gill tissue isolated from FW-adapted juvenile salmon. (X), individual animal; and (○), mean value for a group. Lines connecting the individual points (X) represent the maximum and minimum values for the entire group for that time interval.

C. Biochemical Studies of Salt Secretion

1. Cytodifferentiation of the Salt Secretory Epithelium

Maetz (1968) has shown in preliminary experiments with actinomycin D that injection of this material into seawater-adapted eel, *Anguilla anguilla*, will cause a decrease in the sodium turnover rate from 30%/hr to 1–2%/hr within 4–5 days with the treated fish dying on the sixth day with abnormally high plasma sodium and chloride levels (Fig. 22). Injection into freshwater eels has no measurable effect upon the sodium influx after the same dosage, but the sodium outflux goes up by a large factor (5–10). The fishes appear to survive longer in freshwater after the treatment. During transfer of FW-adapted eels to saltwater, treated fishes do not show any instantaneous regulation; that is, the sodium efflux remains nearly constant, while in normal animals the sodium outflux soars up after a 1–2-hr lag period to reach a very high efflux value characteristic of the SW-adapted animal (Fig. 23). If the injection of actinomycin D is made 2 hr prior to FW → SW transfer, the animal appears

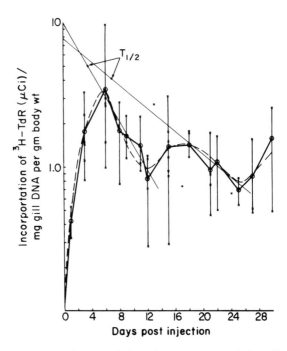

Fig. 21. Incorporation of tritiated thymidine into DNA of the gill tissue isolated from SW-adapted juvenile salmon: (X), individual animal; and (○), mean value for a group. Lines connecting the individual points (X) represent the maximum and minimum values for the entire group of animals for that time interval.

normal for at least 24 hr. Since the known biological action of actinomycin D is to chemically enucleate the cell by eliminating the synthesis of DNA-dependent RNA, it appears that the ion secretory mechanism is dependent upon synthesis of messenger RNA. Experiments utilizing puromycin caused the injected fish to have an immediate increase in the sodium renewal rate of approximately 10% (30 → 40%) followed by slight decrease and then a return to normal within 8–10 hr. Apparently rapid release of protein from ribosomes augments the sodium turnover rate.

Conte and Morita (1968) have reported the existence of specific salt-inducible molecular complexes in the cells of the gill filaments of euryhaline chinook salmon, *O. tshawytscha*. Utilizing immunochemical techniques, they found that in a comparison of ring precipitin titers from homologous antigen–antisera reactions the SW-gill filament epithelium gave a 4–8-fold greater response than that of the FW-gill filament. Immunodiffusion assays of the various cell fractions indicated that the microsomal fraction had at least three distinct and specific antigenic groups. The biochemical nature of the antigens was investigated by pretreatment

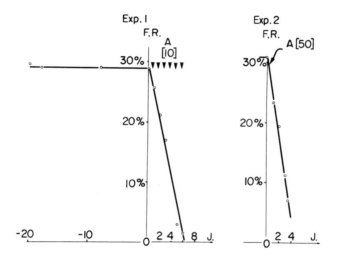

Fig. 22. Inhibition of sodium renewal rate of seawater-adapted eel, *A. anguilla*, following intraperitoneal injection of actinomycin D. Experiment 1: (▼) repeated injection of 10 μg/100 g/day. Experiment 2: (▼) single injection of 50 μg/100 g. Abscissa in days. Ordinate in per cent per hour. Maetz (1968).

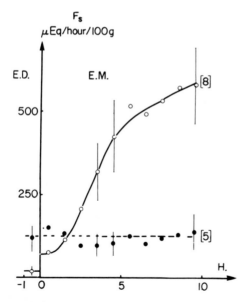

Fig. 23. Inhibition of the instantaneous regulation characteristics of the sodium outflux by actinomycin D. (○) Freshwater-adapted eel returned to seawater (controls). (●) Freshwater-adapted eel which received an intraperitoneal injection 3 days prior to immersion in seawater. Abscissa in hours. Ordinate in μEq Na/hr/100 g. Maetz (1968).

of the antigen with certain enzymes, i.e., trypsin ribonuclease A and α-amylase. Trypsinizing the antigens prior to assaying resulted in complete loss of antigenicity, whereas the other enzymes were without effect. Thus, most of the antigens were thought to be protein in nature or carry a substantial protein moiety. All of the evidence so far suggests that the mechanism of the salt inducible differentiation occurs in the soluble phase of the cytoplasm. Presumably this cytodifferentiation involves development of the ion efflux apparatus and is dependent upon the biosynthesis of proteins necessary for the formation of either the tubular labyrinth and/or the ancillary enzymes required to manufacture the polyanionic substance. In contrast, the ion influx apparatus appears to be preassembled and is activated without the need of a genetic transcription process.

2. Adaptive Enzyme Formation

Earlier cytochemical findings showed qualitative increases in alkaline phosphatases (Pettengill and Copeland, 1948) and succinic dehydrogenase (Natochin and Bocharov, 1962). However, quantitative studies which involved the isolation and assay of enzymes following saltwater adaptation have provided conflicting evidence. No changes in succinate–cytochrome c reductase, NADH–cytochrome c reductase, β-hydroxybutyrate dehydrogenase, glutaminase, and residual adenosinetriphosphatase activities were reported in the studies of *Oncorhynchus* (Tripp, 1967) and of *Fundulus* (Epstein *et al.*, 1967). However, increased Na–K ATPase has been reported in *F. heteroclitus* (Epstein *et al.*, 1967).

D. Endocrinological Studies of Hypoosmoregulation in Teleosts

The role of the adrenal cortical hormones, the caudal neurosecretory system, the urophysis, and other possible endocrine factors in osmoregulation in fishes is reviewed by Bern (1967) and in the chapters by Foster and Goldstein, Hochacka, and Phillips, this volume. Neurohypophyseal hormones have been shown to be involved in the mineral regulation of teleost fishes in several investigations (Acher *et al.*, 1962; Dodd and Perks, 1964; Heller, 1963; Heller and Bentley, 1965; Maetz and Juien, 1961; Rasmussen and Craig, 1961). *In vivo* studies on hypophysectomized fish by Burden (1956), Pickford (1953), Pickford *et al.* (1965), and Maetz *et al.* (1967) have shown that survival in the euryhaline killifish, *F. heteroclitus*, can be achieved for long periods of time in seawater following removal of the pituitary, but death will occur within 6–14 days from a progressive fall in plasma osmolality when fish are transferred into freshwater. Other euryhaline teleosts which appear to require an intact pituitary for freshwater survival are *Poecilia latipinno* and *P. formosa* (Ball, 1962; Ball and Kallman, 1962; Ball and Olivereau, 1964), *Xiphophorus maculatus* and *X. helleri* (Schreibman and Kallman, 1963), *Tilapia mossambica* (Handin *et al.*, 1964; Dharmamba *et al.*, 1967), and *Gambusia* (Chambolle, 1966). However, other teleosts which can survive indefinitely in freshwater without a pituitary are the European eel, *A. anguilla* (Fontaine *et al.*, 1949), and the trout, *Salmo gairdneri* (Donaldson and McBride, 1967). Replacement therapy (Ball *et al.*, 1965; Pickford *et al.*, 1965; Pickford and Phillips, 1959) has shown that the ability of pituitary preparations to protect hypophysectomized fish from failure in freshwater appears to be related to their prolactin activity and not to their content of neuro-hypophysial peptides.

Maetz *et al.* (1967), pursuing the problem of maintaining mineral balance in the hypophysectomized *F. heteroclitus*, found that operated fish are in a negative sodium balance in which the sodium outflux is greater than sodium influx. Injection of prolactin appears to prevent this imbalance by limiting the sodium outflux. The site (or sites) at which this hormone acts to limit the sodium loss is apparently the gill epithelium and not the glomerulus or renal tubule. This has been confirmed by Potts and Evans (1966) and Fleming and Kamemoto (1963). Stanley and Fleming (1966), investigating the euryhaline plains killifish, *Fundulus kansae*, found that hypophysectomy impairs freshwater survival, but these animals showed an increased renal sodium loss rather than an increased extrarenal sodium outflux. Prolactin injections in hypophysec-tomized fish caused a significant reduction in urine electrolyte concen-trations. Therefore, there appears to be different modes of action for this hormone in these two species. *In vitro* studies on isolated gill fila-ments by Bellamy (1961) and Kamiya (1967) have shown that in eels, *A. anguilla* and *A. japonica*, which are adapted to freshwater, the sodium content of the filaments increased following incubation in seawater. The sodium concentration was maintained at a level much lower than the incubation medium. Gills removed from seawater-adapted fish did not show a similar increase in sodium content, but when metabolic inhibitors were added to the incubation medium a rise in sodium concentration occurred. Hirano *et al.* (1967), utilizing the identical technique but in animals which had been hypophysectomized and/or urophysectomized, found that the gill filaments did not salt load from animals which had been exposed to seawater 24 hr prior to the isolation (Table X). How-ever, gill filaments isolated from FW-adapted fish, which had either the pituitary or urophysis extirpated were not assayed. Therefore, it is difficult at this time to state whether prolactin and/or other neurohypo-physial hormones have a direct action upon the cells of the epithelium.

Table X

Effect of Hypophysectomy and/or Urophysectomy on Sodium Transport
of Eel Gills Incubated in Seawater[a,b]

Treatment	No. of eels	Sodium content (mEq/100 mg dry wt)		Net sodium influx during 1 hr of incubation
		Before incubation	1 hr after incubation	
Intact (freshwater)	5	31.4 ± 1.24^c	49.7 ± 0.85	18.3 ± 2.06
Intact (seawater)	5	38.7 ± 1.41	44.5 ± 2.64	5.82 ± 1.35^d
Hypophysectomy	3	34.9 ± 2.76	39.2 ± 2.69	4.33 ± 2.82
Sham-hypophysectomy	3	40.8 ± 0.87	47.1 ± 1.64	6.30 ± 1.79
Hypo-urophysectomy	5	33.4 ± 0.49	38.8 ± 2.58	5.50 ± 2.52
Sham-hypourophysectomy	4	39.2 ± 1.50	45.9 ± 1.76	6.65 ± 0.34
Urophysectomy	5	38.8 ± 1.31	42.7 ± 1.98	3.92 ± 2.13
Sham-urophysectomy	5	36.4 ± 2.02	44.7 ± 0.32	8.32 ± 1.62

[a] Reproduced from Hirano et al. (1967).
[b] Transferred to SW for 24 hr prior to isolation.
[c] Mean $\pm$ standard error.
[d] Significant compared to intact freshwater value ($p = 0.001$).

It appears from this preliminary evidence that induction of salt secretion in the cells can be achieved in the absence of neurohypophysial
hormones. Much more information is needed before definitive conclusions
can be drawn.

For a discussion of embryological studies of hypoosmoregulation in
teleosts, see chapter by Blaxter, Volume III.

E. Summary

(1) Most species of marine teleosts, as well as the migratory forms,
exhibit hypoosmoregulation when living in seawater under steady state
conditions.

(2) Water loss occurs as a result of the dehydration of body tissues
effected by the external salt gradient. To offset water loss, decreased
formation of urine occurs primarily through a reduction of glomerular
filtration rate. Urine usually remains hypotonic to blood.

(3) Hydromineral balance is maintained by replenishing the body
water loss by the ingestion of seawater. The amount swallowed, depending upon the species, is approximately 0.2–0.5% of the body weight per
hour.

(4) The composition of the intestinal fluid following the drinking of
seawater changes during its passage through the gastrointestinal tract.

Approximately 70–80% of the water in the ingested salt solution is absorbed. Monovalent ions readily pass through the intestinal wall and enter the extracellular fluid compartment. Most of the divalent ions remain behind and are lost via the mucous tubes, feces, and, to a lesser extent, by the kidney.

(5) Transitions from the steady state marine environment into environments of lower salinities (or the reverse) provide for marked changes in permeability of the epithelial membranes with regard to electrolytes, especially so for the gill epithelium. Rapid increases or decreases of the blood electrolytes will accompany these transitions until renal and extrarenal regulatory mechanisms are balanced.

(6) An extrarenal mechanism for the regulation of electrolytes does exist in the gill epithelium. Cells with an ultrastructure similar to the secretory cells of salt glands of other animals are found in the filamental epithelium. Inhibition of the cellular regeneration of the gill epithelium destroys the extrarenal mechanism. The major site of cell replication is found in the basal cells which lie under the mitochondria-rich cells which are considered to be the ion secretory cell.

(7) Differentiation of the replacement cell appears to be directly salt inducible and the formation of the ion secretory apparatus is thought to be dependent upon the formation of messenger RNA. Proteins or protein synthesis appears involved in the biosynthesis of the ion efflux system. A polyanionic substance has been suggested as being the electrolyte carrier.

(8) Hormones are involved in hypoosmoregulation, but the exact site of action in regard to the extrarenal mechanism is not known at this time.

ACKNOWLEDGMENTS

The author wishes to express his thanks to Miss L. Ebeling for her devoted assistance in the collection of reference material and to Mrs. D. Hancock and Mrs. S. Conte for their tireless efforts in the typing and preparation of the manuscript. In addition, he is indebted to the many friends and colleagues who have lent him original pictures which are reproduced in this manuscript. This review was supported by funds from the U.S. Atomic Energy Commission under contract number AT(45-1)-RLD-2013-2.

REFERENCES

Acher, R., Chauvet, M., and Crepy, D. (1962). Isolement d'une nouvelle hormone neurophypophysaire, l'isotocine, presente chez les poissons osseux. *Biochim. Biophys. Acta* **58**, 624–625.

Baggerman, B. (1960). Salinity preference, thyroid activity and the seaward migration of four species of Pacific salmon *J. Fisheries Res. Board Can.* **17**, 296–322.

Ball, J. (1962). Blood production after hypophysectomy in the vivoparous teleost *Mollienesia latipinna.* Le Sueur. *Nature* **194**, 787.

Ball, J., and Kallman, K. (1962). Functional pituitary transplants in the all-female gynogenetic teleost, *Mollienesia formosa* (Gerard). *Am. Zoologist* **2**, 389 (abstr.).

Ball, J., and Olivereau, M. (1964). Rôle de la prolactine dans la survie en eau douce de *Poecilia latipinna* hypophysectomise et arguments en faveur de sa synthèse par les cellulis erythrosinaphiles de l'hypophyse des teleosteens. *Compt. Rend.* **259**, 1443–1446.

Ball, J., Olivereau, M., Slicher, A., and Kallman, K. (1965). Functional capacity of ectopic pituitary transplants in the teleost *Poecilia formosa*, with a comparative discussion on the transplanted pituitary. *Phil. Trans. R. Soc. London* **B249**, 69–99.

Bellamy, D. (1961). Movements of potassium, sodium, and chloride in incubated gills from the silver eel. *Comp. Biochem. Physiol.* **3**, 125–135.

Bellamy, D., and Chester Jones, I. (1961). Studies on *Myxine glutinosa.* I. The chemical composition of the tissues. *Comp. Biochem. Physiol.* **3**, 175–183.

Bentley, P. (1962). Permeability of the skin of the cyclostome *Lampetra fluviatilis*, to water and electrolytes. *Comp. Biochem. Physiol.* **6**, 95–97.

Bentley, P., and Follet, B. (1963). Kidney function in a primitive vertebrate, the cyclostome *Lampetra fluviatilis. J. Physiol.* (*London*) **169**, 902–918.

Bergeron, J. (1956). A histophysiological study of osmotic regulation in euryhaline teleost *Fundulus heteroclitus.* Ph.D. thesis, University Microfilms, Ann Arbor, Michigan.

Bern, H. (1967). Horomones and endocrine glands of fishes. *Science* **158**, 455–461.

Bernard, G., and Hartmann, J. (1960). Cytological and histochemical observations on the elasmobranch rectal gland. *Anat. Record* **137**, 340.

Bevelander, G. (1935). Comparative study of the branchial epithelium in fishes with special reference to extrarenal excretion. *J. Morphol.* **57**, 335–348.

Black, V. (1957). Excretion and osmoregulation. In "The Physiology of Fishes" Academic Press, New York. (M. E. Brown, ed.), Vol. 1, pp. 163–205.

Blackstad, T. (1963). The skin and its derivatives. In "Biology of Myxine" (A. Brodal and R. Fänge, eds.), pp. 195–230. Oslo Univ. Press, Oslo.

Bonting, S. (1966). Studies on sodium-potassium-activated adenosine triphosphatase. XV. The rectal gland of the elasmobranch. *Comp. Biochem. Physiol.* **17**, 953–966.

Boroughs, H., Townsley, S., and Hiatt, R. (1957). The metabolism of radionuclides by marine organisms. III. Uptake of ^{45}Ca in solution by marine fish. *Limnol. Oceanog.* **2**, 28–32.

Bulger, R. (1963). Fine structure of the rectal (salt-secreting) gland of *Squalus acanthias. Anat. Record* **147**, 95–127.

Bulger, R. (1965). Electron microscopy of the stratified epithelium lining the excretory canal of the dogfish rectal gland. *Anat. Record* **151**, 589–608.

Burden, C. (1956). The failure of hypophysectomized *Fundulus heteroclitus* to survive in fresh water. *Biol. Bull.* **110**, 8–28.

Burger, J. (1962). Further studies on the function of the rectal gland in the spiny dogfish. *Physiol. Zool.* **35**, 205–217.

Burger, J. (1965). Roles of the rectal gland and the kidneys in salt and water excretion in the spiny dogfish. *Physiol. Zool.* **38**, 191–196.

Burger, J., and Hess, W. (1960). Function of the rectal gland in the spiny dogfish. *Science* **131**, 670–671.

Burger, J., and Tosteson, D. (1966). Sodium influx and efflux in the spiny dogfish *Squalus acanthias*. *Comp. Biochem. Physiol.* **19**, 649–654.

Burian, R. (1910). Funktion der Nierenglomerule und Ultrafiltration. *Arch. Ges. Physiol.* **136**, 741–760.

Busnel, R. (1943). Recherches de physiologie appliquées à la pisciculture: A propos de la migration de la truite arc-en-ciel. *Bull. Franc. Piscicult.* **128**, 108–117.

Chambolle, P. (1966). Recherches dur l'allongement de la durée de survive après hypophysectomie le chez *Gambusia* sp. *Compt. Rend.* **262**, 1750–1753.

Chan, D., and Phillips, J. (1966). The embryology of the rectal gland in the spiny dogfish (*Squalus acanthias*). *J. Anat.* (*London*) **100**, 899–903.

Chan, D., and Phillips, J. (1967). The anatomy, histology and histochemistry of the rectal gland in the lip-shark, *Hemiscyllum plagiosum* (Bennett). *J. Anat.* (*London*) **101**, 137–157.

Chan, D., Phillips, J., and Jones, I. (1967). Studies on electrolyte changes in the lip-shark, *Hemiscyllum plagiosum* (Bennett) with special reference to hormonal influence on the rectal gland. *Comp. Biochem. Physiol.* **23**, 185–198.

Chartier-Baraduc, M. (1960). Etude de quelques aspects du metabolisme hydro-minérale chez *Salmo salar* au cours de la smoltification et la migration catadrome. *Bull. Centre Etude Rech. Sci.* (*Biarritz*) **3**, 19–31.

Columbo, G. (1961). Chloride-secreting cells in the gill of european eels. *Nature* **190**, 19–31.

Conte, F. (1965). Effects of ionizing radiation on osmoregulation in fish *Oncorhynchus kisutch*. *Comp. Biochem. Physiol.* **15**, 293–302.

Conte, F., and Bailey, J. (1968). Unpublished observation.

Conte, F., and Lin, D. (1967). Kinetics of cellular morphogenesis in gill epithelium during sea water adaptation of *Oncorhynchus* (Walb.). *Comp. Biochem. Physiol.* **23**, 945–957.

Conte, F., and Morita, T. (1968). Immunochemical study of cell differentiation in gill epithelium of euryhaline *Oncorhynchus* (Walbaum). *Comp. Biochem. Physiol.* **24**, 445–454.

Conte, F., and Wagner, H. (1965). Development of osmotic and ionic regulation in juvenile steelhead trout, *Salmo gairdneri*. *Comp. Biochem. Physiol.* **14**, 603–620.

Conte, F., Wagner, H., Fessler, J., and Gnose, C. (1966). Development of osmotic and ionic regulation in juvenile coho salmon (*Oncorhynchus kisutch*). *Comp. Biochem Physiol.* **18**, 1–15.

Copeland, D. (1947). The cytological basis of salt excretion from the gills of *Fundulus heteroclitus*. *Biol. Bull.* **93**, 192.

Copeland, D. (1950). Adaptive behavior of the chloride cell in the gill of *Fundulus heteroclitus*. *J. Morphol.* **87**, 369–379.

Copeland, D. (1964a). A mitochondrial pump in the cells of the anal papillae of mosquite larvae. *J. Cell Biol.* **23**, 253–263.

Copeland, D. (1964b). Salt absorbing cells in gills of crabs, *Callinectes* and *Carcinus*. *Biol. Bull.* **127**, 367–368.

Copeland, D. (1967). Fine structure of salt and water uptake in *Gecerinus lateralis*. *Am. Zoologist.* **7**, 765.

Crofts, D. (1925). The comparative morphology of the caecal gland (rectal gland) of selachian fishes. *Proc. Zool. Soc. London* **I**, 101–188.

Datta-Munshi, J. (1964). "Chloride cells" in the gills of fresh-water teleosts. *Quart. J. Microscop. Sci.* **105**, 79–89.

Dekhuyzen, M. (1904). Ergebnisse von osmotischen Studien, nämentlich bei

Knochenfischen, an der Biologischen Station des Bergenser Museums. *Bergens Museums Arbok, Naturv.* 8, 1–7.

Dharmamba, M., Handin, R., Nandi, J., and Bern, H. (1967). Effect of prolactin on freshwater survival and on plasma osmotic pressure of hypophysectomized *Tilapia mossambica. Gen. Comp. Endocrinol.* 9, 295–302.

Dodd, J., Perks, A., and Dodd, M². (1964). Physiological functions of neurohypophysial hormones in sub-mammalian vertebrates. *In* "The Pituitary" (G. Harris and B. Donovan, eds.), Vol. 3, pp. 578–623. Butterworth, London and Washington, D.C.

Donaldson, E., and McBride, J. (1967). The effects of hypophysectomy in rainbow trout *Salmo gairdnerii* (Rich.) with special reference to the pituitary-interrenal axis. *Gen. Comp. Endocrinol.* 9, 93–101.

Doyle, W. (1960a). Fine structure of salt-regulating epithelia. *Anat. Record* 136, 184.

Doyle, W. (1960b). The principal cells of the salt-gland of marine birds. *Exptl. Cell Res.* 211, 386–393.

Doyle, W. (1962). Tubule cells of the rectal salt-gland of *Urolophus. Am. J. Anat.* 111, 223–238.

Doyle, W., and Gorecki, D. (1961). The so-called chloride cell of the fish gill. *Physiol. Zool.* 34, 81–85.

Ellis, K., and Abel, J. (1964). Intercellular channels in the salt-secreting glands of marine turtles. *Science* 144, 1340–1342.

Elson, P. (1957). The importance of size in the change from parr to smolt in Atlantic salmon. *Can. Fish. Culturist* 21, 1–6.

Epstein, F., Katz, A., and Pickford, G. (1967). Sodium and potassium-activated adenosine triphosphatase of gills: Role in adaptation of teleosts to salt water. *Science* 156, 1245–1247.

Evans, D. (1967). Sodium, chloride and water balance of the intertidal teleost, *Xiphister atropurpureus.* II. The role of the kidney and the gut. *J. Exptl. Biol.* 47, 519–542.

Evans, D. (1968). Measurement of drinking rates in fish. *Comp. Biochem. Physiol.* 25, 751–753.

Evropietseva, N. (1957). Transformation to smolt stage and downstream migration of young salmon. *Uch. Zap. Leningr. Gos. Univ., Ser. Biol. Nauk* 44, 117–154.

Fänge, R. K., Schmidt-Nielsen, K., and Robinson, M. (1958). Control of secretion from the avian salt glands. *Am. J. Physiol.* 195, 321–326.

Fleming, W., and Kamemoto, F. (1963). The site of sodium outflux from the gill of *Fundulus kansae. Comp. Biochem. Physiol.* 8, 263–269.

Fleming, W., and Stanley, J. (1965). Effects of rapid change in salinity on the renal function of a euryhaline teleost. *Am. J. Physiol.* 209, 1025–1030.

Fontaine, M. (1930). Recherches sur le mileu intérieur de la lamproie marine (*Petromyzon marinus*). Ses variations en fonction de celles du milieu exterieur. *Compt. Rend.* 191, 680–682.

Fontaine, M. (1960). Quelques problèmes physiologiques poses par le *Salmo salar*. Intérêt de l'étude de la smoltification, type de préparation au comportement migratoire. *Experientia* 16, 433–472.

Fontaine, M., Callaman, O., and Olivereau, M. (1949). Hypophyse et euryhalinite chez l'*Auguille. Compt. Rend.* 288, 513–514.

Galloway, T. (1933). The osmotic pressure and saline content of the blood of *Petromyzon fluviatilis. J. Exptl. Biol.* 10, 313–316.

Getman, H. (1950). Adaptive changes in the chloride cells of *Anguilla rostrata*. *Biol. Bull.* **99**, 439–445.

Gordon, M. (1959a). Osmotic and ionic regulation in scottish brown trout and sea trout (*Salmo trutta* L.). *J. Exptl. Biol.* **36**, 253–260.

Gordon, M. (1959b). Rates of exchange of chloride ions in rainbow trout (*Salmo gairdneri*) acclimated to various salinities. *Anat. Record* **134**, 571–572.

Gordon, M. (1963). Chloride exchanges in rainbow trout (*Salmo gairdneri*) adapted to different salinities. *Biol. Bull.* **124**, 45–54.

Greene, C. (1904). Physiological studies of the chinook salmon. *Bull. U.S. Bur. Fisheries* **24**, 431–456.

Handin, R., Nandi, J., and Bern, H. (1964). Effects of hypophysectomy on survival and on thyroid and interrenal histology of the cichlid teleost *Tilapia mossambica*. *J. Exptl. Zool.* **157**, 339.

Heller, H. (1963). Pharmacology and distribution of neurohypophysial hormones. *Symp. Zool. Soc. London* **9**, 93–106.

Heller, H., and Bentley, P. (1965). Phylogenetic distribution of the effects of neurohypophysial hormones on water and sodium metabolism. *Gen. Comp. Endocrinol.* **5**, 96–108.

Hickman, C. (1968). Ingestion, intestinal absorption and elimination of sea water and salts in the southern flounder *Paralichthys lethostigma*. *Can. J. Zool.* **46**, 457–466.

Hirano, T. (1967). Effect of hypophysectomy on water transport in isolated intestine of the eel *Anguilla japonica*. *Proc. Japan Acad.* **43**, 793–796.

Hirano, T., Kamiya, M., Saishu, S., and Utida, S. (1967). Effects of hypophysectomy and urophysectomy on water and sodium transport in isolated intestine and gills of Japanese eel (*Anguilla japonica*). *Endocrinol. Japon.* **14**, 182–186.

Hodler, J., Heinemann, H., Fishman, A., and Smith, H. (1955). Urine pH and carbonic anhydrase activity in the marine dogfish. *Am. J. Physiol.* **183**, 155–162.

Holliday, F., and Parry, G. (1962). Electron microscopic studies of the acidophil cells in the gills and pseudobranchs of fish. *Nature* **193**, 192.

Hoskins, E. (1917). On the development of the digitform gland and the post-valvular segment of the intestine in *Squalus acanthias*. *J. Morphol.* **28**, 329–360.

Houston, A. (1959). Osmoregulatory adaptation of the steelhead trout (*Salmo gairdneri* Richardson) to sea water. *Can. J. Zool.* **37**, 729–748.

Houston, A. (1960). Variations in the plasma-level of chloride in hatchery-reared yearling Atlantic salmon during pari-smolt transformation and following transfer into sea water. *Nature* **185**, 632–633.

Houston, A. (1964). On passive features of the osmoregulatory adaptation of anadromous salmonids to sea water. *J. Fisheries Res. Board Can.* **21**, 1535–1538.

Houston, A., and Threadgold, L. (1963). Body fluid regulation in smolting Atlantic salmon. *J. Fisheries Res. Board Can.* **20**, 1355–1369.

Hughes, G., and Grimstone, A. (1965). The fine structure of the secondary lamellae of the gills of *Gadus pollachius*. *Quart. J. Microscop. Sci.* **106**, 343–353.

Hughes, G., and Shelton, G. (1962). Respiratory mechanisms and their nervous control in fish. *Advan. Comp. Physiol. Biochem.* **1**, 275–364.

Johnston, C., and Eales, J. (1966). Purines in the integument of the Atlantic salmon (*Salmo salar*) during parr-smolt transformation. *J. Fisheries Res. Board Can.* **24**, 955–964.

Kamiya, M. (1967). Changes in ion and water transport in isolated gills of the cultured eel during the course of salt adaptation. *Annotationes Zool. Japon.* **40**, 123–129.

Kessel, R., and Beams, H. (1960). An electron microscope study of the mitochondria rich "chloride cells" from the gill filaments of fresh water and sea water adapted *Fundulus heteroclitus. Biol. Bull.* **119**, 322.

Kessel, R., and Beams, H. (1962). Electron microscopic studies on the gill filaments of *Fundulus heteroclitus* from the sea water and fresh water with special reference to the ultrastructure organization of the "Cl cell." *J. Ultrastruct. Res.* **4**, 77–87.

Keys, A. (1931a). Chloride and water secretion and absorption by the gills of the eel. *Z. Vergleich. Physiol.* **15**, 364–388.

Keys, A. (1931b). The heart-gill preparation of the eel and its perfusion for the study of a natural membrane *in situ. Z. Vergleich. Physiol.* **15**, 352–363.

Keys, A., and Bateman, J. (1932). Branchial responses to adrenaline and pitressin in the eel. *Biol. Bull.* **63**, 327–336.

Keys, A., and Willmer, E. (1932). "Chloride secreting cells" in the gills of fishes with special reference to the common eel. *J. Physiol. (London)* **76**, 368–378.

Koch, H., Evans, J., and Bergstrom, E. (1959). Sodium regulation in the blood of parr and smolt stages of Atlantic salmon. *Nature* **184**, 283–284.

Komnick, H. (1963). Elektronenmikroskopische Untersuchungen zur funktionellen Morphologie des Ionentransportes in der Salzdrüse von Larus argentatus I. Teil: Bau und Feinstruktur der Salzdrüse. *Protoplasma* **56**, 274–314.

Komnick, H., and Wohlfarth-Bottermann, G. (1966). Zur cytologie der rectaldrusen von Knorpelfischen. I. Teil die Feinstruktur der Tubulusephithelzellen. *Z. Zellforsch. Mikroskop. Anat.* **74**, 123–144.

Krogh, A. (1939). "Osmotic Regulation in Aquatic Animals." Cambridge Univ. Press, London and New York.

Kubo, T. (1953). On the blood of salmonid fishes of Japan during migration. *Bull. Fac. Fisheries, Hokkaido Univ.* **4**, 138–148.

Kubo, T. (1954). Some nitrogenous compounds of blood and metamorphosis of *Oncorhynchus masou. Bull. Fac. Fisheries, Hokkaido Univ.* **5**, 248–252.

Kubo, T. (1955). Changes of some characteristics of blood of smolt of *Oncorhynchus masou* during seaward migration. *Bull. Fac. Fisheries, Hokkaido Univ.* **6**, 201–207.

Liu, C. (1942). Osmotic regulation and chloride secreting cells in the paradise fish *Macropodium opercularis. Sinensia* **13**, 17–20.

Lovern, J. (1934). Fat metabolism in fishes. V. The fat of salmon in its young fresh-water stages. *Biochem. J.* **28**, 1961–1963.

McFarland, W., and Munz, F. (1958). A re-examination of the osmotic properties of the Pacific hagfish, *Polistotrema stouti. Biol. Bull.* **114**, 348–356.

McFarland, W., and Munz, F. (1965). Regulation of body weight and serum composition by hagfish in various media. *Comp. Biochem. Physiol.* **14**, 383–398.

McInerney, H. (1963). Salinity preference and orientation mechanism in salmon migration. *J. Fisheries Res. Board Can.* **21**, 995–1018.

Maetz, J. (1968). Personal communication.

Maetz, J., and Romeu, F. (1964). The mechanism of sodium and chloride uptake by the gills of a fresh-water fish *Carassius auratus.* II. Evidence for NH_4^+/Na^+ and HCO_3^-/Cl^- exchanges. *J. Gen. Physiol.* **47**, 1209–1227.

Maetz, J., and Juien, M. (1961). Action of neurohypophysial hormones on the sodium fluxes of a fresh-water teleost. *Nature* **189**, 152–153.

Maetz, J., and Lahlou, B. (1966). Les échanges de sodium et de chlore chez un elasmobranche, *Scylliorhinus* mesures à l'aide des isotope ^{24}Na et ^{36}Cl. *J. Physiol. (Paris)* **58**, 249.

Maetz, J., and Skadhauge, E. (1968). Drinking rates and gill ionic turnover in relation to external salinities in the eel. *Nature* **217**, 371–373.

Maetz, J., Sawyer, W., Pickford, G., and Mayer, N. (1967). Evolution de la balance minerale du sodium chez *Fundulus heteroclitus* au cours du transfert d'eau de mer en eau douce: Effets de l'hypophysectomie et de la prolactine. *Gen. Comp. Endocrinol.* **8**, 163–176.

Malikova, E. (1957). Biochemical estimation of young salmon in the transition stage before migration to the sea and of smolts in a period of detention in fresh water. *Tryabl Pameuuckoso Omaera BHNPO* **2**, 241–255.

Mashiko, K., and Jozuka, K. (1964). Absorption and excretion of calcium by teleost fishes with special reference to routes followed. *Annotationes Zool. Japon.* **37**, 41–50.

Morris, R. (1956). The osmoregulatory ability of the lampern (*Lampetra fluviatilis*) in sea water during the course of its spawning migration. *J. Exptl. Biol.* **33**, 235–248.

Morris, R. (1957). Some aspects of the structure and cytology of the gills of *Lampetra fluviatilis*. *Quart. J. Microscop. Sci.* **98**, 473–485.

Morris, R. (1958). The mechanism of marine osmoregulation in the lampern (*Lampetra fluviatilis* L.) and the causes of its breakdown during the spawning migration. *J. Exptl. Biol.* **35**, 649–665.

Morris, R. (1965). Studies on salt and water balance in *Myxine glutinosa* (L.). *J. Exptl. Biol.* **42**, 359–371.

Motais, R. (1961a). Les échanges de sodium chez un teleosteen euryhalin, *Platichthys flesus* Linne: Cinétique de ces échanges lors des passages d'eau de mer en eau douce et d'eau douce en eau de mer. *Compt. Rend.* **253**, 724–726.

Motais, R. (1961b). Cinétique des échanges de sodium chez un teleosteen euryhalin (*Platichthys flesus* L. au cours de passages successifs eau de mer, eau douce, eau de mer en fonction du temps de sejour en eau douce. *Compt. Rend.* **253**, 2609–2611.

Motais, R. (1967). Les méchanismes d'échanges ioniques branchiaux chez les teleosteens. *Ann. Inst. Oceanog.* (*Paris*) **45**, 1–84.

Motais, R., and Maetz, J. (1965). Comparison des échanges de sodium chez un teleosteen euryhaline (le Flet) et un teleosteen stenohalin (le Serran) en eau de mer. Importance relative du tube digestif et de la branchie dans ces échanges. *Compt. Rend.* **261**, 532–535.

Motais, R., Romeu, F., and Maetz, J. (1965). Méchanisme de l'euryhalinite. Etude du Flet (euryhalin) et du Serran (stenohalin) au cours du transfert en eau douce. *Compt. Rend.* **261**, 801–804.

Motais, R., Romeu, F., and Maetz, J. (1966). Exchange diffusion effect and euryhalinity in teleosts. *J. Gen. Physiol.* **50**, 391–422.

Mullins, L. (1950). Osmotic regulation in fish as studied with radioisotopes. *Acta Physiol. Scand.* **21**, 303–314.

Munz, F., and McFarland, W. (1964). Regulatory function of a primitive vertebrate kidney. *Comp. Biochem. Physiol.* **13**, 381–400.

Natochin, Y., and Bocharov, G. (1962). Activation of sodium excreting cells in the gills of pink and chum salmon during adaptation to life in salt water. *Vopr. Ikhtiol.* **2**, 687–692.

Newstead, J. (1967). Find structure of the respiratory lamellae of teleostean gills. *Z. Zellforsch. Mikroskop. Anat.* **79**, 396–428.

Newstead, J., and Conte, F. (1969). Cytological and ultrastructural studies on the mitochondria-rich cells in the gills of *Oncorhynchus* (Walbaum). In press.

Ogawa, M. (1962). Chloride cells in Japanese common eel, *Anguilla japonica*. *Sci. Rept. Saitama Univ.* **134**, 131–137.

Oguri, M. (1964). Rectal glands of marine and fresh-water sharks: Comparative histology. *Science* **144**, 1151–1152.

Oide, M. (1967). Effects of inhibitors on transport of water and ion in isolated intestine and Na^+-K^+ ATPase in intestinal mucosa of the eel. *Annotationes Zool. Japon.* **40**, 130–135.

Oide, M., and Utida, S. (1967). Changes in water and ion transport in isolated intestines of the eel during salt adaptation and migration. *Intern. J. Life Oceans Coastal Waters* **1**, 102–106.

Østlund, E., and Fänge, R. (1962). Vasodilation by adrenaline and nonadrenaline, and the effect of some other substances on perfused fish gills. *Comp. Biochem. Physiol.* **5**, 307–309.

Palmer, R. (1966). Unpublished observation.

Parry, G. (1958). Size and osmoregulation in fishes. *Nature* **181**, 1218.

Parry, G. (1960). The development of salinity tolerance in the salmon *Salmo salar* (L.) and some related species. *J. Exptl. Biol.* **37**, 425–434.

Parry, G. (1961). Osmotic and ionic changes in the blood and muscle of migrating salmonids. *J. Exptl. Biol.* **38**, 411–427.

Parry, G. (1966). Osmotic adaptation in fishes. *Biol. Rev.* **41**, 392–444.

Parry, G., Holliday, F., and Blaxter, J. (1959). "Chloride secretory" cells in the gills of teleosts. *Nature* **183**, 1248–1249.

Pettengill, O., and Copeland, D. (1948). Alkaline phosphatase activity in the chloride cell of *Fundulus heteroclitus* and its relation to osmotic work. *J. Exptl. Zool.* **108**, 235–241.

Philpott, C. (1966). The use of horseradish peroxidase to demonstrate functional continuity between the plasmalemma and the unique tubular system of the chloride cell. *J. Cell Biol.* **31**, 88a (abstr.).

Philpott, C., and Copeland, D. (1963). Fine structure of chloride cells from three species of *Fundulus*. *J. Cell Biol.* **18**, 389–404.

Pickford, G. (1953). Disturbances of mineral metabolism and osmoregulation in hypophysectomized *Fundulus*. *Anat. Record* **115**, 409.

Pickford, G., and Phillips, J. (1959). Prolactin, a factor in promoting survival of hypophysectomized killifish in fresh water. *Science* **130**, 454–455.

Pickford, G., Robertson, E., and Sawyer, W. (1965). Hypophysectomy, replacement therapy, and the tolerance of the euryhaline killifish, *Fundulus heteroclitus*, to hypotonic media. *Gen. Comp. Endocrinol.* **5**, 160–180.

Potts, W. (1968). Osmotic and ionic regulation. *Ann. Rev. Physiol.* **30**, 73–104.

Potts, W., and Evans, D. (1966). The effects of hypophysectomy and bovine prolactin on salt fluxes in fresh-water adapted *Fundulus heteroclitus*. *Biol. Bull.* **131**, 362–368.

Potts, W., and Evans, D. (1967). Sodium and chloride balance in the killifish, *Fundulus heteroclitus*. *Biol. Bull.* **133**, 411–424.

Potts, W., and Parry, G. (1964). "Osmotic and Ionic Regulation in Animals." Pergamon Press, Oxford.

Potts, W., Foster, M., Rudy, P., and Parry-Howells, G. (1967). Sodium and water balance in the cichlid teleost, *Tilapia mossambica*. *J. Exptl. Biol.* **47**, 461–470.

291

Prosser, C., and Brown, F. (1962). In "Comparative Animal Physiology," 2nd ed., pp. 27–31. Saunders, Philadelphia, Pennsylvania.

Rall, D., and Burger, J. (1967). Some aspects of hepatic and renal excretion in Myxine. Am. J. Physiol. 212, 354–356.

Rasmussen, H., and Craig, L. (1961). Isolation of arginine vasotocin from fish pituitary glands. Endocrinology 86, 1051–1055.

Robertson, J. (1963). Osmoregulation and ionic composition of cells and tissues. In "Biology of Myxine" (A. Brodal and R. Fänge, eds.), pp. 503–515. Oslo Univ. Press, Oslo.

Romeu, F., and Maetz, J. (1964). The mechanism of sodium and chloride uptake by the gills of a fresh-water fish, Carassius auratus. I. Evidence for an independent uptake of sodium and chloride ions. J. Gen. Physiol. 47, 1195–1207.

Rosenthal, H. (1961). Uptake and turnover of ^{35}S by Lebistos. Biol. Bull. 120, 183–191.

Schmidt-Nielsen, B., and Schmidt-Nielsen, K. (1923). Beitrage zur Kenntnis des osmotischen Druckes der Fische. Kongol. Norske Videnskab. Selskabs, Skrifter No. 1, pp. 1–24.

Schreibman, M., and Kallman, K. (1963). Effects of hypophysectomy on Xiphophorus. Am. Zoologist 3, 556 (abstr.).

Schulz, H. (1960). Die submikroskopische Morphologie des Kiemenepithels. 4th Intern. Conf. Electron Microscopy, Berlin, 1958 2, 421–426.

Shaw, J. (1960). Mechanisms of osmoregulation. Comp. Biochem. 7, 504–518.

Shehadeh, Z., and Gordon, M. (1969). The role of the intestine in salinity adaptation of the rainbow trout, Salmo gairdneri. In press.

Skou, J. (1957). Influence of some cations on an ATPase from peripheral nerves. Biochim. Biophys. Acta 23, 394–401.

Skou, J. (1960). Further investigations on a Mg-Na-activated ATPase, possibly related to the active, linked transport of Na and K across the nerve membrane. Biochim. Biophys. Acta 42, 6–23.

Smith, D. (1956). The role of the endocrine organs in the salinity tolerance of trout. Mem. Soc. Endocrinol. 5, 83–101.

Smith, H. (1930). The absorption and excretion of water and salts by marine teleosts. Am. J. Physiol. 93, 480–505.

Smith, H. (1931). The absorption and excretion of water and salts by the elasmobranch fishes. I. Fresh-water elasmobranches. Am. J. Physiol. 98, 279–295.

Smith, H. (1932). Water regulation and its evolution in the fishes. Quart. Rev. Biol. 7, 1–26.

Smith, L. (1966). Blood volumes of three salmonids. J. Fisheries Res. Board Can. 23, 1439–1446.

Stanley, J., and Fleming, W. (1966). The effect of hypophysectomy on sodium metabolism of the gill and kidney of Fundulus kansae. Biol. Bull. 131, 155–165.

Steen, J., and Kruysse, A. (1964). The respiratory function of teleostean gills. Comp. Biochem. Physiol. 12, 127–142.

Straus, L. (1963). A study of the fine structure of the so-called chloride cell in the gill of the guppy Lebistes reticulatus P. Physiol. Zool. 36, 183–198.

Sullivan, M. (1908). The physiology of the digestive trace of elasmobranchs. Bull. U.S. Bur. Fisheries 27, 1–27.

Threadgold, L., and Houston, A. (1961). An electron microscope study of the "chloride secretory cell" of Salmo salar L., with reference to plasma electrolyte regulation. Nature 190, 612–614.

Threadgold, L., and Houston, A. (1964). An electron microscope study of the "chloride cell" of *Salmo salar* L. *Exptl. Cell Res.* **34,** 1–23.

Tosteson, D. (1962a). Salt transport by eel gill epithelium. II. Attempts to define the site of active transport. *Bull. Mt. Desert Isl. Biol. Lab.* p. 82.

Tosteson, D. (1962b). The electrical potential difference across the eel gill membrane. *Bull. Mt. Desert Isl. Biol. Lab.* p. 80.

Tosteson, D., Nelson, D., Spivack, S., and Schmidt-Nielsen, A. (1962a). Salt transport by eel gill epithelium. I. Demonstration of active transport of both Na and Cl. *Bull. Mt. Desert Isl. Biol. Lab.* p. 80.

Tosteson, D., Nelson, D., Spivack, S., and Schmidt-Nielsen, A. (1962b). Salt transport by eel gill epithelium. I. Apparent active transport of both Na and Cl by the eel heart-gill preparation. *Bull. Mt. Desert Isl. Biol. Lab.* p. 81.

Tosteson, D., Blaustein, M., and Schmidt-Nielsen, B. (1962c). Salt transport by eel gill epithelium. III. The role of blood pressure and flow in salt transport in the perfused eel gill. *Bull. Mt. Desert Isl. Biol. Lab.* pp. 82–83.

Tosteson, D., Blaustein, M., and Schmidt-Nielsen, B. (1962d). Salt transport by eel gill epithelium. IV. The effect of sodium concentration on sodium and water transport in perfused gills from fresh and salt water adapted eels. *Bull. Mt. Desert Isl. Biol. Lab.* p. 83.

Tripp, M. (1967). Enzymatic patterns of ionocytes in *Oncorhynchus tschawytscha*. Master's thesis, Oregon State University.

Utida, S., Hirano, T., Oide, M., Kamiya, M., Saisyu, S., and Oide, H. (1966). Na⁺–K⁺ activated adenosinetriphosphatase in gills and Cl⁻-activated alkaline phosphatase in intestinal mucosa with special reference to salt adaptation of eels. Reprinted from *Abstr. Papers, Proc. 11th Pacific Sci. Congr., Tokyo, 1966.* Vol. 7, Symp. 55, p. 5.

Utida, S., Isono, N., and Hirano, T. (1967). Water movement in isolated intestine of the eel adapted to fresh water or sea water. *Zool. Mag.* (*Tokyo*) **76,** 203–204.

Van Dyck, M. (1966). DeKeys-Willmercellen in de kieuwen. *Koninkl. Vlaam Acad. Wetenschap., Letter. Schone Kunsten Belg., Kl. Wetenschap.* **28,** 1–99.

Vickers, T. (1961). A study of the so-called chloride secretory cells of the gills of the teleosts. *Quart. J. Microscop. Soc.* **104,** 507–518.

Wagner, H., Conte, F., and Fessler, J. (1969). Development of osmotic and ionic regulation in two races of juvenile chinook salmon, *O. tschawytscha*. *Comp. Biochem. Physiol.* **29,** 325–342.

Wikgren, B. (1953). Osmotic regulation in some aquatic animals with special reference to the influence of temperature. *Acta Zool. Fennica* **71,** 1–102.

Wilson, A., and Arnheim, N. (1968). A small molecule in hagfish tissues, possibly related to the cardiac agent, eptatreten. *Comp. Biochem. Physiol.* **25,** 359–362.

Zaks, M., and Sokolova, M. (1962). On the mechanisms of adaptation to changes in water salinity by Sockeye salmon (*Oncorhynchus nerka* Walbaum). *Vopr. Ikhtiol.* **1,** 333–346, translation in *Fisheries Res. Board Can.* Ser. No. 372 (1966).

Zaks, M., and Sokolova, M. (1966). Changes in the type of osmoregulation at different periods of the migration cycle in sockeye salmon (*Oncorhynchus nerka* Walbaum). *Vopr. Ikhtiol.* **5,** 331–337.

THE EFFECTS OF SALINITY ON THE
EGGS AND LARVAE OF TELEOSTS

F. G. T. HOLLIDAY

I. INTRODUCTION

Salinity is an important factor in the survival, metabolism, and distribution of many fish. The eggs of some marine teleosts are often spawned close inshore, near large estuaries where fluctuations of salinity are likely to occur, and the larvae of some species of fish that spawn in freshwater migrate relatively early in their life histories to marine or estuarine conditions. Recently there has been an increased amount of work on the effects of salinity on these early stages of the life history. It is useful physiological material for an experimental analysis of the mechanisms of regulation, and work on the culture and general exploitation of teleost species that spend all or part of their lives in saltwater has demanded a great amount of information on the effects of such an important environmental variable as salinity.

The effects of water of a particular salinity on eggs and larvae may be the result of one or more of a number of factors. There are the effects of the total osmotic concentration, the incidence and concentration of particular ions, the availability of oxygen (the higher the salinity the lower the oxygen content of the water, other factors being equal), and the specific gravities of different salinities may exert an effect through the different buoyancies that the organisms will show. A further factor which needs to be taken into consideration is the effect that different salinities have on the competitors, diseases, and predators of the eggs and larvae; for example, the effect of saltwater on certain freshwater bacteria and fungi may well mean that a slightly enhanced salinity is favorable to the developing egg of the freshwater spawning fish. It is impossible to consider any one of these effects of salinity in isolation. Indeed, as will be seen, it is only by considering the combination of salinity with, for example, temperature, that a full understanding of a particular situation may be arrived at.

It is not intended to present a compendium of all the observed effects of salinity on eggs and larvae; but by reference to selected works it is intended to indicate some of the more significant structural and functional responses and their ecological significance. The word "larva" will be used to include stages from hatching up to and including metamorphosis (see chapter by Blaxter, Volume III). The physiological and structural adaptations that constitute metamorphosis are often of great significance in relation to salinity responses. Some long-term experiments will be referred to in which the environment of the embryo produced effects in the preadult or adult fish.

II. THE GAMETES AND FERTILIZATION

Before spawning the gametes are generally isosmotic with, or slightly hypoosmotic to, the body fluids of the parent fish (Hayes, 1949; Holliday, 1965). However, it has recently been shown that the eggs, even at this stage, can be affected by the salinity of the water. Solemdal (1967) found that the eggs of *Pleuronectus flesus* were larger and had a lower osmotic pressure if the females were transferred to water of low salinity some weeks prior to spawning. The low osmotic pressure of the parent blood had probably contributed to an increased water content of the eggs by way of the ovarian fluids. This is an important finding indicating that parental responses to salinity may result in an altered specific gravity of the eggs, which may influence their chances of survival in brackish waters.

At spawning the gametes are often subjected to an abrupt salinity shock which might be expected to result in death, or at least considerable impairment of the ability to produce fertile eggs, but in fact the gametes are in many species remarkably tolerant to salinity change.

Yanagimachi (1953) found that the sperm of *Clupea pallasii*, which normally spawns close inshore (Schaeffer, 1937), will remain fertile to some extent after 12 hr in full strength seawater, and in 50% seawater retain a level of fertility for at least 24 hr. Helle *et al.* (1964) showed that *Oncorhyncus gorbuscha* and *Oncorhyncus keta* frequently spawn in the gravel of the intertidal zone, even when freshwater spawning grounds are available; and Rockwell (1956) showed that provided the salinity of the water during the process of fertilization did not exceed 18‰, then normal fertilization and development of the egg would occur. Salinities above 24‰ did inhibit fertilization in *Oncorhyncus*, but some eggs were fertilized at 30‰, showing that at least some of the gametes produced by some of the fish were very tolerant. This general finding has also been shown, for example, by Rutter (1902), Ellis and Jones (1939), and Rukker (1949).

Work by Holliday and Blaxter (1960) and Holliday (1965) showed that the gametes of the herring, *Clupea harengus*, and the plaice, *Pleuronectes platessa*, were especially tolerant to high salinities. The gametes were placed in a range of salinities and mixed after 1–2 min. The percentage fertilization was used as the criterion of tolerance, and the results are summarized in Fig. 1.

It has been suggested that certain salinities will act as a stimulus to the development of the unfertilized egg and hence the parthenogenetic production of larvae. Volodin (1956) reported parthenogenesis induced

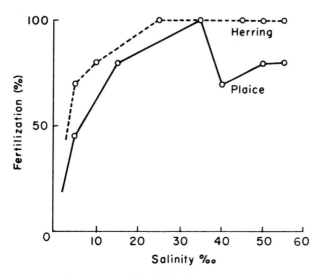

Fig. 1. Percentage fertilization of herring, *Clupea harengus*, and plaice, *Pleuronectes platessa*, eggs. From Holliday, 1965; data for herring from Holliday and Blaxter, 1960.

in *Clupea harengus* by water of low salinity, and Galkina (1957) found that parthenogenesis and low salinity were associated in *Clupea pallasii*.

III. EARLY DEVELOPMENT AND HATCHING

There have been a number of works describing the effects of salinity on the general rate of development and success of hatching of teleost eggs; fewer workers have attempted a causal analysis of the effects. Kinne and Kinne (1962) and Kinne (1964) found that in *Cyprinodon macularius* there was a progressive retardation of development as the salinity increased. However, in this case the effect was not directly resulting from the ionic and osmotic strength of the water but the amount of dissolved oxygen present, which was less in the higher salinities. When the oxygen available was artificially raised for each salinity, the effect was no longer apparent. There are reports of development taking less time in high salinities, e.g., Holliday and Blaxter (1960) in *Clupea harengus* and Forrester and Alderdice (1966) in *Gadus macrocephalis*. Heuts (1947) found that the rate of development of the eggs of *Gasterosteus aculeatus* in different salinities was dependent (among other things) on the race

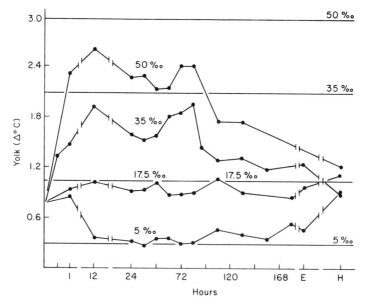

Fig. 2. Osmotic concentration of the yolk of developing herring, *Clupea harengus*, eggs which had been fertilized and incubated in the given salinity; E stands for eyed stage and H for hatching. Salinities in ‰ NaCl. From Holliday and Jones (1965).

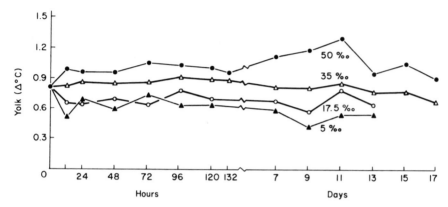

Fig. 3. Osmotic concentration of the yolk of the eggs of the plaice, *Pleuronectes platessa*, fertilized and incubated in given salinities. From Holliday and Jones (1967).

of the fish. This was a genotypic effect; in some cases development was faster in high salinities, in other cases it was slower. Cross fertilization between the races indicated that the response to salinity was characteristic of the race of the female parent. He also found that the effects of salinity on development and hatching could be profoundly modified by the temperature during incubation. Alderdice and Forrester (1968) showed how salinity and temperature acting together produced the most rapid rate of development to hatching in *Parophrys vetulus*, at a salinity of about 25‰ with temperatures between 6 and 12°C.

The chorion of teleost eggs appears to be freely permeable to water, and changes in the salinity of the water are followed by changes in the perivitelline fluid in 4–6 hr (Shanklin, 1954; Lasker and Theilacker, 1962; Holliday, 1963; Weisbart, 1968). In consequence the embryonic tissues are in direct contact with a perivitelline fluid very similar to the medium in which the eggs are being incubated. The perivitelline space is formed by the imbibition of water through the chorion (Hayes and Armstrong, 1942; Zotin, 1965), and this space is not formed in many freshwater spawning species if the imbibing process is inhibited by the presence of relatively small amounts of NaCl (Zotin, 1965). The process of "hardening" of the chorion of *Oncorhyncus* is also inhibited in water of a salinity above 3‰ (Moore, in Black, 1951). Despite the permeability of the chorion to the medium, its presence does afford significant resistance both to increased salinity, demonstrated by decreased survival time of dechorionated embryos in *Oncorhyncus* sp. by Weisbart (1968), and to reduced salinity in which the dechorionated eggs of *Sardinops caerulea* showed rapid swelling and bursting (Lasker and Theilacker, 1962).

Eggs tend to be larger in low salinities and this is almost certainly a

298

F. G. T. HOLLIDAY

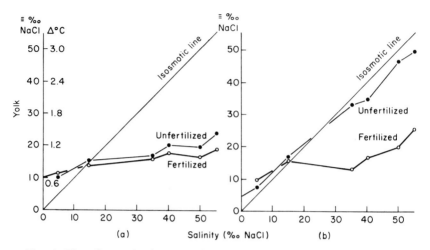

Fig. 4. The effects of salinity on the osmotic concentration of the yolk of unfertilized and fertilized eggs of the plaice, *Pleuronectes platessa*, after (a) 1 hr and
(b) 24 hr. From Holliday and Jones (1967).

result of increased water content (Holliday, 1965; Solemdal, 1967). The
cells of the blastula cap of *Gadus callarius* and *Pleuronectes platessa*
respond rapidly to changes in salinity of the medium, swelling by up to
20% in low salinities and shrinking by about 8% in high salinities (Holliday, 1965; Holliday and Jones, 1967). Changes are also evident in the
yolk; these changes are very marked in some species, e.g., in *Clupea
harengus* (Fig. 2) where the ability to regulate the yolk appears to be
closely linked with the overgrowth of the cells of the blastodisc, and full
regulation is not achieved until the blastopore is closed. In contrast to
this, the fertilized eggs of *Pleuronectes platessa* (Fig. 3) show differences
in the osmotic pressure of the yolk between salinities, but they show
little or no change from the time of fertilization to the time of hatching in
any single salinity, indicating that the power of regulation is present from
the time of fertilization; unfertilized eggs do show changes. (Figs. 4a
and b.) Weisbart (1968) showed that changes in serum osmotic concentration occurred in the embryo of five species of *Oncorhyncus* kept in
seawater, and that *O. gorbuscha* and *O. keta* had a greater ability to
regulate in the embryonic stage than *O. kisutch, O. tshawytscha,* and
O. nerka. The greater regulatory ability resulted in a higher resistance to
and better survival in seawater.

The response to salinity of the most susceptible stage of development
will determine the survival of the embryo up to hatching, and high
mortalities at specific stages of development have been recorded by a
number of workers, for example, at gastrulation (McMynn and Hoar,
1953; Holliday, 1963; Alderdice and Forrester, 1968) and at hatching

(Ford, 1929; Battle, 1930; McMynn and Hoar, 1953; Alderdice and Forrester, 1968). It is difficult to be certain as to the cause of death at particular stages; the overgrowth of cells that accompanies gastrulation may result in a less vulnerable embryo after this process is complete, especially when the regulatory function of these cells is recalled. Battle (1930) thought that the increased mortality at hatching in *Enchelyopus cimbrius* reared in low salinities resulted from poorly developed tail musculature; however, it may be that the low specific gravity of such salinities makes it more difficult for the larvae to free themselves from the chorion, and larvae are often found dead in a partly emerged state. Forrester and Alderdice (1966) found that in *Gadus macrocephalus* the duration of the hatching period was greater in lower salinities.

The effect of salinity on the percentage hatching of eggs has been recorded by a number of workers, and while it is not surprising that successful hatching has been found in salinities up to 70‰ in *Cyprinodon macularius* (Kinne, 1962), which normally inhabits highly saline desert springs and creeks, it is more surprising to find successful development and hatching in salinities as high as 70‰ in *Enchelyopus cimbrius* (Battle, 1930), 50‰ in *Pleuronectes flesus* (Zaitsev, 1955), and 60‰ in *Clupea harengus, Pleuronectes platessa,* and *Gadus callarius,* (Holliday and Blaxter, 1960; Holliday, 1965). It is also surprising to find that freshwater spawning species such as *Abramis brama, Lucioperca lucioperca,* and *Caspiolosa volgensis* will hatch in salinities of 10–20‰ and that survival and hatching are better in 2.5 and 5‰ than in freshwater (Oliphan, 1940, 1941).

IV. THE LARVAE

At hatching the skin of the teleost larva is a thin two-layered epithelium the outer surface of which is increased by a series of ridges (Jones et al., 1966; Threadgold and Lasker, 1967; Lasker and Threadgold, 1968; Fig. 5). Osmotic and ionic movement takes place through this surface to an extent that depends on the salinity of the water. The newly hatched larva may have no gill filaments, the kidney represented only by a pronephric glomerulus, and the gut may not be open. Clearly the regulatory mechanisms of the adult are not available to it.

A. Survival

The ability of the larvae to survive changes of salinity will depend on either or both of two factors, first, the ability of the body fluids to function

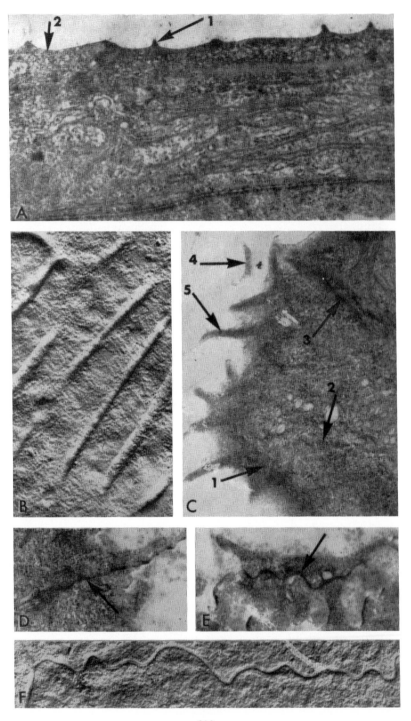

at least for a short time in an abnormal range of internal osmotic and ionic concentrations, and, second, the ability of the larvae to regulate the body fluids in order to restore the levels of osmotic pressure to near normal.

The yolk-sac larvae of many teleost species can survive in a very wide range of salinities, perhaps the most extreme example is that newly hatched *Cyprinodon variegatus* can live in a salinity of 110‰ (Renfro, 1960). Kurata (1959), Holliday and Blaxter (1960), and Holliday (1965) studied the survival of the larvae of *Clupea pallasii, Clupea harengus, Pleuronectes platessa*, and *Gadus calarias* and showed that they would tolerate salinities of 60 to 65‰. (In most of these experiments, tolerance was defined as 50% of the larvae tested remaining active for at least 24 hr after abrupt transfer from seawater.) *Clupea harengus* and *Clupea pallasii* would also tolerate salinities as low as 1–2‰; in *Pleuronectes platessa* the lower limit was 5‰ and in *Gadus calarias* 10‰. Kurata (1959) and Holliday (1965) showed that with increasing age there was a gradual change in tolerance to extremes of salinity in *Clupea pallasii, Clupea harengus*, and *Pleuronectes platessa* so that at metamorphosis tolerance levels were the same as in the adult (see Section V). Conte *et al.* (1966) showed that there were age-linked changes in the ability to survive water of high salinity in *Oncorhyncus kisutch*. The eggs hatch in freshwater, and the ability to tolerate high salinities begins soon after the yolk sac has been absorbed. The increased ability to survive precedes the seaward migration of the juvenile stages by about 6 months. A change with age of salinity preference by the fry and under-yearlings species of *Oncorhyncus* has been found by Baggermann (1960) and correlated with survival data. *Oncorhyncus keta* fry cannot live for long periods in freshwater, for example, the larvae hatched in freshwater in March started to die in June and were all dead by November. When presented with a salinity choice situation, the larvae at the end of the yolk sac period (April) in general preferred freshwater, within about 4 weeks, however, there had been a change to a saltwater preference. The yolk sac stage of *O. gorbuscha* also showed a freshwater preference, but soon after the

Fig. 5. (A) Transverse section of epidermis from herring larva in seawater (35‰) (×20,000). Arrow 1—ridge on surface of outer cell; arrow 2—cell from inner layer appearing at the surface. The letter "A" is on the boundary membrane of the inner cell, which is in contact with the yolk. (B) Carbon replica of surface of epidermis (×16,000). (C) Section cut almost parallel to and obliquely through the epidermis (×16,000). Arrows 1, 2, and 3—granular osmiophilic substance; arrows 4 and 5—longitudinal sections through the surface ridges. (D) and (E) Sections through the junctions of two epidermal cells (×16,000). (F) Carbon replica of surface showing the junctions of epidermal cells (×16,000). From Jones *et al.* (1966); reproduced with the permission of *J. Marine Biol. Assoc. U.K.*

yolk sac was absorbed this changed to a seawater preference. In contrast
to these results *Oncorhyncus kisutch* showed an almost entire preference
for freshwater throughout the test situation, and it is difficult to correlate
this with the findings of Conte *et al.* (1966) described earlier. It is clear
that many species possess powers of regulation and survival that are not
exploited in nature.

B. Physiological Effects

There are a number of aspects of the physiology of teleost larvae
which are influenced by salinity, the most obvious being the osmotic and
ionic concentrations of the body fluids. As already discussed, survival is
based on a combination of tissue tolerance and regulation, and Weisbart
(1968) showed that the alevins of *Oncorhyncus tschawytscha* survived
longer in seawater than the alevins of *O. kisutch* and *O. nerka* by virtue
of a high tissue tolerance but that *O. gorbuscha* and *O. keta* survived
longer because of a higher ability to regulate serum sodium and chloride
concentrations, and blood osmotic pressure (see Figs. 6a and b).

Holliday and Blaxter (1960) showed that the yolk-sac larvae of
Clupea harengus living in seawater of salinity 34‰ have body fluids of
a concentration equivalent to about 12‰ NaCl. Following abrupt transfer
to water of 50‰, the osmotic concentration of the body fluids was equal
to an internal salinity of about 22.5‰, and this level is tolerated by the
tissues for 3–6 hr. Regulation then commences and, after 24 hr, returns
the fluids to a level equivalent to approximately 15‰. Holliday (1965)
and Holliday and Jones (1967) found that the yolk sac larvae of *Pleuronectes platessa* when similarly transferred to water of 50‰ showed a
change in body fluids that resulted in a concentration equivalent to 35‰.
This was again followed by regulation, restoring the level to near its
normal (seawater) value. When transferred to low salinities body fluid
concentrations fell, but again some form of regulatory process operated
to maintain a level of concentration close to normal. Lasker and Theilacker (1962) obtained similar results for *Sardinops caerulea*. These
changes in the osmotic concentration of the body fluids were reflected in
changes in body weight of the larvae. In salinities greater than seawater
the larvae of *Clupea harengus* and *Pleuronectes platessa* could lose up
to 25% of their weight. In low salinities they would gain up to 30%. These
changes were at least partially reversible and probably indicated the
amount of water movement into and out of the larvae.

Some attempts have been made to measure the metabolic cost of
this regulatory ability. Lasker and Theilacker (1962) measured the

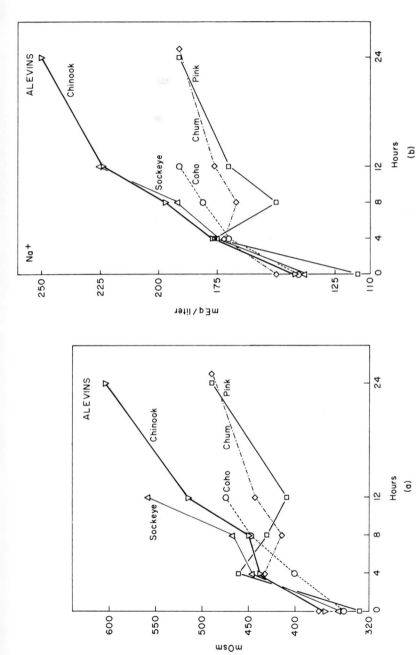

Fig. 6. Serum concentrations of Pacific salmon, *Oncorhyncus* sp., alevins plotted against exposure times to 31.8‰ seawater. From Weisbart (1968). (a) Serum osmotic concentrations, and (b) serum sodium concentrations.

304

oxygen uptake of *Sardinops caerulea* in seawater, half seawater, and double strength seawater. Holliday *et al.* (1964) measured the oxygen uptake of the larvae of *Clupea harengus* in 5, 15, 35, and 50‰. All of these workers concluded that fully adapted larvae showed no differences of oxygen uptake in the salinities tested, but some differences of an irregular nature were present during the period immediately after the transfer when osmotic changes were taking place in the body fluids; after regulation was complete oxygen uptake levels returned to normal. Shanklin (1954) showed that if the glycolytic pathways of metabolism were inhibited, then the embryonic stages of *Fundulus* could not osmoregulate and died in salinities from tap water to double strength seawater in which they normally survived.

Some measurements have also been made on the rate of heartbeat in teleost embryos and larvae subjected to a range of salinities; for example, Kryzhanovsky (1956) found that heartbeat of *Clupea harengus membras* was less rapid (86 beats/min) in salinities of 4–5‰ than in 25‰ (90 beats/min). Holliday and Blaxter (1960) recorded a reduced rate of heartbeat (from 66 to 42 beats/min) in larvae of *Clupea harengus* suffering osmotic death in distilled water. Helle and Holliday (1965) found heartbeat to be more rapid in low salinities in the alevins of *Salmo salar* which normally hatch and live in freshwater (see Table I). The effects possibly result from a combination of the direct response of heart muscle to unfavorable conditions and the changes in viscosity of the blood owing to osmotic effects.

Kinne (1962) recorded the effects of salinity on growth and food intake of larval *Cyprinodon macularius* that had been incubated in different salinities (freshwater, 15‰, and 35‰). He found that if the eggs were allowed to remain in the salinity of spawning throughout development then the larvae exhibited higher growth rates and greater efficiency of food conversion than larvae hatching from eggs that were transferred 3–6 hr after fertilization to another salinity. He concluded that the

Table I

Salmo salar Larvae Held in Salinities for 1 Week after Hatching in Freshwater

Salinity	Heartbeat per minute
0.0	98
2.5	92
5.0	87
10.0	76
17.5	58
24.0	31

water in which the parent fish lived, and to which the eggs were exposed during the first 3–6 hr after oviposition, caused adjustments that persisted throughout the lives of fish hatched from the eggs, and that these were irreversible adjustments by the developing organism to the salinity of the environment. The adjustments were not transmitted to the next generation, i.e., they were nongenetic. In considering how the adaptations came about, Kinne (1962) reviews the literature on the structure and function of the egg membranes of euryhaline fish and concludes that the spawning medium affects the physical properties of the perivitelline fluid and hence the embryo.

Kinne (1960) also has shown that young *Cypinodon macularius* exhibited differences in food intake at different salinities; for example, at a temperature of 30°C most food was eaten in 35‰, less in 15‰ and 55‰, and least in freshwater. Growth was similarly affected. However, conversion efficiency of the food was at a maximum in 15‰, less in 35‰, and less still in freshwater. Kinne clearly demonstrated that salinity and temperature were interrelated in their effects on growth and metabolism and could not easily be considered separately.

C. Structural Effects

There are a number of accounts of salinity influencing the structure of teleost larvae, and many are descriptions of abnormalities induced as a result of the effects of the salinity in which the eggs were incubated and the larvae hatched. Battle (1930) reported deformities of the tail and cardiac regions in *Enchelyopus cimbrius* hatching in salinities up to 70‰. Kryzhanovsky (1956) incubated the eggs of the Baltic herring, *Clupea harengus membras,* in water of up to 25‰ (the eggs normally develop in salinities of 4–5‰), and he reported abnormalities of the cardiac region, otic region, yolk sac, alimentary canal, and associated organs such as the liver. Alderdice and Forrester (1968) studied the early development and hatching of *Parophrys vetulus* in a range of salinities and temperatures and related various abnormalities of development to approximate areas of the salinity–temperature range; this is illustrated in Fig. 7.

Other differences in body structure often difficult to define as deformities are also associated with salinity. For example, Holliday and Blaxter (1960) and Holliday (1965) reported that the larvae of *Clupea harengus* and *Pleuronectes platessa* hatching in salinities of 5–25‰ were up to 23% longer and 33% heavier than those hatching in salinities of 35–55‰, and that the yolk sacs of the larvae of *Clupea harengus* hatching in

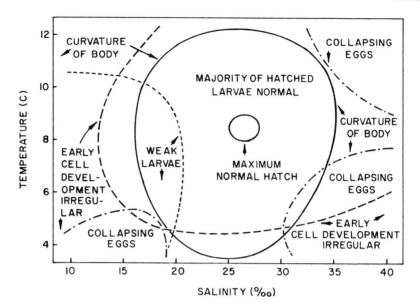

Fig. 7. Approximate areas over the salinity–temperature range examined, in which various qualitative differences in development were observed. *Early cell development irregular*—usually a gross irregularity of cell sizes in the blastodermal cap. At salinities below 20‰ for all temperatures from 2 to 12°C, and at all salinities at temperatures below 6°C. *Collapsing eggs*—invagination of the chorion. At salinities above 30‰ predominately at 6 and 12°C, and below 20‰ S at 4°C. *Death of embryo at early developmental stage*—generally at the stage of blastodermal cap formation. At all salinities (10, 20, 30‰) at 2°C. *Death at advanced stage of egg development*—from the stage of blastopore closure to near full term. At salinities from 15 to 35‰ at 4°C, at 10‰ at 6°C. *Weak larvae*—larvae hatching either head- or tail-first but unable to shed chorion, inactive. Generally below 20‰ S from 4 to 10°C. *Pronounced curvature of body*—At salinities below 20‰ and above 30‰. Most prominent below 20‰ at 6–12°C and above 30‰ at 4–8°C. *Normal larvae*—Highest numbers obtained at 20–30‰ S and 8–10°C. From Alderdice and Forrester, 1968, reproduced with the permission of the *J. Fisheries Res. Board Can.*

low salinities were turgid and pale yellow in color compared with the firm and bright yellow yolk sacs of those hatching in salinities above 35‰. A size difference in larvae related to salinity has been reported by many other workers (Forrester and Alderdice, 1966; Alderdice and Forrester, 1968; Kinne and Kinne, 1962).

Some studies have been made on cell size in larvae reared in different salinities. Jones *et al.* (1966) could find no differences in either size or ultrastructure of the epidermal cells of the larvae of *Clupea harengus* that had been incubated and hatched in salinities of 5, 17.5, 35, and 50‰. However, Lasker and Threadgold (1968), in a study of the cells of the skin of the larvae of *Sardinops caerulea*, found that there was an initial swelling of epidermal cells when larvae were transferred from water of 35‰ to 5 and 50‰; most of the skin cells returned to near normal within 6 hr of the transfer, but some of the cells (designated "chloride cells" by these authors) continued to swell even after the epidermal cells returned to normal. The "chloride cells" did not return to near normal dimensions until between 6 and 24 hr after transfer. The "chloride cells" also showed differences in ultrastructure after 24 hr in 50‰; the cell was packed with microtubules which increased in diameter, and the cytoplasm of the cell appeared crammed. In larvae subjected to low salinities the tubules became thinner in diameter but after 24 hr returned to the same size as in the larvae living in normal seawater. These observations on changes in size and structure of larval epidermal tissues, with a change in salinity, agree well with the observations discussed earlier of the effects of salinity on the size and shape of cells of the blastula cap of the embryo. It would appear that in the early stages of development at least, regulation is a property of all the cells.

Sweet and Kinne (1964), in a series of carefully designed experiments on *Cyprinodon macularius*, described some of the effects of various combinations of salinity and temperature on the length, depth, and width of the whole body and various regions of the body. By using salinities of freshwater, 35 and 75‰ at temperatures within the range 26–36°C, it was found that in general body lengths decreased with increase in salinity, and body depths and widths increased with decreasing salinity. Hempel and Blaxter (1961) showed that the mean myotome counts of the larvae of *Clupea harengus* hatched from eggs incubated in salinities ranging from 5 to 50‰ were highest in higher salinities.

It is clear that salinity often in combination with other factors has marked reversible and irreversible effects on the structure of teleost larvae. Most of these changes are adaptive in nature, but some are gross abnormalities resulting from the influence of the salinity on developmental processes.

V. METAMORPHOSIS

It is not intended to discuss in detail the general effects of salinity upon metamorphosis, which has been defined as the rapid and striking physiological and morphological changes that develop as part of the adaptive processes of the young fish to its environment. Barrington (1961) has reviewed the subject and pointed out the significance of the link between metamorphosis, thyroid gland activity, salinity preference, and osmoregulatory ability in fish at this stage. It would appear that in some species, at least, changes of both behavior and physiology in response to salinity are a part of the process of metamorphosis. Special attention has been paid to *Salmo* sp. and *Oncorhyncus* sp. where the transformation from parr to smolt occurs close to the time when there is a change in environment from freshwater to saltwater. Houston and Threadgold (1963) conclude that some of the physiological changes at this time are preadaptations to a saltwater existence, but other workers (Conte and Wagner, 1965; Conte et al., 1966) suggest that the mechanisms involved in adaptation to seawater are independent of the parr-smolt transformation. Holliday and Jones (1967) discussed the changes in the regulatory ability that occur at metamorphosis in *Pleuronectes platessa* reared in seawater. At this time there is an increased ability to survive in low salinities (down to $1\%_0$), and survival in high salinities is reduced from $60\%_0$ to an upper tolerance level of $45–50\%_0$. There is a pronounced change in body form at metamorphosis, the larva becomes laterally flattened and adopts a benthic way of life. The epidermis thickens as connective tissue is laid down and the skin becomes pigmented. This thickening renders the skin unsuitable for the functions of ionic and gaseous exchange that it possessed in the early larva. These functions are confined to the gills and if they provide a relatively less extensive regulatory surface this may well determine the upper salinity limit of survival. The change in the ability to tolerate low salinities would appear to be related to the increased development of the kidney at this stage, certainly metamorphosed *Pleuronectes platessa* living in low salinities are able to maintain the body fluids at a higher relative concentration than the body fluids of the yolk-sac larvae in the same salinities.

VI. DISTRIBUTION

In the egg and early larval stages most teleosts are at the mercy of the environment, either drifting in the plankton or being attached or buried in the substrate of a spawning bed. For most species there has been little experimental work on the salinity optima of these early stages, although some ecological studies have correlated the distribution of eggs and larvae with particular hydrographic conditions; only a few, e.g., Alderdice and Forrester (1968), have attempted to correlate laboratory and field data, and Bishai (1961) gives an extensive bibliography. Certainly passive factors such as differences in specific gravity may influence the vertical distribution of eggs and larvae, especially in areas such as the Baltic, where marine fish often spawn in water of low salinities; and Shelbourne (1956) discusses the link between buoyancy, body fluid distribution, and embryonic osmoregulation in marine fish.

Bishai (1961) found that the larva of *Clupea harengus* survived longer in salinities of 10–15‰ than in higher or lower salinities, and this has also been found in *Pleuronectes platessa* (Holliday, 1965) and *Brevoortia tyrannus* (Lewis, 1966); McMynn and Hoar (1953) showed that the optimum salinity for rearing the eggs of *Clupea pallasii* was between 11.53 and 16.24‰. These salinities are approximately isosmotic with the body fluids.

The beneficial effects of isosmotic salinities may derive at least as much from the aid that their specific gravity gives to swimming activity as to the saving in energy by reducing the osmotic and ionic effects. Activity levels are often lower in low salinities (Holliday, 1965), and energy expenditure is therefore less and the ability to survive and achieve rapid growth rates may thus be increased. Considerations such as these have important economic implications to the fish culturist.

REFERENCES

Alderdice, D. F., and Forrester, C. R. (1968). Some effects of salinity and temperature on early development and survival of the English sole (*Parophrys vetulus*). *Fisheries Res. Board Can.* **25**, 495–521.

Baggermann, B. (1960). Salinity preference, thyroid activity and the seaward migration of four species of Pacific salmon. *Fisheries Res. Board Can.* **17**, 296–322.

Barrington, E. J. W. (1961). Metamorphic processes in fishes and lampreys. *Am. Zoologist* **1**, 97–106.

Battle, H. I. (1930). Effects of extreme temperatures and salinities on the development of *Enchelyopus cimbrius* L. *Contrib. Can. Biol. Fisheries* [N.S.] **5**, 107–192.

Bishai, H. M. (1961). The effect of salinity on the survival and distribution of larvae and young fish. *Conseil, Conseil Perm. Intern. Exploration Mer* **26**, 292–311.

Black, V. S. (1951). Osmotic regulation in teleost fishes. *Univ. Toronto Biol. Ser.* **59**, 53–89; *Publ. Ontario Fisheries Res. Lab.* **71**.

Conte, F. P., and Wagner, H. H. (1965). Development of osmotic and ionic regulation in juvenile steelhead trout *Salmo gairdneri*. *Comp. Biochem. Physiol.* **14**, 603–620.

Conte, F. P., Wagner, H. H., Fessler, J., and Gnose, G. (1966). Development of osmotic and ionic regulation in juvenile coho salmon *Oncorhyncus kisutch*. *Comp. Biochem. Physiol.* **18**, 1–15.

Ellis, W. G., and Jones, J. W. (1939). The activity of the spermatozoa of *Salmo salar* in relation to osmotic pressure. *J. Exp. Biol.* **16**, 530–534.

Ford, E. (1929). Herring investigations at Plymouth VII. On the artificial fertilization and hatching of herring eggs under known conditions of salinity with some observation on the specific gravity of the larvae. *Marine Biol. Assoc. U.K.* **16**, 43–48.

Forrester, C. R., and Alderdice, D. F. (1966). Effects of salinity and temperature on embryonic development of the Pacific cod (*Gadus macrocephalus*). *Fisheries Res. Board Can.* 23, 319–340.

Galkina, L. A. (1957). Effect of salinity on the sperm, eggs and larvae of the Okhotsk herring. *Izv. Tikhookeansk. Nauchn.-Issled. Inst. Rybn. Khoz i Okeanogr.* 45, 37–50.

Hayes, F. R. (1949). The growth, chemistry and temperature relations of Salmonoid eggs. *Quart. Rev. Biol.* 24, 281–308.

Hayes, F. R., and Armstrong, F. H. (1942). Physical changes in the constituent parts of developing eggs. *Can. J. Res.* 20, 99–114.

Helle, J. H., and Holliday, F. G. T. (1965). Unpublished data.

Helle, J. H., Williamson, R. S., and Bailey, J. E. (1964). Intertidal ecology and life history of Pink salmon at Olsen Creek, Prince William Sound, Alaska. *U.S. Fish Wildlife Serv., Spec. Sci. Rept., Fisheries* 483, 1–26.

Hempel, G., and Blaxter, J. H. S. (1961). The experimental modification of meristic characters in herring (*Clupea harengus* L.). *Conseil, Conseil Perm. Intern. Exploration Mer* 26, 336–346.

Heuts, M. J. (1947). Experimental studies on adaptive evolution in *Gasterosteus aculeatus* L. *Evolution* 1, 89–102.

Holliday, F. G. T. (1965). Osmoregulation in marine teleost eggs and larvae. *Calif. Coop. Oceanic Fisheries Invest. Rep.* 10, 89–95.

Holliday, F. G. T., and Blaxter, J. H. S. (1960). The effects of salinity on the developing eggs and larvae of the herring (*Clupea harengus*). *Marine Biol. Assoc. U.K.* 39, 591–603.

Holliday, F. G. T., and Jones, M. P. (1965). Osmotic regulation in the embryo of the herring (*Clupea harengus*). *Marine Biol. Assoc. U.K.* 45, 305–311.

Holliday, F. G. T., and Jones, M. P. (1967). Some effects of salinity on the developing eggs and larvae of the plaice (*Pleuronectes platessa*). *Marine Biol. Assoc. U.K.* 47, 39–48.

Holliday, F. G. T., Blaxter, J. H. S., and Lasker, R. (1964). Oxygen uptake of the developing eggs and larvae of the herring (*Clupea harengus*). *Marine Biol. Assoc. U.K.* 44, 711–723.

Houston, A. H., and Threadgold, L. T. (1963). Body fluid regulation in smolting Atlantic salmon. *Fisheries Res. Board Can.* 20, 1355–1369.

Jones, M. P., Holliday, F. G. T., and Dunn, A. E. G. (1966). The ultra-structure of the epidermis of larvae of the herring (*Clupea harengus*) in relation to the rearing salinity. *Marine Biol. Assoc. U.K.* 46, 235–239.

Kinne, O. (1960). Growth, food intake, and food conversion in a euryplastic fish exposed to different temperatures and salinities. *Physiol. Zool.* 33, 288–317.

Kinne, O. (1962). Irreversible non-genetic adaptation. *Comp. Biochem. Physiol.* 5, 265–282.

Kinne, O., and Kinne, E. M. (1962). Rates of development in embryos of cyprinodont fish exposed to different temperature-salinity-oxygen conditions. *Can. J. Zool.* 40, 231–253.

Kryzhanovsky, S. G. (1956). Development of *Clupea harengus membras* in water of high salinity. *Vopr. Ikhtiol.* 6, 100–104.

Kurata, H. (1959). Preliminary report on the rearing of the herring larvae. *Bull. Hokkaido Reg. Fisheries Res. Lab.* 20, 117–138.

Lasker, R., and Theilacker, G. (1962). Oxygen consumption and osmoregulation by single Pacific sardine eggs and larvae (*Sardinops caerulea* Girard). *Conseil, Conseil Perm. Intern. Exploration Mer* 27, 25–33.

Lasker, R., and Threadgold, L. T. (1968). 'Chloride cells' in the skin of the larval sardine. *Exptl. Cell Res.* **52,** 582–590.

Lewis, R. M. (1966). Effects of salinity and temperature on survival and development of larval Atlantic Menhaden, *Brevoortia tyrannus. Trans. Am. Fisheries Soc.* **95,** 423–426.

McMynn, R. G., and Hoar, W. S. (1953). Effects of salinity on the development of the Pacific herring. *Can. J. Zool.* **31,** 417–432.

Oliphan, V. I. (1940). Contributions to the physiological ecology of the eggs and larvae of fishes. I. The effect of salinity on early developmental stages of *Abramis brama* L., *Lucioperca lucioperca,* L. and *Caspialosa volgensis* Berg. *Zool. Zh.* **19,** 73–98.

Oliphan, V. I. (1941). Effect of salinity on the eggs and larvae of carp, vobla and bream. *Vses. Nauchn.-Issled. Inst. Morsk. Nybnogo. Khoz. i Okeanogr. Tr.* **16,** 159–172.

Renfro, N. C. (1960). Salinity relations of some fishes in the Aransus River, Texas. *Tulane Studies Zool.* **8,** 85–91.

Rockwell, J. (1956). Some effects of sea-water and temperature on the embryos of Pacific salmon, *Oncorhyncus gorbuscha* (Walb.) and *Oncorhyncus keta* (Walb.). Ph.D. thesis, University of Washington, Seattle, Washington.

Rukker, R. R. (1949). Facts and fiction in spawn taking. *Progressive Fish Culturist* **11,** 75–77.

Rutter, C. (1902). Studies on the natural history of the Sacremento salmon. *Popular Sci. Monthly* **61,** 195–211.

Schaeffer, M. B. (1937). Notes on the spawning of the Pacific herring (*Clupea pallasii*). *Copeia* **1,** 57.

Shanklin, D. R. (1954). Evidence for active transport in Fundulus embryos. *Biol. Bull.* **107,** 320.

Shelbourne, J. E. (1956). The effect of water conservation on the structure of marine fish embryos and larvae. *Marine Biol. Assoc. U.K.* **35,** 275–286.

Solemdal, P. (1967). The effect of salinity on buoyancy, size and development of Flounder eggs. *Sarsia* **29,** 431–442.

Sweet, J. G., and Kinne, O. (1964). The effect of various temperature-salinity combinations on the body form of newly-hatched *Cyprinodon macularius* (Teleostei). *Helgolaender Wiss. Meeresuntersuch.* **11,** 49–69.

Threadgold, L. T., and Lasker, R. (1967). Mitochondriogenesis in integumentary cells of the larval sardine (*Sardinops caerulea*). *J. Ultrastruct. Res.* **19,** 238–249.

Volodin, V. M. (1956). Embryonic development of the autumn Baltic herring and their oxygen requirement during the course of development. *Vopr. Ikhtiol.* **7,** 123–133; *Fisheries Res. Board Can.* (*Transl.*) **252.**

Weisbart, M. (1968). Osmotic and ionic regulation in embryos, alevins, and fry of the five species of Pacific salmon. *Can. J. Zool.* **46,** 385–397.

Yanagimachi, R. (1953). Effect of environmental salt concentration on fertilizability of herring gametes. *J. Fac. Sci., Hokkaido Univ., Ser. VI* **11,** 481–486.

Zaitsev, J. P. (1955). Influence of the salinity of the water on the development of the eggs of the flounder (*Pleuronectes flesus luscus* Pallas). *Dokl. Akad. Nauk SSSR* **105,** 1364.

Zotin, A. (1965). The uptake and movement of water in embryos. *Symp. Soc. Exptl. Biol.* **19,** 365–384.

5

FORMATION OF EXCRETORY PRODUCTS

ROY P. FORSTER and LEON GOLDSTEIN

I. INTRODUCTION*

Excretory patterns of nitrogen excretion in fishes have been explored in considerable detail, and the partitioning of the separate roles of the gills and kidneys in various environmental situations is fairly well established. However, only quite recently with the development of isotopic labeling and other biochemical enzymic techniques has it been possible to evaluate the relative importance of specific synthetic pathways implicated in the formation of excretory products. The conversion from

---

* Abbreviations used in this chapter include: ATP, adenosine triphosphate; ADP, adenosine diphosphate; AMP, adenosine monophosphate; NAD⁺, oxidized nicotinamide-adenine dinucleotide; NADP⁺, oxidized nicotinamide-adenine dinucleotide phosphate; NADH, reduced nicotinamide-adenine dinucleotide; NADPH, reduced nicotinamide-adenine dinucleotide phosphate; and DDT, 1,1,1,-trichloro-2,2-bis(p-chlorophenyl)ethane.

313

ammoniotelism to ureotelism when certain lungfishes estivate, and the role of urea and trimethylamine oxide retention in cartilaginous fishes as a means of attaining osmotic balance in an otherwise hypertonic seawater environment, for example, constitute some of the most fascinating adaptive devices in comparative physiology. Exploration of the means by which these biochemical shifts are accomplished is equally interesting, and, in addition, these may lead to an explanation of how metabolic events generally are controlled and modified in response to environmental stimuli. This chapter will consider recent studies which have contributed to an understanding of biochemical sources, especially of nitrogenous end products, and emphasize particularly their adaptive and evolutionary significance.

II. PATTERNS OF EXCRETORY END PRODUCTS

Values for urinary nonprotein nitrogen appearing as amino N, ammonia, creatine, creatinine, urea, and uric acid in a wide variety of marine and freshwater fishes can be found in "Blood and Other Body Fluids" in the Biological Handbook series (Altman, 1961) and in Prosser and Brown (1961). In considering patterns of nitrogen excretion we will be concerned mainly with ammonia, urea, and trimethylamine oxide; and the participation of liver, gills, kidneys, and certain other tissues in their synthesis and selective handling.

A. Ammonia

Only a small fraction of the total nitrogen excreted by fishes appears in the urine. Early studies by Denis (1913–1914), Marshall and Grafflin (1928), Grollman (1929), and Edwards and Condorelli (1928) all showed that urine of freshwater and marine fish, both fasted and fed, had very low nitrogen values. Urinary nitrogen is lower in freshwater than in marine fish, and, additionally in both, there is a peculiar distribution of nitrogen, with the principal constituent being creatine, not because of its unusually high values but because the total nitrogen content is so very low. These observations led Homer Smith to suspect that nitrogen, possibly as ammonia or urea, was being excreted by some route other than the kidneys, and the gills appeared to him to offer the most likely avenue for this escape. Smith's classic divided box experiments on the carp and goldfish disclosed within minutes the presence of ammonia in the front chamber around the gills, where the ammonia concentration steadily increased, while no trace of ammonia appeared in the back chamber around the body of the fish for several hours. Comparison of the branchial and urinary excretion showed that 6–10 times as much nitrogen was excreted in ammonia by the gills as in all the nitrogenous compounds together by the kidneys. The branchial excretion consisted also of some other highly diffusible substances such as urea and amine or amine oxide derivatives, whereas the less diffusible nitrogenous end products, creatine, creatinine, and uric acid, were excreted solely by the kidneys.

Delaunay (1931), in his review of patterns of excretion of nitrogenous compounds in invertebrates, showed that ammonia is the chief end product of nitrogen metabolism in all aquatic animals, freshwater and marine, from protozoa to the most complexly organized. Of the main forms of nitrogenous compounds excreted by aquatic organisms—ammonia, urea, amino acids, creatine, creatinine, uric acid, and purine bases—ammonia is the smallest and simplest. It is extremely soluble in water. With a pK_a of approximately 9.24 at 25°C ammonia is about 99% dissociated at neutrality, and as a weak base this means that in blood and most other body fluids the ratio of ionized to the nonionized form is about 100/1. Weak acids and bases diffuse across biological membranes mainly in their lipid soluble nonionized forms which accounts for their ready movement into and out of all cells, plant and animal alike (Jacobs, 1940).

With fish gills and other excretory surfaces the pH difference that exists between the internal and external environment of aquatic animals may be a significant factor in determining rates of ammonia elimination. Alkaline seawater would tend to retard and acidic freshwater would accelerate diffusion of the free base into the external aqueous environment. Euryhaline forms that can withstand wide ranges of salinity, either by conformity or regulation, would have the excretion of ammonia accelerated by the transition from the characteristically alkaline seawater to neutral or acidic freshwater.

Except for its relative toxicity, ammonia has many advantages over urea and uric acid as the chief end product of nitrogen metabolism. In contrast to the latter, no expenditure of energy is required for the conversion of protein nitrogen to ammonia. Actually some of the reactions involved in the production of ammonia, such as the deamination of glutamate, ultimately lead to the production and capture of free energy (Fig. 1). Coupling of the glutamic acid dehydrogenase reaction with transamination systems is of considerable importance in generating ATP. Glutamic acid dehydrogenase requires NAD^+ or $NADP^+$, which in the reduced form can enter the chain of oxidative phosphorylation (Braunstein, 1951; Krebs and Kornberg, 1957; P. P. Cohen and Brown, 1960).

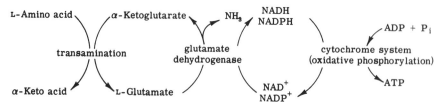

Fig. 1. Deamination of amino acids and production of ATP via coupled trans-amination and glutamate dehydrogenase.

Another advantage derived from excreting nitrogen in the form of ammonia lies in the small size and highly lipid soluble nature of the free base which permits its easy elimination by diffusion without an accompanying obligatory loss of water from the fish or other aquatic organism. At the pH of body fluids only 1% of the compound is in the form of free base, but the conversion of $NH_4^+ \rightarrow NH_3$ is instantaneous and hence probably not a rate-limiting step in its elimination.

Maetz and García Romeu (1964), in following up a suggestion made earlier by Krogh (1939), demonstrated the ability of NH_4^+ to exchange with Na^+ absorption by the gills of freshwater fish. Thus a further advantage of excreting nitrogen in the form of ammonia lies in this exchange of cations where the active absorption of Na^+ is critically important in maintaining salt and water balance. Hence, in these freshwater species the exchange of NH_4^+ for Na^+ serves the dual purpose of eliminating the nitrogenous end product and in facilitating the accumulation of Na^+. Injection of ammonium sulfate into goldfish doubled the rate of Na^+ uptake from the external freshwater environment without simultaneously affecting the rate of Na^+ loss, and the addition of ammonium ions to the external medium inhibited sodium influx and net uptake. A separate exchange process across the gill involving bicarbonate and chloride ions was also demonstrated. The tentative scheme representing these ionic absorption mechanisms in the branchial cell of freshwater teleosts is shown in Fig. 2. Confirmation of the Na^+/NH_4^+ exchange process was obtained in the freshwater eel where it was shown that injection of ammonium sulfate increased sodium influx sixfold without affecting sodium efflux (García Romeu and Motais, 1966).

Whereas ammonia may be the ideal end product of nitrogen metabolism for purely aquatic organisms, for animals living in more limited environments its toxicity may become a critical disadvantage. It sometimes appears in rivers as a pollutant, and elevated levels of ammonia in well-aereated water have been shown to be acutely toxic to bream, perch, roach, rudd, and rainbow trout, for example (Ball, 1967). Certain fish that leave the water for extended periods of time, as during estivation, or that rear their young in the limited environments of protective self-containing egg cases must somehow provide for the detoxication of ammonia, and adaptive alternatives in synthetic pathways have provided for the conversion of toxic ammonia to other nontoxic nitrogenous end products.

B. Urea

The shift from ammoniotelism to ureotelism and uricotelism among both the invertebrates and vertebrates in response to water deprivation and the need to carry on developmental metabolic processes in a cleidoic environment is one of the most interesting and widely applicable generalizations in biochemical evolution (Needham, 1938). There is no simple biochemical method known for the detoxication of ammonia; only by complex energy demanding synthetic pathways can ammonia nitrogen be incorporated into such less toxic end products as urea. The urea molecule is a dipole which in aqueous solution behaves in many ways as do the water molecules themselves. It has a low oil–water partition

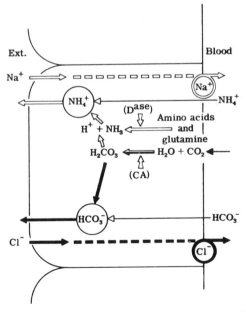

Fig. 2. Schematic representation of ionic exchanges in the branchial cell of *Carassius*. Deamidation and deamination enzymes (Dase) and carbonic anhydrase (CA). Reproduced from Maetz and García Romeu (1964) by permission of Rockefeller University Press.

coefficient and penetrates cell membranes through aqueous pores rather than through the lipid-protein component. In a very few instances in fish, most notably the elasmobranchs, urea may be actively transported against a chemical gradient by a carrier-mediated reabsorptive process which is presumably dependent upon the availability of free energy, as is urea secretion in proximal renal tubules of the frog (Forster, 1954), and is likewise subject to competitive inhibition, in this case, by such analogous compounds as acetamide and methylurea (Schmidt-Nielsen and Rabinowitz, 1964). Despite the similarity in diffusion coefficients of ammonia and urea, the former passes through most biological membranes faster than urea, probably because of the high lipoid solubility of the NH_3^0 form of ammonia and the low oil–water partition coefficient of urea. In the fish gill of the shorthorn sculpin, *Myoxocephalus scorpius*, for example, where afferent and efferent gill blood was simultaneously sampled, more than two-thirds of the blood ammonia was lost in passing through the branchial circulation whereas no detectable drop in urea concentrations could be discerned (Goldstein *et al.*, 1964).

Conversion to urea as the predominant nitrogenous end product has been of adaptive value in at least two groups of fishes. First, cartilaginous sharks, skates, rays, and chimaeroids since very early in their evolution seem all to have shared the internal fertilization feature, and paleontological evidence has disclosed that male fossils always possessed claspers. As with present forms of the Chondrichthyes, this suggests that throughout their entire evolutionary history a protected embryonic environment was obtained either by ovoviviparity or by the intrauterine deposition of leatherlike protective egg cases. In these encapsulated environments during the characteristically long periods of embryonic development typical of cartilaginous fishes (the longest of any of the vertebrates, sometimes with gestation periods up to 2 years), it would be necessary to obviate the accumulation of toxic ammonia and converting it to urea could provide this protection. In the elasmobranch, *Squalus*, for example, during the first year of the almost 2-year pregnancy, the embroys develop in the cleidoic environment of egg cases bathed by a few milliliters of intrauterine fluid similar in composition to a plasma exudate. During this period the sparsely vascularized uteri are sealed off from the exterior, and only after the egg cases break, and the pups are capable of living essentially independently, does the seal break. Seawater is then periodically flushed into and out of the uteri in volumes up to several hundred milliliters per female (Burger and Loo, 1959; Forster, 1967a; Price and Daiber, 1967). The presence of urea-synthesizing enzymes in early embryos of *Squalus* and the skate, *Raja* (Read, 1968), can be correlated with the need to detoxicate ammonia during the intrauterine period or while encapsulated in the characteristically tough keratinized egg cases, respectively.

The relatively naked egg of ammoniotelic teleosts, on the other hand, can readily dispose of ammonia by diffusing it into their aqueous environment, or, as in the relatively few viviparous teleosts, pass it across elaborate fetal membranes into the maternal circulation for its eventual disposal by her gills. In all of the latter, the young develop not in the uterus but in the ovary, and here nourishment is supplied for the embryos. Even in those teleost eggs with enough yolk to sustain embryological development ammonia can readily be disposed of through the ovarian follicular sac which is highly vascularized and brought into intimate association with expanded and equally well-vascularized embryonic pericardial, peritoneal, intestinal, or urinary bladder membranes. Hence, with these adaptations, which establish close relationships between the maternal and fetal circulations, the elimination of embryonic ammonia via the mother's gills is just about as efficient as the excretion of preformed ammonia of maternal origin. In the elasmobranchs retention of urea has also provided them with a method of maintaining osmotic balance with the otherwise hypertonic seawater environment (Smith, 1936). Chimaeroids, the other surviving group of cartilaginous fishes, also retain urea as an osmoregulatory device, and the ratfish, *Chimaera monstrosa*, takes a position between that of the typical elasmobranch and the cyclostomes (e.g., *Myxine glutinosa*), however, more closely resembling the former. Hence, plasma urea levels are not quite as high as in the elasmobranch, and the sodium chloride levels are correspondingly higher, but they do not rely entirely on salt retention to maintain approximate isotonicity with seawater as do the purely ammoniotelic cyclostomes (Fänge and Fugelli, 1962). The relatively higher salt levels in chimaeroid body fluids might reflect an inability to excrete salt efficiently via rectal glands or some other extrarenal route. Varying degrees of development of extrarenal salt glands among the cartilaginous fishes might be related to the availability of estuarial waters or possibly to their freshwater phylogenetic ancestry. Modern elasmobranchs occurring in freshwater but with access to the sea have significantly lower urea concentrations in their body fluids than do the purely marine forms (Smith, 1931; Price and Creaser, 1967; Price, 1967), and purely freshwater elasmobranchs such as the South American stingrays of the Orinoco and Amazon drainage systems (e.g., *Potamotrygon*) appear to have abandoned urea retention entirely (Thorson *et al.*, 1967).

Estivating lungfish such as the African *Protopterus* provide another example of fish deriving an advantage from converting ammonia production to ureotelism during water deprivation that enables them to estivate for long periods during which the relatively inert urea accumulates in body fluids to very high levels (Smith, 1930; Janssens, 1964; Forster and Goldstein, 1966). Members of the two surviving families of lungfishes, Lepidosirenidae and Ceratodontidae, differ from one another in ecological behavior. The African *Protopterus* and the South American *Lepidosiren*, belonging to the former family, must breathe air at rather frequent intervals, and they estivate in burrows out of water during dry seasons. The sole surviving member of the latter family, the Australian *Neoceratodus forsteri* uses its lung as an accessory respiratory organ when its poorly developed gills are inadequate to sustain gas exchange during periods of high activity; however, it does not estivate and it cannot survive out of water as can the African and South American dipnoans. In contrast to *Protopterus* which eliminates up to 50% of its nitrogenous end products as urea when in water, little or no urea is excreted by the Australian lungfish which, with its totally aquatic existence, has no need to obviate the production of toxic ammonia (Goldstein et al., 1967b). Details of the biosynthetic pathways will be presented later in this chapter.

It has recently been shown that urea occurs in the blood of the marine coelacanth, *Latimeria chalumnae*, at a concentration comparable with that of elasmobranches (Pickford and Grant, 1967), and enzymes have been identified in its liver which suggest that urea is synthesized here via the same ornithine–urea cycle that occurs in all the predominantly ureotelic vertebrates (Brown and Brown, 1967). In referring to the origin and evolution of fishes, Smith pointed out soon after the living coelacanths were first discovered that if these primitive fish closely related to the Devonian crossopterygians had originated in freshwater, in accordance with a hypothesis he favored, analysis of their body fluids would reveal whether the mechanism of osmotic regulation were that of retaining urea or some other osmotically active organic solute, as with the cartilaginous fishes, or by the typically teleost method of drinking and then subsequently desalinating seawater at the gills and kidney. Disclosure of a well-developed glomerular kidney induced Smith to suggest that the former mechanism was more likely (Smith, 1953), and, of course, these results bear out his prediction.

Mud skippers, such as the amphibious goboid, *Periophthalmus* sp., are interesting teleosts that have developed respiratory adaptations that enable them to spend most of their time out of water, darting about in the intertidal zone of the hot tropical mangrove areas in which they thrive. In addition, they seem to have adjusted to a euryhaline existence, and Gordon et al. (1965) showed that *Periophthalmus sobrinus* could tolerate direct transfers successively from 100% seawater to 80, 60, 40, and 20%, with survival in good condition for extended periods of time. While in 100% seawater Gordon found that they excreted almost as much urea

(0.36) as ammonia (0.49 mmole/kg/hr), and then they exhibited a shift toward ammoniotelism in lower salinities. In 40% seawater, for example, urea excretion amounted only to 0.06 mmole/kg/hr, whereas ammonia production was 0.77 mmole/kg/hr. The higher fraction of ammonia excretion when favorable gradients for water influx are established in a dilute environment does not necessarily mean that a biochemical suppression of ureogenesis, as such, occurs. As in the lungfish, *Protopterus* (Forster and Goldstein, 1966; Janssens and Cohen, 1968a), the activity of the ornithine–urea cycle or purine pathway enzymes may be unaltered whether or not the fish are experiencing water deprivation, be it osmotic or evaporative. With high water influx and a correspondingly vigorous branchial circulation, ammonia could be extracted from blood and eliminated into the aqueous environment so efficiently that there would be relatively little available in plasma for capture by the ureosynthetic systems in liver, kidney, or other body tissues.

C. Trimethylamine Oxide

Another nitrogenous compound that occurs generally in fish is trimethylamine oxide (TMAO) (Groniger, 1959). A relatively large fraction of the total nitrogen excreted by marine fish is in the form of this slightly basic, soluble and nontoxic trimethylamine oxide, $(CH_3)_3N \rightarrow O$. Freshwater teleosts have characteristically low levels of TMAO in their urine and body fluids, and euryhaline forms such as young salmon excrete very little nitrogen in this form until they are transferred to a marine environment. Marine teleosts such as the aglomeruler goosefish, *Lophius*, may excrete up to 50% of their total nitrogen in the form of TMAO (Grollman, 1929). The renal tubules of elasmobranchs actively reabsorb filtered TMAO, as they do urea, and its accumulation contributes approximately one-third of the total organic osmolarity of body fluids (J. J. Cohen et al., 1958; Forster et al., 1958). In contrast to the elasmobranchs where TMAO is avidly retained by the kidney, the renal tubules of certain marine teleosts such as *Lophius americanus* actively secrete TMAO into urine via an energy-dependent organic base-secreting mechanism that occurs generally in nephrons throughout the vertebrates (Forster et al., 1958).

Trimethylamine oxide may also play a role in intracellular regulation in euryhaline teleosts where cell volumes are maintained constant in both seawater- and freshwater-adapted fishes, respectively, despite the fact that the freezing point of serum is consistently lower when the fish lives in seawater than when it lives in freshwater. Lange and Fugelli (1965) have emphasized the active roles trimethylamine oxide and free ninhydrin-positive substances play in attaining this kind of isosmotic intracellular regulation. Thus, TMAO may participate actively in regulating the volume and composition of the fish's extracellular fluids by means of the kidneys, as with the cartilaginous fishes, and also in this intracellular system. In the flounder, for example, there is a linear correlation between the total activity of intracellular solutes and the osmolarity of the blood resulting in a drop of TMAO content in muscle tissue water from 30 mmoles/kg to 20 when the fish is transferred from seawater ($-1.88°C$) and then subsequently adapted to freshwater ($-0.01°C$).

The TMAO contents in blood and tissues of various marine fishes have been compiled by Nicol (1960). Presence of the noxious free amine in spoiled fish food and in the feces of birds and other animals that prey on fish results from the action of bacteria capable of reducing the oxide to the volatile form of trimethylamine. Endogenous volatile amines detected by diffusion analyses are present in small amounts in the urine of some marine fishes, but these have not been definitely identified as trimethylamine.

D. Other Nonprotein Nitrogenous End Products

Specific pathways of nitrogen metabolism in fishes have not been systematically explored, but available information suggests that the metabolic pathways occurring here are similar to those in higher vertebrates. This is indicated by the occurrence of the usual main forms of nonprotein nitrogen constituents in urine (Altman, 1961). As pointed out earlier, relatively low urinary nitrogen excretion rates prevail in teleosts because of the importance of the branchial route of ammonia elimination, the nitrogen levels being generally lower in the urines of freshwater fish than of marine forms. Also, compared with the more familiar terrestrial species, there is an unusual distribution of nitrogen with the principal nitrogenous constituent usually being the organic base creatine. A large fraction of that usually designated "undetermined" nonprotein nitrogen, at least in marine species, is TMAO. Both of these weak bases are actively secreted by renal tubules of marine teleosts, and this is reflected in their characteristically high urine/plasma concentration ratios (Forster et al., 1958). The uric acid present is excreted in traces, presumably resulting from the oxidative deamination of purines.

E. Inorganic Excretory Products

Urinary electrolytes in marine teleosts reflect the separate roles of the kidneys and gills in desalinating seawater subsequent to its alimentary ingestion and absorption. High concentrations of urinary magnesium, sulfate, and phosphate are accounted for by secretory mechanisms in renal tubules that actively remove these divalent ions from blood. As in all vertebrates, the renal tubules of both marine and freshwater fishes play a salt-saving role with respect to sodium and chloride ions which accounts for the very low concentrations of these univalent ions in urine of freshly captured and nontraumatized fish (Forster, 1953; Forster and Berglund, 1956; Berglund and Forster, 1958). Extrarenal routes, including the rectal gland in cartilaginous fish (Burger, 1967) and the gills of all fishes, are involved in regulating the univalent ion composition of body fluids (Krogh, 1939).

III. CONJUGATION AND DETOXICATION

Fish may be expected to differ from terrestrial animals with respect to the formation of detoxified end products of drugs and other foreign compounds that find their way into the organisms. In mammals certain specific biochemical transformations including conjugation, N-dealkylation, deamination, aromatic hydroxylation, ether cleavage, sulfoxide formation, alkyl chain oxidation, nitro-group reduction, azo-link cleavage, and glucuronide formation result in more rapid excretion of the foreign agent because of its conversion to a less lipid soluble and, hence, more readily excretable form. This could be accomplished either by enabling it to be actively secreted by the renal tubules or by making it less likely to diffuse back into plasma for recycling subsequent to glomerular filtration. Brodie and his co-workers (1955, 1958) showed that biogenesis of these products in higher forms occurred largely by enzymically controlled reactions localized in liver microsomes. Maickel et al. (1958) observed, however, that fish (goldfish and perch) did not have the ability to detoxicate phenols by the otherwise ubiquitous formation of conjugation derivatives with glucuronides and sulfates. Doses of these phenols as low as 0.5 mg/kg were often fatal, and fish placed in solutions of 10 ppm of phenols absorbed enough through their gills to be toxic in 4–8 hr. Also, tadpoles of both frogs and toads were unable to conjugate phenols, whereas the adults were capable of forming the usual glucuronide and sulfate conjugates.

Brodie and Maickel (1962) extended their observations on the comparative biochemistry of drug metabolism and speculated further that, as with the evolutionary development of mechanisms to convert ammonia to the less toxic urea and uric acid, the development of adaptive detoxication processes might be considered similarly. It was postulated that lipid-soluble organic compounds not readily excretable by the kidneys of terrestrial forms may be no problem to the fish because here they could be rapidly eliminated through the lipoidal gill epithelial membranes into the effectively limitless aqueous external environment. They considered also the possibility of evolutionary regression in that enzymic functions might disappear, presumably because genes are lost, when they no longer confer an advantage in natural selection. Aquatic turtles were cited as a possible example where such a regression might have taken place accompanying their migration back to the sea. When underwater they use their highly vascularized pharyngeal cavity as a sort of accessory branchial structure to facilitate respiratory gas exchange, and this might obviate previous needs for detoxication reactions.

Studies on the biogenesis of conjugation and detoxication products in a wide variety of fish, however, have revealed many exceptions to this provocative hypothesis. For example, trout, salmon, and cod contain uridine diphosphate-glucuronic acid, and glucuronide formation can be demonstrated in homogenates of trout liver. Sulfate esters have been identified in the bile of carp and hagfish, and elasmobranch bile contains a sulfate ester derivative of a steroid alcohol (Williams, 1963). Trout can hydroxylate biphenyl (Creaven et al., 1965).

Buhler (1966) suggests that differences in pesticide toxicity among various species of fishes may result from differences in their hepatic drug-metabolizing activity. Furthermore, he showed that pretreatment of rainbow trout with DDT or phenylbutazone resulted in a selective induction of the drug-metabolizing enzymes. Studies on shad, perch, carp, sucker, steelhead trout, rainbow trout, sockeye salmon, chinook salmon, and silver salmon disclosed NADPH-dependent hepatic hydroxylation of aniline, reduction of nitrobenzoic acid, and N-dealkylation of aminopyrine. The more primitive lamprey eel and sturgeon showed only nitroreductase activity.

Acetylation of the p-amino group of p-aminohippuric acid (PAH) by kidney tissue occurs in varying degrees among fishes, and this has practical significance in doing renal function studies because, although the acetylated form is secreted as avidly as free PAH, this will give false "clearance" values and spurious measurements of minimal effective renal plasma flow unless in the Bratton-Marshall procedure the amino group is uncovered by acid hydrolysis before diazotization (Taggart et al., 1953; Forster, 1967b). Phenol red in the uterus of the pregnant dogfish, Squalus acanthias, can be converted in an utterly unique biological halogenation to bromophenol blue by a process which may be analogous to the biogenesis of thyroxin where tyrosine residues undergo iodination

(Burger and Loo, 1959). Huang and Collins (1962) showed that three types of marine fishes—the elasmobranch Squalus, the aglomular teleost Lophius, and the glomerular flounder Pseudopleuronectes—all were capable of conjugating p-, m-, and o-aminobenzoic acid isomers with glycine and glucuronic acid. Acetylation of the m- and p-isomers occurred in all three fishes, but only Lophius was found capable of acetylating o-aminobenzoic acid.

Conversion of the highly volatile base trimethylamine to TMAO is an oxidation detoxication, and Baker et al. (1963) have shown in an exhaustive study including representatives of marine elasmobranchs and teleosts, as well as anadromous and freshwater fishes, that the enzyme catalyzing the conversion of trimethylamine to the oxide is present in the livers of some but not in others, and with no apparent relation to the presence or absence of osmoregulatory or any other physiological feature of adaptive significance. Trimethylamine, as such, is rapidly eliminated by the elasmobranch, Squalus (Goldstein et al., 1967a), and presumably by all the fishes. This would not support the hypothesis that a relationship exists between the metabolic fate of foreign compounds and the degree to which they are readily excreted by the organism.

Dixon et al. (1967) have recently extended the list of exceptions to the hypothesis that animals living in water environments do not need special detoxication mechanisms to aid in the disposal of lipid-soluble substances because of the availability of gill surfaces for elimination by dialysis into their aqueous environment. They reported on azo- and nitro-reductase activity in liver of Squalus, the lemon shark Negaprion, the stingray Dasyatis, the skate, barracuda, and yellow tail snapper Ocyurus. In agreement with Brodie and Maickel (1962) they showed that liver homogenates of Squalus fortified with a proven NADPH-generating system were unable to oxidize hexobarbital (side-chain oxidation) or chlorpromazine (sulfoxide formation), and also incapable of converting aminopyrine to 4-aminoantipyrine (N-demethylation). However, all of the elasmobranch and teleost fishes studied were able to reduce the azolinkage of neoprontosil (azoreductase) to form sulfamilamide. Sharks and rays lacked nitroreductase, but the skate and the two teleosts studied were able to reduce p-nitrobenzoic acid to p-aminobenzoic acid (nitroreductase). These aromatic azo- and nitro-compounds are used as drugs, food additives, and pesticides; some of the azo-compounds are potent carcinogens. Their metabolic fate in fishes and various other species that might be exposed to them by way of industrial waste or other contaminants entering the environments warrants continued investigation.

Direct studies on half-times in plasma by Dixon et al. (1967) of lipid-soluble and readily diffusible drugs such as 4-aminoantipyrine and sulfanilamide do not suggest an extremely rapid diffusion from plasma to seawater in the dogfish, *Squalus*, for example. Thus although it may appear to be reasonable that in fish, with the wide avenues of an extensive branchial circulation available for the elimination of drugs and other foreign compounds by diffusion, no need may exist for their detoxication or conversion to more excretable forms, such does not appear to be the case. Conjugation and detoxication mechanisms in fishes generally resemble those of terrestrial forms, and indeed the gill does not seem to provide a very efficient method even of eliminating many of the more readily diffusible foreign substances tested to date. The kidneys of both marine and freshwater fishes generally share with vertebrate kidneys the ability to actively excrete by specific tubular transport mechanisms many of the conjugated and detoxicated derivatives of drugs and foreign compounds (Forster, 1961, 1967b).

IV. SOURCES OF NITROGENOUS END PRODUCTS

Proteins within all living organisms are in a constant state of turnover, continuously being synthesized and degraded, and the metabolic pathways in fish are probably similar to those of mammals and other vertebrates. However, the terminal production of specific nitrogenous end products varies widely from species to species and according to the environmental status of the organism. The various fishes have two main pathways of nitrogen elimination leading to the production of ammonia, on the one hand, and to the synthesis of urea, on the other. The ultimate source of nitrogen in both cases is presumably the amino and amide groups of amino acids. Other minor end products of nitrogen metabolism include creatine and TMAO. Metabolic pathways involving the former are apparently similar to those occurring in higher vertebrates, although the predominance of creatine in fishes over that of its internal anhydride, creatinine, is of interest since the latter is the main form in blood and urine of the higher vertebrates. As will be pointed out later, despite many recent attempts to ascertain the source of TMAO in fishes, the problem still is unsolved, even with respect to determining whether the compound is synthesized endogenously or originates exogenously in the diet.

A. Deamination and Transamination

Speculations concerning the source of branchially excreted ammonia, which comprises from 60 to 90% of the nitrogen excreted by fish (Smith,

1929; Wood, 1958; Fromm, 1963), began with Smith's suggestion that branchially excreted ammonia was derived by diffusion directly from blood ammonia preformed in other tissues of the body and delivered, as such, to the gills. Later he abandoned this view in favor of its formation *de novo* at the gill rather than at some central site, such as the liver or kidney, as in most vertebrates. Evidence favoring the peripheral origin of ammonia came from his observation that in the estivating lungfish, *Protopterus*, when there was a complete cessation of branchial excretion ammonia did not accumulate in blood or other body fluids (Smith, 1930). Subsequently, in his studies on the freshwater elasmobranch, *Pristis microdon*, evidence was obtained that suggested to him that the excretion of ammonia at the gills was under physiological control (Smith, 1931, 1936).

More recently it was shown that branchially excreted ammonia may be derived both from preformed sources and from the extraction of plasma α-amino acid N at the gill, with most of it coming from ammonia delivered, as such, to the gills after having been formed in the liver and perhaps to some extent in other central organs (Goldstein *et al.*, 1964). Ammonia may diffuse across the branchial epithelium as free base or, as cited previously, by an exchange of NH_4^+ in blood for Na^+ in the external environment of freshwater fishes. The amount of ammonia excreted by the kidneys is very small compared to that eliminated at the gills (Smith, 1929), even when the urine is highly acidic. Hence, it does not appear that the renal production of ammonia participates in acid–base regulation as it does in mammals.

The main route for the formation of ammonia by the deamination of amino acids proceeds via the system of transamination first proposed by Braunstein (1939) and Braunstein and Byechkov (1939). This pathway couples the transamination of various L-amino acids with α-ketoglutarate to form L-glutamate which is subsequently deaminated by L-glutamate dehydrogenase (Fig. 1).

Indirect evidence favoring the source of branchially excreted ammonia as the result of the enzymically controlled deamination of glutamine and other amino acids in the gill was found in studies that showed significant glutaminase I and glutamic acid dehydrogenase activities in the gill tissue of the shorthorn sculpin, *Myoxochephalus scorpius*, and other representative teleosts (Goldstein and Forster, 1961). The rate of branchial ammonia excretion by sculpins in oxygenated seawater was measured directly and found to be 260 μmoles/kg body weight per hour. The maximal glutaminase I and glutamic acid dehydrogenase activities as assayed in gill homogenates and calculated by summing the activity per gram of both gills, and multiplying this figure by the average gill weight (5.0 g/kg body weight), was 355 μmoles NH_3/kg body weight per hour; more than that needed to account for the measured rate of ammonia production by gills in intact sculpins. However, the activity of an enzyme as measured *in vitro* does not necessarily correspond with its activity *in vivo*; substrate availability and other environmental factors can modify rates, and its localization *in situ* as part of an integrated bio-chemical system modifies its kinetics. Pequin's observations (1962) on the carp in which measurements were made of the ammonia content in venous blood entering the heart and in the ventricle itself, and simultan-eously in blood leaving the gill circulation via the dorsal aorta, showed that, in this instance at least, branchial extraction of preformed ammonia at the gills could account for all of the ammonia excreted. Cardiac output (gill blood flow) was measured as the rate determined by electrocardiog-raphy multiplied by the estimated ventricular volume. It was also noticed that ammonia concentrations were highest in venous blood drain-ing the liver. Later Pequin and Serfaty (1963) showed that the liver plays the essential role in the formation of ammonia, but that the kidney's participation was not negligible, being approximately one-third that produced by the liver. Mixed blood taken from the ventricle contained 2.89 μg NH_3-N per milliliter and blood sampled in the three main veins delivering blood back to the heart in the carp contained 6.96, 4.60, and 1.04 μg/ml in the hepatic vein, posterior cardinals, and caudal vein, re-spectively. Significant branchial extraction of this preformed ammonia was indicated by the relatively low level of ammonia in dorsal aorta blood leaving the gill circulation, 0.68 μg/ml, compared with 2.89 μg/ml in ven-tricular blood on its way to the branchial circulation.

Furthermore, Pequin and Serfaty (1936) devised a method for per-fusing the liver and gut which demonstrated that L-asparagine, L-gluta-mine, and some purines could act as substrates for hepatic ammoniogene-sis in the freshwater carp. Perfusion with hydroxylamine also led to an increase in ammonia production, although the biological significance of this source was questioned. They also noted seasonal differences in blood ammonia levels that corresponded with variations in the amounts of ammonia excreted via the gills.

To evaluate the relative importance of branchial extraction of pre-formed ammonia, on the one hand, and its formation *de novo* at the gill on the other, Goldstein *et al.* (1964) used the Fick principle to measure rates of gill blood flow while simultaneously determining branchial ammonia excretion. Total oxygen consumption in milliliter O_2 per kilo-gram per hour by the marine teleost, *Myoxocephalus scorpius*, was meas-ured with an oxygen electrode while the fish were maintained in a specially designed sealed container. The amount of oxygen taken up per milliliter of blood as it traversed the gill circulation (milliliter O_2 per milliter) was taken as the difference in oxygen content of efferent and afferent samples taken from the dorsal aorta and the bulbus arteriosus, respectively. Ammonia excretion was measured simultaneously under conditions identical with those used in determining oxygen consumption by the fish. Under these circumstances gill blood flow (cardiac output) ranged from 1100 to 2200 ml/kg/hr, with an average flow of 1665 ml/kg/hr. Extraction of preformed ammonia from blood flowing through the gills accounted for about 60% of the ammonia excreted, and the remainder could be accounted for by the extraction of α-amino acid N in plasma. The lack of a net extraction of glutamine from plasma observed in these experiments indicates that unlike the mammalian kidney glutaminase does not play a major role in ammonia formation by the gills. Thus, it appears that in this marine teleost branchially excreted ammonia has its source both in that delivered in blood to the gill after being derived from transamination and deamination steps going on centrally in the liver and to a lesser extent in kidneys and other tissues, and also in that formed peripherally at the gills by the enzymic extraction of α-amino acid N. The relative importance of these two sources probably varies between species and with the physiological or environmental status of the fish.

Another possible pathway of amino acid deamination (Fig. 3) originally proposed by Braunstein (1957) has been suggested as a source of ammonia in the gills of carp and other freshwater fishes by Zydowo (1960), Makarewicz and Zydowo (1962), and Makarewicz (1963). Branchial tissues were found to have much higher activities of adenosine monophosphate aminohydrolase than of glutaminase, and it was suggested that this route might be the major source of ammonia excreted by intact lower aquatic vertebrates generally. This pathway involves the amination of inosine monphosphate by aspartate and then the subsequent deamination of the product, adenosine monophosphate, by AMP-deaminase. The importance of this system as an actual producer of ammonia *in vivo* must be questioned on the basis of the observation made by Pequin (1962) on the carp that extraction of preformed blood ammonia can account for all the ammonia excreted by the gills. In gill homogenates of the marine teleost, *Myoxocephalus scorpius*, adenylic acid deaminase activity is approximately twice that of glutaminase (Goldstein *et al.*, 1964). However, glutamine is not extracted as blood traverses the branchial circulation, and again it must be kept in mind that the activity of an enzyme measured *in vitro* does not necessarily reflect its normal kinetics in a structurally and functionally integrated cell.

Concerning the central formation of ammonia, Pequin and Serfaty (1963) showed a low rate of deamination of adenosine monophosphate by perfused carp livers, but adenosine and adenosine triphosphate were

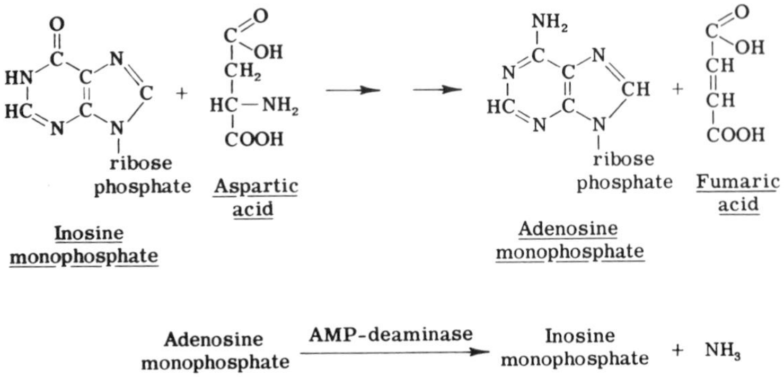

Fig. 3. Scheme for the deamination of amino acids via adenosine monophosphate aminohydrolase.

quickly deaminated. The suggestion that adenosine nucleotides can be deaminated by fish liver supports the possibility that amino acids could be deaminated by transamination to aspartate which could subsequently be deaminated via the purine cycle.

There are other specialized deamination pathways present in higher animals that may occur in fishes. Homogenates of fish liver, kidney, and brain can deaminate aspartate, cysteine, and histidine (Salvatore *et al.*, 1965). Deaminases for cysteine and histidine are known to occur in higher vertebrates. The presence of aspartase has not been reported, but if it is found in fish tissues, then a fourth pathway may exist in which amino acids are transaminated to aspartate and subsequently deaminated by aspartase.

The importance of extrabranchial sites of ammonia formation was again shown in the study of McBean *et al.* (1966) concerned with the assay and characterization of glutamate dehydrogenase in various fish tissues. The finding that glutamate dehydrogenase activity was lower in the gills than in the liver and kidney of eels agrees with observations on the site of formation of branchially excreted ammonia by Pequin (1962) on the freshwater carp and by Goldstein *et al.* (1964) on the marine sculpin. The occurrence of glutamate dehydrogenase was demonstrated in the livers of five species, including two marine teleosts, two elasmobranchs, and a freshwater teleost. L-Alanine, a major amino acid in fish plasma, was not deaminated directly by the eel enzyme but was transaminated to glutamate before deamination.

Crystalline glutamate dehydrogenase prepared from *Squalus* liver has a molecular weight of 330,000 ± 20,000, and an electrophoretic pattern that differs significantly from the chicken and beef enzyme (Corman *et al.*, 1967). Nucleotides such as adenosine diphosphate and guanosine triphosphate activated the dogfish enzyme, and excess nicotinamide-adenine dinucleotide inhibited it without altering its molecular weight. This is of general significance because it indicates that in contrast to the enzyme of higher vertebrates nucleotide can affect the reaction kinetics without altering the aggregation state of the enzyme. Hagfish, *Myxine glutinosa*, liver enzymic activity is stimulated by adenosine diphosphate and diethylstilbesterol without changing its affinity for oxidized nicotinamide-adenine dinucleotide or for glutamate when assayed in tris buffer, but not when tested in physiological saline.

Indirect evidence for the presence of glutamine synthetase in carp was provided by Pequin and Serfaty's observation (1966) that prolonged intravenous injections of L-glutamate decreased the branchial excretion of ammonia and simultaneously increased amino acid excretion, while raising the levels of L-glutamic acid in blood, liver, and kidney. Concentrations of glycine, alanine, serine, and histidine also increased in blood, liver, and kidney, presumably as a result of transamination occurring on a massive scale. Glutamine synthetase could account for diminished ammonia excretion because of its combination with the massive levels of administered L-glutamate to form the amide. The authors do not interpret their observations as supporting the role of L-glutamate dehydrogenase in forming ammonia *de novo* in the gills. They noted that the content of amino acids in gill tissue did not vary during and after glutamate injection, and the excess glutamate in blood was excreted without deamination. However, since glutamate levels were not measured, it is not possible to separate the direct effect glutamate may have had on glutamine synthesis from other effects indirectly affecting the excretion of ammonia. Pequin (1967) has recently found that elevation of blood ammonia leads to increase in plasma glutamine and decrease in hepatic glutamate levels in the carp, suggesting the presence of glutamine synthetase in the liver of this species. Wu (1963) had previously failed to find glutamine synthetase activity in homogenates of liver from the blue gill, *Lepomis macrochirus*, and the black crappie, *Pomoxis nigromaculatus*.

B. Biosynthesis of Urea

Protein metabolism is purely ammoniotelic in the protochordate ancestors of the fishes. Even the uricolytic enzymes are missing in modern ascidian sea squirts, for example, and purine metabolism stops with the formation of urates and xanthine which are disposed of by storage in the form of concretions in special excretory vesicles or renal sacs (Good-body, 1965). Similarly, the uricolytic enzymes (urate oxidase, allanto-inase, and allantoicase) are absent in the cyclostome fishes (Florkin and Duchateau, 1943), and there is also no evidence for the presence of the ornithine–urea cycle in these primitive forms. The small amount of urea in their body fluids might be accounted for by the hydrolysis of arginine by arginase, and accumulation may result from their characteristically low excretion rates. By the time amphibians arose from the crossoptery-gian stem of ancient fishes, complex mechanisms of urea synthesis presumably had evolved which helped prepare them and the higher verte-brates for terrestrial life. Modern frogs in the course of their meta-morphosis undergo a sudden switch from ammoniotelism to the produc-tion of urea, and simultaneously to its active renal tubular excretion when the fishlike tadpole acquires the structural and functional changes charac-teristic of the amphibious adult (Brown and Cohen, 1958; Forster *et al.*, 1963). Increased biosynthesis of urea at this time is by some type of in-duction resulting in enhancement of all enzymic activities of the ornithine cycle. A similar induction of function might account for the simultaneous activation of the carrier-mediated urea transport process in the renal tubule.

1. Ornithine–Urea Cycle and the Purine Pathway

The classification of animals as ammoniotelic, ureotelic, or uricotelic according to the major end product of protein catabolism is somewhat arbitrary because the body fluids and excretory products of most animals contain a mixture of ammonia, urea, and uric acid in varying proportions. Teleost fishes, of course, are primarily ammoniotelic but their blood con-tains significant amounts of urea (Denis, 1913–1914), and indeed, in some teleosts, it may account for 20% or more of the total nitrogen ex-creted (Wood, 1958). In the freshwater carp there is an increase in the excretion of urea during prolonged fasting, but the origin of this supple-mentary urea is difficult to explain (Vellas and Serfaty, 1967).

Urea cannot be synthesized in teleosts via the Krebs ornithine–citrul-line–arginine cycle (Krebs and Henseleit, 1932) which is the main route of ureogenesis in mammals and all the other nonuricotelic higher verte-brates. Arginase is present in the livers of teleost fishes (Hunter and Dauphinee, 1924–1925; Hunter, 1929), but early studies by Manderscheid (1933) suggested that the rest of the ornithine cycle mechanism was in-complete or lacking. Brown and Cohen (1960) have shown that two en-zymes involved in the first two steps of the ornithine cycle, carbamoyl phosphate synthetase and ornithine transcarbamylase (Fig. 4), are not

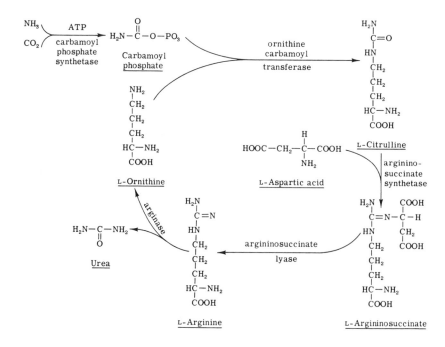

Fig. 4. Ornithine–urea cycle.

present in teleost livers. Dietary arginine potentially could be a source of urea, but it is unlikely that the degradation of an essential amino acid which cannot be synthesized by teleosts could provide more than a very minor fraction of the nitrogen excreted as urea. Purines, however, are a likely source of urea in fish. The early studies on the degradation of uric acid by Przylecki (1925), Stansky (1933), Brunel (1937a,b), and Florkin and Duchateau (1943) revealed the presence of urate oxidase, allantoinase, and allantoicase in the livers of the many fish and amphibians, and Brunel (1937a) proposed that urea was produced in a three-step process: urate → allantoin → allantoicate → urea. More recently, it has been proposed that amino acids could be funneled into the pathway during formation of the purine ring and then converted to uric acid eventually to be transformed into urea (Fig. 5).

2. TELEOSTS

The activity of the uricolytic pathway (Fig. 5) was assayed in liver slices of five representative species of marine and freshwater teleost fishes

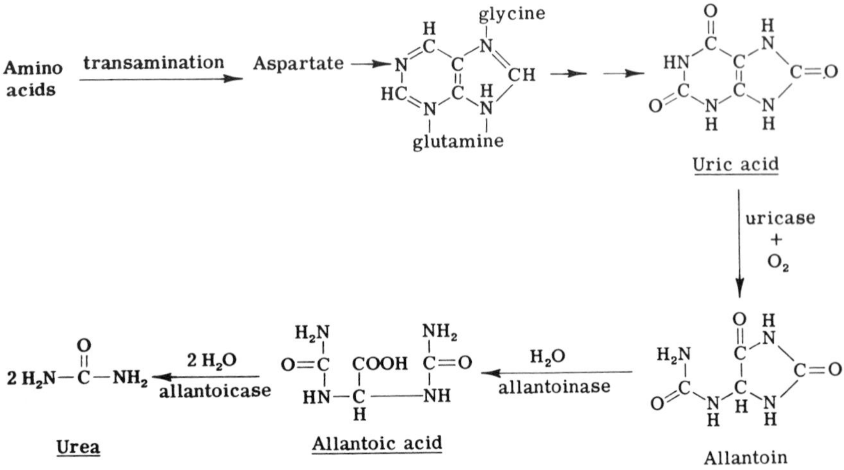

Fig. 5. Urea synthesis via the purine pathway.

by Goldstein and Forster (1965) to evaluate the role this route might play in the production of urea. The fact that urate, allantoin, and allantoicate were all readily converted to urea confirmed Brunel's proposed pathway for the conversion of urate to urea (1937a). Rates of conversion of urate to urea in liver slices ranged from 5 μmoles urea per gram per hour in the goosefish, *Lophius americanus*, up to 23 umoles/g/hr in the winter flounder, *Pseudopleuronectes americanus*. Assays of kidney slices showed that the rates of conversion of urate to urea were only 10% or less that observed in liver slices. Rates of uricolytic urea production *in vitro* corresponded generally with actual rates of urea excretion by the intact fish.

No evidence was found to support Brunel's classification of teleost fishes into two groups (1937b), based on those which degrade uric acid to urea and those which supposedly lack allantoicase and instead convert urate to allantoicate as the end product of purine catabolism. To determine whether the conversion of urate to urea takes place only in certain families of teleosts or whether this pathway is more widespread, allantoicase activity was assayed in liver homogenates prepared from 18 species of teleosts, including eel, cod, and flatfishes which had previously been reported as lacking the enzyme. Allantoicase activity was present in the livers of all 18 species of fish assayed, and the relative rates are shown in Fig. 6.

Since urea is produced and excreted, although at a relatively low rate, in tadpoles (Munro, 1953; Brown and Cohen, 1958) and *Necturus* (Walker and Hudson, 1937; Fanelli and Goldstein, 1964) in which relatively low activities of ornithine–urea cycle enzymes prevail (Brown and

Cohen, 1960), it is of interest to note that liver slices of these aquatic amphibians showed significant rates of urea production from urate. In contrast, there was no detectable conversion of urate to urea by liver slices of large adult frogs. Thus, it appears that uricolysis is the major source of urea produced by teleosts and other primarily "ammoniotelic" vertebrates generally. Inasmuch as nitrogens in the purine ring may come from the amide N of glutamine and the α-amino groups of aspartate and glycine (Buchanan and Hartman, 1959), a pathway may exist also for the synthesis of purine that may provide a route for the disposal of α-amino acid N via transamination, formation of urate, and then the subsequent production of urea via the uricolytic pathway (Goldstein and Forster, 1965).

3. CARTILAGINOUS FISHES

The very high concentration of urea and its osmoregulatory role in the body fluids of marine elasmobranchs and holocephaleans makes the search for possible pathways of ureogenesis here of special interest (Smith, 1953). All the enzymes necessary for urea synthesis via the ornithine cycle have been demonstrated in elasmobranchs (Baldwin, 1960; Brown and Cohen, 1960; Campbell, 1961; Brown, 1964; Schooler,

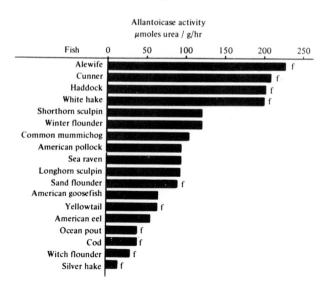

Fig. 6. Allantoicase activity of teleost fish livers. Activities were assayed in homogenates of either fresh or frozen (f) livers. Values are the average of 2–6 assays performed on 1–3 individuals of each species. Reproduced from Goldstein and Forster (1965) by permission of Pergamon Press.

1964; Read, 1968) and in the chimaeroid, *Hydrolagus colliei* (Read, 1967). Arginase occurs in various tissues of elasmobranchs (Hunter and Dauphinee, 1924–1925; Campbell, 1961) and in chimaeroids (Hunter, 1929) but, again, as in teleosts the production of urea from dietary arginine here is probably of little quantitative significance. Elasmobranchs also have the enzymes needed to convert uric acid to urea–urate oxidase, allantoinase, and allantoicase (Brunel, 1947a). Thus, all the enzymes needed for operation of both the ornithine cycle and the purine pathway in ureogenesis are present in elasmobranchs.

Schooler *et al.* (1966) used isotopic labeling techniques in the dogfish, *Squalus acanthias*, to assess the relative importance of the two major routes known for urea synthesis. These studies were carried out both *in vivo* with sodium bicarbonate-^{14}C and uric acid-2-^{14}C, and *in vitro* with isolated liver slices using these substrates and also serine-3-^{14}C. Earlier studies on other vertebrates had shown that CO_2 is incorporated into urea via the ornithine cycle (Brown and Cohen, 1960) and that the 3-carbon of serine is a precursor for the ureido-carbons (2 and 8) of uric acid (Elwyn and Sprinson, 1950). The 3-carbon of serine is transferred to tetrahydrofolic acid by serine hydroxymethyltransferase to form N^5, N^{10}-methylenyltetrahydrofolate and is subsequently incorporated in a series of steps into the purine ring (Buchanan and Hartman, 1959). In these *in vitro* experiments with a large pool of serine available in the flasks (2 mM) it was assumed that most of the 2- and 8-carbons of purine came from serine and, hence, that the conversion rate of serine-3-^{14}C to urea-^{14}C approximated the rate of urea production via urate synthesis and then its subsequent degradation in the uricolytic pathway.

These studies on *Squalus* (Schooler *et al.*, 1966) showed that the rate of urea formation via the ornithine cycle was approximately 50–100 times faster than that produced in the purine pathway. Isolated liver slices synthesized urea from urate-2-^{14}C substrate at the rate of 0.023 μmole/g/hr and incorporated isotopically labeled serine as a precursor into the purine pathway to produce urea at the rate of 0.040 μmole/g/hr, whereas the synthesis of urea-^{14}C from bicarbonate-^{14}C via the ornithine cycle proceeded at the rate of 2.25 μmoles/g/hr. The addition of "cold" ornithine increased urea synthesis 23%, whereas the addition of citrulline reduced the synthesis of urea-^{14}C to 65% of control values by diluting the fraction of isotopically labeled urea (see Fig. 5), thus confirming the specific involvement of the ornithine cycle as the pathway of urea synthesis via incorporation of bicarbonate-^{14}C.

In the intact dogfish also, Schooler *et al.* found the ornithine cycle to be much more active in forming urea than was the purine pathway. The rate of urea synthesis from either bicarbonate or uric acid was calculated by dividing the total counts of urea produced during a 30-minute period following the intravenous injection of either bicarbonate-^{14}C and urate-2-^{14}C, by the mean specific activity of the respective precursor in the plasma. Total counts in urea were obtained from the distribution of urea-^{14}C in blood, liver, and other body fluids. Urea was found to be synthesized by the purine pathway at a rate of 7.0 μmoles/kg body weight per hour and via the much faster ornithine cycle at the rate of 135 μmoles/kg per body weight per hour.

4. THE LUNGFISHES

Biochemical sources of nitrogenous end products in the *Dipnoi* are of unusual interest because lungfish such as the African form, *Protopterus*, shifts from a primarily ammoniotelic status to a completely ureotelic habitus while undergoing estivation (Smith, 1930). During its aquatic phase *Protopterus* has urine flows and glomerular filtration rates that are quite comparable with those of freshwater teleosts. It then obtains about half of its total metabolic energy from protein, and, resembling some aquatic amphibians, the nitrogenous end products are excreted both as ammonia and urea (Smith, 1930; Sawyer, 1966). During estivation when the fish is enveloped in a leathery protective cocoon with a small aperture to allow breathing it may survive for years without food or water while urine formation is completely suspended and the metabolic rate drops to very low levels. During this period when the metabolism of endogenous protein is the main source of energy, ammonia does not accumulate in body fluids and nitrogen is almost entirely incorporated into urea, which within a year may amount to as much as 1% of the animal's total body weight (Smith, 1930; Janssens, 1964).

All of the enzymes of the ornithine–urea cycle are present in *Protopterus* livers (Janssens and Cohen, 1966) and, in addition, all of the enzymes of the uricolytic pathway (Florkin and Duchateau, 1943; Brown *et al.*, 1966) that in teleosts appears to be solely responsible for the relatively small amounts of urea synthesized (Goldstein and Forster, 1965).

Forster and Goldstein (1966) evaluated the relative importance of the purine pathway and the ornithine cycle by determining actual rates of urea synthesis via these routes in livers taken from *Protopterus dolloi*, both while in their aquatic phase and during estivation. Rates of urea formation via the ornithine cycle were determined by incubating liver slices in balanced isotonic media containing ammonium ion, bicarbonate-^{14}C, ornithine, and an energy source. To evaluate rates of ureogenesis via the purine pathway a similar *in vitro* preparation contained L-serine labeled with ^{14}C in the third position as the source of the "formate" carbon atoms 2 and 8 in uric acid according to the method described above for similar studies with *Squalus* by Schooler *et al.* (1966). To ascertain specificity unlabeled citrulline, one of the intermediates in the ornithine cycle was added to note its damping effect on the rate of production of ^{14}C-labeled urea. Similarly, "cold" uric acid was used to observe its damping effect on the incorporation of ^{14}C into urea derived from serine-^{14}C via the purine pathway. Both routes of ureogenesis were found to be operating in *Protopterus* livers, but the rate of urea synthesis by the ornithine cycle was at least 100 times greater than that of the purine pathway under the conditions of these experiments. These observations on the African lungfish correspond with the results of similar comparisons made by Carlisky *et al.* (1966) in the frog kidney and by Schooler *et al.* (1966) in the elasmobranch, *Squalus*, liver.

Radical alterations in biochemical pathways during estivation may not be needed to account for the shift from ammonia to urea production. Ammonia which normally escapes rapidly from lungfish by branchial diffusion might instead be captured during estivation by carbamoyl phosphate synthetase to be incorporated into carbamoyl phosphate along with carbon dioxide and then fed into the cycle, which was found to be operative in both the aquatic and estivating livers of *Protopterus*. Forster and Goldstein (1966) calculated that, at the rate of urea synthesis observed in nonestivating lungfish livers (approximately 1 μmole/g/hr), a typical 100-g fish whose liver weighs about 1 g could form approximately 0.5 g of urea in a year, a value for urea accumulation within the range actually found in the tissues of African lungfish estivating for this period of time (Smith, 1930).

The Australian lungfish, *Neoceratodus forsteri*, the only surviving member of the family *Ceratodontidae*, differs from the African dipnoan *Protopterus* and the South American *Lepidosiren* in its ecological behavior. The latter, sole survivors of the family Lepidosirinidae, must breathe air at rather frequent intervals and they estivate in burrows out of water during dry seasons in their tropical environment. On the other hand, the nontropical Australian lungfish uses its single lung only as an accessory respiratory organ during periods when it is very active (Grigg, 1965). In contrast to the African and South American dipnoans, it dies if removed from water, and it has no adaptive provisions for entering a state of estivation by burrowing and breathing air. Having no need to detoxify ammonia because of its purely aquatic status it was of interest to determine whether the Australian form had the capability of synthesizing urea via the ornithine cycle as had previously been demonstrated in livers of the African lungfish, *Protopterus*. Goldstein *et al.* (1967b) compared actual rates of incorporation of bicarbonate-^{14}C into urea in the Australian and African forms and evaluated relative activities of enzymes of the ornithine–urea cycle. Studies on two adult *Neoceratodus* shipped to Boston from Australia showed them to be much more "ammoniotelic" than *Protopterus* which eliminates up to 50% of its nitrogenous end products as urea when in water (Smith, 1930). Rates of ammonia production in the two fish were 98 and 156 μmoles/kg/hr when simultaneous rates of urea excretion were 5.5 and zero, respectively.

The percentage incorporation of bicarbonate-^{14}C into urea-^{14}C by *Neoceratodus* liver slices was very low, being only 0.01–0.02% of the radioactivity present in the *in vitro* medium, (about 100 counts/min above background). Addition of unlabeled citrulline, which in similar studies on *Protopterus* reduced the ^{14}C incorporated into urea by competing with citrulline-^{14}C formed from bicarbonate-^{14}C, had little effect on the rate of ^{14}C-urea formation. The addition of unlabeled arginine, however, significantly reduced the rate of synthesis of urea-^{14}C by about one-third. Hence, it appears likely that bicarbonate-^{14}C was incorporated to urea via the ornithine cycle, but the actual rate of synthesis was only about a hundredth of that observed in liver slices of the African *Protopterus*. This finding is consistent with the low rate of excretion (production) of urea in the totally aquatic *Neoceratodus*.

Activities of all the enzymes involved in the ornithine cycle were also assayed by Goldstein *et al.* Carbamoyl phosphate synthetase, ornithine carbamoyltransferase, argininosuccinate synthetase, argininosuccinate lyase and arginase all had activities much lower in *Neoceratodus* than in *Protopterus*, a finding consistent with the low rates of excretion, the low level of incorporation of bicarbonate into urea by liver slices *in vitro*, and with the exclusively aquatic nature of the Australian lungfish.

C. Sources of Trimethylamine Oxide

The origin of trimethylamine oxide, a weakly basic nontoxic end product of nitrogen metabolism found in blood and body fluids of fish and other aquatic animals, has been a matter of dispute ever since the early investigations of Hoppe-Seyler (1930). He demonstrated its occurrence in various fishes and also suggested an osmoregulatory role along with that of urea in marine elasmobranchs. The hypothesis that these organic solutes were actively retained was based upon observed plasma concentrations of TMAO of from 100 to 120 mmoles/liter when urine concentrations were only about one-tenth this amount. The TMAO content of tissues of freshwater teleosts is generally quite low compared with that of marine forms, and it fluctuates correspondingly in euryhaline forms such as the salmon when they migrate between fresh- and salt-water. The increased TMAO in tissues during the marine phase may be of endogenous origin and associated somehow with biochemical or physiological osmoregulatory adaptations, or it might be derived from exogenous dietary sources in the marine environment.

Benoit and Norris (1945) based their experiments concerning sources of TMAO on the reasoning that if it originates exogenously, then transferring young salmon which have been maintained on a TMAO-free diet from fresh- to saltwater should raise the levels of the base in their muscle tissues. However, young salmon maintained in excellent condition on a TMAO-free liver diet during a period of increasing salinity showed no significant variation in TMAO content; and even after 5 weeks in full strength seawater there was no increase. Young control salmon fed TMAO-containing food such as salmon meal and ground scallop muscle rapidly accumulated the compound and within 3 weeks on the diet reached levels characteristic of normal marine adults. These results strongly suggest that, in salmon at least, high levels of TMAO in muscle result from the accumulation of ingested TMAO rather than from an endogenous metabolic source.

Bilinski (1964) used isotopic labeling techniques to identify precursors of TMAO in marine teleosts (lemon sole, *Parophrys vetulus,* and starry flounder, *Platichthys stellatus*). ^{14}C-labeled compounds were administered mainly intraperitoneally, and incorporation of tracer into TMAO was determined using the whole body of the fish as starting material. Of the compounds tested the free amine, trimethylamine-^{14}C, was found to be the best precursor of the oxide. Limited labeling of TMAO occurred after administration of γ-butyrobetaine-methyl-^{14}C, betaine-methyl-^{14}C, and methionine-methyl-^{14}C. Little or no incorporation of ^{14}C into TMAO occurred after administration of methylamine-^{14}C, carnitine-methyl-^{14}C, glycine-2-^{14}C, sodium formate-^{14}C, sodium acetate-1-^{14}C, sodium acetate-2-^{14}C, and $NaHCO_3$-^{14}C. Intraperitoneal injections of choline-methyl-^{14}C gave higher conversion rates to TMAO than the intramuscular route, and only trace amounts of radioactivity were detected after intravenous injection of precursor. Dependence of conversion to TMAO on the route of administration of the precursor might be because choline given intravenously or intramuscularly is rapidly excreted in the characteristically acid urine of marine teleosts, and thus unavailable for conversion to TMAO, whereas when administered by the intraperitoneal route the slower rate of precursor uptake would result in a more complete conversion to TMAO. However, in the dogfish, *Squalus acanthias,* when choline was administered by the intravenous route only about 1% of the base was excreted in 23 hr (Goldstein *et al.,* 1967a). The slow rates of conversion of the methyl donors to TMAO in teleosts and elasmobranchs as compared with the lobster and other marine crustaceans, for example (Bilinski, 1960, 1961, 1962), raise some doubts as to the relative importance of endogenous synthesis by fish tissues as the source of TMAO. Also, the possibility of synthesis by microbial activity in the digestive tract has not been ruled out as a source of this limited conversion of choline and other precursors, especially when administered intraperitoneally.

The fact that plasma levels of TMAO remain relatively constant in dogfish, *Squalus acanthias,* that were maintained without feeding for as long as 40 days suggested an endogenous source of the base (J. J. Cohen *et al.,* 1958). Goldstein *et al.* (1967a) used isotopically labeled precursors to test the ability of intact dogfish and *in vitro* liver preparations to synthesize TMAO. There was no incorporation of trimethylamine-^{14}C or choline-methyl-^{14}C into TMAO *in vivo* or *in vitro*. A test of possible incorporation of L-methionine-methyl-^{14}C by liver slices was also negative. Sustained maintenance of relatively steady plasma levels of TMAO may be achieved by drawing on the large store of the oxide in the muscles of these elasmobranchs and then minimizing loss by its active tubular reabsorption in the kidneys. Trimethylamine oxide levels in muscle are higher than in other tissues of these cartilaginous fishes, and there is no indication of any particular role the compound might play in muscle contraction or metabolism.

The problem of the biochemical origin of TMAO in fishes is still unresolved. The osmoregulatory advantage in its presence in the body fluids of marine forms is clear. Possible explanations for the difference in TMAO levels between marine and freshwater species could include higher TMAO content in the diets of marine fishes, the contributions by intestinal microorganisms, differences in retention capabilities of gills and kidneys, as well as the possibility of some endogenous synthesis of TMAO.

V. EVOLUTIONARY CONSIDERATIONS

The recent disclosure that the coelacanth, *Latimeria,* uses high concentrations of urea to maintain a serum osmolarity approximately equal to that of its marine environment raises fresh speculations concerning the evolutionary origin of the "ureotelic" habitus in amphibians and other terrestrial vertebrates. Pickford and Grant (1967) showed that the urea concentration in serum was 355 mmoles/liter when the mean value for total osmolarity was 1181 mOsm/liter. This urea level is comparable to that in the elasmobranch, *Squalus*, for example (322–364 mmoles/liter), where 81–94% of the urea in the glomerular filtrate is actively reabsorbed by the renal tubules (Forster and Berglund, 1957) and then retained in body fluids as part of the osmoregulatory device used by all marine cartilaginous fishes to maintain approximate isosmolarity with their environment. Arginase and ornithine carbamoyltransferase have been identified in homogenates of coelacanth liver (Brown and Brown, 1967); hence, it appears quite likely that the ornithine cycle plays a predominant role in urea synthesis in this ancient representative of the proto-amphibian crossopterygian fishes that were the evolutionary progenitors of land vertebrates.

Fossils of primitive coelacanths are present in Devonian deposits, and all indications are that early members of the group were, in general, freshwater types. As Romer (1966) points out, by the Triassic they had already invaded the sea, and were common and varied. No fossil records of coelacanths were found beyond the cretaceous, and until living *Latimeria* were discovered in 1938, it was generally assumed that they and the *Crossopterygii* as a whole had become extinct at the end of the Mesozoic. The lungfishes of the order Dipnoi in the late Paleozoic and Triassic were abundant in the freshwater environment in which they arose and had adapted themselves. The third group of "ureotelic" fishes, the Chondrichthyes, appear to have been ocean dwellers since their first appearance in the Devonian. Romer (1966) states, "Possibly their remote ancestors may have dwelt in inland waters; but the advent of sharklike fishes into the sea was certainly early, and was accomplished by special structural and functional adaptations to a saline environment quite different from those evolved (presumably at a later date) by marine bony fishes." Possession of the ornithine–urea cycle by both the chimaeras and the elasmobranchs suggests that these two main groups of cartilaginous fishes shared a marine environment at the time their separate evolutionary lines were established. It seems more plausible that absence of the urea-retention habitus in freshwater riverrays of the genus *Potamotrygon* (Thorson *et al.*, 1967) represents the deletion or repression of the ornithine–urea cycle apparatus rather than that those forms descended from ancient Chondrichthyes that had never developed urea retention as an osmoregulatory device.

Individual enzymes involved in the ornithine–urea cycle and the purine pathway are widely distributed in microorganisms, plants, and animals, and here, as in most fishes with the ammonia-producing habitus, they may serve a variety of nonureogeneic functions. As Ratner (1954) pointed out, linked enzymically controlled reactions in the ornithine–urea cycle may be regarded as a diversion of more general mechanisms that provide the arginine and other amino acids of cellular protein. Microbial and other enzyme systems are also known that can degrade arginine and citrulline by routes that do not result in urea formation. Similarly, individual enzymes of the purine pathway, that incorporated amino acid nitrogen into uric acid resulting eventually in the production of urea, have widespread functions related to the general biosynthesis and metabolism of purines quite apart from ureogenesis. As a requisite for the development of the ureotelic habitus by cartilaginous fishes, the African and South American lungfishes, and by ancestral crossopterygians on their way to acquiring terrestrial status as amphibians, new genetically transmittable patterns of enzymic organization had to be assembled. What specific inductions, mutations or other genetic processes brought this about is, of course, unknown.

VI. CONTROL MECHANISMS

The transition to terrestrial life demands either suppression of ammonia production or its incorporation into urea. The rate of urea synthesis in African lungfish, *Protopterus*, is similar in estivating and nonestivating fish (Janssens and Cohen, 1968a). Furthermore, there are no differences in the activities of the ornithine–urea cycle enzymes in isolated slices and homogenates of livers taken from fish in both conditions (Forster and Goldstein, 1966; Janssens and Cohen, 1968a). These observations suggest that there is no adaptive increase in urea synthesis during estivation in *Protopterus;* rates of urea synthesis normally are high enough to account for the actual amount of urea accumulated during estivation. In contrast the switch from ammoniotelism to ureotelism in the African frog, *Xenopus laevis*, during water deprivation (owing to estivation or osmotic stress) is accompanied by an increase in urea production and adaptation of ornithine–urea cycle enzymes (McBean and Goldstein, 1967; Janssens and Cohen, 1968b). It is probable that ammonia production is suppressed during estivation in lungfish. Conversion of ammonia to some nitrogenous compound other than urea appears unlikely in view of Smith's observation (1930) that the accumulation of other nitrogenous compounds accounts for less than 30% of the nonprotein nitrogen in muscle of estivating *Protopterus*. Furthermore, Janssens and Cohen (1968a) found no significant changes in the levels of the major amino acids in the livers of estivating *Protopterus* (Janssens and Cohen, 1968b).

More information is needed concerning the specific pathways of ammonia synthesis in lungfish before an explanation is forthcoming about the nature of the control mechanism which shuts off ammonia synthesis during estivation. If the pathway is transdeamination via glutamate dehydrogenase then there are several possible control sites. Ammonia could block its own synthesis from glutamate by some sort of feedback inhibition. For example, McBean et al. (1966) found that ammonia was a competitive inhibitor of glutamate dehydrogenase in eel liver homogenates. A tenfold increase in ammonia concentration was needed, however, to produce a 50% inhibition in enzymic activity, and inasmuch as the lungfish does not accumulate detectable amounts of ammonia during estivation, it does not appear that ammonia is controlling its own rate of production in this instance.

Janssens (1964) found that glutamic-alanine transaminase levels during estivation were one-tenth those in the nonestivating lungfish, Protopterus. If alanine is the predominant amino acid of extracellular fluid in lungfish, as it is in teleosts and elasmobranchs, then reducing the conversion rate of alanine to glutamate could significantly lower the rate of glutamate formation and, thus, ammonia production. The relative rates of glutamate deamination vs its synthesis via glutamate dehydrogenase is partially dependent on the ratio of $NAD^+/NADH$ in cells. Tissues may be somewhat hypoxic during estivation, thereby increasing the fraction of reduced nucleotide. This would result in depressing deamination and favoring the formation of glutamate. The α-amino group of glutamate, in turn, could be transaminated to aspartate via glutamate-aspartate transaminase (Janssens, 1964) and then into the ornithine–urea cycle.

In both invertebrate and vertebrate preparations glutamate dehydrogenase is sensitive to NaCl concentrations in the assay medium (Chaplin et al., 1965). Although both the forward and reverse reactions: α-oxoglutarate $+ NADH_2 + NH_4^+ \rightleftharpoons$ glutamate $+ NAD^+ + H_2O$ were accelerated by increasing the ionic concentrations particularly of Cl^-, the reaction rate in the direction toward glutamate synthesis was affected more. The hypothesis is worth tesing that it may not be water deprivation per se that is the effective stimulus in suppressing ammonia production by fishes as they assume a semiterrestrial habitus, but rather corresponding increases in concentrations of inorganic ions in liver and other tissues involved in ammonio- and ureogenesis. The fact that osmotic stress (hypertonic saline) is as effective as estivation in inducing a switch from ammoniotelism to ureotelism in Xenopus and the amphibious mudskippers supports this hypothesis. Certainly one of the most challenging problems in relation to the biochemical sources of nitrogenous excretory products concerns the identification of environmentally influenced control mechanisms that bring about reciprocal alterations in ammonia production and ureogenesis as fish adapt to varying degrees of water deprivation in a hyperosmotic or semiterrestrial environment.

REFERENCES

Altman, P. L. (1961). "Blood and Other Body Fluids." Fed. Am. Soc. Exptl. Biol., Washington, D.C.

Baker, J. R., Struempler, A., and Chaykin, S. (1963). A comparative study of tri-methylamine-N-oxide biosynthesis. Biochim. Biophys. Acta 71, 58–64.

Baldwin, E. (1960). Ureogenesis in elasmobranch fishes. Comp. Biochem. Physiol. 1, 24–37.

Ball, I. (1967). The relative susceptibilities of some species of fresh-water fish to poisons. I. Ammonia. Water Res. 1, 767–775.

Benoit, G. J., and Norris, E. R. (1945). Studies on trimethylamine oxide. II. The origin of trimethylamine oxide in young salmon. J. Biol. Chem. 158, 439–442.

Berglund, F., and Forster, R. P. (1958). Renal tubular transport of inorganic divalent ions by the aglomerular marine teleost, Lophius americanus. J. Gen. Physiol. 41, 429–440.

Bilinski, E. (1960). Biosynthesis of trimethylammonium compounds in aquatic animals. I. Formation of trimethylamine oxide and betaine from C^{14}-labelled compounds by lobster (Homarus americanus). J. Fisheries Res. Board Can. 17, 895–902.

Bilinski, E. (1961). Biosynthesis of trimethylammonium compounds in aquatic animals. II. Role of betaine in the formation of trimethylamine oxide by lobster (Homarus americanus). J. Fisheries Res. Board Can. 18, 285–286.

Bilinski, E. (1962). Biosynthesis of trimethylammonium compounds in aquatic animals. III. Choline metabolism in marine crustacea. J. Fisheries Res. Board Can. 19, 505–510.

Bilinski, E. (1964). Biosynthesis of trimethylamine compounds in aquatic animals. IV. Precursors of trimethylamine oxide and betaine in marine teleosts. J. Fisheries Res. Board Can. 21, 765–771.

Braunstein, A. E. (1939). The enzyme system of trans-amination, its mode of action and biological significance. Nature 143, 609–610.

Braunstein, A. E. (1951). Transamination and the integrative functions of the di-carboxylic acids in nitrogen metabolism. Advan. Protein Chem. 3, 1–52b.

Braunstein, A. (1957). Les voies principales de l'assimilation et dissimilation de l'azote chez les animaux. Advan. Enzymol. 19, 335–377.

Braunstein, A. E., and Byechkov, S. M. (1939). A cell-free enzymatic model of 1-amino-acid dehydrogenase ('1-deaminase'). Nature 144, 751–752.

Brodie, B. B., and Maickel, R. P. (1962). Comparative biochemistry of drug me-tabolism. Proc. 1st Intern. Pharmacol. Meeting, Stockholm, 1961 Vol. 6, pp. 299–324. Pergamon Press, Oxford.

Brodie, B. B., Axelrod, J., Cooper, J. R., Gaudette, L. E., LaDu, B. N., Mitoma, C., and Udenfriend, S. (1955). Detoxication of drugs and other foreign compounds by liver microsomes. Science 121, 603–605.

Brodie, B. B., Gillette, J. R., and LaDu, B. N. (1958). Enzymatic metabolism of drugs and other foreign compounds. Ann. Rev. Biochem. 27, 427–454.

Brown, G. W., Jr. (1964). Urea synthesis in elasmobranchs. In "Taxonomic Bio-chemistry and Serology" (C. A. Leone, ed.), p. 407. Ronald Press, New York.

Brown, G. W., Jr., and Brown, S. G. (1967). Urea, and its formation in Coelacanth liver. Science 155, 570–573.

Brown, G. W., Jr., and Cohen, P. P. (1958). Biosynthesis of urea in metamorphosing tadpoles. In "The Chemical Basis of Development" (W. D. McElroy and B. Glass, eds.), pp. 495–513. Johns Hopkins Press, Baltimore, Maryland.

Brown, G. W., Jr., and Cohen, P. P. (1960). Activities of urea cycle enzymes in various higher and lower vertebrates. *Biochem. J.* 75, 82–91.

Brown, G. W., Jr., James, J., Henderson, R. J., Thomas, W. N., Robinson, R. O., Thompson, A. L., Brown, E., and Brown, S. G. (1966). Uricolytic enzymes in liver of the dipnoan *Protopterus aethiopicus*. *Science* 153, 1653–1654.

Brunel, A. (1937a). Catabolisme de l'azote d'origine purique chez les Sélaciens. *Bull. Soc. Chim. Biol.* 19, 805–826.

Brunel, A. (1937b). Catabolisme de l'azote d'origine purique chez Téléostéens. *Bull. Soc. Chim. Biol.* 19, 1027–1036.

Buchanan, J. M., and Hartman, S. C. (1959). Enzymatic reactions in the synthesis of purines. *Advan. Enzymol.* 21, 199–261.

Buhler, D. R. (1966). Hepatic drug metabolism in fishes. *Federation Proc.* 25, 343.

Burger, J. W. (1967). Problems in the electrolyte economy of the spiny dogfish, *Squalus acanthias. In* "Sharks, Skates and Rays" (P. W. Gilbert, R. F. Mathewson, and D. P. Rall, eds.), pp. 177–185. Johns Hopkins Press, Baltimore, Maryland.

Burger, J. W., and Loo, T. L. (1959). Bromination of phenol red by the dogfish, *Squalus acanthias. Science* 129, 778–779.

Campbell, J. W. (1961). Studies on tissue arginase and ureogenesis in the elasmobranch *Mustelis canis. Arch. Biochem. Biophys.* 93, 448–455.

Carlisky, N. J., Jard, S., and Morel, F. (1966). *In vivo* tracer studies of renal urea formation in the bullfrog. *Am. J. Physiol.* 211, 593–599.

Chaplin, A. E., Huggins, A. K., and Munday, K. A. (1965). Ionic effects on glutamate dehydrogenase activity from beef liver, lobster muscle and crab muscle. *Comp. Biochem. Physiol.* 16, 49–62.

Cohen, J. J., Krupp, M. A., and Chidsey, C. A., III (1958). Renal conservation of trimethylamine oxide by the spiny dogfish, *Squalus acanthias. Am. J. Physiol.* 194, 229–235.

Cohen, P. P., and Brown, G. W., Jr. (1960). Ammonia metabolism and urea biosynthesis. *Comp. Biochem.* 2, 161–244.

Corman, L., Prescott, L. M., and Kaplan, N. O. (1967). Purification and chemical characteristics of dogfish liver glutamate dehydrogenase. *J. Biol. Chem.* 242, 1383–1390.

Creaven, P. J., Parke, D. V., and Williams, R. T. (1965). A fluorimetric study of the hydroxylation of biphenyl *in vitro* by liver preparations of various species. *Biochem. J.* 96, 879–885.

Delaunay, H. (1931). L'excretion azotée des invertébrés. *Biol. Rev.* 6, 265–301.

Denis, W. (1913–1914). Metabolism studies on cold-blooded animals. II. The blood and urine of fish. *J. Biol. Chem.* 16, 389–393.

Dixon, R. L., Adamson, R. H., and Rall, D. P. (1967). Metabolism of drugs by elasmobranch fishes. *In* "Sharks, Skates and Rays" (P. W. Gilbert, R. F. Mathewson, and D. P. Rall, eds.) pp. 547–552. Johns Hopkins Press, Baltimore, Maryland.

Edwards, J. G., and Condorelli, L. (1928). Studies on aglomerular and glomerular kidneys. II. Physiological. *Am. J. Physiol.* 86, 383–398.

Elwyn, D., and Sprinson, D. B. (1950). Role of serine and acetate in uric acid formation. *J. Biol. Chem.* 184, 465–474.

Fanelli, G. M., and Goldstein, L. (1964). Ammonia excretion in the neotenous newt, *Necturus maculosus* (Rafinesque). *Comp. Biochem. Physiol.* 13, 193–205.

Fänge, R., and Fugelli, K. (1962). Osmoregulation in chimaeroid fishes. *Nature* 196, 689.

Florkin, M., and Duchateau, G. (1943). Les formes du système enzymatique de l'uricolyse et l'évolution du catabolisme purique chez les animaux. *Arch. Intern. Physiol.* **53**, 267–307.

Forster, R. P. (1953). A comparative study of renal function in marine teleosts. *J. Cellular Comp. Physiol.* **42**, 487–510.

Forster, R. P. (1954). Active cellular transport of urea by frog renal tubules. *Am. J. Physiol.* **179**, 372–377.

Forster, R. P. (1961). Kidney cells. In "The Cell" (J. Brachet and A. E. Mirsky, eds.), Vol. 5, pp. 89–161. Academic Press, New York.

Forster, R. P. (1967a). Osmoregulatory role of the kidney in cartilaginous fishes (*Chondrichthyes*). In "Sharks, Skates and Rays" (P. W. Gilbert, R. F. Mathewson, and D. P. Rall, eds.), pp. 187–195. Johns Hopkins Press, Baltimore, Maryland.

Forster, R. P. (1967b). Renal transport mechanisms. In "Proceedings of an International Symposium on Comparative Pharmacology" (E. J. Cafruny, ed.), pp. 1008–1019. Fed. Am. Soc. Exptl. Biol., Bethesda, Maryland.

Forster, R. P., and Berglund, F. (1956). Osmotic diuresis and its effect on total electrolyte distribution in plasma and urine of the aglomerular teleost, *Lophius americanus*. *J. Gen. Physiol.* **39**, 349–359.

Forster, R. P., and Berglund, F. (1957). Contrasting inhibitory effects of Probenecid on the renal tubular reabsorption of p-aminohippurate and on the active reabsorption of urea in the dogfish, *Squalus acanthias*. *J. Cellular Comp. Physiol.* **49**, 281–286.

Forster, R. P., and Goldstein, L. (1966). Urea synthesis in the lungfish: Relative importance of purine and ornithine cycle pathways. *Science* **153**, 1650–1652.

Forster, R. P., Berglund, F., and Rennick, B. R. (1958). Tubular secretion of creatine, trimethylamine oxide, and other organic bases by the aglomerular kidney of *Lophius americanus*. *J. Gen. Physiol.* **42**, 319–327.

Forster, R. P., Schmidt-Nielsen, B., and Goldstein, L. (1963). Relation of renal tubular transport of urea to its biosynthesis in metamorphosing tadpoles. *J. Cellular Comp. Physiol.* **61**, 239–244.

Fromm, R. O. (1963). Studies on renal and extra-renal excretion in a freshwater teleost, *Salmo gairdneri*. *Comp. Biochem. Physiol.* **10**, 121–128.

García Romeu, F., and Motais, R. (1966). Mise en évidence d'échanges Na^+/NH^+_4 chez l'anguille d'eau douce. *Comp. Biochem. Physiol.* **17**, 1201–1204.

Goldstein, L., and Forster, R. P. (1961). Source of ammonia excreted by the gills of the marine teleost, *Myoxocephalus scorpius*. *Am. J. Physiol.* **200**, 1116–1118.

Goldstein, L., and Forster, R. P. (1965). The role of uricolysis in the production of urea by fishes and other aquatic vertebrates. *Comp. Biochem. Physiol.* **14**, 567–576.

Goldstein, L., Forster, R. P., and Fanelli, G. M., Jr. (1964). Gill blood flow and ammonia excretion in the marine teleost, *Myoxocephalus scorpius*. *Comp. Biochem. Physiol.* **12**, 489–499.

Goldstein, L., Hartman, S. C., and Forster, R. P. (1967a). On the origin of trimethylamine oxide in the spiny dogfish, *Squalus acanthias*. *Comp. Biochem. Physiol.* **21**, 719–722.

Goldstein, L., Janssens, P. A., and Forster, R. P. (1967b). Lungfish *Neoceratodus forsteri*: Activities of ornithine-urea cycle and enzymes. *Science* **157**, 316–317.

Goodbody, I. (1965). Nitrogen excretion in Ascidiacea. II. Storage excretion and the uricolylic enzyme system. *J. Exptl. Biol.* **42**, 299–305.

Gordon, M. S., Boëtius, J., Boëtius, I., Evans, D. H., McCarthy, R., and Oglesby, L. C.

(1965). Salinity adaptation in the mudskipper fish, *Periophthalmus sobrinus*. *Hvalradets Skrifter Norske Videnskaps-Akad. Oslo* **48,** 85–93.

Grigg, G. C. (1965). Studies on the Queensland lungfish, *Neoceratodus forsteri* (Krefft). III. Aereal respiration in relation to habits. *Australian J. Zool.* **13,** 413–421.

Grollman, A. (1929). The urine of the goosefish (*Lophius piscatorius*): Its nitrogenous constituents with special reference to the presence in it of trimethylamine oxide. *J. Biol. Chem.* **81,** 267–278.

Groniger, H. S. (1959). The occurrence and significance of trimethylamine oxide in marine animals. *U.S. Fish Wildlife Serv., Spec. Sci. Rep., Fisheries* **333,** 1–22.

Hoppe-Seyler, F. W. (1930). Die Bedingungen und die Bedetung biologischer Methylierungsprozesse. *Z. Biol.* **90,** 433–466.

Huang, K. C., and Collins, S. F. (1962). Conjugation and excretion of amino-benzoic acid isomers in marine fish. *J. Cellular Comp. Physiol.* **60,** 49–52.

Hunter, A. (1929). Further observations on the distribution of arginase in fishes. *J. Biol. Chem.* **81,** 505–511.

Hunter, A., and Dauphinee, J. A. (1924–1925). Quantitative studies concerning the distribution of arginase in fishes and other vertebrates. *Proc. Roy. Soc.* **B97,** 227–242.

Jacobs, M. H. (1940). Some aspects of cell permeability to weak electrolytes. *Cold Spring Harbor Symp. Quant. Biol.* **8,** 30–39.

Janssens, P. A. (1964). The metabolism of the aestivating African lungfish. *Comp. Biochem. Physiol.* **11,** 105–117.

Janssens, P. A., and Cohen, P. P. (1966). Ornithine-urea cycle enzymes in the African lungfish, *Protopterus aethiopicus*. *Science* **152,** 358–359.

Janssens, P. A., and Cohen, P. P. (1968a). Biosynthesis of urea in the estivating African lungfish and in *Xenopus laevis* under conditions of water shortage. *Comp. Biochem. Physiol.* **24,** 887–898.

Janssens, P. A., and Cohen, P. P. (1968b). Nitrogen metabolism in the African lungfish. *Comp. Biochem. Physiol.* **24,** 879–886.

Krebs, H. A., and Henseleit, K. (1932). Untersuchungen über die Harnstoffbildung in Tierkörper. *Z. Physiol. Chem.* **210,** 33–66.

Krebs, H. A., and Kornberg, H. L. (1957). A survey of the energy transformations in living matter. *Ergeb. Physiol., Biol. Chem. Exptl. Pharmakol.* **49,** 212–298.

Krogh, A. (1939). "Osmotic Regulation in Aquatic Animals." Cambridge Univ. Press, London and New York (republished by Dover, New York 1965).

Lange, R., and Fugelli, K. (1965). The osmotic adjustment in the euryhaline teleosts, the flounder, *Pleuronectes flesus* L. and the three-spined stickleback, *Gasterosteus aculeatus* L. *Comp. Biochem. Physiol.* **15,** 282–292.

McBean, R. L., and Goldstein, L. (1968). Ornithine-urea cycle activity in *Xenopus laevis*: Adaptation in saline. *Science* **157,** 931–932.

McBean, R. L., Neppel, M. J., and Goldstein, L. (1966). Glutamate dehydrogenase and ammonia production in the eel (*Anguilla rostrata*). *Comp. Biochem. Physiol.* **18,** 909–920.

Maetz, J., and García Romeu, F. (1964). The mechanism of sodium and chloride uptake by the gills of a fresh-water fish, *Carassius auratus*. III. Evidence for NH_4^+/Na^+ and HCO_3^-/Cl^- exchanges. *J. Gen. Physiol.* **47,** 1209–1227.

Maickel, R. P., Jondorf, J. R., and Brodie, B. B. (1958). Conjugation and excretion of foreign phenols by fish and amphibia. *Federation Proc.* **17,** 390.

Makarewicz, W. (1963). AMP-aminohydrolase and glutaminase activities in the kidneys and gills of some freshwater vertebrates. *Acta Biochim. Polon.* 10, 363–369.

Makarewicz, W., and Zydowo, M. (1962). Comparative studies on some ammonia-producing enzymes in the excretory organs of vertebrates. *Comp. Biochem. Physiol.* 6, 269–275.

Manderscheid, H. (1933). Über die Harnstoffbildung bei den Wirbeltieren. *Biochem. Z.* 263, 245–249.

Marshall, E. K., Jr., and Grafflin, A. L. (1928). The structure and function of the kidney of *Lophius piscatorius. Bull. Johns Hopkins Hosp.* 43, 205–236.

Munro, A. F. (1953). The ammonia and urea excretion of different species of amphibia during their development and metamorphosis. *Biochem. J.* 54, 29–36.

Needham, J. (1938). Nitrogen excretion in relation to environment. *Biol. Rev.* 13, 224–251.

Nicol, J. A. C. (1960). "The Biology of Marine Animals." Wiley (Interscience), New York.

Pequin, L. (1962). Les teneurs en azote ammoniacal du sang chez la Carpe (*Cyprinus carpio* L.). *Compt. Rend.* 255, 1795–1797.

Pequin, L. (1967). Dégradation et synthèse de la glutamine chez la Carpe (*Cyprinus carpio* L.). *Arch. Sci. Physiol.* 21, 193–203.

Pequin, L., and Serfaty, A. (1963). L'excretion ammoniacale chez un Téléostéen dulcicole: *Cyprinus carpio* L. *Comp. Biochem. Physiol.* 10, 315–324.

Pequin, L., and Serfaty, A. (1966). Acide glutamique et excretion azotee chez la Carpe commune, *Cyprinus carpio* L. *Comp. Biochem. Physiol.* 18, 141–149.

Pickford, G. E., and Grant, F. B. (1967). Serum osmolarity in the coelacanth, *Latimeria chalumnae:* Urea retention and ion regulation. *Science* 155, 568–570.

Price, K. S., Jr. (1967). Fluctuations in two osmoregulatory components, urea and sodium chloride, of the clearnose skate, *Raja eglanteria Bosc 1802.* II. Upon natural variation of the salinity of the external medium. *Comp. Biochem. Physiol.* 23, 77–82.

Price, K. S., Jr., and Creaser, E. P., Jr. (1967). Fluctuations in two osmoregulatory components, urea and sodium chloride, of the clearnose skate, *Raja eglanteria Bosc 1802.* I. Upon laboratory modification of external salinities. *Comp. Biochem. Physiol.* 23, 65–76.

Price, K. S., Jr., and Daiber, F. C. (1967). Osmotic environments during fetal development of dogfish, *Mustelus canis* (Mitchell) and *Squalus acanthias* Linnaeus, and some comparisons with skates and rays. *Physiol. Zool.* 40, 248–260.

Prosser, C. L., and Brown, F. A. (1961). "Comparative Animal Physiology." Saunders, Philadelphia, Pennsylvania.

Przylecki, S. J. (1925). La dégradation de l'acide urique chez les vertébrés. *Arch. Intern. Physiol.* 24, 238–264.

Ratner, S. (1954). Urea synthesis and metabolism of arginine and citrulline. *Advan. Enzymol.* 15, 319–387.

Read, L. J. (1967). Enzymes of the ornithine-urea cycle in the Chimaera *Hydrolagus colliei. Nature* 215, 1412–1413.

Read, L. J. (1968). Ornithine-urea cycle enzymes in early embryos of the dogfish *Squalus suckleyi* and the skate *Raja binoculata. Comp. Biochem. Physiol.* 24, 669–674.

Romer, A. S. (1966). "Vertebrate Paleontology." Univ. of Chicago Press, Chicago, Illinois.

Salvatore, F., Zappia, V., and Costa, C. (1965). Comparative biochemistry of deamination of L-amino acids in elasmobranch and teleost fish. *Comp. Biochem. Physiol.* **16**, 303–309.

Sawyer, W. H. (1966). Diuretic and natriuretic responses of lungfish (*Protopterus aethiopicus*) to arginine vasotocin. *Am. J. Physiol.* **210**, 191–197.

Schmidt-Nielsen, B., and Rabinowitz, L. (1964). Methylurea and acetamide: Active reabsorption by elasmobranch renal tubules. *Science* **146**, 1587–1588.

Schooler, J. M. (1964). Ph.D. thesis, University of Wisconsin, Madison, Wisconsin.

Schooler, J. M., Goldstein, L., Hartman, S. C., and Forster, R. P. (1966). Pathways of urea synthesis in the elasmobranch, *Squalus acanthias. Comp. Biochem. Physiol.* **18**, 271–281.

Smith, H. W. (1929). The excretion of ammonia and urea by the gills of fish. *J. Biol. Chem.* **81**, 727–742.

Smith, H. W. (1930). Metabolism of the lung-fish, *Protopterus aethiopicus. J. Biol. Chem.* **88**, 97–130.

Smith, H. W. (1931). The absorption and excretion of water and salts by the elasmobranch fishes. I. Fresh-water elasmobranchs. *Am. J. Physiol.* **98**, 279–295.

Smith, H. W. (1936). The retention and physiological role of urea in the elasmobranchii. *Biol. Rev.* **11**, 49–82.

Smith, H. W. (1953). "From Fish to Philosopher." Little, Brown, Boston, Massachusetts (rev. ed., Ciba Pharmaceutical Products Co., Summit, New Jersey, 1959).

Stransky, E. (1933). Untersuchungen über den Purinhaushalt bei Fischen und Amphibien. *Biochem. Z.* **266**, 287–300.

Taggart, J. V., Forster, R. P., Schachter, D., Kaplan, S. A., and Trayner, E. M. (1953). Studies on the tubular excretion of creatinine and p-aminohippurate in thin slices of dogfish kidney (*Squalus acanthias*). *Bull. Mt. Desert Is. Biol. Lab.* **4**, 54–60.

Thorson, T. B., Cowan, C. M., and Watson, D. E. (1967). *Potamotrygon* spp.: Elasmobranchs with low urea content. *Science* **158**, 375–377.

Vellas, F., and Serfaty, A. (1967). Sur l'excrétion uréique de la Carpe (*Cyprinus carpio* L). *Arch. Sci. Physiol.* **21**, 185-192.

Walker, A. M., and Hudson, C. L. (1937). The role of the tubule in the excretion of urea by the amphibian kidney. *Am. J. Physiol.* **118**, 153–166.

Williams, R. T. (1963). The biogenesis of conjugation and detoxication products. In "Biogenesis of Natural Compounds" (P. Bernfeld, ed.), p. 427. Pergamon Press, Oxford.

Wood, J. D. (1958). Nitrogen excretion in some marine teleosts. *Can. J. Biochem. Physiol.* **36**, 1237–1242.

Wu, C. (1963). Glutamine synthetase, I. A comparative study of its distribution in animals and its inhibition by DL-allo-δ-hydroxylysine. *Comp. Biochem. Physiol.* **8**, 335–351.

Zydowo, M. (1960). Dezaminazy Kwasu adenilowego i adenozyny w nerce szczura. *Acta Biochim. Polon.* **7**, 215–226.

6

INTERMEDIARY METABOLISM IN FISHES

P. W. HOCHACHKA

GLOSSARY

AMP	adenosine 5′-monophosphate	ITP	inosine triphosphate
ADP	adenosine diphosphate	αKGA	α-ketoglutarate
ATP	adenosine triphosphate	LDH	lactate dehydrogenase
CNS	central nervous system	MDH	malate dehydrogenase
FDP	fructose diphosphate	NAD	nicotinamide-adenine dinucleo-
F1P	fructose 1-phosphate		tide
F6P	fructose 6-phosphate	NADP	nicotinamide-adenine dinucleo-
FDPase	fructose diphosphatase		tide phosphate
G1P	glucose 1-phosphate	OXA	oxalacetate
G6P	glucose 6-phosphate	PEP	phosphoenolpyruvate
GDH	glutamate dehydrogenase	PGM	phosphoglucomutase
GDP	guanosine diphosphate	PFK	phosphofructokinase
GTP	guanosine triphosphate	PYK	pyruvate kinase
HMP	hexose monophosphate	SDH	succinate dehydrogenase
IDH	isocitrate dehydrogenase	TDH	triose phosphate dehydrogenase
IDP	inosine diphosphate	UDPG	uridine diphosphoglucose

I. INTRODUCTION

The metabolism of a cell or organism may be defined as the totality of chemical processes that it is capable of performing. It is evident that all aspects of the metabolism of even simple organisms will not be expressed at any instant in time. Those aspects which are expressed, and these need not be the same in different organisms, are determined by a delicate set of interrelated checks and balances which have both intrinsic (i.e., genetic) and extrinsic (i.e., physiological or environmental) components. These effects must be kept in mind since it will be convenient in this review to focus on particular reactions and their participating components, i.e., on highly dissected, rather than on integrated, biochemical systems.

Classically, metabolic pathways have been divided into two types—catabolic and anabolic. The former define essentially degradative processes in which large organic molecules are broken down to simple cellular constituents with the release of chemical free energy. The latter define synthetic processes that produce complex organic cellular components from simpler precursors and frequently involve reductive energy-requiring reactions. Research into the metabolism of a wide variety of organisms has shown that a remarkable diversity or individuality of metabolites handled is balanced by an equally remarkable order and simplicity in metabolic patterns. Of these patterns, none is more significant than the elucidation of a central area of metabolism that provides a link with catabolic and anabolic routes.

During initial phases of catabolism, large molecules are broken down to yield, apart from CO_2 and H_2O, a quite restricted group of small organic molecules, liberating about one-third of the available free energy in the process. For carbohydrates these product molecules are triose phosphates and/or pyruvate; for fats, they are acetyl CoA, propionyl CoA, and glycerol; for proteins, acetyl CoA, oxalacetate, α-ketoglutarate, fumarate, and succinate.

A major and unifying theme in biochemistry is the finding that the same set of reactions are involved in three crucial phases of metabolism: (1) the interconversion of various products of catabolism mentioned above; (2) their complete combustion to CO_2 and H_2O, which releases to the organism the remaining two-thirds of its available free energy supply; and (3) the supply of crucial intermediates for biosynthetic, anabolic processes. These central pathways are composed of relatively few reactions. In essence, they consist of these steps: triose phosphate $\rightleftharpoons$ pyruvate; pyruvate $\rightarrow$ acetyl CoA; oxalacetate $\rightleftharpoons$ aspartate; α-ketoglutarate $\rightleftharpoons$ glutamate and the citric acid cycle reactions designed to catalyze the complete combustion of acetyl CoA to CO_2 and H_2O.

An important distinction between anabolic and catabolic routes is that they rarely follow the same enzyme pathways in detail. This is readily apparent when the product of catabolism is not identical with the carbon source in anabolism, as is the case with many amino acids. In the case of fatty acids, catabolism leads to acetyl CoA as the end product, and biosynthesis commences with the same intermediate. However, the enzyme reactions involved are different and also are located in different cellular compartments. Even in the case of the biosynthesis of glucose, which proceeds in large part by a reversal of a number of glycolytic reactions, biosynthesis and degradation differ at the two most critical points in the enzyme sequence, that is, at either end. Thus, glucose is converted to G6P during catabolism in an ATP requiring reaction, but it is formed in anabolism by simple hydrolysis of the phosphate ester. Pyruvate is produced in catabolism from PEP by a transphosphorylation to ADP; it is utilized in gluconeogenesis in most organisms by two linked reactions, which first carboxylate pyruvate to oxalacetate and then transform the latter to PEP. Two important implications arise: (1) The two pathways cannot be maximally active simultaneously for this would lead to a kind of short circuit in metabolism, and (2) controls on the two pathways to some extent must be integrated.

This connection between anabolism and catabolism is a very close one and is absolutely requisite for maintenance of metabolic control. It is manifest on at least three levels:

(1) On the level of carbon sources: As discussed above, the products of catabolism, through interconversions in the central routes, become the substrates of anabolism.

(2) On the level of energy supply exchange and demand: Catabolism produces useful energy in the form of ATP (or compounds easily convertible to ATP), while anabolism requires energy and consumes ATP.

(3) On the level of reducing power: Catabolism, being essentially oxidative, consumes oxidizing power and produces reducing power. The reverse is true for anabolism. Much of the reducing power generated in catabolism and consumed in anabolism is provided by the pyridine (nicotinamide) nucleotides, NAD/NADH$^+$ and NADP/NADPH$^+$; catabolism produces both NADH$^+$ and NADPH$^+$, while anabolism requires and consumes the latter almost exclusively.

II. GLUCOSE METABOLISM

A. Studies of Multienzyme Systems

The major route of glucose catabolism in most systems including fish tissues is encompassed by the series of reactions, termed the Embden-Meyerhof-Parnas (EMP) pathway, that convert glucose to pyruvate. A second pathway of glucose catabolism, the HMP shunt, diverges from glycolysis at the level of G6P. Since any cell, in theory, can oxidize the NADPH produced by the first two steps of the shunt and can thus regenerate the NADP, the shunt provides for the complete oxidation of G6P to CO_2 and H_2O. This entails the catalytic intervention of five additional molecules of G6P and a cyclic set of reactions and intermediates. Significantly, the molecules of CO_2 produced have their origin in C-1 of glucose (Katz and Wood, 1960). One way of testing the occurrence of this metabolic sequence consists of comparing the rate of appearance of label in the respired CO_2 derived from C-1-labeled glucose with labeled CO_2 produced from other positions. The rate of G6-^{14}C oxidation frequently has been used as a standard for comparison with G1-^{14}C oxidation since in glycolysis, followed by oxidation of pyruvate, the C-1 and C-6 carbons appear in CO_2 at equal rates. Any deviation of the C-6/C-1 ratio from unity, then, indicates some participation of the nonglycolytic pathway.

The first report of this approach to ascertain the relative contributions of these two pathways of glucose metabolism in fishes is that of Hoskin (1959). In in vitro studies of electric eel, the radiochemical yield of $^{14}CO_2$ from G6-^{14}C compared to G1-^{14}C (C-6/C-1 ratio) shows pronounced tissue specificity. The ratios for electric tissues and muscle, both highly anaerobic tissues in these species (Williamson et al., 1967a), are about 0.1–0.3 and suggest an important HMP contribution to metabolism. Glucose oxidation rates by brain preparations are about 50-fold higher than that of the electric organ and the C-6/C-1 ratio approaches one, indicating that initial phases of glucose dissimilation in this tissue occur through the EMP pathway alone.

Brown (1960) arrived at similar interpretations of in vivo tests with carp. In this species, initial rates of glucose-^{14}C oxidation (after intraperitoneal injections of the labeled substrate) are about the same regardless of whether the C-1 or the C-6 position is labeled. In 2-hr experiments, C-6 incorporation into liver fatty acids and glycerol is greater than that of C-1, but after 48 hr C-6 and C-1 incorporation into fatty acids, glycerol, glutamate, and alanine are virtually identical. These results were taken to indicate that glucose dissimilation occurs primarily by the EMP pathway with only a minor contribution of the shunt. Low, but demonstrable, levels of G6P dehydrogenase and 6-phosphogluconate dehydrogenase support this interpretation (Brown, 1960). Using comparable techniques in tests with liver and muscle slices of the brook trout, Salvelinus fontinalis, Hochachka (1961) observed an exclusive EMP participation during anoxia; during aerobic metabolism, both the EMP and the HMP pathways were operative.

The above criteria for estimating the participation of these two pathways are open to serious question (Katz and Wood, 1960). A more quantitative and theoretically acceptable method for evaluating the degree of HMP participation is based upon the ^{14}C distribution in glucose following administration of such components as $^{14}CO_2$, acetate-1-^{14}C, or proprionate-1-^{14}C. The view is that ^{14}C is introduced into C-3,4 of glucose by effective reversal of the EMP pathway. Recycling through the shunt introduces label into C-1,2; equilibration of triose phosphates produced by both pathways introduces label into C-5,6. A preponderance of ^{14}C in C-3,4 indicates a predominant EMP participation, and the activity spread into the C-1,2 positions can be used as a measure of shunt activity. This approach was utilized in examining glucose metabolism in the brook trout (Hochachka and Hayes, 1962), but since we were looking for large changes, we did not separate the C-1,2 and the C-5,6 carbons in these studies. In liver slices with acetate-1-^{14}C only one-tenth to one-thirtieth as much ^{14}C appeared in the C-1,2 positions of glucose (glycogen) as appeared in the C-3,4 positions (Hochachka, 1962) indicating a relatively minor participation of the shunt. However, the relative activity of the shunt apparently increases during cold acclimation (Hochachka and Hayes, 1962) or following treatment with high (pharmocological) doses of thyroid hormones (Hochachka, 1962).

Recently, Liu (1968) studied the *in vivo* metabolism of C-1, C-2, C-3,4, and C-6 labeled glucose in the cichlid, *Cichlasoma bimaculatum*, using a modified radiorespirometric method. His data likewise are explicable on the basis of a major EMP contribution to glucose metabolism, with minor participation of the HMP shunt and the glucuronate pathway.

A detailed kinetic analysis of the changes of glycolytic intermediates has been made in the main electric organ of *Electrophorus* during periods of discharge and recovery (Williamson *et al.*, 1967a,b).

The electric organ of *Electrophorus* offers unique advantages for the study of metabolic regulation in an electrically excitable tissue, since it is available in bulk, and metabolic changes can be measured readily during and after the rapid performance of relatively large amounts of electrical work. The homogeneity of the tissue makes it ideally suitable for kinetic experiments, where emphasis is placed on the transient state rather than the steady state levels of metabolites. On the other hand, differences between organ sections from either the same or different eels resulting from unavoidable variations in the condition of the eel and preparation of the tissue make it unsuitable for evaluating data on a statistical basis.

Previous studies with the electrical organ of *Electrophorus electricus* have indicated that electrical discharge of the organ is associated with an entry of Na^+ ions into the electroplates and an exit of K^+ ions. The ionic gradients inside and outside the electroplates are restored by increased activity of the sodium pump. Expenditure of high energy phosphate (as ATP and creatine phosphate) is involved in this process, and the tissue levels of ADP and inorganic phosphate rise. The energy stores are replenished during recovery, principally by increased glycolytic activity, which is indicated by the substantial formation of lactate and the poor capability of the organ for oxidative phosphorylation.

Aubert *et al.* (1964) studied the changes of fluorescence emission in slices of the electric organ after periods of discharge and recovery and observed a cyclic increase and decrease of fluorescence. These results were interpreted as indicating an increased reduction of NAD+ by glyceraldehyde-3-P dehydrogenase, caused by an activation of glycolysis. Maitra *et al.* (1964) observed large increases in the tissue levels of FDP and triose phosphates at the end of a 30-sec discharge period, which supported this interpretation of the fluorescence changes, and showed that PFK activity was enhanced following the increased electrical activity. By analogy with studies on brain PFK (Passoneau and Lowry, 1962), control of PFK in eel was presumed due to relief of ATP inhibition by the elevated levels of ADP and inorganic phosphate, although other control substances were also thought to be involved. The kinetics of the DPN+ reduction, as measured fluorometrically, and the results of metabolite assays in the organ showed, in addition, a delayed onset of glycolytic activity. However, the cause of this induction period in the metabolic response was not ascertained.

Williamson *et al.* (1967a,b) identified the principal sites of control in glycolysis of the electric organ at the phosphorylase and PFK reactions. Mobilization of glycogen is brought about by an increase in phosphorylase b kinase activity. Increases of phosphorylase a and G1P follow this activation after a brief delay, which corresponds to that observed by Aubert *et al.* The chemical mediator of the phosphorylase kinase activation is probably Ca^{2+}, which is known to enhance the conversion of the inactive to active kinase in other systems, and which is thought to be released during membrane depolarization. Cyclic 3'5'-AMP, which can activate the inactive kinase, is not involved in this control in electric organ.

The increased rate of input of hexose phosphates to the glycolytic pathway is accommodated by concurrent activation of PFK, and the levels of G6P and F6P either remain approximately constant or diminish while FDP levels increase greatly. Kinetic measurements of adenine nucleotides, creatine phosphate, and inorganic phosphate show that a disappearance of creatine phosphate occurs within a few seconds of the onset of discharge. The levels of ATP are maintained approximately constant by the high activity of creatine phosphokinase, and marked increases of ADP and AMP are not observed until after the end of a 60-sec discharge period when the creatine phosphate reserves are depleted. Phosphofructokinase activity, as depicted by changes in the PFK mass action ratio [FDP] $\times$ [ADP]/[F6P] $\times$ [ATP], increases during discharge and shows a further large but transient increase during recovery, especially after discharge times greater than 40 sec. The initial activation of PFK is caused by the early rise of inorganic phosphate, but thereafter the change of the PFK mass action ratio closely follows the changes of ADP and AMP.

Phosphorylation of glucose by hexokinase provides an alternative to glycogenolysis for the increased supply of substrate to the glycolytic pathway. However, no change in the glucose level occurs either during or after electrical discharge. Furthermore, measurement of the activities of glycolytic enzymes in electric organ homogenates show that hexokinase is present in by far the lowest amount of all the glycolytic enzymes. Consequently, the rate of formation of G6P from glucose is considerably slower than its rate of formation from G1P via glycogenolysis.

Once the ATP/ADP ratio is restored to normal during recovery, PFK activity decreases, G6P levels increase, and phosphorylase a is converted to the b form by phosphorylase phosphatase. In addition, G6P activates the glycogen synthetase system and these events contribute to a fall in net glycogenolysis.

The action of temperature on the flow of carbon through the G6P branch point in glucose metabolism was investigated using liver slices of two Amazon fishes, *Lepidosiren* and *Symbranchus* (Hochachka, 1968). At elevated temperatures, the pathway to glycogen and the HMP shunt become increasingly effective in competing for the common substrate, G6P; carbon flow through glycolysis consequently is held constant or is slightly diminished when the temperature is raised. A high (40–45°C) thermal optimum for G6P dehydrogenase probably contributes to HMP activation at high temperatures. An increase in the affinity of glycogen synthetase for G6P at high temperatures with concomitant inhibition of phosphorylase can account for increased glycogenesis at elevated temperatures.

A detailed consideration of hormonal control of glucose metabolism in fishes is beyond the scope of this review. Some of the main control circuits in fishes are patently similar to their homologs in mammals. Thus, in both groups, blood glucose is decreased by insulin and is increased by glucocorticoids (Falkmer, 1961; Bentley and Follett, 1965). As in mammals, adrenaline raises blood glucose levels and depletes glycogen stores in muscle; however, in contrast to mammals, liver glycogen appears to be unaffected by adrenaline (Bentley and Follett, 1965). Glucagon does not increase the blood glucose concentration in cyclostomes (Bentley and Follett, 1965), but its action in other fishes has not been extensively examined. Studies of the mechanisms by which hormones mediate adjustments in intermediary metabolism in fishes are notably lacking (Bern, 1967), whereas such studies are predominant in the field of mammalian endocrinology (see, e.g., Weber *et al.*, 1965; Young *et al.*, 1964; Szego, 1965; Toft and Gorski, 1966).

B. Studies with Single Enzyme Systems

1. Conversion of Glucose to G6P

Glucose 6-phosphate is the mobilized form of glucose and it is the substrate common not only to the glycolytic pathway but to all other pathways of glucose metabolism. Hence, the enzyme (hexokinase) catalyzing the conversion of glucose to G6P is found widely distributed in nature. In fish tissues, MacLeod *et al.* (1963) first identified the enzyme in tissue extracts of muscle, brain, liver, and kidney. Activity in heart is some 60-fold higher than in muscle. Ronald *et al.* (1964), using a sensitive coupled assay for hexokinase based on release of $^{14}CO_2$ from 6-phosphogluconic acid formed enzymically from glucose, showed that pancreatic islet tissue of toadfish possesses rather high hexokinase activity (30–50 μmole glucose/g N_2/min). Glucose, mannose, and fructose all could serve as substrate. The Michaelis (K_m) constants of glucose and mannose are similar (about $1 \times 10^{-4} M$) while the K_m of fructose is about 10-fold higher.

In mammalian systems, the level of hexokinase can be regulated by hormones (Weber *et al.*, 1965); its activity can be and probably normally is regulated by such metabolites as G6P, ADP, and ATP (Uyeda and Racker, 1965) and its cellular distribution (between mitochondria and cytosol) can be affected directly by Mg^{2+} or indirectly by hormones (Fomina, 1963; Rose and Warms, 1967). Because of its importance in metabolic control, we have initiated a study of the regulatory and catalytic properties of this enzyme in the CNS of fishes (Somero and Hochachka, 1968). In rainbow trout brain, the ratio of mitochondrial (M) bound to soluble (Sol) hexokinase (Hk) varies between about 0.2 and 2.0 depending on isolation procedures. Both bound and unbound hexokinases are stable for several hours. The equilibrium, M-Hk $\rightleftharpoons$ Sol-Hk shows a negative temperature coefficient of about two, suggesting that *in vivo* the cellular localization of the enzyme may be affected by the outer temperature. We do not yet appreciate the significance of these effects, for the regulatory properties of bound and unbound hexokinase appear to be similar.

2. HYDROLYSIS OF G6P

The hydrolysis of G6P is a key reaction both in gluconeogenesis and in the conversion of liver glycogen to blood glucose. Glucose-6-phosphatase occurs in liver and kidney but not in muscle and is localized in the endoplasmic reticulum. In mammalian systems, G6Pase activity is both hormonally (Weber et al., 1965) and metabolite (Arion and Nordlie, 1964) regulated. The enzyme occurs at a potential control point in glucose metabolism since simultaneous function of hexokinase and G6Pase in a single cell should lead to ATP hydrolysis and thus to a short circuit in metabolism. Separate cellular localizations of the two enzymes suggest one means of circumventing this problem. Because of its importance in metabolic control, the enzyme should receive further and careful analysis. At this writing, the enzyme from fish sources has not been characterized, although Jannsens (1965) has identified the enzyme in lungfish extracts and has suggested (unfortunately on rather inadequate data) that G6Pase, and not glycogen phosphorylase, is rate limiting in liver glycogen metabolism.

3. CONVERSION OF GLYCOGEN TO G1P

The conversion of glycogen to G1P is catalyzed by phosphorylases (more correctly referred to as α1,4-glucan phosphorylases) which are widely distributed in nature. In mammalian tissues, the enzyme exists in 2 forms, a and b. Phosphorylase a is active in the absence of AMP, whereas the b form requires the presence of this nucleotide for activity. Other factors which determine, and at times control, the activity of phosphorylase include phosphorylation or dephosphorylation of the protein (i.e., conversion of b to a by phosphorylase kinase and a to b by phosphorylase phosphatase), presence of pyridoxal 5'-phosphate, and state of aggregation of the enzyme. In all cases, specific regions of the enzyme are involved, and regulation of enzymic activity can be visualized as resulting through a number of site–site interactions. Phosphorylation of the enzyme during b to a conversion does not proceed in an all-or-none reaction as originally postulated but in a stepwise fashion in which partially phosphorylated intermediates are produced. These "hybrids" show catalytic and physicochemical behavior different from either the b or a form of the enzyme: They possess high enzymic activity when measured at high G1P levels, but this activity is greatly depressed by low levels of G6P. Likewise, G1P increases the proportion of the enzyme occurring in tetrameric form, while G6P favors formation of the dimeric form (see Fischer *et al.*, 1966, for recent review of this area).

Phosphorylase was first identified in fishes (*Scorpaena* and *Scyllium*) by Cordier and Cordier (1957) and in sea bass by Ono *et al.* (1957). Nagayama (1961a, b, c) purified the enzyme of sea bass about 20-fold by ammonium sulfate fractionation. Both the a and b forms occur in fish muscle (tested by sensitivity to AMP activation). Phosphorylase from sea bass has an optimum temperature, under Nagayama's assay conditions, of 37°C, while the enzyme from trout muscle shows optimum activity at 25°C. Since the temperature optimum of phosphorylase is dependent on substrates and modulators (Lowry *et al.*, 1964; Helmreich and Cori, 1964), the significance of Nagayama's thermal studies are not entirely clear.

The only detailed study of phosphorylase of fish tissues is that of Yamamoto (1968). Yamamoto, using 70–90-fold purified phosphorylase b of trout muscle, examined kinetic properties at 37°C. The choice of temperature is unfortunate, for trout lethal temperatures are in the vicinity of 25°C (Fry and Gibson, 1953) and, as mentioned above, the kinetic and conformational properties of phosphorylase a are markedly temperature-sensitive (Helmreich and Cori, 1964; Lowry *et al.*, 1964) as is the b → a conversion (Danforth *et al.*, 1962). Under Yamamoto's assay conditions, the trout muscle phosphorylase b exhibits maximal activity near pH 6.8, the b form of the enzyme is activated by AMP, the Michaelis constants of G1P and AMP are 10–15 mM and 0.2–0.4 mM, respectively, and the enzyme is inhibited by high levels of glucose and ATP. Glucose 6-phosphate, inhibitory in other systems, does not affect the fish enzyme. *In vivo* the b form is converted to the a form during physical exercise (observed also by Nakano and Tomlinson, 1967). The b → a conversion is catalyzed *in vitro* with purified rabbit muscle phosphorylase b kinase in the presence of ATP and Mg^{2+}. The total activity per milligram protein is highest in heart and white muscle; substantial activities occur in red muscle and brain but the activities in liver, gills, and gut tissues are low.

Although the enzyme does not appear to function at a control site in lungfish or goldfish (at least under conditions of starvation) (Jannsens, 1965; Stimpson, 1965), it appears to be a rate-limiting step in electric eel, and it is strongly facilitated during discharge of the electric organ (Williamson *et al.*, 1967a). In the latter case, the enzyme is probably activated by Ca^{2+} accumulation during membrane depolarization. The level of the enzyme system in cardiac muscle shows pronounced species differences (Nayler and Merrillees, 1966). The enzyme should be further examined in fishes in order to clear up its role in metabolic control in these animals.

4. THE SYNTHESIS OF GLYCOGEN

Since the phosphorylase reaction is reversible, it was originally postulated that glycogen synthesis and degradation were catalyzed by the same enzyme system. However, Leloir and his colleagues discovered only in 1957 that animal tissues possess an enzyme, glycogen synthetase, which makes glycogen by a reaction analogous to that catalyzed by phosphorylase except that UDPG (formed by a specific pyrophosphorylase and serving as a glycosyl donor) rather than G1P is employed and the products are polysaccharide and UDP. The reaction requires a primer polysaccharide (glycogen itself) as glucose acceptor and provides only for chain lengthening; i.e., the reaction leads to a linear polymer. The branches in naturally occurring glycogen ($\alpha 1 \rightarrow 6$ bonds, which occur about every 8–12 residues in the $\alpha 1 \rightarrow 4$ chain) arise by action of a glycogen branching enzyme, which cleaves off small fragments from the $1 \rightarrow 4$ linkages and transfers them to polyglucose chains, but now at $1 \rightarrow 6$ linkages.

An interesting feature of glycogen synthetase is that it, like phosphorylase, can exist in two forms. An I form is active without G6P and can be converted upon phosphorylation to a D form, which is active only when G6P is bound to an allosteric site. Thus for glycogen synthetase, G6P appears to fulfill a role similar to that of AMP for phosphorylase, except that here G6P is effective with the fully phosphorylated D form and that G6P is a direct, stoichiometric precursor of the eventual product [see Larner (1966) and Mersmann and Segal (1967) for references in this area].

Except for the observation that in lungfish liver slices glucose-^{14}C incorporation into glycogen appears to be sensitive to G6P levels (Hochachka, 1968), nothing is known of the enzymes of glycogen synthesis in fishes.

5. INTERCONVERSION OF G1P AND G6P

The reaction is catalyzed by PGM, which has been obtained in pure form from a variety of sources. Martin and Tarr (1961) extracted and partially purified the enzyme from muscle of Pacific lingcod, *Ophiodon*. The dephospho enzyme is inactive unless glucose diphosphate is added. Ribose diphosphate and deoxyribose diphosphate can replace glucose diphosphate but are less effective. The enzyme requires Mg^{2+} and cysteine. The optimum pH is about 7.5, and the K_{eq} at pH 7.5 and 30°C (G6P/G1P) is 19.

The most complete study of this enzyme from fishes is that of Handler's group who have purified PGM from flounder and shark. These PGMs, like the rabbit muscle enzyme, exhibit a molecular weight of about 63,000 by ultracentrifugal and phosphate analysis, require glucose diphosphate and Mg^{2+} for activity, and are stimulated by pre-incubation with metal chelating agents and Mg^{2+}. The enzyme is phosphorylated by glucose diphosphate and dephosphorylated by G1P. Unlike the mammalian enzyme, flounder and shark PGMs are extremely labile below pH 5 and at pH 7.4 exhibit a significant level of glucose diphosphatase activity. The enzyme does not catalyze the interconversion of G1P and G6P by the consecutive operation of two reactions; in contrast, its kinetic behavior is of the shuttle type, designated by Cleland as "ping-pong." (Plots of $1/V$ vs $1/S_1$ at varying levels of S_2 give a family of parallel lines.) At high concentrations, G1P and GDP are each competitively inhibitory with respect to the other. For the flounder PGM, the K_m of G1P is 4×10^{-4} M and the K_m of GDP is 3×10^{-7} M, while the respective K_i constants are 1×10^{-3} M and 7×10^{-4} M. For the shark enzyme, the 2 K_m constants are 4×10^{-5} M and 4×10^{-7} M, while the K_i constants are 1.2×10^{-3} M and 5×10^{-4} M. The K_{eq} for the overall reaction is about 17.5 for both enzymes. Both shark and flounder PGMs are activated by short (5 min) heat (58°C) exposures or by low (1.66 M) urea levels (Hashimoto and Handler, 1966). Phosphoglucomutase from all sources examined appears to be a single polypeptide chain. In this connection, it is interesting that Tsuyuki and Roberts (1963) observed five isozymic forms of rainbow trout PGM. More recently, reports of mammalian PGM isozymes have also been made (Joshi et al., 1967).

6. THE INTERCONVERSION OF G6P AND F6P

This reaction, catalyzed by phosphoglucoisomerase was first studied in the fish, *Ophiodon*, by Martin and Tarr (1961). The enzyme occurs in isozymic forms, at least two isozymes being separable on DEAE-cellulose (Martin and Tarr, 1961) and several bands being separable on gel electrophoresis (Roberts and Tsuyuki, 1963). The enzyme of lingcod is relatively stable with an alkaline pH optimum and a K_{eq} at pH 8.2 and 30°C (G6P/F6P) of 0.6.

7. THE FORMATION OF FDP FROM F6P

The formation of FDP is catalyzed by a specific enzyme, PFK, and shares some of the properties of other kinases (requirements for ATP and Mg^{2+} and a negative ΔG). Since alternative pathways of hexose metabolism diverge from all other hexose phosphates thus far discussed, this reaction may be regarded as the first characteristic of the glycolytic sequence proper. This argument can be carried further, for the steps from triose phosphate to pyruvate are again common to several pathways and they, plus aldolase, must be utilized in carbohydrate synthesis. Thus, the PFK reaction is the only one truly unique to glycolysis and therefore constitutes an important control point, subject to strong metabolic regulation in fishes and other organisms (Atkinson, 1966; Williamson *et al.*, 1967a).

In fishes, MacLeod *et al.* (1963) first assayed the enzyme in crude extracts of trout muscle, heart, brain, liver, and kidney. Specific activity of brain PFK, under their assay conditions, was five times higher than that of muscle. Freed (1968) has observed that goldfish muscle PFK is inhibited by high levels of ATP. Cyclic AMP reverses the inhibition at 25°C but not at 5°C. Specific activity of goldfish muscle PFK is comparable to the activity of rabbit muscle. During discharge of the electric organ of *Electrophorus*, PFK is initially activated by P_i but further regulation is dependent upon the adenylates, activation by AMP, and inhibition by ATP (Williamson *et al.*, 1967a,b).

The enzyme is difficult to purify, and therefore detailed studies are only currently appearing in the literature. It is generally recognized as the major control point in glycolysis (Atkinson, 1966). For a fuller understanding of glycolytic control in fishes, further studies of PFK are essential.

8. The Hydrolysis of FDP

During gluconeogenesis, the physiologically irreversible PFK reaction is bypassed by the FDPase catalyzed hydrolysis of FDP to yield F6P and P_i. It is generally recognized that close regulation of this enzyme also is required since operation of both PFK and FDPase simultaneously would lead to ATP hydrolysis and a short circuit in energy metabolism. (In the case of hexokinase and G6Pase, a similar problem is partly relieved by a spatial separation of the two enzymes within the cell; however, both FDPase and PFK are apparently localized in the cytosol.) In mammalian systems, the level of FDPase is hormonally regulated (Weber *et al.*, 1965) and its activity is inhibited by high levels of AMP and FDP (see Atkinson, 1966; Horecker *et al.*, 1966). The observation of a remarkable temperature dependence of the FDPase–AMP interaction (Taketa and Pogell, 1965) led us to examine in detail the kinetic properties of the enzyme in a species which may routinely encounter fluctuating environmental temperatures. We have extracted trout liver FDPases (rainbow, brook, and lake trout) and partially purified the enzyme by ammonium sulfate fractionation (Behrisch and Hochachka, 1969). The FDP saturation curve is sigmoidal at all temperatures and the K_m of FDP

$(5 \times 10^{-5}\ M)$ is essentially independent of temperature. Fructose di-phosphate inhibits at concentrations over about $10^{-4}\ M$. The enzyme shows an absolute requirement for a divalent cation, which can be satis-fied by Mg^{2+} or Mn^{2+}. The Mg^{2+} saturation curve is also sigmoidal and the K_a (apparent) is about $5 \times 10^{-4}\ M$. These kinetics suggest that in addition to serving as substrate and cofactor, FDP and Mg^{2+} serve also as positive modulators of the enzyme. The Mn^{2+} saturation curve, on the other hand, is hyperbolic with a K_a about 100-fold lower than the K_a of Mg^{2+}; Ca^{2+} and Zn^{2+} inhibition is competitive with respect to both Mg^{2+} and Mn^{2+}. Trout liver FDPase shows a complex pH profile, with a general increase in activity at alkaline pH values. The K_a of Mg^{2+} falls and the V_{max} of the enzyme increases when the pH increases. Adenosine 5-monophosphate inhibition is marked but variable from preparation to preparation, possi-bly because of the presence of AMP-insensitive conformations. The K_i of AMP shows a complex temperature dependency (K_i at 15°C being mini-mal). It is evident that modulation of cellular Ca^{2+}, Mg^{2+}, Mn^{2+}, AMP, and FDP levels can modify FDPase activity in a remarkable manner. This is of particular relevance to fishes since AMP and FDP levels are un-doubtedly unstable (Williamson et al., 1967a), and cation levels are known to change during temperature acclimation (Hickman et al., 1964).

9. The Conversion of FDP to Two Triose Phosphates and the
   Interconversion of the Aldo- and Ketotriose Phosphates

These two reactions are catalyzed by aldolase and triose phosphate isomerase, respectively. The combined equilibrium for the formation of FDP from two molecules of glyceraldehyde 3-phosphate lies in favor of the hexose (FDP:GAP:dihydroxy acetone phosphate is 89:0.5:10.5). Therefore for cleavage to occur, provision must be made for the efficient removal of the product trioses. Aldolase prepared from carp muscle (Shibita, 1958) is less stable than the rabbit enzyme and the pH optimum is between pH 8.3 and 9.4 depending on the buffer system used. At low concentrations Mg^{2+} and Zn^{2+} have no effect on activity but inhibit at high concentrations. The K_m of FDP is about $5 \times 10^{-4}\ M$ at pH 7.5–8.6 and increases to $1 \times 10^{-3}\ M$ at pH 9.0. The specific activity in carp muscle is about one-sixth to one-half that of rabbit muscle.

Aldolase makes up about 3% of the total extractable protein of tuna muscle (Kwon and Olcott, 1965). This is about one-third the amount found in rabbit muscle, and the specific activity of the enzyme is about one-half that of comparable rabbit preparations. The tuna enzyme occurs as a single homogeneous protein electrophoretically and during ultra-centrifugal sedimentation. The optimum temperature for maximum activ-ity is about 42°C. The activation energy, calculated on the basis of activity between 20 and 42°C, is 4.2 kcal, a value about one-third that of rabbit muscle aldolase. The activity appears to be fairly constant between pH 7 and pH 9. The K_m of FDP is about 2×10^{-3} M, a value comparable to that for the carp enzyme.

In shark, perch, and trout, aldolase occurs in multimolecular form, each isozyme consisting of four subunits. In general, the profiles suggest that three kinds of aldolases, A, B, and C, along with appropriate hybrid heteropolymers, can occur in shark and perch and vertebrates in general (Lebherz and Rutter, 1968; Penhoet et al., 1967). In shark muscle and spleen only aldolase A was detected. The A–C hybrid set was detected in heart and brain. Aldolase A is present in liver and kidney. In addition, aldolase B is present in these latter tissues as reflected by the low FDP/F1P ratios of liver and kidney extracts. The liver nonaldolase A activities are not part of the A–C set since these activities are resolved from those of brain by co-electrophoresis. The liver and kidney profiles are more complex than those of the higher vertebrates, but they can be explained on a three-gene basis (aldolases A, B, and C and their hybrids).

In contrast, the aldolase patterns of trout and salmon cannot be explained with a three-gene model. Aldolase A is detected in muscle. Aldolases A and B are detected in liver. The FDP/F1P ratio of liver is low indicating that the anodic band in liver is aldolase B. A five-membered set is present in heart, the parental aldolases of which are aldolase A and what Rutter and co-workers have termed "aldolase C." A different five-membered set is detected in brain. The set is formed by the interaction of aldolase C and aldolase D. Coelectrophoresis of heart and brain extracts results in nine activity bands suggesting that aldolase C is present in both five-membered sets.

Aldolase C is the predominant activity of kidney and spleen although upon prolonged incubation of the strips, additional bands appear (probably A–C and/or C–D hybrids). Aldolase B has not been detected in kidney. However, this aldolase may be present in low concentration.

The aldolase profiles of perch and shark are more similar to those of other vertebrates than to those of trout and salmon. This observation suggests that the four parental aldolases of trout and salmon represent a diversion from the main line of aldolase evolution since the profiles of both primitive (shark) and more advanced (perch and higher vertebrates) species can be explained with a three-gene model (aldolases A, B, and C).

10. DEHYDROGENATION OF GLYCERALDEHYDE 3-PHOSPHATE

In most tissues, the major fate of the glyceraldehyde 3-phosphate is its conversion to 1,3-diphosphoglycerate. The reaction catalyzed by TDH requires NAD⁺ and P_i and leads to a product (1,3-diphosphoglycerate) that shows a very high negative free energy of hydrolysis. It is this fact that accounts for the driving force for the TDH and phosphoglycerate kinase reactions, and indeed for the whole conversion of FDP to phosphoglycerate.

In a detailed comparative approach, Allison and Kaplan (1964) purified TDH from a variety of sources, including halibut and sturgeon. The sedimentation constants of TDH from each of these organisms are similar ($s_{20,w}$ being about 7.44) and indicate a molecular weight of about 120,000. The turnover numbers of the fish TDHs, indeed of all TDHs examined, are identical when measured at pH 8.5. The two fish enzymes show minor differences in reactivity toward NAD analogs. In each case, starch gel electrophoresis at four different pH values indicates a single protein component, but differences do occur in migration distance and direction. Amino acid compositions of the sturgeon and halibut enzymes are quite different from each other, suggesting that during evolution important and unique substitutions have been incorporated into each of these TDHs. Quantitative complement fixation tests, yielding results parallel to the amino acid analyses, suggest that the structure of TDH has changed more markedly during the evolution of bony fishes than during the evolution of birds and mammals (see also Allison and Kaplan, 1964). Both of the latter groups regulate their body temperatures and cellular salt levels, which, within narrow limits, are constant from species to species, whereas fishes appear to have adapted metabolism to the conditions of the waters in which they live. A part of this process appears to involve adaptation of enzyme structure for function under unique physiological conditions.

Triose phosphate dehydrogenase from muscle of mirror carp occurs as a single protein component of about 120,000 MW and a pH optimum of about pH 8.5 (Ludovicy-Bungert, 1961). Cod muscle TDH (Cowey, 1967) also occurs as a single protein, with an activation energy of 14.5 kcal (compared to 19 kcal for the rabbit enzyme). The K_m of P_i at 5°C is 0.7 mM and increases with temperature to 4.5 mM at 35°C.

In perch and trout, TDH occurs in electrophoretically distinct multiple molecular forms (Lebherz and Rutter, 1967). Although the TDH patterns in each species are tissue specific and, particularly in trout, quite complex, the enzymes appear to occur in five-membered sets, and probably are generated by random combination of two kinds of subunits into tetramer holoenzymes. If this interpretation is correct, each kind of subunit must be coded separately. In both perch and trout, over 10 different TDHs can be generated, and their electrophoretic properties suggest that at least three, and more probably four, subunits are required for their biosynthesis.

In mammalian muscle, TDH occurs as a single protein species which appears to be distinct from a single liver isozyme (Papadopoulos and Velick, 1967). The kinetic properties of mammalian muscle TDH appear to be adapted to glycolytic function, while the liver TDH is better adapted for gluconeogenic function. It will be of considerable interest to determine whether the kinetic properties of the various TDH variants in fishes also are modulated according to the metabolic functions of the various tissues.

11. CONVERSION OF 1,3-DIPHOSPHOGLYCERATE TO 3-PHOSPHOGLYCERATE

The reaction catalyzed by phosphoglyceric kinase requires ADP and Mg^{2+}, the K_{eq} for the mammalian enzyme is 3000. The enzyme has not been examined from fish sources.

12. THE SYNTHESIS OF SERINE

It is generally believed that the synthesis of serine occurs from 3-phosphoglycerate through phosphorylated intermediates (such as phosphohydroxypyruvate and phosphoserine) or through nonphosphorylated intermediates (i.e., glycerate and hydroxypyruvate). In perch and eel liver, where a specific phosphoserine phosphatase is absent (Grillo and Coghe, 1966), glucose carbon is not incorporated into phosphoserine, although ^{14}C of glucose appears in serine (Grillo et al., 1966). Grillo et al. suggest that serine synthesis in these species occurs through the nonphosphorylated pathway, and this postulate is supported by the presence of unusually active hydroxypyruvate-alanine transaminase activity in the liver.

13. INTERCONVERSION OF 3-PHOSPHOGLYCERATE TO 2-PHOSPHOGLYCERATE

The reaction, catalyzed by phosphoglyceromutase, requires Mg^{2+} and 2,3-phosphoglycerate. The K_{eq} of the mammalian enzyme is 4.0. MacLeod et al. (1963) demonstrated the occurrence of this enzyme in fish tissues, but its properties have not been examined.

14. CONVERSION OF 2-PHOSPHOGLYCERATE TO PEP

The reaction, catalyzed by enolase, requires Mg^{2+} or Mn^{2+}. The K_{eq} is 1.4 for the mammalian enzyme. Tsuyuki and Wold (1964) showed that the enzyme occurs in at least three different forms in muscle tissue of eight species of the genus Salmo. Later, Cory and Wold (1965) isolated and crystallized the three isozymes from skeletal muscle of rainbow

368 P. W. HOCHACHKA

trout. The three isozymes sedimented together upon ultracentrifugation, indicating a uniform molecular weight. Enolase is known to consist of two subunit chains. A model assuming the occurrence of two kinds of subunits, A and B, which can randomly self-assemble into three compositionally distinct dimers (A_2, A_1B_1, and B_2) can account for the enolase isozymes in trout.

15. Conversion of PEP to Pyruvate

In fish and mammalian tissues none of the above three glycolytic enzymes is likely to be rate limiting to glycolysis (Williamson et al., 1967a); such, however, is not the case for PyK, which catalyzes the production of pyruvate from PEP by a transphosphorylation to ADP. Pyruvate kinase, which requires Mg^{2+} and K^+ and catalyzes a reaction favoring pyruvate and ATP formation (K_{eq} for the mammalian enzyme is about 2000), occurs at an important branch and control point in glycolysis and gluconeogenesis. In fishes (Somero and Hochachka, 1968b) PyK occurs in at least two isozymic forms, a M-PyK occurring in muscle and sensitive to feed-forward activation by FDP, and a L-PyK, which is found along with M-PyK, in fish liver and which is insensitive to FDP activation. Fructose diphosphate activation of trout M-PyK is independent of pH between 6.5 and 7.5 and independent of temperature between 7 and 25°C. The K_a of FDP is about 2×10^{-6} M. Although FDP affects both K_m and V_{max}, decreasing the K_m of PEP and increasing V_{max}, the effect on K_m is more dramatic. Thus, at low PEP levels, 16-fold activation of the enzyme can be achieved by FDP. As in the case of rabbit M-PyK, trout muscle PyK is subject to feedback inhibition by ATP. Adenosine triphosphate inhibition, presumably competitive with respect to both PEP and ADP, is independent of temperature between 7 and 25°C. Activation energy for the trout M-PyK is about 30 kcal. Arrhenius plots (log V_{max} vs $1/T$) for muscle Pyk of an Antarctic fish, Trematomus, are discontinuous. At temperatures within the environmental range, the activation energy of the Trematomus enzyme is about 10 kcal. The K_m values of PEP for the fish PyKs depend upon temperature in a complex manner, passing through a minimum at temperatures corresponding to those the organism encounters in nature. Thus, the K_m values of PEP for the trout PyK isozymes are minimal at about 15°C, while the K_m of PEP for the Trematomus enzyme is minimal between 0 and 5°C (Somero and Hochachka, 1968b).

These experiments suggest that in the case of PyK evolutionary adaptation in fishes has favored the production of enzymes with (1) maximum affinity for unique substrate at temperatures corresponding to environmental ranges, (2) low activation energy (high catalytic efficiency) to compensate for low temperatures, and (3) temperature-independent regulatory properties. Presumably, these kinds of mechanisms contribute toward temperature independence of this particular reaction.

16. THE SYNTHESIS OF PEP

The thermodynamic and kinetic block against the use of the PyK reaction in PEP formation (for carbohydrate synthesis) is bypassed by two reactions. In a CO_2 fixation, energy requiring reaction, pyruvate conversion to OXA is catalyzed by a mitochondrial-bound enzyme, pyruvate carboxylase. A second enzyme, PEP carboxykinase, which occurs in the cytosol, can catalyze, in the presence of GTP (or ITP), the production of PEP, CO_2, and GDP (or IDP). The first of these is known to occur in carp liver mitochondria (Gumbmann and Tappel, 1962a,b), and the second is presumed to occur since pyruvate stimulates glycogen synthesis in lungfish liver (Hochachka, 1968). These enzymes occur at major control sites in glucose production from pyruvate (Gevers and Krebs, 1966; Gevers, 1967); for this reason, a better understanding of their properties in fish systems should be useful and should lead to greater insight into control mechanisms in fishes.

17. THE REDUCTION OF PYRUVATE TO LACTATE

The reversible reduction of pyruvate to lactate is the terminal step that characterizes glycolysis in vertebrate muscle. The NADH-linked reaction is catalyzed by LDHs which are abundant and relatively easy to handle experimentally. Hence, a large amount of information is available. Lactate dehydrogenases from vertebrate sources all appear to be tetramers containing four independent catalytic sites. They can be dissociated into four subunits of 35,000 MW. In fishes, the number of LDH isozymes can vary from 1 to about 20 (Hochachka, 1966; Markert and Faulhaber, 1965; Goldberg, 1965). A common pattern of five isozymes, often observed, is generated by the random self-assembly of two kinds of subunits (usually termed M and H in mammalian systems) into five compositionally distinct LDH forms: M_4, M_3H, M_2H_2, M_1H_3, and H_4. The only complex LDH isozyme system which is adequately described in terms of subunit structure is that of the salmonids (Hochachka, 1966, 1967; Morrison and Wright, 1966). In skeletal muscle of these species at least five kinds of subunits are involved in a nonrandom assembly of a complex of 14 tetramer LDHs. A group of three (A, B, and C) assemble to form nine isozymes (A_4, A_3B_1, A_2B_2, A_1B_3, B_4, B_3C_1, B_2C_2, B_1C_3, C_4). A second set (D and E) assemble to form five isozymes migrating near the origin at neutral pH. Interactions between the two sets (A, B, C and the D, E subunits) have not been observed. However, in brain and heart, A–C and A–B interactions appear to be favored (Hochachka, 1966a), and *in vitro* dissociation–reassociation studies likewise indicate that the A–C and A–B heterotetramers self-assemble more readily than do the B–C forms (Hochachka, 1966, 1967). Fairly complete descriptions of LDHs are available for many other fish species, including goldfish (Hochachka, 1965, 1967), dogfish, lamprey, and halibut (Pesce *et al.*, 1967); tuna, an Antarctic fish, *Trematomus*, and lungfish (Hochachka and Somero, 1968); and hagfish (Arnheim *et al.*, 1967; Ohno *et al.*, 1967). Markert and Faulhaber (1965) have described LDH patterns in 30 species of fish, but their data do not allow unequivocal assignments of subunit formulas for the various isozymes. Similarly, Wilson *et al.* (1967) described the electrophoretic mobility and inactivation temperature of the major muscle LDH from some 26 species of fishes.

In a number of fish species, the catalytic properties of different LDHs have been compared. Thus, Pesce *et al.* (1967) observed that the K_m values of pyruvate for the muscle LDHs of dogfish, halibut, and tuna are similar (about 0.3–0.9 mM at 25°C and pH 7.5), and quite comparable to the value for mammalian M_4 LDHs. Different LDH isozymes appear to vary in sensitivity to activation by OXA and other Krebs cycle intermediates (Hochachka, 1967; Fritz, 1965), and in pH sensitivity (Hochachka, 1968a; Fritz, 1967). Kinetics of reactions catalyzed by single isozymes are Michaelis–Menten at physiological pH, although at alkaline pH the pyruvate saturation curve for some LDHs can become sigmoidal (Hochachka, 1965). The turnover numbers of the fish LDHs (Pesce *et al.*, 1967) are about two times larger than the analogous mammalian M_4 enzymes and nearly four times larger than the mammalian H_4 LDHs. None of the physicochemical characteristics examined (sedimentation constants, optical rotatory dispersion, tryptic peptide maps, amino acid compositions, and temperature of thermal inactivation) appears to correlate with the unusually high turnover number of the fish enzymes. Of all the LDHs examined, estimates of percentage helicity were lowest for tuna and halibut M_4 LDH; but the differences are small, probably within the error limits of the estimate.

In most mammalian species examined, H_4 is the most abundant of the five isozymes in heart, kidney, brain, and erythrocytes; whereas M_4 is predominant in liver, skeletal muscle, and leukocytes. A popular theory developed to account for this distribution maintains that M_4 is the major isozyme in anaerobically metabolizing tissues because it is less sensitive than the H_4 LDH to inhibition by high substrate levels [see Kaplan (1965) and Markert (1963, 1965) for literature in this area]. In aerobic tissues, such as the heart, a steady supply of energy is required and is maintained by the complete oxidation of pyruvate in the mitochondria; hence, inhibition of H_4 LDH by excess pyruvate would favor channeling the keto acid into the Krebs cycle. In skeletal muscle, the energy requirements usually arise suddenly and are supplied by rapid glycolysis and pyruvate formation. To maintain glycolysis the M_4 LDH is adapted to catalyze NADH oxidation even when pyruvate levels are very high. Thus, in these conditions lactate serves as a temporary H_2 storage reservoir until O_2 supplies become available.

Vessel (1965) and Vessel and Pool (1966) reexamined the above problem since the experimental basis for the anaerobic–aerobic theory are investigations of purified isozymes at 25°C with pyruvate and lactate concentrations much higher than any levels thus far shown to exist *in vivo*. They observed that lactate and pyruvate do not reach inhibitory levels in canine skeletal muscle stimulated anaerobically to exhaustion. Moreover, when assayed at the physiological temperature (39°C in the dog), differences in substrate inhibition are quite small. These observations are incompatible with the aerobic-anaerobic theory of LDH function.

In mammals a great similarity of isozyme pattern among homologous tissues of different species has tended to reinforce the conviction that the characteristic proportion of isozymes in a given cell is related to aerobic/anaerobic ratios. That conviction is weakened by the observation that in fishes, homologous tissues need not possess homologous LDH isozymes (Hochachka, 1965, 1966; Markert and Faulhaber, 1965). In salmonids, for example, muscle tissue generally contains many (14) isozymes; in tuna, muscle tissue characteristically possesses a single isozyme (Hochachka and Somero, 1968; Pesce *et al.*, 1967). Both groups of fishes are highly aerobic. In goldfish, which belong to a highly anaerobic group of fishes (Blažka, 1958), muscle likewise possesses multiple (9–12) LDHs and these are also found in heart and brain. Within three closely related species of trout (between which interspecies hybrids can be formed), the liver of each possesses a single predominant LDH isozyme, A_4 in rainbow trout, B_4 in lake trout, and C_4 in brook trout (Hochachka, 1966). Interspecies hybrids possess the appropriate LDH heteropolymers and appear to be as viable as the parental types. Moreover, the same LDHs occur in brain and heart tissues exclusively, and, along with the D–E series, also in skeletal muscle. Thus, it has been difficult to argue from tissue distribution to biological functions of these isozymes.

A kinetic analysis of the various salmonid LDHs, however, yielded some insight into one probable physiological function of isozymes in fishes. Whereas neither thermal optima based on maximal velocities nor activation energies correlate with environmental temperature, LDH affinity for substrate (measured by K_m) is maximal at these temperatures. The minimum K_m (maximum affinity) of pyruvate for trout muscle LDH activity, which does not change noticeably during temperature acclimation, is about 15–20°C. During cold acclimation A_4, B_4, and C_4 are increased and for these the minimum K_m occurs between 10 and 15°C. The LDHs induced during cold acclimation are isozymes with low K_m (A_4 and B_4) and with low thermal optima (A_4, B_4, and C_4). These results suggest that an important site of natural selection, during evolutionary adaptation to temperature, is enzyme affinity for substrate and predict that in poikilotherms enzyme affinity for substrate should be maximal at the environmental temperature. On tests of LDHs from lungfish, tuna, the Antarctic fish, *Trematomus*, and of PyK from trout and *Trematomus*, this prediction is realized (Hochachka and Somero, 1968; Somero and Hochachka, 1968b).

18. THE OXIDATION OF PYRUVATE TO ACETYL CoA

The predominant fate of pyruvate in most animal cells is its oxidation to acetyl CoA mediated by the multienzyme pyruvate dehydrogenase complexes associated with mitochondria. These complexes are multi-component aggregates, with different subunits exhibiting separate enzymic activities. Descriptions of these complexes have been made (see Koike *et al.*, 1963; Hayakawa *et al.*, 1964); although undoubtedly present in fish mitochondria (Gumbmann and Tappel, 1962a,b), this enzyme system from fish sources has not been examined.

III. LIPID METABOLISM

A. Fatty Acid Utilization

In 1959, Brown and Tappel described an enzyme system in carp liver mitochondria that oxidizes fatty acids in the presence of ATP, cytochrome c, and Mg^{2+}. The addition of a "sparker" intermediate of the citric acid cycle greatly enhanced the rate of fatty acid oxidation. The rate of degradation of saturated and unsaturated fatty acid of similar chain length are similar. Cysteine, Ca^{2+}, and supernatant derived from the tissue preparation all slightly stimulate fatty acid oxidation. Exogenously added CoA, FAD, GSH, Fe^{2+}, Mn^{2+}, and Zn^{2+} are without effect. Lipoic acid is inhibitory to fatty acid oxidation as are high levels of fatty acid.

Adenosine diphosphate or AMP can replace ATP in the system, presumably because of active adenylate kinase equilibration. Acetoacetate is produced in less than theoretical amounts from the oxidation of octanoate and linoleate because of a channeling of a portion of the fatty acid carbon into the citric acid cycle. These requirements for optimal fatty acid oxidation by enzymes of fish mitochondria are similar to the corresponding enzyme systems of mammalian liver, in which the presence of an enzymic mechanism for β-oxidation of fatty acids has been established. The major characteristics of β-oxidation (1) that the fatty acyl chain is shortened by two carbons at a time, (2) that it is the CoA derivatives and not the free fatty acids that are involved in these reactions, and (3) that the liberated two-carbon fragment is acetyl CoA, thus appear to be common to both fish and mammalian mitochondria.

Experiments with ^{14}C-labeled fatty acids yielded similar results in studies of red (lateral line) muscle mitochondria of salmonids (Bilinski and Jonas, 1964). These preparations, however, are not sensitive to cytochrome c but are activated by CoA. Jonas and Bilinski (1964) also examined the rates of fatty acid oxidation in various tissues. In general, liver, kidney, heart, and red muscle preparations tend to have highest fatty acid oxidizing activities, while preparations of brain and dorsal (white) muscle are relatively inactive.

Carnitine plays a major role in the transport of activated fatty acids into the mitochondria where β-oxidation of the fatty acids takes place. Activated fatty acids are probably carried through the mitochondrial membranes as acylcarnitines. The acylcarnitines are formed according to the reaction, acyl CoA + carnitine $\rightleftharpoons$ acylcarnitine + CoA, which is catalyzed by two different enzyme systems. Carnitine acetyltransferase is active in the transfer of short-chain fatty acids, whereas carnitine palmityltransferase catalyzes the transfer of long-chain fatty acids (Norum, 1964). Norum and Bremer (1966) examined the distribution of the latter enzyme in a variety of organisms and tissues. In cod, the specific activity of the enzyme is highest in liver, heart, and kidney and is least in muscle, spleen, gonad, and intestinal tissue. These activities seem to correlate with the tissue ability to oxidize fatty acids in fishes (Jonas and Bilinski, 1964) as well as in other animals (Norum and Bremer, 1966).

B. Lipogenesis

Each step in the oxidation of fatty acids is reversible. Consequently, the postulate was advanced that fatty acid biosynthesis might occur by a simple reversal of the oxidative pathway (Lynen, 1952). Indeed, several workers were able to demonstrate the incorporation of acetyl CoA into fatty acids in mitochondrial preparations supplemented with NADH, NADPH, and ATP (see Wakil, 1961). However, it now seems that at least two pathways for fatty acid biosynthesis exist in the mitochondrion; one catalyzes the elongation of preformed fatty acids and is similar to the reverse of fatty acid oxidation, while the other is similar to that occurring in the cytosol and catalyzes the *de novo* synthesis of fatty acids (Harlan and Wakil, 1963).

One of the first suggestions that fatty acid synthesis may occur via a pathway distinct from that of oxidation was the demonstration by Van Baalen and Gurin (1953) and Tietz and Popják (1955) that fatty acids could be synthesized by soluble extracts of avian liver and mammary gland. Cofactor requirements for fatty acid synthesis by soluble cell extracts determined by Wakil and his co-workers (1961) include an absolute dependence upon bicarbonate, ATP, Mn^{2+}, and NADPH. These cofactor requirements clearly distinguish the synthetic and oxidative pathways since bicarbonate has no effect on fatty acid oxidation and NAD, rather than NADP, is the oxidation–reduction coenzyme in the oxidative pathway. Fatty acid synthesizing systems have been investigated from *E. coli* (Alberts *et al.*, 1964), from yeast (Lynen, 1961), from pigeon liver (Wakil, 1961), and from a number of invertebrate forms (see Beames *et al.*, 1967). In fishes, except for the fact that acetate incorporation into lipid appears to be strongly activated by citrate (Hochachka, 1968) as in other systems (Kallen and Lowenstein, 1962), nothing is known of the enzyme systems involved in lipogenesis. This represents an unfortunate gap in our knowledge for changes in lipid metabolism are probably basic to temperature adaptation in fishes and other poikilotherms (Hochachka, 1967).

In this connection, it was previously shown that, in the case of trout, acetate and glucose carbon appear in lipid at rates which vary between tissues and depend upon acclimation temperature (Hochachka and Hayes, 1962; Hochachka, 1961, 1967). Predictably, G6-^{14}C incorporation exceeds that of G1-^{14}C. In *Electrophorus* preparations, the channeling of acetate carbon into lipid (vs the Krebs cycle) appears to be favored, at high temperatures and under conditions of high citrate, ATP, and OXA (Hochachka, 1968). The probable sites of action of these intermediates are the citrate cleavage enzyme and acetyl CoA carboxylase.

Knipprath and Mead (1968) found that the *in vivo* incorporation of ^{14}C acetate into fatty acids of the goldfish is much greater during low temperature acclimation, with a tendency toward greater incorporation into unsaturated fatty acids at low temperatures and toward preferential incorporation into certain components. Thus, ^{14}C incorporation into palmitic acid is increased 2-fold during cold acclimation, whereas incorporation into palmitoleic is about 12-fold higher. The total incorporation into stearic acid is about 16-fold higher in cold- than in warm-acclimated fish whereas incorporation into oleic acid, by far the major constituent of the fatty acid mixture, is about 4-fold higher. The changes occurring in lipogenesis presumably are related causally, or at least intimately, to adjustments in gluconeogenesis, glycolysis, the HMP shunt, and the Krebs cycle (Hochachka, 1967) and to adjustments in the lipid composition of fishes (Roots, 1968; Caldwell, 1967) which are known to occur during thermal acclimation.

Inasmuch as carbohydrate is a major source of acetyl CoA for lipogenesis, it is of interest to consider the overall process of fatty acid synthesis from glucose. The process of glycolysis and pyruvate decarboxylation results in formation of two moles of CO_2 and two moles of acetyl CoA for each mole of glucose. Thus four of every six carbons of glucose are available for fatty acid synthesis under ideal (energy-saturating) conditions. For palmitic acid synthesis, for example, the process may be represented stoichiometrically as

$$4C_6H_{12}O_6 + O_2 \rightarrow C_{16}H_{32}O_2 + 8CO_2 + 8H_2O$$

Glucose   Palmitic acid

The theoretical energetic efficiency of this process is relatively high. Although it cannot be stated precisely, it is interesting to note that 4 moles of glucose approximate 2744 kcal, whereas 1 mole of palmitate approximates 2400 kcal when these are completely oxidized. As mentioned above, incorporation of glucose carbon into lipid has been observed in fishes (Hochachka and Hayes, 1962). The process of lipogenesis from carbohydrate may be unusually important in intermediary metabolism of fishes, for Blažka (1958) has shown that lipid accumulates extensively as an end product of carbohydrate metabolism, particularly in sluggish, anaerobic-type fishes.

IV. THE CITRIC ACID CYCLE AND ASSOCIATED REACTIONS

A. Studies with Multienzyme Systems

Although it is often coined as the hub of metabolism of all cells, the citric acid cycle in fish tissues has received very little critical study. The most extensive work remains that of Gumbmann and Tappel (1962a,b), who first showed that the reactions of the citric acid cycle occur in fish mitochondria. Carp liver mitochondria oxidize pyruvate and alanine without a requirement for a sparker (malate, fumarate, etc.) intermediate. The carboxyl carbon of alanine appears as CO_2 at a rate over 50 times that of carbon-2. From the stoichiometry of alanine oxidation, the carbon entering the citric acid cycle appears to be completely oxidized, with the exception of that which accumulates as citrate. Although citrate is readily oxidized by carp liver mitochondria, citrate levels rise 2-fold with alanine as substrate and rise over 4-fold with alanine and OXA as substrates. Fluoroacetate increases citrate accumulation in a manner in accord with the formation of fluorocitrate, the active inhibitor of aconitase. All intermediates of the citric acid cycle become labeled upon incubation with alanine-2-^{14}C in a manner which indicates the sequential flow of carbon from pyruvate through each intermediate in the cycle. Malonate blocks carbon flow at the site of succinate oxidation. Similar studies with ^{14}C metabolites (acetate, pyruvate, glutamate, glyoxylate, and glucose) are consistent with the operation of the citric acid cycle in salmonid liver and muscle (Hochachka and Hayes, 1962; Hochachka, 1961; Dean, 1969), in tissues of Lepidosiren, Symbranchus, and Electrophorus (Hochachka, 1968), in salmon and cod testis (Mounib, 1967a,b; Mounib and Eisan, 1968), and in Cichlasoma bimaculatum (Liu, 1968). In in vivo studies of Cichlasoma, Liu compared the appearance of carbons 1, 2, 3, 4, and 5 of glutamate in respiratory [14]CO_2 using radiorespirometric techniques.

In this species, carbon-1 and -5 appear in respiratory CO_2 at essentially similar rates that are somewhat higher than that of carbon-2. These carbons (1, 2, and 5) all appear in respiratory CO_2 at rates substantially higher than do carbon-3 and -4. Liu's results are not incompatible with the operation of the citric acid cycle. However, if glutamate metabolism occurs via the citric acid cycle alone, carbon-2 and -5 should appear in CO_2 at identical rates; hence, the data suggest a preferential and selective decarboxylation of carbon-5 of glutamate.

Phosphorylation occurring with the oxidation of various intermediates of the citric acid cycle in fish systems have been examined by Gumbmann and Tappel (1962a) in carp liver mitochondria. Phosphorylation efficiencies showed general agreement with theoretical values. Thus, respiratory–chain-linked phosphorylation in the oxidation of NAD-dehydrogenase-dependent substrate yields values approaching 3, while in the case of succinate, a flavoprotein-dependent substrate, the P/O ratio is only 1.4 (compared to a theoretically expected value of 2). Dinitrophenol almost completely inhibits the uptake of P_i in carp liver mitochondria oxidizing αKGA, demonstrating that phosphorylation is coupled with electron transport. Oxidative activity and hence electron transport is dependent to some degree on the presence of an external phosphate acceptor. In the absence of a phosphate acceptor system (glucose plus hexokinase), O_2 uptake of carp liver mitochondria drops by about 20–40%, a value somewhat lower than that observed in comparable mammalian preparations. All of the vitamins and cofactors of the vitamin B complex occur in carp liver mitochondria at concentrations again somewhat lower than in mammalian tissues. Difference spectra of carp liver mitochondria show striking similarity to those of mammalian tissues. The α peaks of cytochromes c, c_1, and b are separable by low temperature difference spectra. The positions of absorbancy maxima show that carp cytochromes $a + a_3$, b, c, and c_1 are spectrally identical to those of rat liver and beef heart. (In contrast, marine invertebrates show some fundamental variations.) Ratios of $a_3 : a : c + c_1$ are equal to unity for both carp and mammalian mitochondria. Absolute amounts of these cytochromes in carp liver mitochondria are comparable to the levels in rat liver mitochondria. Similar results were obtained by Kanungo and Prosser (1959) in studies of goldfish liver mitochondria. In addition, a number of inhibitors (amytal, antimycin, azide, cyanide, and CO), specific to the electron transport system of intact, tightly coupled mitochondria, effectively block respiration of the goldfish liver mitochondrial preparations.

Investigations of changes at the subcellular level following thyroid hormone administration in mammals revealed a partial uncoupling of oxidation and phosphorylation in isolated mitochondrial preparations. It was proposed that this effect may be the basis for the calorigenic action of the hormone. This concept, however, has been criticized since in mammalian preparations the thyroxine levels used have been nonphysiological. Massey and Smith (1968) investigated this problem in trout liver mitochondria. Prolonged treatment of trout with low thyroxine levels (3.5 μM) decreases phosphorylative activity of liver mitochondria and increases the specific activity of oxidative enzyme systems. Thiourea tends to reverse these results. Upon addition of thyroxine to mitochondrial preparations *in vitro*, uncoupling occurs at thyroxine levels between 5×10^{-7} M and 5×10^{-8} M (comparable values for mammalian systems are about 5×10^{-5} M).

Richardson and Tappel (1962) compared the physical properties of liver mitochondria of catfish and bluegills to similar preparations from rat liver. Fish mitochondria appear to swell at a more rapid rate over a 0–33°C temperature range, whereas the rates of swelling are comparable at 30–40°C. Discharge of the energized state of mitochondrial membranes can be linked to any of the mitochondrial work performances, including swelling and translocation of monovalent ions, translocation of divalent metal ions plus P_i, and energized transhydrogenation (Harris *et al.*, 1968). The swelling rates of both fish and rat mitochondria vary with temperature according to the Arrhenius law. Apparent activation energies for swelling average about 26 and 13 kcal for rat and fish, respectively. Although fish mitochondria appear to be generally less stable during osmotic variation than do rat preparations, the above observations suggest that mitochondrial functions associated with swelling and contraction cycles may be more efficient in fish liver than in comparable mammalian preparations. To some extent, the unique swelling properties of fish mitochondria may be related to unique composition of mitochondrial membranes. In this connection, it is interesting that polyunsaturated fatty acids in fish mitochondria tend to be longer chained and more highly unsaturated (Richardson et al., 1962).

Caldwell (1967a,b) examined the effect of temperature acclimation on terminal electron transport in mitochondria from gills and other tissues of the goldfish. Cytochrome oxidase activity in brain, gill, and muscle homogenates and in mitochondrial preparations from cold-acclimated animals are consistently higher than in comparable preparations from warm-acclimated animals (observed also by Freed, 1965). Similarly, the specific activities of succinate cytochrome c reductase and of NADH cytochrome c reductase are higher in mitochondria from cold-adapted animals. These changes could result from changes in enzyme levels or changes in enzymic activity. The mitochondrial cytochromes are integral oxidation–reduction components of the electron transport chain and as such can be used to estimate the concentration of the electron transfer system in mitochondria. Cytochromes a $+$ a$_3$ are believed to be the major catalytic proteins of cytochrome oxidase and can be used to estimate the concentration of this enzyme. Cytochrome b and c$_1$ in turn are part of the coenzyme Q–cytochrome c reductase segment of the chain. This latter segment is common to both succinate and NADH cytochrome c reductase. Caldwell's estimates indicate that cytochrome concentrations in fish mitochondria do not change significantly with temperature acclimation and therefore cannot account for the observed changes in enzymic activities; hence, a temperature-controlled modulation of the respiratory enzymes appears likely. A major means by which respiratory control is maintained in tightly coupled mitochondria is through the levels of ADP, ATP, and P$_i$. Although Caldwell did not assay these metabolites, their involvement is considered unlikely since the extent of uncoupling by dinitrophenol is similar for mitochondrial preparations from both warm- and cold-acclimated fish. Caldwell observed a decrease in concentration of saturated and monounsaturated fatty acids and a general increase in the concentration of polyunsaturated fatty acids in mitochondria from cold-acclimated goldfish (see also Roots, 1968; Richardson et al., 1962). This observation raises the possibility that the activity of the electron transfer enzymes might be modulated by temperature-induced changes in lipid composition. However, the reactivation properties of gill mitochondrial lipids from cold- and warm-acclimated fish (tested on lipid-deficient beef heart cytochrome oxidase) are entirely similar. Equally important, the proportions of the various polyunsaturated fatty acids are relatively unimportant in reactivating the lipid-deficient cytochrome oxidase. Since Caldwell did not test the reactivation of goldfish lipid-deficient cytochrome oxidase, these results are not unequivocal, but certainly appear to be inconsistent with the postulate that changes in mitochondrial lipids during thermal acclimation influence electron transfer enzymes. The basic observation, then, of increased activity of the electron transfer chain in preparations from cold-acclimated fish remains unexplained. An obvious point of departure, thus far not considered, is an examination of the kinetics of changes in metabolites which are important in the modulation of this system.

B. Studies with Single Enzyme Systems

1. CONDENSING ENZYME

The production of citrate from acetyl CoA and OXA has been demonstrated in minnow muscle, carp muscle, and liver (Gumbmann and Tappel, 1962) and in muscle of tuna and swordfish (Yamada and Suzuki, 1950). Liver preparations show highest activity. The enzyme functions at an important branch and control point in metabolism (Atkinson, 1966); interest in metabolic control in fishes will undoubtedly lead to further characterization of this reaction.

2. ACONITASE

The conversion of citrate to cis-aconitate, catalyzed by aconitase, has been demonstrated in liver and muscle of minnow. Activity in liver is about six times greater on a wet weight basis than in muscle (Gumbmann and Tappel, 1962a).

3. ISOCITRATE DEHYDROGENASE

The activity of NADP-linked IDH in minnow liver is twice the muscle activity (Gumbmann and Tappel, 1962). The NADP-linked enzyme occurs in the supernatant. NAD-dependent IDH, which occurs in the mitochondria and is a key control site in channeling carbon between citric acid and fatty acid cycles (Atkinson, 1966; Hochachka, 1968), has not been carefully examined in fish systems.

4. α-Ketoglutaric Dehydrogenase Complexes

The oxidation of α-ketoglutaric acid is entirely analogous to that of pyruvic acid, leads to the production of succinyl CoA, and requires the concerted action of two enzymes. α-Ketoglutarate dehydrogenases from fishes have not been characterized.

5. Succinic Thiokinase

The predominant fate of succinyl CoA is hydrolysis of the thioester in the presence of GDP and P_i to yield succinate and GTP + CoA. Succinic thiokinase, catalyzing this reaction, has not been characterized from fish sources.

6. Dehydrogenation of Succinic Acid

The oxidation of succinic to fumaric is the only dehydrogenation in the citric acid cycle which is not NAD-linked. The enzyme, SDH, occurs as a ferri flavoprotein, is specific for the *trans* form of succinate and is competitively inhibited by malonate. Malonate can therefore be used to block the citric acid cycle at this point. Operationally, this feature often has been taken as evidence for the occurrence of SDH and even of the citric acid cycle. In muscle mitochondria of minnows, malonate inhibits succinate oxidation almost completely, whereas αKGA oxidation falls by only 20% (Gumbmann and Tappel, 1926b). Thus, under these conditions most of the O_2 uptake arises from the single step oxidation of αKGA; only about 20% can be attributed to the newly formed succinate. Succinate dehydrogenase activities in goldfish gill (Sexton and Russell, 1955) and minnow muscle (Gumbmann and Tappel, 1962a) are substantially lower than in mammalian muscle. The activation energy of the enzyme in muscle of the Antarctic fish, *Trematomus,* is unusually low (Somero et al., 1968).

7. Hydration of Fumaric to L-Malic Acid

The reversible hydration of fumaric acid to yield L-malic acid is catalyzed by fumarase. The enzyme is abundant in minnow muscle and liver (Gumbmann and Tappel, 1962a) at levels comparable to those in mammalian muscle. Fumarases from fish sources have not been characterized.

8. Regeneration of OXA by Malate Oxidation

Malate undergoes oxidation in the presence of MDH and NAD to yield OXA, NADH, and H^+. By this reaction the citric acid cycle is completed. The enzyme has been assayed in minnow muscle (Gumbmann and Tappel, 1962a) and in mitochondrial and supernatant fractions of various tissues of Amazon fishes (South American lungfish, *Electrophorus* and *Symbranchus*) (Hochachka, 1967a). Purified supernatant and mitochondrial MDHs of tuna heart tissue (Kitto and Lewis, 1967) are relatively small proteins (67,000 MW) which can be electrophoretically resolved at alkaline pH (the mitochondrial MDH migrates more rapidly toward the anode).

The two MDHs can be differentiated on kinetic grounds: mitochondrial MDH is more sensitive to inhibition by high substrate (OXA) levels and the K_m of malate appears to be about 2–3-fold higher. Both enzymes are equally sensitive to inhibition by high malate levels. The mitochondrial MDH is more thermolabile than the supernatant enzyme and shows different reactivities to coenzyme analogs. Large differences in primary structure of the two MDHs are evident since rabbit antiserum directed against the tuna mitochondria MDH fails to inhibit or cross-react with supernatant enzyme. The same antiserum cross-reacts strongly with mitochondria MDHs of species closely related to the tuna (mackerel and bass), less strongly with more distantly related fish (herring and trout), and only very weakly with MDHs of elasmobranchs.

9. GLUTAMATE DEHYDROGENASE AND AMMONIUM PRODUCTION

A predominant fraction (one-half to three-fourths) of the waste N_2 of teleost fishes is excreted as NH_4^+ at the gills. Although the gills of some species may show significant glutaminase and deaminating activities, it is generally believed that most of the NH_4^+ excreted at the gills originates from blood NH_4^+ which is in turn supplied by the liver. A major pathway of liberation of amino nitrogen involves the transamination of amino acids with αKGA to form glutamate, followed by the deamination of the glutamate to yield αKGA and NH_4^+ (see McBean et al., 1966, for literature in this area). The latter reaction, catalyzed by NAD-dependent GDHs has been examined in eel, pollock, sculpin, and two elasmobranchs, the skate and the dogfish (McBean et al., 1966; Corman and Kaplan, 1967; Corman et al., 1967). In the case of eel liver GDH the K_m of glutamate is $1.5 \times 10^{-2} M$ and the K_i of NH_4^+ is about $1 \times 10^{-2} M$. In the eel, GDH activity is highest in liver (nearly 20 times higher than in muscle) and kidney, suggesting that these tissues are major sources of blood NH_4^+. The K_m of glutamate and the K_i of NH_4^+ are high ($1.5 \times 10^{-2} M$ and $1.1 \times 10^{-2} M$, respectively) and at physiological concentrations of these two compounds, GDH activity can readily account for all of the NH_4^+ excreted by the eel. However, it is difficult to argue from potential GDH activity under these conditions to NH_4^+ excretion since there may be other important deaminating reactions (Salvatore *et al.*, 1965) and since GDH activity also is closely regulated by other compounds (Corman and Kaplan, 1967).

Corman *et al.* (1967) purified GDH from dogfish liver. The $s_{20,w}$ value for the enzyme is 13.0 (molecular weight of about 330,000) and, unlike comparable mammalian enzymes, this sedimentation value for dogfish GDH does not change in the presence of NADH or GTP or these two compounds together. Similar experimental conditions have been shown to lead to changes in polymerization state of mammalian GDHs. Michaelis constants for the reaction (NADPH$^+$ + NH$_4^+$ + αKGA $\rightleftarrows$ glutamate + NADP) are 4×10^{-4} M, 8×10^{-2} M, 4.5×10^{-3} M, 8×10^{-2} M, and 8×10^{-5} M for each of the above compounds, respectively. The dogfish GDH is activated by high levels of NAD and the kinetic evidence is consistent with both catalytic and regulatory NAD binding sites on the enzyme. High NADH levels inhibit the enzyme, and in this case also the evidence is consistent with an inhibitory NADH binding site and a catalytic NADH site. In analogy with other GDHs, GTP is a noncompetitive inhibitor with respect to αKGA and it appears that GTP must bind at a site other than the active site. As with other GDHs, ADP activates dogfish GDH, but the activation occurs only at low NADH levels; at high NADH levels, ADP activation is overcome by NADH inhibition of the enzyme. Added anions such as chloride tend to minimize these regulatory effects. In the case of NADH, anions do not alter the V_{max} but increase the apparent K_m of NADH and effectively prevent inhibition at high levels of the coenzyme. Added anions can entirely reverse both the GTP inhibition and the ADP activation. It is evident that a number of mechanisms are available for the close regulation of GDH in fishes and that the argument of McBean *et al.* concerning the role of GDH in NH$_4^+$ secretion is not on a firm foundation.

These considerations are of some relevance to the problem of NH$_4^+$ excretion in lungfish during estivation (Janssens and Cohen, 1968). Nitrogenous waste, excreted largely as NH$_4^+$ while the animal is in the aquatic phase, accumulates during estivation in the tissues as urea; after long periods of estivation, urea has been reported to reach an amount as high as 3% of the body weight. Since urea biosynthesis rate does not change noticeably during estivation, a major problem for the estivating lungfish is the maintenance of low tissue levels of NH$_4^+$. The possibility that the NH$_4^+$ is held in a "storage" form as glutamine is ruled out by the observation that glutamine synthetase is undetectable in lungfish liver. The level of NH$_4^+$ could also be controlled by a close regulation of GDH activity. Kinetically, lungfish liver GDH resembles that of other organisms in that it is activated by high NAD levels and by ADP and is inhibited by NADH and by quite low levels of GTP. However, the levels of these metabolites have not been measured in estivating lungfish. Hence, the implications of these observations to regulation of NH_4^+ levels remain untested (see chapter by Forster and Goldstein, this volume, for further discussion of these points).

REFERENCES

Alberts, A. W., Majerus, P. W., Talamo, B., and Vagelos, P. R. (1964). Acyl-carrier protein. II. Intermediary reactions of fatty acid synthesis. *Biochemistry* 3, 1563–1571.
Allison, W. S., and Kaplan, N. O. (1964). The comparative enzymology of triose-phosphate dehydrogenase. *J. Biol. Chem.* 239, 2140–2152.
Arion, W. J., and Nordlie, R. C. (1964). Liver microsomal glucose 6-phosphatase, inorganic pyrophosphatase, and pyrophosphate-glucose phosphotransferase. II. Kinetic studies. *J. Biol. Chem.* 239, 2752–2757.
Arnheim, N., Jr., Cocks, G. T., and Wilson, A. C. (1967). Molecular size of hagfish muscle lactate dehydrogenase. *Science* 157, 568–569.
Atkinson, D. E. (1966). Regulation of enzyme activity. *Ann. Rev. Biochem.* 35, 85–124.
Aubert, X., Chance, B., and Keynes, R. D. (1964). Optical studies of biochemical events in the electric organ of *Electrophorus*. *Proc. Royal Soc. (London) Ser. B*, 160, 211–245.
Beames, C. J., Jr., Harris, B. G., and Hopper, F. A., Jr. (1967). The synthesis of fatty acids from acetate by intact tissue and muscle extract of *Ascaris lumbricoides* Suum. *Comp. Biochem. Physiol.* 20, 509–521.
Behrisch, H. W., and Hochachka, P. W. (1969). Temperature and the regulation of enzyme activity in poikilotherms: Properties of rainbow trout fructose diphosphatase. *Biochem. J.* 111, 287–295.
Bentley, P. J., and Follett, B. K. (1965). The effects of hormones on the carbohydrate metabolism of the Lamprey *Lampreta fluviatilis*. *J. Endocrinol.* 31, 127–137.
Bern, H. A. (1967). Hormones and endocrine glands of fishes. *Science* 158, 455–462.
Bilinski, E., and Jonas, R. E. E. (1964). Utilization of lipids by fish. II. Fatty acid oxidation by a particulate fraction from lateral line muscle. *Can. J. Biochem.* 42, 345–352.
Blažka, P. (1958). The anaerobic metabolism of fish. *Physiol. Zool.* 31, 117–128.
Brown, W. D. (1960). Glucose metabolism in carp. *J. Cellular Comp. Physiol.* 55, 81–85.
Brown, W. D., and Tappel, A. L. (1959). Fatty acid oxidation by carp liver mitochondria. *Arch. Biochem. Biophys.* 85, 149–158.
Caldwell, R. S. (1967). Temperature acclimation in the goldfish (*Carassius auratus* L): Studies on terminal electron transport and the role of lipids. Ph.D. dissertation, Duke University, Durham, North Carolina.
Caldwell, R. S. (1967b). Effects of temperature acclimation on respiratory enzyme activity in goldfish. *Am. Zoologist* 7, 134.
Cordier, D., and Cordier, M. (1957). Phosphorolyse du glycogène du muscle strié chez les Poissons marins. *Compt. Rend. Soc. Biol.* 151, 1909–1911.

Corman, L., and Kaplan, N. O. (1967). Kinetic studies of dogfish liver glutamate dehydrogenase with Diphosphopyridine Nucleotide and the effect of added salts. J. Biol. Chem. 242, 2840–2846.

Corman, L., Prescott, L. M., and Kaplan, N. O. (1967). Purification and kinetic characteristics of dogfish liver glutamate dehydrogenase. J. Biol. Chem. 242, 1383–1390.

Cory, R. P., and Wold, F. (1965). Multiple forms of rainbow trout enolase. Federation Proc. 24, 594.

Cowey, C. B. (1967). Comparative studies on the activity of D-glyceraldehyde-3 phosphate dehydrogenase from cold and warm-blooded animals with reference to temperature. Comp. Biochem. Physiol. 23, 969–976.

Danforth, W. H., Helmrich, E., and Cori, C. F. (1962). The effect of contraction and of epinephrine on the phosphorylase activity of frog Sartorius muscle. Biochem. J. 48, 1191–1199.

Dean, J. M. (1966). Unpublished studies.

Dean, J. M. (1969). The metabolism of tissues of thermally acclimated trout (Salmo gairdneri). Comp. Biochem. Physiol. 29, 185–196.

Falkmer, S. (1961). Experimental diabetes research in fish. Acta Endocrinol. Suppl. 59–122.

Fischer, E. H., Hurd, S. S., Koh, P., and Teller, D. (1966). The activation and inactivation of muscle phosphorylase. Natl. Cancer Inst. Monograph 27, 47–59.

Fomina, M. P. (1963). Hexokinase activity in liver cell fractions from normal rabbits and rabbits with Alloxan Diabetes. Biokhimiya 28, 185–189.

Freed, J. (1968). Personal communication.

Freed, J. (1965). Changes in activity of cytochrome oxidase during adaptation of goldfish to different temperatures. Comp. Biochem. Physiol. 14, 651–659.

Fritz, P. J. (1965). Rabbit muscle lactate dehydrogenase 5: A regulatory enzyme. Science 150, 364–366.

Fritz, P. J. (1967). Rabbit lactate dehydrogenase isozymes: Effect of pH on activity. Science 156, 82–83.

Fry, F. E. J., and Gibson, M. B. (1953). Lethal temperature experiments with speckled trout × lake trout hybrids. J. Heredity 44, 56–57.

Gevers, W. (1967). The regulation of phosphoenolpyruvate synthesis in pigeon liver. Biochem. J. 103, 141–152.

Gevers, W., and Krebs, H. A. (1966). The effects of adenine nucleotides on carbohydrate metabolism in pigeon liver homogenates. Biochem. J. 98, 720–735.

Goldberg, E. (1965). Lactate dehydrogenases in trout: Evidence for a third subunit. Science 148, 391–392.

Grillo, M. A., and Coghe, M. (1966). Phosphoserine phosphatase of vertebrates. Comp. Biochem. Physiol. 17, 169–173.

Grillo, M. A., Fossa, T., and Coghe, M. (1966). Synthesis of serine in the liver of vertebrates. Comp. Biochem. Physiol. 19, 589–596.

Gumbmann, M., and Tappel, A. L. (1962a). The tricarboxylic acid cycle in fish. Arch. Biochem. Biophys. 98, 262–270.

Gumbmann, M., and Tappel, A. L. (1962b). Pyruvate and alanine metabolism in carp liver mitochondria. Arch. Biochem. Biophys. 98, 502–575.

Harlan, W. R., Jr., and Wakil, S. J. (1963). Synthesis of fatty acids in animal tissues. I. Incorporation of C^{14} Acetyl Coenzyme A into a variety of long chain fatty acids by subcellular particles J. Biol. Chem. 238, 3216–3223.

Harris, R. A., Penniston, J. T., Asai, J., and Green, D. E. (1968). The conformational basis of energy conservation in membrane systems. II. Correlation between conformational change and functional states. *Proc. Natl. Acad. Sci. U.S.* **59**, 830–837.

Hashimoto, T., and Handler, P. (1966). Phosphoglucomutase. III. Purification and properties of phosphoglucomutases from flounder and shark. *J. Biol. Chem.* **241**, 3940–3948.

Hayakawa, T., Muta, H., Hirashima, M., Ide, S., Okabe, K., and Koike, M. (1964). Isolation and properties of pyruvate and α-ketoglutarate dehydrogenation complexes from pig heart muscle. *Biochem. Biophys. Res. Commun.* **17**, 51–56.

Helmreich, E., and Cori, C. F. (1964). The effects of pH and temperature on the kinetics of the phosphorylase reaction. *Proc. Natl. Acad. Sci. U.S.* **52**, 647–654.

Hickman, C. P., Jr., McNabb, R. A., Nelson, J. S., Van Breeman, E. D., and Comfort, D. (1964). Effects of cold acclimation on electrolyte distribution in rainbow trout (*Salmo gairdneri*). *Can. J. Zool.* 577–597.

Hochachka, P. W. (1961). Glucose and acetate metabolism in fish. *Can. J. Biochem. Physiol.* **39**, 1937–1941.

Hochachka, P. W. (1962). Thyroidal effects on pathways for carbohydrate metabolism in a teleost. *Gen. Comp. Endocrinal.* **2**, 499–505.

Hochachka, P. W. (1965). Isoenzymes in metabolic adaptation of poikilotherm: Subunit relationships in lactic dehydrogenases of goldfish. *Arch. Biochem. Biophys.* **111**, 96–103.

Hochachka, P. W. (1966a). Unpublished data.

Hochachka, P. W. (1966b). Lactate dehydrogenases in poikilotherms: Definition of a complex isozyme system. *Comp. Biochem. Physiol.* **18**, 261–269.

Hochachka, P. W. (1967). Organization of metabolism during temperature compensation. *In* "Molecular Aspects of Temperature Adaptation," Publ. No. 84, pp. 177–203. Am. Assoc. Advance. Sci., Washington, D. C.

Hochachka, P. W. (1967a). Unpublished data.

Hochachka, P. W. (1968). Action of temperature on branch points in glucose and acetate metabolism. *Comp. Biochem. Physiol.* **25**, 107–118.

Hochachka, P. W. (1968a). Lactate dehydrogenase function in *Electrophorus* swimbladder and in the lungfish lung. *Comp. Biochem. Physiol.* **27**, 613–615.

Hochachka, P. W., and Hayes, F. R. (1962). The effect of temperature acclimation on pathways of glucose metabolism in the trout. *Can. J. Zool.* **20**, 261–270.

Hochachka, P. W., and Somero, G. N. (1968). The adaptation of enzymes to temperature. *Comp. Biochem. Physiol.* **27**, 659–668.

Horecker, B. L., Pontremoli, S., Rosen, O., and Rosen, S. (1966). Structure and function in fructose diphosphatase. *Federation Proc.* **25**, 1521–1528.

Hoskin, F. C. G. (1959). Intermediate metabolism of electric tissue in relation to function. III. Oxidation of substrates by tissues in *Electrophorus electricus* as compared to other vertebrates. *Arch. Biochem. Biophys.* **85**, 141–148.

Janssens, P. A. (1965). Phosphorylase and glucose-6-phosphatase in the African lungfish. *Comp. Biochem. Physiol.* **16**, 317–319.

Janssens, P. A., and Cohen, P. P. (1968). Nitrogen metabolism in the African lungfish. *Comp. Biochem. Physiol.* **24**, 879–886.

Jonas, R. E. E., and Bilinski, E. (1964). Utilization of lipids by fish. III. Fatty acid oxidation by various tissues from Sockeye salmon (*Onchorhynchus nerka*). *J. Fisheries Res. Board Can.* **21**, 653–656.

Joshi, J. G., Hooper, J., Kuwaki, T., Sakurada, T., Swanson, J. R., and Handler, P. (1967). Multiple forms of phosphoglucomutase. *Federation Proc.* **26**, 557.

Kallen, R. G., and Lowenstein, J. M. (1962). The stimulation of fatty acid synthesis by isocitrate and malonate. *Arch. Biochem. Biophys.* **96**, 188–190.

386   P. W. HOCHACHKA

Kanungo, M. S., and Prosser, C. L. (1959). Physiological and biochemical adaptation of goldfish to cold and warm temperatures. II. Oxygen consumption of liver homogenate; oxygen consumption and oxidative phosphorylation of liver mitochondria. *J. Cellular Comp. Physiol.* **54**, 265–274.

Kaplan, N. O. (1965). Evolution of dehydrogenases. In "Evolving Genes and Proteins" (V. Bryson and H. J. Vogel, eds.), pp. 243–277. Academic Press. New York.

Katz, J., and Wood, H. G. (1960). The use of glucose-C¹⁴ for the evaluation of the pathways of glucose metabolism. *J. Biol. Chem.* **235**, 2165–2177.

Kitto, G. B., and Lewis, R. G. (1967). Purification and properties of tuna supernatant and mitochondrial malate dehydrogenases. *Biochim. Biophys. Acta* **139**, 1–15.

Knipprath, W. G., and Mead, J. F. (1968). The effect of the environmental temperature on the fatty acid composition and on the *in vivo* incorporation of 1-¹⁴C-acetate in goldfish (*Carassius auratus* L.). *Lipids* **3**, 121–128.

Koike, M., Reed, L. J., and Carroll, W. R. (1963). α-Keto acid dehydrogenation complexes. IV. Resolution and reconstitution of the *Escherichia coli* pyruvate dehydrogenation complex. *J. Biol. Chem.* **238**, 30–39.

Kwon, T. W., and Olcott, H. S. (1965). Tuna muscle aldolase. I. Purification and properties. *Comp. Biochem. Physiol.* **15**, 7–16.

Larner, J. (1966). Hormonal and non-hormonal control of glycogen metabolism. *Trans. N.Y. Acad. Sci.* [2] **29**, 192–209.

Lebherz, H. G., and Rutter, W. J. (1968). Personal communication.

Lebherz, H. G., and Rutter, W. J. (1967). Glyceraldehyde-3-phosphate dehydrogenase variants in phyletically diverse organisms. *Science* **157**, 1198–1200.

Liu, D. (1968). Personal communication.

Lowry, O. H., Schulz, D. W., and Passonneau, J. V. (1964). Effects of adenylic acid on the kinetics of muscle phosporylase a. *J. Biol. Chem.* **239**, 1947–1953.

Ludovicy-Bungert, L. (1961). Isolation et propriétés de la D-3-phosphoglycéraldéhyde-déhydrogénase de carpe. *Arch. Intern. Physiol. Biochim.* **69**, 265–276.

Lynen, F. (1952). Acetyl coenzyme A and the fatty acid cycle. *Harvey Lectures* **48**, 210–244.

Lynen, F. (1961). Biosynthesis of saturated fatty acids. *Fed. Proc.* **20**, 941–951.

McBean, R. L., Neppel, M. J., and Goldstein, L. (1966). Glutamate dehydrogenase and ammonia production in the eel (*Anguilla rostrata*). *Comp. Biochem. Physiol.* **18**, 909–920.

MacLeod, R. A., Jonas, R. E. E., and Roberts, E. (1963). Glycolytic enzymes in the tissues of a Salmonid fish (*Salmo gairdnerii gairdnerii*). *Can. J. Biochem. Physiol.* **41**, 1971–1981.

Maitra, P. K., Ghosh, A., Schoener, B., and Chance, B. (1964). Transients in glycolytic metabolism following electrical activity in *Electrophorus*. *Biochim. Biophys. Acta* **88**, 112–119.

Markert, C. L. (1963). Epigenetic control of specific protein synthesis in differentiating cells. *Symp. Soc. Study Develop. Growth* **21**, 64–84.

Markert, C. L. (1965). Development genetics. *Harvey Lectures* **59**, 187–218.

Markert, C. L., and Faulhaber, I. (1965). Lactate dehydrogenase isozyme patterns of fish. *J. Exptl. Zool.* **159**, 319–332.

Martin, G.-B., and Tarr, H. L. A. (1961). Phosphoglucomutase, phosphoribomutase and phosphoglucoisomerase of lingcod muscle. *Can. J. Biochem. Physiol.* **39**, 297–308.

Massey, B. D., and Smith, C. L. (1968). The action of thyroxine on mitochondrial respiration and phosphorylation in the trout (*Salmo trutta fario* L.). *Comp. Biochem. Physiol.* **25**, 241–255.

Mersmann, H. J., and Segal, H. L. (1967). An on-off mechanism for liver glycogen synthetase activity. *Proc. Natl. Acad. Sci. U.S.* **58**, 1688.

Morrison, W. J., and Wright, J. E. (1966). Genetic analysis of three lactate dehydrogenase isozyme systems in trout: Evidence for linkage of genes coding subunits A and B. *J. Exptl. Zool.* **163**, 259–270.

Mounib, M. S. (1967a). Metabolism of pyruvate, acetate and glyoxylate by fish sperm. *Comp. Biochem. Physiol.* **20**, 987–992.

Mounib, M. S. (1967b). Metabolism of pyruvate in testes of fish and rabbits with particular reference to p-nitrophenol and 2,4-dinitrophenol. *Comp. Biochem. Physiol.* **22**, 539–548.

Mounib, M. S., and Eisan, J. S. (1968). Carbon dioxide fixation by spermatozoa of cod. *Comp. Biochem. Physiol.* **25**, 703–709.

Nagayama, F. (1961a). Enzymatic studies on the glycolysis in fish muscle. I. Activity of phosphorylase. *Bull. Japan. Soc. Sci. Fisheries* **23**, 260–264.

Nagayama, F. (1961b). Enzyme studies on the glycolysis in fish muscle. IV. Some specificities of muscle phosphorylase. *Bull. Japan. Soc. Sci. Fisheries* **27**, 1018–1021.

Nagayama, F. (1961c). Enzymatic studies on the glycolysis in fish muscle. V. Differences of phosphorylase activity due to the part and freshness of muscle. *Bull. Japan. Soc. Sci. Fisheries* **27**, 102–1025.

Nakano, T., and Tomlinson, N. (1967). Catecholamine and carbohydrate concentrations in rainbow trout (*Salmo gairdneri*) in relation to physical disturbance. *J. Fisheries Res. Board Can.* **24**, 1701–1715.

Nayler, W. G., and Merrillees, N. C. R. (1966). Species determined differences in cardiac phosphorylase activity. *Comp. Biochem. Physiol.* **18**, 931–936.

Norum, K. R. (1964). Palmityl-CoA: Carnitine palmityltransferase. Purification from calf liver mitochondria and some properties of the enzyme. *Biochim. Biophys. Acta* **89**, 95–108.

Norum, K. R., and Bremer, J. (1966). The distribution of Palmityl-CoA: Carnitine palmityltransferase in the animal kingdom. *Comp. Biochem. Physiol.* **19**, 483–487.

Ohno, S., Klein, J., Poole, J., Harris, C., Destree, A., and Morrison, M. (1967). Genetic control of lactate dehydrogenase formation in the hagfish (*Eptatretus stoutii*). *Science* **156**, 96–98.

Ono, T., and Nagayama, F. (1957). Enzymatic studies on the glycolysis in fish muscle. I. Activity of phosphorylase. *Bull. Japan. Soc. Sci. Fisheries*, **23**, 260–264.

Ono, T., and Nagayama, F. (1959). Studies on the fat metabolism of fish. 3. Relations between fat and phosphorus in rainbow trout. *J. Tokyo Univ. Fisheries* **45**, 153–162.

Papadopoulos, C. S., and Velick, S. F. (1967). An isozyme of glyceraldehyde-3-phosphate dehydrogenase in rabbit liver. *Federation Proc.* **26**, 557.

Passonneau, J. V., and Lowry, O. H. (1962). Phosphofructokinase and the Pasteur effect. *Biochem. Biophys. Res. Commun.* **7**, 10–15.

Penhoet, E., Kochman, M., Valentine, R., and Rutter, W. J. (1967). The subunit structure of mammalian fructose diphosphate aldolase. *Biochemistry* **6**, 2940–2949.

Pesce, A., Fondy, T. P. Stolzenbach, F., Castillo, F., and Kaplan, N. O. (1967). The comparative enzymology of lactic dehydrogenases. III. Properties of the H4 and M4 enzymes from a number of vertebrates. *J. Biol. Chem.* **242**, 2151–2167.

Richardson, T., and Tappel, A. L. (1962). Swelling of fish mitochondria *J. Cell Biol.* 13, 43–53.

Richardson, T., Tappel, A. L., Smith, L. M.,and Houle, C. R. (1962). Polyunsaturated fatty acids in mitochondria. *J. Lipid Res.* 3, 344–350.

Roberts, E., and Tsuyuki, H. (1963). Zone electrophoretic separation of five phosphoglucomutase activities from fish muscle. *Biochim. Biophys. Acta* 73, 673–676.

Ronald, A. E., Di Pietro, D. L., and Williams, A. K. (1964). The structure and metabolism of the pancreatic islets. *Proc. 3rd Intern. Symp., Stockholm, 1963* pp. 269–279. Pergamon Press, Oxford.

Roots, B. I. (1968). Phospholipids of goldfish (*Carassius auratus* L.) brain: The influence of environmental temperature. *Comp. Biochem. Physiol.* 25, 457–466.

Rose, I. A., and Warms, J. V. B. (1967). Mitochondrial hexokinase release, rebinding and location. *J. Biol Chem* 242, 1635–1645

Salvatore, F., Zappia, V., and Costa, C. (1965). Comparative biochemistry of deamination of L-amino acids in elasmobranch and teleost fish. *Comp. Biochem. Physiol.* 16, 303–309.

Sexton, A. W., and Russell, R. L. (1955). Succinic dehydrogenase activity in the goldfish gill. *Science* 121, 342–343.

Shibata, T. (1958). Enzymatic studies on the muscle of aquatic animals. 1. On the properties of carp muscle aldolase. *Bull. Fac. Fisheries, Hokkaido Univ.*, 9, 218–226.

Somero, G. N., and Hochachka, P. W. (1968a). Unpublished data.

Somero, G. N., and Hochachka, P. W. (1968b). The effect of temperature on the catalytic and regulatory properties of pyruvate kinases from the rainbow trout and the Antarctic fish *Trematomus bernacchii. Biochem. J.* 110, 395–400.

Somero, G. N., Giese, A. C., and Wohlaschlag, D. E. (1969). Cold adaptation of the Antarctic fish *Trematomus bernacchii. Comp. Biochem. Physiol.* 26, 223–233.

Stambaugh, R., and Post, D. (1966). Substrate and product inhibition of rabbit muscle lactic dehydrogenase heart (H4) and muscle (M4) isozymes. *J. Biol. Chem.* 24, 1462–1467.

Stimpson, J. H. (1965). Comparative aspects of the control of glycogen utilization in vertebrate liver. *Comp. Biochem. Physiol.* 15, 187–197.

Szego, C. M. (1965). Role of histamine in the mediation of hormone action. *Federation Proc.* 24, 1343–1352.

Taketa, K., and Pogell, B. M. (1965). Allosteric inhibition of rat liver fructose-1, 6-diphosphatase by adenosine-5' monophosphate. *J. Biol. Chem.* 240, 651–662.

Tietz, A., and Popják, G. (1955). Biosynthesis of fatty acids in cell-free preparations. 3. Coenzyme A dependent reactions in a soluble enzyme system of mammary gland. *Biochem. J.* 60, 155–165.

Toft, D., and Gorski, J. (1966). A receptor molecule for estrogens: Isolation from the rat uterus and preliminary characterization. *Proc. Natl. Acad. Sci. U.S.* 55, 1574–1581.

Tsuyuki, H., and Wold, F. (1964). Enolase: Multiple molecular forms in fish muscle. *Science* 146, 535–537.

Uyeda, K., and Racker, E. (1965). Regulatory mechanisms in carbohydrates metabolism. VIII. Hexokinase and phosphofructokinase. *J. Biol. Chem.* 240, 4682–4688.

Van Baalen, J., and Gurin, S. (1953). Cofactor requirements for lipogenesis. *J. Biol. Chem.* 205, 303–308.

Vesell, E. S. (1965). Lactate dehydrogenase isozymes: Substrate inhibition in various human tissues. *Science* 150, 1590–1592.

Vesell, E. S., and Pool, P. E. (1966). Lactate and pyruvate concentrations in exercised ischemic canine muscle. Relationship of tissue substrate level to lactate dehydrogenase isozyme pattern. *Proc. Natl. Acad. Sci. U.S.* 55, 756–762.

Wakil, S. J. (1961). Mechanism of fatty acid synthesis. *J. Lipid Res.* 2, 1–24.

Weber, G., Singhal, R. L., and Srivastava, S. K. (1965). Action of glucocorticoid as indicer and insulin as suppressor of biosynthesis of hepatic gluconeogenic enzymes. *Advan. Enzyme Regulation* 4, 43–75.

Williamson, J. R., Cheung, W. Y., Coles, H. S., and Herczeg, B. E. (1967a). Glycolytic control mechanisms. IV. Kinetics of glycolytic intermediate changes during electrical discharge and recovery in the main organ of *Electrophorus electricus*. *J. Biol. Chem.* 242, 5112–5118.

Williamson, J. R., Herczeg, B. E., Coles, H. S., and Cheung, W. Y. (1967b). Glycolytic control mechanisms. V. Kinetics of high energy phosphate intermediate changes during electrical discharge and recovery in the main organ of *Electrophorus electricus*. *J. Biol. Chem.* 242, 5119–5124.

Wilson, A. C., Kitto, G. B., and Kaplan, N. O. (1967). Enzymatic identification of fish products. *Science* 157, 82–83.

Yamada, K., and Suzuki, T. (1950). Citric acid formation in fish. *Bull. Japan. Soc. Sci., Fisheries* 15, 765–770.

Yamamoto, M. (1968). Fish muscle glycogen phosphorylase. *Can. J. Biochem.* 46, 423–432.

Young, J. W., Shrago, E., and Lardy, H. A. (1964). Metabolic control of enzymes involved in lipogenesis and gluconeogenesis. *Biochemistry* 3, 1687–1692.

7

NUTRITION, DIGESTION, AND ENERGY UTILIZATION

ARTHUR M. PHILLIPS, JR.

I. INTRODUCTION

A. Definition of Nutrition

Nutrition supplies the raw materials for the maintenance of life (metabolism). Some materials are used for the formation of body tissues (anabolism) and some for the production of energy (catabolism). Foods may be classified as energy and growth foods (carbohydrates, fats, and

391

proteins) and nonenergy foods (minerals, vitamins, water, and oxygen). The nonenergy foods, for the most part, serve in a supporting role to the energy foods; and although many of them are required in very small amounts (vitamins and trace minerals), they are essential for life.

The nutrition of fish has received attention for many years. Atkins in 1908 reported the death of trout fed an all-dry diet. Death was prevented by the addition of fresh meat. Although Atkins did not know it (for vitamins were unknown at the time), he had described a vitamin deficiency of trout that many years later was found to be caused by a folic acid deficiency (Phillips, 1963).

The majority of fish nutrition studies have been conducted upon trout and salmon and for the most part with fish held under fish-cultural conditions and fed artificial foods (hatchery diets). Any discussion of fish nutrition, therefore, through necessity must emphasize these species held under these conditions.

B. Nutritional Value of Foods

The nutritional value of a diet is measured by the presence of the necessary elements and catalysts (minerals and vitamins), an abundant supply of the auxiliary foods (water and oxygen), and a proper balance between the energy and growth foods (carbohydrates, fats, and proteins). A proper balance between the energy and growth foods assures an adequate supply of both energy and raw materials for optimum anabolic activity, which, in addition to growth, includes tissue repair, reproduction, and the formation of essential body products (hemoglobin, hormones, enzymes, etc.).

C. Digestion of Foods

The nutritional value of a diet is ultimately determined by the ability of the animal to digest and absorb it. Digestion depends upon both the physical state of the food and the kind and quantity of enzymes in the tract.

There are species differences in the kind and amount of enzymes present. Some animals are more efficient than others in digesting the different food groups. Since food is not useful until absorbed and made available for metabolism, the decision for the inclusion of a food in an animal diet for purposes other than bulk is partially determined by the ability of the animal to digest and absorb the food.

Although a number of studies have been made to establish the

enzymes present in the digestive tract of fish, there have been comparatively few studies upon the digestibility of specific diet ingredients by
fish. These studies are important because the physical condition of diet
ingredients varies and often determines digestibility. An ingredient altered
by excessive heat during manufacture may pass through the digestive
tract relatively intact, although the necessary enzymes may be present.

D. Measurement of Energy

Food energy is estimated by the heat produced during complete
combustion in a calorimeter. The large calorie or kilogram-calorie (defined as the amount of heat required to raise the temperature of 1 kg of
water 1°C and abbreviated kcal) is the unit of measurement used in
nutrition studies.

E. Types of Energy

There are two types of energy in food: (1) heat (ΔH) and (2) free
(ΔF). Only the latter is used for body metabolism; the former is useful
solely for the maintenance of body temperature. However, because all
food energy is eventually expressed as heat, the calorie confounds both
heat and free energy. Free energy is calculated since there is no method
for direct determination.

F. Estimation of Food Energy

Two methods are used to estimate the energy value of foods:
  (1) It is determined directly in a bomb calorimeter (Maynard
      and Loosli, 1962).
  (2) The average caloric values of the food groups are used to
      estimate the total energy content of the diet.
The latter method is usually used because the first is time consuming and
expensive and a bomb calorimeter is not available in most laboratories.

1. For Higher Animals

Through a series of determinations, it was early estimated that proteins contain 5.65 kcal/g, carbohydrates 4.15 kcal/g, and fats 9.40 kcal/g.
Atwater and Bryant (1899) considered the digestibility and availability
of the food groups in human nutrition, and estimated that the physiological values were 4 kcal/g of protein, 9 kcal/g of fat, and 4 kcal/g of carbo-

hydrate. However, Maynard and Loosli (1962) pointed out that these values do not apply to farm animals because the digestibility factors are too high. Although these values were used for all animal species for a number of years (and still are in many studies), recently they have been adjusted according to the digestive capability of the animal under study.

2. For Trout

Phillips and Brockway (1959) determined physiological caloric values for three species of trout (brook, brown, and rainbow), using 85% for fat digestibility, 40% for carbohydrate, and 90% for protein. Fat digestibility was based upon the digestibility of hard and soft fats by trout (McCay and Tunison, 1935) and the amounts of hard and soft fats in most hatchery diets. The digestibility used for carbohydrate is similar to that for raw starch (Phillips et al., 1948a) since most of the carbohydrate in the usual trout diets is raw starch. However, this value is too low for foods that contain significant amounts of the more readily digestible carbohydrates. The values shown in Table I (Phillips et al., 1948a) may be used when diets are fed that contain the more highly digested carbohydrates. Protein digestibility was based upon studies by Tunison et al. (1944) and Wood (1952). Neither of these workers demonstrated significant differences between the digestibility by trout of plant or animal proteins. However, Kitamikado et al. (1965a) reported that there were differences in the digestibilities of different protein sources by rainbow trout. They found fresh protein (fish and fish viscera) was digested 91–97%, dried protein (whitefish meal) 80%, and soybean meal and commercial fish meal 70%. It may be necessary to change the protein digestibility value used by Phillips and Brockway (1959) as additional studies are made.

Using the preceding digestibility levels, and adjusting the protein for its nitrogen content (nitrogen is unavailable for energy), Phillips and

Table I

Digestion of Carbohydrates by Brook Trout[a]

Carbohydrate	% digested
Glucose	99
Maltose	92
Sucrose	73
Lactose	60
Cooked starch	57
Raw starch	38

[a] From Phillips et al. (1948a).

Brockway (1959) determined the physiological values for trout to be 3.9 kcal/g of protein, 8.0 kcal/g of fat, and 1.6 kcal/g of carbohydrate. These values are not valid for other fish species unless it is known that the fish digest the food groups to the same extent as trout.

G. The Measurement of Energy Used by the Body

Armsby developed a large animal respiration calorimeter (described by Braman, 1933) to determine energy balances through the measurement of the heat evolved, the oxygen consumed, and the gases given off by the animal. These were precise experiments that provided most of our present basic knowledge of energy usage in animals.

Other studies have measured the differences between the oxygen consumed during rest and activity as an estimate of the energy required for the activity under study. Many of these studies have been done upon fish and they reached a high degree of refinement in the experiments of Brett (1964).

Energy requirements are often estimated by "slaughter experiments." The calories consumed by the animal and those deposited in the body are chemically determined. The difference between the calories deposited as fat and protein and those absorbed from the food provides a crude measure of the calories used for energy purposes.

II. ENERGY FOODS

A. Carbohydrates

1. UTILIZATION OF CARBOHYDRATES BY FISH

Costwise, carbohydrates are the cheapest source of food energy but they are not all equally well utilized by all animals. Phillips *et al.* (1948a) found differences in digestibility of the different types of carbohydrates by trout (Table I). They also reported that the level in the diet, under their experimental conditions, was limited to 12% digestible carbohydrate because additional amounts caused a deposition of excess liver glycogen. A severe mortality occurred.

Phillips *et al.* (1948a) concluded that, in terms of human beings, trout were normally diabetic since after feeding a sugar meal the blood glucose increased 110%; the increase followed a curve that was like that of diabetic human beings. The injection of insulin lowered the blood sugar of the carbohydrate fed trout in a manner similar to insulin injected diabetic human beings. [Gray (1928) had earlier demonstrated that injected insulin lowered the blood sugar of fish.] It was suggested that trout were physiologically unable to utilize high levels of dietary carbohydrates, a likelihood further indicated by the diffuse character of the pancreas and its few insulin producing islets of Langerhans (Hess, 1935) (see also chapter by Liley, Volume III).

Phillips et al. (1948a) concluded that carbohydrate was of limited usefulness in trout diets. Trout are not expected to encounter high levels of carbohydrate in their natural diet and apparently they are ill equipped to utilize it.

Phillips et al. (1959) showed that low levels of digestible raw starch in meat-dry meal hatchery diets caused an increase in trout blood sugar that remained at the higher levels during the period the diets were fed. The physiological effect of liver glycogen and blood glucose increases have not been assessed.

Buhler and Halver (1961) found that chinook salmon tolerated relatively high levels of dietary carbohydrate without the development of abnormal conditions ("high glycogen" livers). They suggested that the results reported in the earlier trout studies were caused by dietary imbalance rather than by an inability of trout to utilize carbohydrate.

Species differences may explain the differences between the utilization of carbohydrate by salmon and trout. In addition, the diets fed to salmon by Buhler and Halver (1961) were expressed on a dry weight basis. If the water of the diet is taken into account the salmon consumed the equivalent of between 5 and 24% of the diet as carbohydrate. Allowing for digestibility by the fish, these levels are below those causing difficulty in trout diets.

Schaeperclaus (1933) suggested carbohydrate as a source of energy for carp and reported digestibilities of from 30 to 92%.

2. SPARING ACTION OF CARBOHYDRATE ON PROTEIN

Recent studies by the Cortland workers (Phillips et al., 1966, 1967) showed that carbohydrates were utilized for energy by trout and thus spared protein for protein purposes in the body. However, they were unable to feed levels of maltose as low as 6% of the diet without producing an increase in both liver glycogen and liver size.

Although carbohydrates may have a sparing action on protein metabolism, Kitamikado et al. (1965b) found that large amounts of starch in the diet of rainbow trout decreased the digestion of protein and therefore decreased the amount available for metabolism.

Tiemeier *et al.* (1965) found a sparing action of carbohydrate on protein when fed to channel catfish.

In higher animals carbohydrates are used for energy, stored as glycogen as an immediate reserve of energy, or stored as fat as a future source of energy. In trout and salmon there is evidence that carbohydrates are used immediately for energy as demonstrated by their sparing action on protein (Phillips *et al.*, 1966, 1967; Buhler and Halver, 1961). There is also evidence that they are stored as an immediate reserve of energy as shown by increased liver and muscle glycogen (Phillips *et al.*, 1948a; Wendt, 1964). However, the reported studies that were conducted over relatively short periods have not shown increased body fat after feeding surplus carbohydrate.

3. Beneficial Effects from Glycogen Storage

Wendt (1964) expressed the opinion that increased liver and muscle glycogen, resulting from feeding carbohydrates in hatchery diets, benefited salmon by preventing delayed fish mortalities after stocking by serving as reserve food during the period of adjustment to the new environment. Black *et al.* (1966) found that starvation of trout for 84 hr before shipment significantly lowered liver glycogen. They suggested that starvation prior to transportation and liberation may place the fish at a disadvantage because of reduced energy reserves.

4. Carbohydrate-Digesting Enzymes

Sources of carbohydrate-digesting enzymes in fish are the pyloric ceca, pancreas, and intestinal mucosa.

Because the carbohydrates listed in Table I are absorbed by trout it is assumed that the digestive enzymes sucrase, maltase, lactase, and amylase are present in the tract. The disaccharide maltose was absorbed 92%, sucrose 73%, and lactose 60%. Apparently not all three disaccharide enzymes were present in the same amount, and there was considerably more maltase activity than lactase.

Phillips *et al.* (1948a) found a lactase activity in the pyloric ceca of brook trout although lactose is not normally included in their diet. There was no correlation between the lactase activity of the trout and the inclusion of dried skim milk in the diet.

Kenyon (1925) found a saccharase (invertase) activity in the ceca and intestine of bluegill, carp, and pickerel. Carp and bluegill possessed relatively large amounts but pickerel contained only a small amount. Kenyon noted that both bluegill and carp ingest a considerable amount of vegetable matter but pickerel ingest practically none. He also found a great abundance of maltase in the hepatopancreas of carp but no evidence of lactase.

McGeachin and Debnam (1960) demonstrated a relatively high amylase activity in the digestive tract of a number of freshwater fish. Kitamikado and Tachino (1961a) found an amylase activity in the digestive tract of rainbow trout that was less than that of carp but more than that of the eel, reflecting differences in the feeding habits between fish species. Kenyon (1925) found a similar relationship in that an abundance of the starch-digesting enzyme amylase was present in the intestinal mucosa of carp but almost none in pickerel. Fish (1962) stated that the nature of the relative activity of the digestive enzymes in fish correlates with the nature of the fish's normal diet. In the predominantly herbivorous *Tilapia* amylase activity was distributed throughout the gastrointestinal tract, but in carnivorous perch the pancreas was the only source of amylase activity. Ushiyama *et al.* (1966) found an amylase activity in the pyloric ceca of salmon that was 1/411 that of carp intestine, 1/29.5 of cod pyloric ceca, and 1/95 of flounder intestine. The amylase had an optimum activity at a pH of 8.5 and an optimum temperature of 20°C.

B. Proteins

1. ROLE OF PROTEIN IN FISH NUTRITION

Apparently both plant and animal proteins satisfy, at least in part, the protein requirement of most fish. Even trout and salmon, normally carnivorous under natural conditions, satisfactorily utilize plant products when held under artificial conditions.

The protein requirement changes with changes in the fish's life cycle. Small, fast growing fish need more protein than larger fish that grow at a slower rate. During the prespawning period, a generous supply of protein is required for the formation of viable reproductive products (eggs and sperm).

As water temperatures rise, the protein requirement increases because of accelerated fish growth. Falling water temperatures depress growth rates and less protein is required by the fish.

Protein may serve as a source of energy for fish, but approximately 16% is nitrogen that cannot be used for energy. Protein is not an efficient energy source. Protein will be used for energy if insufficient energy is available from other sources (fats and carbohydrates), if protein is fed in surplus of the needs, or if poor quality protein (either lacking in or having an inefficient ratio of essential amino acids) is fed.

Table II

The Amino Acid Requirement of Chinook and Sockeye Salmon and of Rainbow Trout

Amino acid	Rainbow trout[a]	Sockeye salmon[b]	Chinook salmon (tentative requirement in % of the diet)[c]
Arginine	X	X	2.5
Histidine	X	X	0.7
Isoleucine	X	X	1.0
Leucine	X	X	1.5
Lysine	X	X	2.1
Methionine	X	X	0.5
Phenylalanine	X	X	2.0
Threonine	X	X	0.8
Tryptophan	X	X	0.2
Valine	X	X	1.5

[a] From Shanks et al. (1962).
[b] From Halver and Shanks (1960).
[c] From Halver et al. (1957); Halver (1961).

2. ESSENTIAL AMINO ACIDS

Table II lists the essential amino acids for trout and salmon (Halver et al., 1957; Halver, 1961; Halver and Shanks, 1960; Shanks et al., 1962). The 10 amino acids required by higher animals are required by these fish and, for the most part, at approximately the same levels. These authors found that cystine has a sparing action on methionine and can replace part of the methionine in the diet of salmon, a relation in fish between the sulfur-containing amino acids that is similar to that in the higher animals.

3. PROTEIN QUALITY

The value of protein for growth is governed by its quality. The most efficient dietary protein provides the exact quantitative and qualitative amino acid requirements of the animal.

Block (1959) compared the amino acid contents of natural foods with those of hatchery and wild brook trout and a meat-dry meal diet fed to hatchery trout. Little difference was found in the amino acid content of natural and artificial trout foods, and no difference was found between the amino acid content of wild and hatchery brook trout. Ogino (1963) determined the amino acid content of natural foods. The analytical values derived by these workers provide a means for assessing the protein quality of hatchery diets by comparison with the amino acid content of natural foods.

During growth the body protein, except for glycine, of fingerling rainbow trout did not vary appreciably in amino acid content (Ogino and Suyama, 1957). These authors suggested that the amino acid content of the body offers a means for determining the amino acid requirement of rainbow trout.

Shcherbina et al. (1964) suggested that hatchery-reared carp and rainbow trout should be fed a combination of plant and animal materials because plant products contain 5–8 times less methionine and 2–3 times less lysine (both essential amino acids) than fish tissues. A high percentage of plant products might result in an improper amino acid balance of the diet.

4. PROTEIN EFFICIENCY

Invertebrate fish food averages approximately 11.5% protein. Under experimental conditions brook trout fed natural food required 143 g of food protein for each pound (315 g/kg) of fish produced (Phillips and Brockway, 1959). This compared with approximately 300 g of protein per pound (660 g/kg) for trout fed a meat-dry meal type of hatchery diet. This difference in protein efficiency may be indirect evidence that the protein quality of the hatchery diet was not high and much of the diet protein was utilized for energy or deposited as fat. Recent experiments with ingredient manipulation have resulted in protein conversion efficiencies that approach those of natural foods (Phillips et al., 1964b). Phillips et al. (1966) improved the protein efficiency of trout diets by the addition of supplemental amino acids.

About 70% of the calories of most artificial trout diets and natural foods is protein. This percentage has been lowered to approximately 48%, without loss of fish growth, by the substitution of fat (corn oil) and/or carbohydrate (maltose) for some of the dietary protein (Phillips et al., 1966).

Fowler et al. (1964) found that a protein-to-calorie ratio of 1:1 was optimum for salmon fed a 20% protein diet. The ratio increased to 1:1.35 as the fish grew, indicating a higher energy requirement for the larger salmon.

A successful trout diet containing 18% protein in which 48% of the total calories is protein has a protein-to-calorie ratio of 1:1.4, similar to that found optimal for the larger salmon.

Gerking (1952) found a lessening of protein utilization by long-eared sunfish as the fish grew larger although the same level of protein was absorbed by the small and large sunfish. This was thought to be a result of aging.

5. Protein-Digesting Enzymes

The stomach, intestinal mucosa, pancreas, and the pyloric ceca are sources of protein-digesting enzymes in fish.

Kenyon (1925) reported a peptic activity of the stomach tissues of carp, bluegill, crappie, pickerel, and white bass. He found tryptic activity from extracts of the pancreas and heptopancreas of carp, from the intestine of the pickerel, and from the ceca of the crappie. Protease activity was found in the intestinal mucosa of carp and pickerel. Togasawa *et al.* (1959) prepared a purified protease from the pyloric ceca of bonita and Togasawa and Katsumata (1961) a purified protease from the pyloric ceca of tuna.

Some differences between fish species have been reported in their protein-digesting enzymes. Kitamikado and Tachino (1961b) found a protease activity in the tract of rainbow trout that was higher than that of carp, reflecting, perhaps, the greater intake of protein foods by rainbow trout than by carp. However, Chesley (1934) found that in general the protein-digesting enzymes present paralleled in quantity the general activity of the fishes studied. Morishita *et al.* (1966) found no marked difference in protease activity of the digestive tract of five species of fish. However, enzymes from *Salmo* were more active at lower temperatures than were the enzymes from any other species studied.

Johnston (1937) found that the proteolytic enzymes of the pyloric ceca of cod exerted their greatest effect at about the same pH as trypsin. Croston (1961) found two tryptic activities in extracts of salmon ceca that were similar to those of mammalian trypsin and chymotrypsin. The optimum pH was 9.0 and the optimum temperature 49°C.

Kitamikado and Tachino (1961a) found the optimum activity of stomach protease in rainbow trout was between 35° and 40°C and that of the ceca and intestine was 45°C. Both proteases lost activity at low water temperature. The activity at all temperatures, however, was more powerful than that of carp and eel.

The optimum temperature for the proteases of these cold-blooded animals is well above any possible physiological temperature for their survival. These observations support similar data that have been reported for a number of years. The expected differences in the optimum activity for the digestive enzymes of poikilothermal and homoiothermal animals does not exist.

Babkin and Bowie (1928) found no pepsin in the alkaline environment of the intestine of *Fundulus,* which does not possess a stomach. Every phase of protein digestion takes place in an alkaline medium in these fish. Kitamikado and Tachino (1961a) found an acid protease in the stomach of rainbow trout and an alkaline protease in the intestine. Slightly alkaline media are required for trypsin activity, whereas peptic activity occurs in acid media. This does not differ from the higher animals. Fish without stomachs would not be expected to exhibit peptic activity but, because of the alkaline environment of the intestine, would be expected to exhibit tryptic activity.

Norris and Elam (1940) found differences between the pepsin of king salmon and that of higher animals. Crystallized pepsin from the gastric mucosa of king salmon was effective at lower temperatures and over a wider temperature range than the pepsin of higher animals. Differences were found in the crystalline structure and chemical composition between the two sources of pepsin. In later studies, Norris and Mathies (1953) prepared a crystalline pepsin from tuna that showed higher activity than any previously reported pepsin extract.

Seasonal variation in proteolytic enzymic activity has been reported. Kashiwada (1952) found the proteolytic activity in tuna rose in the spring, fell in the summer, and rose again in the fall. The period of abundant commercial fish catch coincided with periods of strong enzyme activity. Chepik (1966) found an increase in protein digestion in carp in the spring and summer and a decrease by 47–66% in the winter. These observations can be associated with periods of abundant food supply and/or increased fish activity.

C. Fats

1. Site of Fat Absorption

According to Greene (1913), although fats are absorbed through the epithelium of all portions of the alimentary tract of the king salmon, the primary function of the numerous pyloric ceca is fat absorption.

2. Effect of Melting Point on Fat Digestion

McCay and Tunison (1935) found that as in the higher animal the digestion of fat by trout was dependent upon the melting point (Table III). There was no difference between the digestibility of low melting point oils from plants or fish.

3. Sparing Action of Fats on Protein

Fats fed at appropriate levels in trout diets serve as a source of energy, sparing dietary protein for other protein purposes in the body (Phillips et al., 1964a, 1965, 1966). In experimental diets the protein level was

Table III
Effect of Melting Point on the Digestibility of Fat by Brook Trout[a]

Fat	Melting point	% of the diet	% digested
Cottonseed oil	Liquid	25.0	90.0
Salmon oil	Liquid	25.0	93.5
Hydrogenated cottonseed oil	43°C	25.0	78.0
Cottonseed oil	Liquid	7.0	87.0
Salmon oil	Liquid	7.0	84.0
Hydrogenated cottonseed oil	43°C	7.0	56.0

[a] From McCay and Tunison (1935).

reduced to 18% without loss of fish growth, provided the calories removed with the protein were replaced by fat (corn oil). It was not possible in these studies to add supplemental fat to the diet without causing a subsequent increase in body fat. These results may be caused by poor quality protein, a surplus of total diet calories, or simply a reaction to supplemental dietary fat.

Combs et al. (1962) and Fowler et al. (1966) demonstrated a sparing action of peanut oil for protein in the diet of chinook salmon.

4. Essential Fats

Nicolaides and Woodall (1962) showed that fat is essential for chinook salmon. Salmon fed a fat-free diet from hatching showed a marked depigmentation. The fat trilinolein prevented depigmentation from occurring. Trilinolein or linolenic acid, or both, when substituted isocalorically for sucrose in a fat-free diet, resulted in a positive growth response of the salmon. Lee et al. (1967) found a favorable growth response and a mortality reduction of rainbow trout fed unsaturated fatty acids with the $\omega 3$ configuration. As in the higher animals, some of the fatty acids are essential for the well being of trout and salmon.

5. Harmful Effect of Dietary Fats

The superimposing of supplemental fat in high protein diets (Phillips et al., 1951) resulted in a large increase in body fat and, in some experiments, death of the fish. Fatty infiltrated livers were reported by Davis (1953) following the feeding of excess fat calories.

Ono et al. (1960) found that peroxide compounds associated with oxidized dietary fats caused a lipid degeneration of trout liver. Dietary levels as high as 15% fat were not harmful if the food fat was protected from oxidation.

The use of fat as an energy source by fish is not without limitation, but in balanced diets, at appropriate levels, supplemental fats are desirable and may increase fish growth, provide essential vitamins, and spare the protein of the diet for protein purposes.

6. THE FAT CONTENT OF HATCHERY AND WILD FISH

Phillips et al. (1957a) found that hatchery brook trout contained a higher level of body fat (5.5%) than wild brook trout (3.4%). The hatchery food contained a higher level of fat (3.2%) than did the wild food (2.2%) and, perhaps more importantly, a higher level of total calories, 722 and 336 kcal/lb (1588 and 739 kcal/kg), respectively (Phillips et al., 1954).

The depot fat in the fish's body is generally formed from a surplus of calories in the diet. The metabolic rate of the body determines the caloric requirement for energy and growth. Seasonal changes in the environment may alter the fish's metabolic rate and therefore its energy requirement. A diet that exactly satisfies the energy needs during the warm summer months may cause a deposit of fat with the onset of cooler water because of reduced body metabolism and therefore a decreased calorie need. The caloric content of the diet also determines the amount of fat deposition in the body. Brook trout feeding upon a high calorie, oil-enriched food deposited more body fat than the same species feeding under similar condition on a lower calorie diet (Phillips and Brockway, 1959).

Changes in physiological activity affect the fat content of fish. Bailey (1952) found an increase in body fat storage prior to gonad maturation. In salmon and herring the storage was found in the muscle tissue, but in cod it was found in the liver. The increased fat was linked with the period of nonfeeding that usually occurs during the spawning season. The reserve fat provided energy for metabolism of the nonfeeding fish.

As in the higher animals, as fish grow larger and older there is an increase in body fat (Lovern, 1938; Phillips et al., 1960b), probably as a result of reduced physiological (growth) and physical activity of the older animal and thus a reduced metabolic rate.

Although the amount of body fat in wild fish is normally dependent upon the caloric content of its food, the fat content of some wild fish is surprisingly high. On a dry weight basis, mature and apparently healthy Siscowet lake trout contained as much as 89% fat (Eschmeyer and Phillips, 1965). Such extraordinarily high body fat contents are probably not only the result of surplus dietary fat calories but also are governed by genetic differences in the physiology of the trout.

7. THE CHARACTER OF THE BODY FAT OF FISH

The character of the body fat of fish assumes that of the food fat. A liquid fat (cod-liver oil or corn oil) fed to brook trout results in the deposition of a less saturated body fat (higher iodine number) than that deposited in fish fed either a surplus of protein or hard fat (beef tallow or hydrogenated cottonseed oil) (Phillips *et al.*, 1950, 1951). Toyomizu *et al.* (1963) found that the fatty acid composition of the body fat of rainbow trout was influenced by dietary fat.

Hoar and Cottle (1952) reported that goldfish with a hard body fat are less tolerant to temperature changes than those with soft body fats. Irvine *et al.* (1957) reported that the temperature resistance of goldfish was increased by the addition of cholesterol or phospholipid to the basic diets.

Phillips *et al.* (1957a) found that wild trout have a softer body fat (iodine number of 135) than hatchery trout (iodine number of 100). The fat of the hatchery food fed to these fish was harder than that of natural food (iodine numbers of 89 and 125, respectively).

The possible relationship between the melting point of the body fat of hatchery fish and their survival after stocking should be investigated.

Factors other than food alter the physical characteristics of the body fat. Lovern (1938) found that the body fat of eels is less saturated at higher water temperatures than at lower. Privol'nev and Brizinova (1965) found that the melting point of fats of cold water fish was lower than that of warm water fish and they were able to determine the relation of a fish to its environment temperature by the melting point of the body fat.

8. FAT-DIGESTING ENZYMES

Fat is digested by the enzyme lipase into fatty acids and glycerine before absorption. There are at least two sources of lipase in teleost fish, the pyloric ceca and the intestinal mucosa.

Lipase has been reported from the intestinal mucosa of fish by Babkin and Bowie (1928) and MacKay (1929); Johnston (1937) found a lipase activity from the dried preparation of the pyloric ceca. Brockerhoff (1966) described a digestion of fat by cod that was similar to that of pancreatic lipases of the higher animals. He did not know the source of this "pancreatic lipase" of teleosts but thought it could be from the diffuse pancreatic tissue of the fish. The lipase was similar in its action (nonstereospecific) to the pancreatic lipase in mammals and in the skate.

Kitamikado and Tachino (1961c) reported a strong esterase activity in the liver, spleen, and bile of rainbow trout. It was also detected, but much weaker, in the intestine, pyloric ceca, and the stomach. Optimum activity was between a pH of 6.8 and 7.6 and optimum temperature was 25°C for the enzyme from the intestine. Morishita *et al.* (1966) reported that there was no marked difference in the optimum pH for maximum esterase activity of five species of fish. However, the esterase from rainbow trout digestive tract was more active at lower temperatures than those from the other fish species.

Chepik (1966) found a seasonal change in the lipolytic activity of carp intestine. In the spring the activity increased 15–45% over the winter activity, reflecting, perhaps, seasonal changes in food abundance.

III. NONENERGY FOODS

A. Minerals

1. MINERAL REQUIREMENTS

Until the availability of radioactive isotopes, the dual source of minerals in food and water and the inability to obtain both a mineral-free diet and water limited studies upon fish mineral nutrition.

Marine (1914) early demonstrated an iodine deficiency (goiter) in Atlantic salmon.

Krogh (1939) found that after holding in distilled water ("washing out") a variety of freshwater fish (catfish, perch, rainbow trout, roach, and goldfish) absorbed chloride from a solution of solium chloride.

Berg and Gorbman (1953) demonstrated that ^{125}I was utilized by the thyroid of the platyfish. Chavin and Bouwman (1965) traced ^{125}I from the food into the thyroid gland of goldfish and back into the bloodstream as a component of thyroxine, an iodine cycle similar to that of higher animals. Woodall and LaRoche (1964) established the iodine need of chinook salmon fingerlings as 0.6 μg/g of dry diet and that of advanced parr as 1.1 μg (see also chapter by Gorbman, Volume II).

Frolova (1964) found that supplemental cobalt in the diet (0.08 mg/kg of fish) of carp increased the erythrocyte and hemoglobin content of the blood. Tominatik and Batyr (1967) reported that supplemental cobalt decreased the mortality of carp fingerlings and increased the erythrocyte and the hemoglobin content of the fish's blood. Farberov (1965) reported that dietary supplementation with 0.08 mg of $CoCl_2$ per kilogram of fish weight resulted in increased growth, productivity, and food consumption of carp and increased their level of vitamin B_{12}. This latter observation demonstrates a role of cobalt in vitamin B_{12} nutrition of carp that is similar to that in higher animals. Shabalina (1967) increased the growth of fingerling rainbow trout by supplementing their diet with cobalt.

Hevesy *et al.* (1964) found that at 18°C intramuscularly injected ^{59}Fe was incorporated into the blood components of tench to a maximum of 70% and at 5°C only 4% was taken up by the erythrocytes.

Dissolved calcium is used by trout for structural purposes (Phillips *et al.*, 1963). The amount of calcium absorbed is dependent upon the level of calcium in the food, and if the food level is decreased the amount of calcium absorbed from the water increases (see also chapter by Epple, Volume II).

Studies by the Cortland workers (Phillips *et al.*, 1958, 1959, 1960a, 1961; Podoliak and McCormick, 1967) have shown that calcium, phosphorus, cobalt, chloride, sulfate, and strontium are taken directly from the water by trout. These workers have also demonstrated the absorption and distribution of these minerals from the food by and to the body tissues. Other workers have found similar results with other fish species (Nelson, 1961; Templeton and Brown, 1963; Smelova, 1962; Ichikawa and Oguri, 1961; Hunn and Reineke, 1964; Schiffman, 1961).

2. OSMOREGULATION BY MINERALS

Krogh (1939) reported that calcium is involved in osmoregulation in some aquatic animals. In detailed studies, Podoliak and Holden (1965, 1966) confirmed the work of Krogh with brook, brown, and rainbow trout and reported that calcium decreases the permeability of the fish's membranes to actively oppose loss of ions to the environment following abrupt ionic changes of the environment.

3. EFFECT OF EXCESS SODIUM CHLORIDE

Phillips (1944) showed that high levels of dietary sodium chloride resulted in a water edema in brook trout that eventually caused death. The absorption of the salt exceeded the possible rate of excretion, and the chloride level of the blood was lowered by storing the excess salt in the body cavities. Excess dissolved sodium chloride caused loss of fish equilibrium that was correlated with greatly increased levels of blood chloride. Upset osmotic relationship of the tissues was thought to be the cause of the equilibrium loss (see chapter by Holmes and Donaldson, this volume, and chapter by Bern, Volume II).

B. Vitamins

1. METHODS FOR DETERMINING THE VITAMIN REQUIREMENTS OF FISH

Studies at the Cortland laboratory (Phillips *et al.*, 1946, 1947, 1948b) established the tentative daily requirements of trout for the B vitamins by determining the dietary level causing maximum vitamin storage in the fish's liver.

Wolf (1951) developed a diet capable of growing trout and from which vitamins could be withdrawn singly to determine their need and deficiency symptoms. Wolf's diet is composed of a purified protein base (casein and gelatin) supplemented with fat and carbohydrate calories (hydrogenated cottonseed oil, dextrin, and cooked potato starch), vitamins (at levels equivalent to approximately twice those of fresh beef liver), and minerals. Wolf's diet should be considered a semipurified diet and not a synthetic diet. It has been a useful tool in advancing fish nutrition and has led to a more thorough understanding of the vitamin needs of fish.

2. VITAMIN REQUIREMENTS

Using either Wolf's semipurified diet or a modification of Wolf's diet, many studies have been made to determine the essential vitamins for fish. Wolf (1951), Phillips et al. (1955, 1956, 1957b, 1958), Coates and Halver (1958), and Halver (1957) determined the need of trout and salmon for 10 members of the vitamin B complex. Ogino (1965, 1967) established the necessity of carp for pyridoxine, riboflavin, and pantothenic acid and Dupree (1966) that of channel catfish for some of the fat- and water-soluble vitamins. Recently, the need of trout for the fat-soluble vitamins K (Poston, 1964) and E (Poston, 1965) and the water-soluble vitamin C (Kitamura et al., 1965; Poston, 1967) was established. Woodall et al. (1964) determined that vitamin E was essential for chinook salmon.

Phillips and Livingston (1966) found a seasonal variation in the requirement of brook trout for pyridoxine. The variation was associated with seasonal light changes and was independent of fish size and water temperature. These experiments suggest changes in metabolic activity that are phototropic in nature.

Table IV has been included to show the essential vitamins for trout and Pacific Coast salmon and Table V for channel catfish and carp.

3. HYPERVITAMINOSIS

Burrows et al. (1952) suspected a hypervitaminosis A in salmon following the feeding whale liver. Poston et al. (1966) found that 1,100,000 units of vitamin A per pound of diet caused a hypervitaminosis A in hatchery-reared brook trout. The toxicity symptoms were reduced growth, lowered microhematocrit, and erosion of the caudal fin and peduncle.

Table IV

The Vitamin Requirements of Trout and Salmon

Vitamin	Trout[a]		Salmon[b]	
	Required	Deficiency symptoms	Required	Deficiency symptoms
Ascorbic acid	Yes	Scoliosis, lordosis, internal hemorrhaging	?	?
Thiamine	Yes	Nervousness, retracted head, high mortality	Yes	Muscle atrophy, loss of equilibrium, convulsions
Riboflavin	Yes	No growth, opaque eyes	Yes	Cloudy lens, hemorrhagic eyes, incoordination
Pyridoxine	Yes	Complete mortality in 6–12 weeks	Yes	Nervous disorders, anemia, edema of peritoneal cavity
Vitamin B₁₂	Yes	Reduced growth, increased severity of anemia in folic acid deficiency	Yes	Erratic hemoglobin and erythrocyte counts
Biotin	Yes	No growth, "blue-slime" disease, high mortality	Yes	Muscle atrophy, spastic convulsions
Choline	Yes	Reduced growth	Yes	Poor growth, hemorrhagic kidney and intestine
Folic acid	Yes	Anemia, reduced growth	Yes	Poor growth, anemia
Inositol	Yes	Reduced growth	Yes	Poor growth, distended stomach
Nicotinic acid	Yes	Reduced growth, increased susceptibility to suburn	Yes	Increased susceptibility to suburn, colon lesions, muscle spasms
Pantothenic acid	Yes	Poor growth, nonbacterial gill disease, high losses	Yes	Clubbed gills, prostration cellular atrophy
p-Aminobenzoic acid	Yes	Increased severity of anemia during folic acid deficiency	No	None
Tocopherol	Yes	Increased mortality, reduced microhematocrit	Yes	Anemia, poor growth, exophthalmia
Vitamin K	Yes	Retarded blood coagulation reduced microhematocrit	?	?

[a] From Phillips and Brockway (1957), Poston (1964, 1965, 1967), Phillips (1963), Phillips et al. (1964a), and Kitamura et al. (1965).

[b] From Halver (1957), DeLong et al. (1958), and Coates and Halver (1958).

Table V

The Vitamin Requirement of Channel Catfish and Carp

Vitamin	Channel catfish[a]		Carp[b]	
	Required	Deficiency symptoms	Required	Deficiency symptoms
Pyridoxine	Yes	Erratic swimming, tetany, gyrations, muscular spasms, mortality	Yes	Nervous disorders
Pantothenic acid	Yes	Clubbed gill filaments, reduced weight, mortality, "mummy" textured skin	Yes	Poor growth, exophthalmus, anemia
Riboflavin	Yes	Opaqueness of eye, mortality	Yes	Nervousness, photophobia, retarded growth
Thiamine	Yes	Reduced weight gain, difficulty maintaining equilibrium	?	?
Folic acid	Yes	Lethargy, mortality	?	?
Nicotinic acid	Yes	Tetany and death, lethargy, reduced coordination	?	?
Vitamin B_{12}	Yes	Reduced weight gain	?	?
Choline	Yes	Hemorrhagic areas in kidneys, reduced weight gain	?	?
Vitamin A	Yes	"Pop-eye," fluid in body cavity, edema in body cavity, hemorrhagic kidney	?	?
Vitamin K	Yes	Hemorrhages on body surface	?	?

[a] From Dupree (1966).
[b] From Ogino (1965, 1967).

C. Water

Maynard and Loosli (1962) estimated that an animal may lose practically all of its fat and half of its protein and live, but a loss of only 10% of its water causes death.

1. WATER CONTENT OF FISH

Hatchery trout contain between 75 and 80% water (Phillips *et al.*, 1964a, 1965, 1966) and, as in higher animals, the amount of water decreases with age. Wild trout contain less body water than hatchery trout (Phillips *et al.*, 1957a). Generally fat is deposited at the expense of water, and trout that contain high levels of body fat usually have lower levels of body water (Phillips *et al.*, 1951).

The amount of water in the fish's body is higher than that of birds or mammals (Maynard and Loosli, 1962).

2. SOURCES OF WATER

Water may be taken directly into the body by drinking, as a component of other foods, or supplied by both catabolic and anabolic metabolism.

Smith (1930) found that marine fish swallowed large amounts of seawater. Schuster-Woldan (1936) believed that carp and tench drank actively if nutrients were present in the water. X-ray photographs after immersion of the fish in a barium sulfate suspension with and without glucose showed considerable amounts of barium suspension in the intestinal tract if glucose was in the solution but none when glucose was absent (see chapters by Hickman and Trump, and Conte, this volume).

Curtis (1949) stated that freshwater fish do not drink but absorb water through the gills. Saltwater fish, however, drink saltwater. The blood of freshwater fish is ionically stronger than freshwater and, therefore, freshwater passes through the semipermeable membranes of the gills into the blood. There is no need for them to drink. Saltwater fish have blood ionically weaker than seawater. Water does not pass through the gill membranes into the blood. Saltwater fish must drink to obtain water which is then absorbed through the semipermeable membranes of the tract (see chapter by Conte, this volume).

D. Oxygen

Because of the limited supply, oxygen is more critical for aquatic than terrestrial life. An adequate oxygen supply depends upon contact of the water with the air and upon plant phytosynthesis. Neither of these processes occurs at a constant rate but varies with environmental conditions. The supply of dissolved oxygen may change from abundant to critical as environmental conditions abruptly change.

According to Davis (1953) trout should not be held for extended periods in water containing less than 5.0 ppm oxygen. Trout may survive at lower oxygen levels but they will not thrive in a normal manner.

The amount of dissolved oxygen depends upon the movement of air over the water's surface, the movement of the water itself, the population of aquatic plants, the water temperature, the amount of sunshine, and the population and activity of aquatic animals. Water temperature is critical because less oxygen is dissolved as the temperature increases and, equally important, as the temperature increases, the metabolic rate of aquatic animals accelerates increasing the oxygen demand. Leitritz (1960) reported that at 7.2°C yearling rainbow trout consumed 3 cm³ of oxygen/hr and at 20°C 12 cm³/hr—a 300% increase in oxygen utilization over the 12.8°C change in water temperature. This increased oxygen demand coupled with a decreased supply of dissolved oxygen (reduced by approximately 25%) at the higher temperature would limit the fish carrying capacity of the water.

An overload may be placed upon the oxygen supply because of temporary increases in the oxygen demand of the fish. These increases may follow feeding or excitement and stress caused by environmental changes. Davis (1953) showed an increased oxygen demand of trout after feeding, and Schaeperclaus (1933) reported a threefold increase in oxygen consumption by tench after transference from a pond to a barrel showing the effect of stress or excitement upon the fish's oxygen requirement.

Phillips (1947) showed that the erythrocyte content of the blood of blueback salmon and brook trout increased as the fish used the oxygen in a closed, nonaerated system. Similar results were found by Chiba (1965) with carp. In these experiments the oxygen-transporting capacity of the fish's blood was increased to compensate for the reduced oxygen supply in the environment. These results are similar to the increases in the erythrocyte content of human blood following continual living at high elevations (J. B. Sumner, 1929).

The opportunity for aeration (tumbling or surface content) and the abundance of plant life govern the rate of replacement of oxygen removed by aquatic animals.

IV. ENERGY REQUIREMENTS

A. Gross Energy Requirements

Winberg (1960) stated that between 14.1 and 33.1% of the calories consumed were deposited in the tissues of pike, indicating a usage for energy of between 67 and 86%. Slaughter experiments by Phillips and Brockway (1959) determined that trout used approximately 70% of the food calories for energy. Ivlev (1939) found that in sheatfish 65.7% of the yolk calories were recovered in the fish's body after complete yolk absorption; 34.3% of the calories were used for energy. This latter efficiency would be difficult to duplicate in older fish that must seek their food and maintain themselves in the environment, activities not required of fish during the period in which they live in a protected environment and are dependent upon the yolk for food.

Maynard and Loosli (1962) estimated that not more than 40% of the energy consumed by the muscle is actually transformed into work. The rest of the energy appears as heat and is lost to the body. The gross efficiency of the body of man and horse while working was estimated at 25%. This efficiency compares favorably with that of the steam engine (7.5%) and the gas engine (14 to 18%) but is lower than that of the diesel engine (29 to 35%).

The studies reported by Maynard and Loosli (1962) considered heat and other energy losses by the body. The values of Phillips and Brockway (1959) with trout and those of Winberg (1960) with pike and Ivlev (1939) with sheatfish were determined by differences between the calories consumed and those deposited in the body (gross energy). Heat and other losses were not taken into account.

Total metabolism was described mathematically in a series of papers (Paloheimo and Dickie, 1965, 1966a,b) that discussed the food and growth of fishes. These studies evaluated some of the factors that affect the growth (total metabolism and therefore total energy requirements) of fishes.

The relationship between total metabolism (total energy requirement) and body weight under standard conditions was expressed as follows (Paloheimo and Dickie, 1966a):

$$T = \alpha W^r \tag{1}$$

where T = total metabolism measured by the volume of oxygen consumed per hour per fish, W = weight of the fish in grams, α = the level of metabolism, and r = the weight exponent.

These authors found that the level of metabolism (α) apparently corresponded to the "routine" metabolic level found in oxygen consumption studies upon fish fed a "maintenance diet." A higher level of metabolism resulted from higher levels of food availability and ad libitum feeding appeared to produce metabolic levels known in oxygen consumption studies as active metabolic levels. The metabolic levels increased with increasing water temperature. Gross energy requirements varied with the amount of available food and with water temperature.

B. Net Energy Requirements

The energy requirement of fish increases or decreases with body metabolic activity. Because of an incomplete understanding of fish physiology and of the effect of the mechanics and chemistry of the environment upon the fish, many factors that alter the metabolic rate are unknown. However, some are known and experiments have shown their effect on the energy requirement of fish.

For convenience the fish's energy requirements are discussed under the headings of basal (maintenance) metabolism, growth, reproduction, and physical activity.

1. Energy Required for Basal Metabolism

Basal (maintenance) metabolism is the minimum level required to support the animal body under resting conditions. Basal metabolism has not been measured in fish, but standard metabolism (estimated by oxygen uptake) has been measured and defined as the approximate equivalent to basal metabolism in man (Fry and Hart, 1948). Beamish (1964a) stated that "In accord with the general practice, the metabolic rate of fishes is ordinarily measured in terms of oxygen consumption." This may be resting or active metabolism. The difference between the two is taken as a measure of energy requirement for the activity under study. The oxygen consumed by fish is a valid measure of the energy requirement since an increase in metabolic rate requires additional oxygen. Oxygen consumption has been used in numerous fish studies to measure the effect of variables upon the energy requirement.

a. Effect of Water Temperature. The metabolic rate of warm-blooded animals increases 10% for each degree Celsius rise in body temperature (J. B. Sumner, 1929). Schaeperclaus (1933) supported Sumner's findings in fish in that he found the rate of their metabolic activities doubled with a 10°C rise in water temperature. Phillips *et al.* (1960b) showed that during starvation brook trout increased their weight loss by approximately 10% for each degree Celsius rise in water (body) temperature, confirming the observations of both Sumner in higher animals and Schaeperclaus in fish. However, Brett (1965b) found that at a water temperature of 24°C, the basal metabolic rate (measured by oxygen uptake) of salmon was only six times higher than at 5°C.

Fry and Hart (1948) found that over a range of 5–35°C standard metabolism of goldfish increased to its highest value at about 30°C. It then remained steady or decreased slightly at temperatures higher than 35°C.

F. B. Sumner and Lanham (1942) reported that fish (*Crenuchthys baileyi*) living in warm spring water (35–37°C) had a greater oxygen consumption than those living in cool spring water (21°C). The energy required for maintenance increased as the water temperature increased.

Apparently there is an optimum temperature for fish metabolism, and if given a choice fish will select the most "comfortable" environment. Goldfish placed in an environment with a temperature gradient spent most of their time in a chosen temperature range (Fry, 1947). Rozin and Mayer (1961) trained goldfish to alter the temperature of their environment by depressing a lever. The fish maintained the water temperature between 33.5 and 36.5°C.

b. Effect of Light. Vinberg and Khartova (1953) showed there was a greater consumption of oxygen by young carp under illumination than by carp held in darkness.

c. Effect of Water Flow. Increased water flow increases the energy required for environmental maintenance. This is not necessarily swimming energy but may be energy increases required to maintain position in the environment. Washbourn (1936) found that trout fry reared for 3 months in two tanks at the same water temperature consumed more oxygen (measured under standard conditions) at the end of the experimental period when held in swift than when held in slower water.

d. Effect of Season. According to Swift (1964) brown trout were most active in the summer (maximum in June and August) and least active in the winter. Beamish (1964a) found a seasonal variation in the standard rate of oxygen consumption for brook and brown trout acclimated to 10°C and exposed to natural daylight. The consumption was lowest during March and April and at a maximum during the late fall spawning period.

e. Effect of Fish Size. Schaeperclaus (1933) reported that the metabolic rate of small fish was greater than that of large fish because of differences in body surfaces. The metabolic rate of a 12-g carp was 24.48 kcal in 24 hr/kg of body weight and that of a 600-g fish only 7.97 kcal. Based upon a square decimeter, Schaeperclaus calculated that at a water temperature of 15°C the caloric need per hour of the small size carp was 27 cal and that of the larger fish 23 cal. He emphasized that even with small differences in fish length there are great differences in caloric requirements. He pointed out that the greater oxygen consumption of the smaller fish per unit weight was evidence of the relatively greater caloric need of smaller fish.

Woynarovich (1964) found that the oxygen consumption related to 1 g of dried body substance of feeding fish larvae decreased with increasing body weight.

Beamish and Mookherjii (1964) found the proportionate change in the standard oxygen consumption was independent of fish weight for temperature changes in the range of 10–35°C. Beamish (1964b) showed that the proportionate change in standard oxygen consumption for a given change in water temperature was independent of body size within each of five species of freshwater fish.

f. Effect of Fish Species. Belding (1929) showed that under similar conditions brook trout required more oxygen than carp. Beamish (1964b) measured the standard oxygen consumption of five species of freshwater fish and found their standard oxygen uptake increased linearly (logarithmic grid) with weight for all species. The mean slope values of the regressions found for brown trout, brook trout, common white sucker, brown bullhead, and carp were 0.877, 1.052, 0.864, 0.925, and 0.894, respectively. Moss and Scott (1961) reported that at 25°C the standard metabolic rate of bluegill sunfish was lower than that of either largemouth bass or channel catfish. Within species there was little difference between the standard metabolic rates at water temperatures of 25, 30, and 35°C.

g. Effect of Starvation. There was a reduction in the energy requirement of brook trout (standard oxygen consumption) for the first 3 days of starvation, after which the energy requirement reached a minimum and then remained constant for the remainder of the 10-day experimental period (Beamish, 1964c). Similar changes were found in the blood glucose of brook trout (Phillips *et al.*, 1953). These authors believed the long period of constant blood sugar, after the rapid initial drop, represented conservation metabolism in the fish.

The standard oxygen consumption of white suckers decreased rapidly during starvation to a minimum within 2 or 3 days in experiments run in June, September, and December. However, in May the standard oxygen consumption reduced for 5 days, after which it remained constant to the seventh day. A secondary decrease then occurred on the ninth day of starvation and again remained constant for the last 13 days of the experiment (Beamish, 1964c).

2. ENERGY REQUIRED FOR GROWTH

Phillips and Brockway (1959) estimated that brook trout fed "high" calorie meat-dry meal or all-dry meal hatchery diets containing in excess of 700 kcal/lb (1540 kcal/kg) required 2100 kcal to produce a pound (4600 kcal/kg) of flesh. Only 900 kcal were required per pound (2000 kcal/kg) of trout produced after feeding natural foods containing 336 kcal/lb (640 kcal/kg) and 1200 kcal/lb (2600 kcal/kg) after feeding all-meat diets containing 450 kcal/lb (990 kcal/kg). Differences in the efficiencies between these foods may be caused by differences in diet quality or improper ingredient balance since Phillips *et al.* (1964a,b) were able to reduce the required calories to between 1500 and 1800 kcal/lb (3300 and 3960 kcal/kg) of fish produced by altering the formula of the hatchery diet.

Using the method of Phillips and Brockway, Tiemeier *et al.* (1965) determined that 1700 kcal were required to produce a pound (3740 kcal/kg) of channel catfish. Fowler *et al.* (1966) found that a diet containing 2350 kcal/kg (1065 kcal/lb) reared chinook salmon at a rate of 1.81 kg of food per kilogram of fish produced, equivalent to 1850 kcal/lb (4700 kcal/kg) of fish produced.

Although for practical application these results are used as measures of net energy for growth, the experiments actually measured gross energy utilization since the calorie requirement for growth was determined by the conversion of food into flesh and the calories present in a pound of diet. They are not precise studies that measure the energy requirement for the various body activities, but they are useful for comparing the gross energy efficiencies of food sources by different fish species.

Pyle (1966) found seasonal changes in the growth energy requirement of brook trout held in constant water temperature. The fish increased in length at a constant rate in inches during the summer months, but in the fall the rate decreased and continued to do so for a 16-week period from October to February. Pyle (1969) continued these experiments with brook trout held under either constant light, constant darkness, or simulated natural light. Those fish held under constant light increased at a constant rate in length in inches over a 15-month period. Those held under either simulated natural light (duplicating natural light cycles) or under constant darkness increased at a constant rate in inches from April until late October, paralleling the growth of trout held under constant light. However, for the following 16 weeks their growth rate dropped, after which it again paralleled that of fish held under constant light.

Gross *et al.* (1965) found changes similar to those described by Pyle in the growth of green sunfish. These fish generally grew at the highest rate in increasing light, the lowest in decreasing light, and intermediate in two constant photoperiods.

Altered light cycles may offer a means for controlling fish growth under artificial conditions, an intriguing field for investigation.

Phototropically triggered changes in growth rate alter the energy requirement of fish. Perhaps this should be expected for such changes act as a mechanism to lower the food requirement of the fish during periods of short food supply (fall and winter) in the temperate zones.

3. Energy Required for Reproduction

Much of the energy required for fish reproduction is deposited in the reproductive products (eggs and sperm).

Brown and Kamp (1941) found that at the time of spawning approximately 10% of the weight of female brown trout was in the gonads. Eschmeyer (1954) found that during 1950 approximately 12% of the body weight of female lake trout and about 1.7% of the body weight of males was gonads, and, during 1951 and 1953, an average of 18.5% of the weight of females and 3.1% of males was gonads. The range over the 3 years was from 11.7 to 33.0% in females and from 1.5 to 3.7% in males.

Calculations based upon the data of Kelley (1962) showed that the gonads of female largemouth black bass varied from a low of 2% to a high of 15% of the body weight. There was a large variation between and within age groups of the fish.

Based upon the data of Otsu and Uchida (1959) the percentage of the body weight of albacore as female gonads varied from 1 to 3%.

In addition to the measurable store of energy within the eggs and sperm, energy is required to form them. Calories stored in the eggs and sperm and those required for their formation represents consumed food energy that is lost to the body for other energy purposes.

Additional energy is expended during the spawning act and during the recovery period following spawning.

4. Energy Required for Physical Activity

Energy is required by animals, beyond that needed for maintenance, growth, and reproduction, to respond to the environment. A number of investigators have measured the energy required by fish for these responses.

Black (1958) reviewed experiments upon fish energy stores and metabolism and defended these studies by suggesting that a knowledge of the nature of muscular fatigue serves to reduce fish losses in every phase of handling (capture, transportation, tagging, etc.). He further suggested that a knowledge of the usage of body chemical stores for energy by adult migrating fish affords information needed to evaluate success of the fish in reaching their spawning beds in a condition satisfactory for successful spawning at a level capable of maintaining the fishery.

Since the energy used for physical activity is not available for growth, a knowledge of the factors that control this use of energy is valuable in the management of fish populations under both artificial and natural conditions.

a. Methods for Measurement. Fry and Hart (1948) described an apparatus for measurement of the oxygen uptake of goldfish under standard (resting) and maximum (swimming) metabolism. The fish were held in a closed system consisting of two-liter Erlenmeyer flasks. The standard uptake of oxygen was measured in stationary flasks and the maximum uptake of oxygen in rotating chambers that generated a current that forced the fish to swim. These authors were able to alter the water temperature and thus measure differences in energy requirements resulting from differences in water temperature.

In a review article Ivlev (1964) described the methods and apparatus used by Spoor (1946), Fry and Hart (1948), Basu (1959), Wohlschlag (1957), and Kovalevskaya (1956) to determine the metabolism of active fish, all of whom used oxygen uptake under changing conditions and activities to determine energy requirements.

Brett (1962) suggested that comparisons between different conditions of energy expenditures in fish be measured by estimating the energy required for a swimming rate of one body length per second (BL/sec). He pointed out that the energy cost of locomotion at all levels of performance requires exact definition since oxygen consumption may not in itself offer a precise measure of the actual energy expended for the activity under consideration.

Brett (1964, 1965a,b) described a series of experiments which measured the energy expenditures of swimming sockeye salmon under a variety of conditions. Brett (1964) developed a fish respirometer (Fig. 1) and an exercise cage (Brett, 1965b) for his outstanding series of studies. The energy required for physical activity was measured by the oxygen consumption of the fish.

Vincent (1960) evaluated the stamina of hatchery and wild trout in a specially designed apparatus, providing an estimate of the fish's ability to perform work. Thomas et al. (1964) developed a large tunnel for the measurement of salmon stamina under a variety of conditions, and Pyle (1965) reported preliminary experiments in a similar apparatus constructed for trout studies.

b. Effect of Water Temperature. Fry and Hart (1948) concluded that if differences between the oxygen uptake at maximum and standard rates of metabolism approximates the metabolism (energy) available for external work there is an apparent optimum temperature for goldfish in the vicinity of 28°C, at which the fish can perform the most external work. The curve of the relationship between water temperature and the speed at which goldfish can swim steadily showed a maximum between 20° and 30°C. They postulated that the decrease in sustained swimming rate of goldfish at temperatures from 30° to 38°C was probably caused by a decrease in metabolism (energy) available for external work.

Brett (1964) showed that at 5°C the standard metabolic rate of sock-

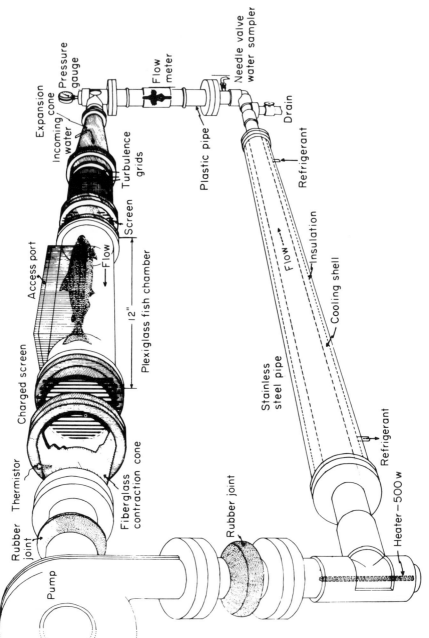

Fig. 1. A fish respirometer. From Brett (1964).

eye salmon, as measured by oxygen consumption, was 41 mg of oxygen per kilogram of fish per hour and at 24°C it was 196. Active metabolism (representing increased energy usage) rose from 514 mg of oxygen per kilogram of fish per hour to a maximum of 895 at 15°C. At temperatures up to 15°C active rates were 10–12 times the standard level, but above 15°C the ratio fell and was only four times the standard rate for salmon acclimated to 24°C. The combined action of temperature and activity elevated respiratory metabolism by a factor of 22.

Brett described a near linear transformation for the effect of water temperature on salmon locomotor metabolism by relating the logarithm of the oxygen swimming requirement (less the standard rate) to acclimation temperature. The slopes of the lines decreased with increasing swimming speed, approaching 0 at 4.21 BL/sec. Brett concluded that burst speeds at this level and above are virtually independent of water temperature. Salmon are similar to homeotherms in this respect.

c. *Effect of Fish Weight and Length.* According to Brett (1964), within the weight range of the salmon studied (25–130 g), there was no correlation between fish weight and either standard or active metabolism at any acclimation temperature, except at 15°C, in whose groupings smaller fish (8 g) were included. Despite a range in condition factor of between 0.64 and 1.04, only the group held at 15°C (which included the smaller fish) showed a significant correlation between plumpness and active metabolism.

Brett (1965a), in studies upon the effect of fish size on the oxygen consumption and swimming speed of sockeye salmon, showed there was a rapid decrease in the relative performance with increasing fish size. For all fish sizes at the highest sustained level of swimming speed the muscular demand for oxygen was met by the maximum rate of oxygen uptake. Burst speeds invoked an oxygen debt and fatigue.

In most cases, the replacement of the oxygen debt was exponential with time. The extent of the debt at the time of fatigue was influenced by acclimation temperature, ranging from 252 mg of oxygen per kilogram per hour at 5°C to 504 mg at 15°C.

d. *Effect of Water Velocity (Swimming Speed).* Pumpkinseed showed an increasing rate of oxygen consumption with water velocity (swimming speed) that was linear on a semilogarithmic basis (Brett and Sutherland, 1965). The results were similar to those of sockeye salmon (Brett, 1965a).

Brett (1965b) discussed experiments in which salmon held at 15°C showed an increase in oxygen consumption similar to a compound interest curve as the swimming speed was increased. The peak of the curve was reached as the fish reached the maximum rate of oxygen consumption and represented the top sustained swimming speed of the fish.

Brett (1965b) measured the actual energy utilized by analyzing the bodies of salmon forced to swim in a swimming apparatus at 1.83 miles/hr for 12.7 days (equivalent to 560 miles). Calculations were based on the known energy yields of fats and proteins by mammals. The results of the analyses were then compared with fish held under control conditions. The results were valid when compared with the rates of energy expenditure by salmon in the respirometer.

In field studies upon migrating salmon, Idler and Clemens (1959) found that female sockeye salmon used 96% of their body fat and 53% of their protein by the end of migration from the ocean to their spawning lake. These authors established that the daily energy output of male and female salmon during migration was 43 kcal/kg of body weight, nearly 80% of the maximum rate they could maintain.

e. Differences between Fish Species. In a study of the energy used for swimming by different fish species at various water temperatures Beamish (1966) found that at a swimming speed of 4 BL/sec in an activity chamber, Atlantic cod swam equally long at 5° and 8°C and redfish and winter flounder each equally long at 5°, 8°, and 11°C. At 6 BL/sec winter flounder swam longer at 14°C than at the lower water temperatures. At swimming speeds less than 4 BL/sec all fish species swam longer at the higher water temperatures. At 8°C, the only water temperature at which all species were tested, endurance at comparable swimming speeds was greatest for winter flounder, followed by cod, redfish, longhorn sculpin, ocean trout, and sea raven. Under comparable experimental conditions the energy available for work varied with species.

f. Fish Stamina. The studies of Vincent (1960) and Thomas *et al.* (1964) depended upon physical exhaustion (stamina) to measure differences in ability to perform work. There were no quantitative measurements or estimates of the energy used.

Vincent (1960) showed that wild trout were able to sustain themselves against a current in a stamina tunnel for longer periods than hatchery trout. These experiments showed differences between hatchery and wild trout in the energy available for outside work.

Thomas *et al.* (1964) showed differences in stamina between groups of hatchery salmon receiving different methods of hatchery care prior to testing, indicating differences in ability for sustained work.

C. Cyclic Changes in Energy Requirements

Many cyclic and seasonal changes in fish growth, body chemistry, and physiological activity have been described in the literature. These changes in metabolic activity alter the energy requirement of the fish.

The spawning of fish is the best known cyclic phototropic response to seasonal light changes (Pickford and Atz, 1957). Schaeperclaus (1933) described a winter rest in fish that was independent of water temperature, fish age, and spawning activity. Chepik (1966) noted that the proteolytic activity of carp intestine was less during fall and winter than during spring and summer. Phillips and Livingston (1966) found that trout survived for longer periods when fed diets deficient in pyridoxine during the fall and winter months than when fed similar diets during the spring and summer. These differences were independent of fish size and water temperature. Booke (1964) found a seasonal variation in the calcium content of trout blood similar to those observed in cod by Hess (1928). McCartney (1966) found seasonal changes in the cholesterol content of trout blood, and Shell (1961) showed cyclic changes in many of the blood constituents of black bass. Saito (1957) found that the total protein and albumin fractions of mackerel and carp blood were lower in February than in May or June when feeding behavior reached a maximum.

Wohlschlag and Juliano (1959) found a seasonal variation in the metabolism of bluegills. There were similar relationships of respiratory rates with swimming movements in the spring, summer, and autumn, but during winter a higher oxygen consumption was required for similar swimming activity.

Pyle (1966) reported seasonal changes in the rate of increases in length of brook trout held in constant temperature water. He was able (1969) to eliminate these changes by exposing the trout to continual light.

Studies concerning the energy requirement of fish should consider the phototropic exposure of the fish and the season of the year.

REFERENCES

Atkins, C. G. (1908). Foods for young salmonoid fishes. *U.S. Bur. Fisheries, Bull.* **28**, 839–851.
Atwater, W. O., and Bryant, A. P. (1899). The availability and fuel value of food materials *Storrs Agr. Exp. Sta. Ann. Rep.* pp. 73–110.
Babkin, B. P., and Bowie, D. J. (1928). The digestive system and its function in *Fundulus heteroclitus* (North American killifish). *Biol. Bull.* **54**, 254–278.
Bailey, B. E. (1952). Marine oils with particular reference to those of Canada. *Bull., Fisheries Res. Board Can.* **89**, 1–413.
Basu, S. P. (1959). Active respiration of fish in relation to ambient concentrations of oxygen and carbon dioxide. *J. Fisheries Res. Board Can.* **16**, 175–212.
Beamish, F. W. H. (1964a). Seasonal changes in the standard rate of oxygen consumption of fishes. *Can. J. Zool.* **42**, 189–194.
Beamish, F. W. H. (1964b). Respiration of fishes with special emphasis on standard oxygen consumption. II. Influence of weight and temperature on respiration of several species. *Can. J. Zool.* **42**, 177–188.

Beamish, F. W. H. (1964c). Influence of starvation on standard and routine oxygen consumption Trans. Am. Fisheries Soc. 93, 103–107.

Beamish, F. W. H. (1966). Swimming endurance of some Northwest Atlantic fishes. J. Fisheries Res. Board Can. 23, 341–347.

Beamish, F. W. H., and Mookherjii, P. S. (1964) Respiration of fishes with special emphasis on standard oxygen consumption. I. Influence of weight and temperature on respiration of goldfish, Carassius auratus. Can. J. Zool. 42, 161–175.

Belding, D. L. (1929). The respiratory movements of fish as an indicator of a toxic environment. Trans. Am. Fisheries Soc. 59, 238–245.

Berg, A., and Gorbman, A. (1953). Utilization of iodine by the thyroid of the platyfish, Xiphophorus (Platypoecilus) maculatus. Proc. Soc. Exptl. Biol. Med. 83, 751–756.

Black, E. C. (1958). Energy stores and metabolism in relation to muscular activity in fishes. "The Investigation of Fish-power Problems," H. R. MacMillan Lecture in Fisheries, pp. 51–67. Univ. of British Columbia.

Black, E. C., Bosomworth, N. J., and Docherty, G. E. (1966). Combined effect of starvation and severe exercise on glycogen metabolism of rainbow trout, Salmo gairdneri. J. Fisheries Res. Board Can. 23, 1461–1463.

Block, R. J. (1959). The approximate amino acid composition of wild and hatchery trout (Salvelinus fontinalis) and some of the principal foods (Gammarus and Hexagenia bilineata). Contrib. Boyce Thompson Inst. 20, 103–105.

Booke, H. E. (1964). Blood serum protein and calcium levels in yearling brook trout. Progressive Fish Culturist 26, 107–110.

Braman, W. W. (1933). The respiration calorimeter. Penn. State Univ., Agr. Expt. Sta., Bull. 302.

Brett, J. R. (1962). Some considerations in the study of respiratory metabolism in fish, particularly salmon. J. Fisheries Res. Board Can. 19, 1025–1038.

Brett, J. R. (1964). The respiratory metabolism and swimming performance of young sockeye salmon. J. Fisheries Res. Board Can. 21, 1183–1226.

Brett, J. R. (1965a). The relation of size to rate of oxygen consumption and sustained swimming speed of sockeye salmon (Oncorhynchus nerka). J. Fisheries Res. Board Can. 22, 1491–1501.

Brett, J. R. (1965b). The swimming energies of salmon. Sci. Am. 213, No. 2, 80–85.

Brett, J. R., and Sutherland, D. B. (1965). Respiratory metabolism of pumpkinseed (Lepomis gibbosus) in relation to swimming speed. J. Fisheries Res. Board Can. 22, 405–409.

Brockerhoff, H. (1966). Digestion of fat by cod. J. Fisheries Res. Board Can. 23, 1835–1839.

Brown, C. J. D., and Kamp, G. C. (1941). Gonad measurements and egg counts of brown trout from the Madison River, Montana. Trans. Am. Fisheries Soc. 71, 195–200.

Buhler, D. R., and Halver, J. E. (1961). Nutrition of salmonoid fishes. IX. Carbohydrate requirement of chinook salmon. J. Nutr. 74, 307–318.

Burrows, R., Palmer, D. D., Newman, H. W., and Azevedo, R. L. (1952). Tests of hatchery foods for salmon, 1951. U.S. Fish Wildlife Serv., Spec. Sci. Rept., Fisheries 86, 1–24.

Chavin, W., and Bouwman, B. W. (1965). Metabolism of iodine and thyroid hormone synthesis in the goldfish, Carassius auratus. Gen. Comp. Endocrinol. 5, 493–503.

Chepik, L. (1966). Activity of carp digestive enzymes at different seasons of the year. Biol. Abstr. 47, 60747.

Chesley, L. C. (1934). The influence of temperature upon the amylases of cold and warm-blooded animals. *Biol. Bull.* **66**, 330–338.

Chiba, K. (1965). A study on the influence of oxygen concentration on the growth of juvenile common carp. *Bull. Freshwater Fisheries Res. Lab.* **15**, 35–47.

Coates, J. A., and Halver, J. E. (1958). Water-soluble vitamin requirements of silver salmon. *U.S. Fish Wildlife Serv., Spec. Sci. Rept., Fisheries* **281**, 1–9.

Combs, B. D., Heinemann, W. W., Burrows, R. E., Thomas, A. E., and Fowler, L. G. (1962). Protein and calorie levels of meat-meal, vitamin-supplemented salmon diets. *U.S. Fish Wildlife Serv., Spec. Sci. Rept., Fisheries* **432**, 1–24.

Croston, C. B. (1961). Tryptic enzymes of chinook salmon. *Arch. Biochem. Biophys.* **89**, 202–206.

Curtis, B. (1949). "The Life Story of the Fish. His Manners and Morals." Republished 1961 by Dover, New York.

Davis, H. S. (1953). "Culture and Disease of Game Fishes." Univ. of California Press, Berkeley, California.

DeLong, D. C., Halver, J. E., and Mertz, E. T. (1958). Nutrition of salmonoid fishes. VI. Protein requirements of chinook salmon at two water temperatures. *J. Nutr.* **65**, 589–599.

Dupree, H. K. (1966). Vitamins essential for growth of channel catfish. *U.S. Fish Wildlife Serv., Tech. Papers* **7**, 1–12.

Eschmeyer, P. H. (1954). The reproduction of lake trout in Southern Lake Superior. *Trans. Am. Fisheries Soc.* **84**, 47–74.

Eschmeyer, P. H., and Phillips, A. M., Jr. (1965). Fat content of the flesh of siscowets and lake trout from Lake Superior. *Trans. Am. Fisheries Soc.* **94**, 62–74.

Farberov, V. G. (1965). Influence of cobalt feeding on fish growth and accumulation of vitamin B_{12} in carp liver. *Chem. Abstr.* **62**, 5617g.

Fish, G. R. (1962). The comparative activity of some digestive enzymes in the alimentary canal of Tilapia and perch. *Chem. Abstr.* **57**, 15623b.

Fowler, L. G., McCormick, J. H., and Thomas, A. E. (1964). Further studies of protein and calorie levels of meat-meal, vitamin-supplemented salmon diets. *U.S. Fish Wildlife Serv., Spec. Sci. Rept., Fisheries* **480**, 1–14.

Fowler, L. G., McCormick, J. H., and Thomas, A. E. (1966). Studies of calorie and vitamin levels of salmon diets. *U.S. Fish Wildlife Serv., Tech. Papers* **6**, 1–13.

Frolova, L. K. (1964). Effect of cobalt on some hematological characteristics of carp. *Biol. Abstr.* **45**, 80900.

Fry, F. E. J. (1947). Effects of environment on animal activity. *Univ. Toronto Biol. Ser.* **55**, 1–62.

Fry, F. E. J., and Hart, J. S. (1948). The relation of temperature to oxygen consumption in the goldfish. *Biol. Bull.* **94**, 66–77.

Gerking, S. D. (1952). The protein metabolism of sunfishes of different ages. *Physiol. Zool.* **25**, 358–372.

Gray, I. E. (1928). The effect of insulin on the blood sugar of fishes. *Am. J. Physiol.* **84**, 566–573.

Greene, C. W. (1913). The fat absorbing function of the alimentary tract of the king salmon. *U.S. Bur. Fisheries, Bull.* **33**, 149–176.

Gross, W. L., Roelofs, E. W., and Fromm, P. O. (1965). Influence of photoperiod on growth of green sunfish, *Lepomis cyanellus. J. Fisheries Res. Board Can.* **22**, 1379–1386.

Halver, J. E. (1957). Nutrition of salmonoid fishes. III. Water-soluble vitamin requirements of chinook salmon. *J. Nutr.* **62**, 225–243.

Halver, J. E. (1961). A big role for vitamins and amino acids. *U.S. Trout News* 6, No. 4, 8–9.

Halver, J. E., and Shanks, W. E. (1960). Nutrition of salmonoid fishes. VIII. Indispensable amino acids for sockeye salmon. *J. Nutr.* 72, 340–346.

Halver, J. E., DeLong, D. C., and Mertz, E. T. (1957). Nutrition of salmonoid fishes. V. Classification of essential amino acids for chinook salmon. *J. Nutr.* 63, 95–105.

Hess, A. (1928). Differences in calcium level of blood between the male and female cod. *Proc. Soc. Exptl. Biol. Med.* 25, 349-359.

Hess, W. (1935). Reduction of the islets of Langerhans in the pancreas of fish by means of diet, overeating, and lack of exercise. *J. Exptl. Zool.* 70, 187–194.

Hevesy, G., Lockner, D., and Sletten, K. (1964). Iron metabolism and erythrocyte formation in fish. *Acta Physiol. Scand.* 60, 256–266.

Hoar, W. S., and Cottle, M. K. (1952). Dietary fat and temperature tolerance of goldfish. *Can. J. Zool.* 30, 41–48.

Hunn, J. B., and Reineke, E. P. (1964). Influence of iodine intake on iodine distribution in trout. *Proc. Soc. Exptl. Biol. Med.* 115, 91–93.

Ichikawa, R., and Oguri, M. (1961). Metabolism of radionuclides in fish. I. Strontium-calcium discrimination in gill absorption. *Bull. Japan. Soc. Sci. Fisheries* 27, 351–356.

Idler, D. R., and Clemens, W. A. (1959). The energy expenditures of Frazer River sockeye salmon during the spawning migration to Chilko and Stuart Lakes. International Pacific Salmon Fisheries Report, Progress Report. 6. 80 pp.

Irvine, D. G., Newman, K., and Hoar, W. S. (1957). Effects of dietary phospholipid and cholesterol on the temperature resistance of goldfish. *Can. J. Zool.* 35, 691–709.

Ivlev, V. S. (1939). The energy balance of growing larvae of *Silurus glanis*. *Dokl. Akad. Nauk SSSR* 25, 87–89.

Ivlev, V. S. (1964). Determinations of active metabolism. "Techniques for the Investigation of Fish Physiology." Translated from Russian and published for The USDI and the National Science Foundation by the Israel Program for Scientific Translations, pp. 80–90.

Johnston, W. W. (1937). Some characteristics of the enzymes of the pyloric caeca of cod and haddock. *J. Biol. Board Can.* 3, 473–485.

Kashiwada, K. (1952). Studies on the enzymes of shipjack entrails. I. On the seasonal variation of proteolytic enzyme activity in pyloric caeca. *Bull. Japan. Soc. Sci. Fisheries* 18, 151–154.

Kelley, J. W. (1962). Sexual maturity and fecundity of the largemouth bass in Maine. *Trans. Am. Fisheries Soc.* 91, 23–28.

Kenyon, W. A. (1925). Digestive enzymes in poikilothermal vertebrates. An investigation of enzymes in fishes with comparative studies on those of amphibians, reptiles, and mammals. *U.S. Bur. Fisheries, Bull.* 41, 181–200.

Kitamikado, M., and Tachino, S. (1961a). Digestive enzymes of rainbow trout. I. Carbohydrases. *Chem. Abstr.* 55, 5789a.

Kitamikado, M., and Tachino, S. (1961b). Digestive enzymes of rainbow trout. II. Proteases. *Chem. Abstr.* 55, 5789b.

Kitamikado, M., and Tachino, S. (1961c). Digestive enzymes of rainbow trout. III. Esterases. *Chem. Abstr.* 55, 5789c.

Kitamikado, M., Morishita, T., and Tachino, S. (1965a). Digestibility of dietary protein in rainbow trout. I. Digestibility of several dietary proteins. *Chem. Abstr.* **62,** 15129d.

Kitamikado, M., Morishita, T., and Tachino, S. (1965b). Digestion of dietary protein in rainbow trout. II. Effect of starch and oil contents in diets and the size of fish on digestibility. *Chem. Abstr.* **62,** 15129e.

Kitamura, S., Ohara, S., Suwa, T., and Nakagawa, K. (1965). Studies on vitamin requirements of rainbow trout *Salmo gairdneri.* I. On the ascorbic acid. *Bull. Japan. Soc. Sci. Fisheries* **31,** 818–826.

Kovalevskaya, L. A. (1956). "The Energy of Fish in Motion." Sbornik Posvyaschenny Pamyati Akad. PP Lazarea, Izd. Akad. Nauk S.S.S.R.

Krogh, A. (1939). "Osmotic Regulation in Aquatic Animals." Cambridge Univ. Press, London and New York.

Lee, D. J., Roehm, J. N., Yu, T. C., and Sinnhuber, R. O. (1967). Effect of 3 fatty acids on the growth rate of rainbow trout, *Salmo gairdnerii.* **J.** *Nutr.* **92,** 93–98.

Leitritz, E. (1960). "Trout and Salmon Culture," State of California Fishery Bull. 107. Dept. of Fish and Game, Sacramento, California.

Lovern, J. A. (1938). Fat metabolism in fishes. XIII. Factors influencing the composition of the depot fat of fishes. *Biochem. J.* **32,** 1214–1224.

McCartney, T. H. (1966). Monthly variations of the serum total cholesterol of mature brown trout. Cortland Hatchery Report 34 for the year 1965. Fisheries Res. Bull. 29, pp. 72–75. State of New York Conservation Department, Albany.

McCay, C. M., and Tunison, A. V. (1935). Report of the experimental work at the Cortland Hatchery for the year 1934. State of New York Conservation Department, Albany.

McGeachin, R. L., and Debnam, J. W., Jr. (1960). Amylase in freshwater fish. *Proc. Soc. Exptl. Biol. Med.* **103,** 814–815.

MacKay, M. E. (1929). The digestive system of the eel-pout (*Zoarces onguitharis*). *Biol. Bull.* **56,** 8–23.

Marine, D. (1914). Further observations and experiments on goiter (so called thyroid carcinoma) in brook trout, *Salvelinus fontinalis.* III. Its prevention and cure. *J. Exptl. Med.* **19,** 70–88.

Maynard, L. A., and Loosli, J. K. (1962). "Animal Nutrition." McGraw-Hill, New York.

Morishita, T., Noda, H., Kitamikado, M., Takahashi, T., and Tachino, S. (1966). The activity of the digestive enzymes in fish. *Chem. Abstr.* **64,** 7093b.

Moss, D. D., and Scott, D. C. (1961). Dissolved oxygen requirements of three species of fish. *Trans. Am. Fisheries Soc.* **90,** 377–393.

Nelson, W. C. (1961). Uptake of P^{32} from the water and from a single oral dose by brook and rainbow trout. *U.S. At. Energy Comm. Rept., T.D.* **13109,**

Nicolaides, N., and Woodall, A. N. (1962). Impaired pigmentation in chinook salmon fed diets deficient in essential fatty acids. *J. Nutr.* **78,** 431–437.

Norris, E. R., and Elam, D. W. (1940). Preparation and properties of crystalline salmon pepsin. *J. Biol. Chem.* **134,** 443–454.

Norris, E. R., and Mathies, J. C. (1953). Preparation, properties and crystallization of tuna pepsin. *J. Biol. Chem.* **204,** 673–680.

Ogino, C. (1963). Studies on the chemical composition of some natural foods of aquatic animals. *Bull. Japan. Soc. Sci. Fisheries* **29,** 459–462.

Ogino, C. (1965). B vitamin requirements of carp, *Cyprinus carpio.* I. Deficiency symptoms and requirements of vitamin B_6. *Bull. Japan. Soc. Sci. Fisheries* **31,** 546–551.

428   ARTHUR M. PHILLIPS, JR.

Ogino, C. (1967). B vitamin requirements of carp. II. Requirements of riboflavin and pantothenic acid. *Bull. Japan. Soc. Sci. Fisheries* 33, 351–354.

Ogino, C., and Suyama, M. (1957). Variations in amino acid composition of protein during the development of egg and the growth of young fish of rainbow trout, *Salmo irideus. Bull. Japan. Soc. Sci. Fisheries* 23, 227–229.

Ono, T., Nagayama, F., and Masuda, T. (1960). Studies on the fat metabolism of fish muscles. 4. Effects of the components in foods on the culture of rainbow trout. *J. Tokyo Univ. Fisheries* 46, 97–110.

Otsu, T., and Uchida, R. N. (1959). Sexual maturity and spawning of albacore in the Pacific Ocean. *U.S. Fish Wildlife Serv., Fishery Bull.* 59, 287–305.

Paloheimo, J. E., and Dickie, L. M. (1965). Food and growth of fishes. I. A growth curve derived from experimental data. *J. Fisheries Res. Board Can.* 22, 521–542.

Paloheimo, J. E., and Dickie, L. M. (1966a). Food and growth of fishes. II. Effect of food and temperature on the relation between metabolism and body weight. *J. Fisheries Res. Board Can.* 23, 869–908.

Paloheimo, J. E., and Dickie, L. M. (1966b). Food and growth of fishes. III. Relation among food, body size, and growth efficiency. *J. Fisheries Res. Board Can.* 23, 1209–1248.

Phillips, A. M., Jr. (1944). The physiological effect of sodium chloride upon brook trout. *Trans. Am. Fisheries Soc.* 74, 297–309.

Phillips, A. M., Jr. (1947). The effect of asphyxia upon the red cell content of fish blood. *Copeia* 3, 183–186.

Phillips, A. M., Jr. (1963). Folic acid as an anti-anemia factor for brook trout. *Progressive Fish Culturist* 25, 132–134.

Phillips, A. M., Jr., and Brockway, D. R. (1957). The nutrition of trout. IV. Vitamin requirements. *Progressive Fish Culturist* 19, 119–123.

Phillips, A. M., Jr., and Brockway, D. R. (1959). Dietary calories and the production of trout in hatcheries. *Progressive Fish Culturist* 21, 3–16.

Phillips, A. M., Jr., and Livingston, D. L. (1966). The effect of season of the year and fish size on the pyriodixine requirement of brook trout. Cortland Hatchery Report 34 for the year 1965. Fisheries Res. Bull. 29, pp. 15–19. State of New York Conservation Department, Albany.

Phillips, A. M., Jr., Tunison, A. V., Schaffer, H. B., White, G. K., Sullivan, M. W., Vincent, C., Brockway, D. R., and McCay, C. M. (1946). Cortland Hatchery Report 14 for the year 1945. Fisheries Res. Bull. 8. State of New York Conservation Department, Albany.

Phillips, A. M., Jr., Brockway, D. R., Rodgers, E. O., Sullivan, M. W., Cook, B., and Chipman, J. R. (1947). Cortland Hatchery Report 15 for the year 1946. Fisheries Res. Bull. 9. State of New York Conservation Department, Albany.

Phillips, A. M., Jr., Tunison, A. V., and Brockway, D. R. (1948a). The utilization of carbohydrate by trout. Fisheries Res. Bull. 11. State of New York Conservation Department, Albany.

Phillips, A. M., Jr., Brockway, D. R., Rodgers, E. O., Robertson, R. L., Goodell, H., Thompson, J. A., and Willoughby, H. (1948b). Cortland Hatchery Report 16 for the year 1947. Fisheries Res. Bull. 10. State of New York Conservation Department, Albany.

Phillips, A. M., Jr., Brockway, D. R., Bryant, M., Rodgers, E. O., and Maxwell, J. M. (1950). Cortland Hatchery Report 18 for the year 1949. Fisheries Res. Bull. 13. State of New York Conservation Department, Albany.

Phillips, A. M., Jr., Brockway, D. R., Kolb, A. J., and Maxwell, J. M. (1951). Cortland

Hatchery Report 19 for the year 1950. Fisheries Res. Bull. 14. State of New York Conservation Department, Albany.

Phillips, A. M., Jr., Lovelace, F. E., Brockway, D. R., and Balzer, G. C., Jr. (1953). Cortland Hatchery Report 21 for the year 1952. Fisheries Res. Bull. 16. State of New York Conservation Department, Albany.

Phillips, A. M, Jr., Neilsen, R. S., and Brockway, D. R. (1954). A comparison of hatchery diets and natural foods. *Progressive Fish Culturist* **16**, 153–157.

Phillips, A. M., Jr., Lovelace, F. E., Podoliak, H. A., Brockway, D. R., and Balzer, G. C., Jr. (1955). Cortland Hatchery Report 23 for the year 1954. Fisheries Res. Bull. 18. State of New York Conservation Department, Albany.

Phillips, A. M., Jr., Lovelace, F. E., Podoliak, H. A., Brockway, D. R., and Balzer, G. C., Jr. (1956). Cortland Hatchery Report 24 for the year 1955. Fisheries Res. Bull. 19. State of New York Conservation Department, Albany.

Phillips, A. M., Jr., Brockway, D. R., Lovelace, F. E., and Podoliak, H. A. (1957a). A chemical comparison of hatchery and wild brook trout. *Progressive Fish Culturist* **19**, 19–25.

Phillips, A. M., Jr., Brockway, D. R., Podoliak, H. A., and Balzer, G. C., Jr. (1957b). Cortland Hatchery Report 25 for the year 1956. Fisheries Res. Bull. 20. State of New York Conservation Department, Albany.

Phillips, A. M., Jr., Podoliak, H. A., Brockway, D. R., and Vaughn, R. R. (1958). Cortland Hatchery Report 26 for the year 1957. Fisheries Res. Bull. 21. State of New York Conservation Department, Albany.

Phillips, A. M., Jr., Podoliak, H. A., Dumas, R. F., and Thoesen, R. W. (1959). Cortland Hatchery Report 27 for the year 1958. Fisheries Res. Bull. 22. State of New York Conservation Department, Albany.

Phillips, A. M., Jr., Podoliak, H. A., Livingston, D. L., Dumas, R. F., and Thoesen, R. W. (1960a). Cortland Hatchery Report 28 for the year 1959. Fisheries Res. Bull. 23. State of New York Conservation Department, Albany.

Phillips, A. M., Jr., Livingston, D. L., and Dumas, R. F. (1960b). Effect of starvation and feeding on the chemical composition of brook trout. *Progressive Fish Culturist* **22**, 147–154.

Phillips, A. M., Jr., Podoliak, H. A., Livingston, D. L., Dumas, R. F., and Hammer, G. L. (1961). Cortland Hatchery Report 29 for the year 1960. Fisheries Res. Bull. 24. State of New York Conservation Department, Albany.

Phillips, A. M., Jr., Podoliak, H. A., Poston, H. A., Livingston, D. L., Booke, H. E., Pyle, E. A., and Hammer, G. L. (1963). Cortland Hatchery Report 31 for the year 1962. Fisheries Res. Bull. 26. State of New York Conservation Department, Albany.

Phillips, A. M., Jr., Podoliak, H. A., Poston, H. A., Livingston, D. L., Booke, H. E., Pyle, E. A., and Hammer, G. L. (1964a). Cortland Hatchery Report 32 for the year 1963. Fisheries Res. Bull. 27. State of New York Conservation Department, Albany.

Phillips, A. M., Jr., Hammer, G. L., and Pyle, E. A. (1964b). Dry concentrates as complete trout foods. *Progressive Fish Culturist* **26**, 21–24.

Phillips, A. M., Jr., Livingston, D. L., and Poston, H. A. (1965). The effect of protein and calorie levels and sources on the growth of brown trout. Cortland Hatchery Report 33 for the year 1964. Fisheries Res. Bull. 28, pp. 11–19. State of New York Conservation Department, Albany.

Phillips, A. M., Jr., Livingston, D. L., and Poston, H. A. (1966). The effect of changes in protein quality, calorie sources and calorie levels upon the growth and chemical composition of brook trout. Cortland Hatchery Report 34 for the year 1965. Fisheries Res. Bull. 29, pp. 6–7. State of New York Conservation Department, Albany.

Phillips, A. M., Jr., Poston, H. A., and Livingston, D. L. (1967). The effects of calorie sources and water temperature upon trout growth and body chemistry. Cortland Hatchery Report 35 for the year 1966. Fisheries Res. Bull. 30, pp. 25–34. State of New York Conservation Department, Albany.

Pickford, G. E., and Atz, J. W. (1957). "The Physiology of the Pituitary Gland of Fishes." N.Y. Zool. Soc., New York.

Podoliak, H. A., and Holden, H. K., Jr. (1965). Distribution of dietary calcium to the skeleton and skin of fingerling brown trout. Cortland Hatchery Report 33 for the year 1964. Fisheries Res. Bull. 28, pp. 64–70. State of New York Conservation Department, Albany.

Podoliak, H. A., and Holden, H. K., Jr. (1966). Calcium ion regulation by fingerling brook, brown, and rainbow trout. Cortland Hatchery Report 34 for the year 1965. Fisheries Res. Bull. 29, pp. 59–65. State of New York Conservation Department, Albany.

Podoliak, H. A., and McCormick, J. H. (1967). Absorption of strontium by brook trout. Cortland Hatchery Report 35 for the year 1966. Fisheries Res. Bull. 30, pp. 5–13. State of New York Conservation Department, Albany.

Poston, H. A. (1964). Effect of dietary vitamin K and sulfaguanidine on blood coagulation time, microhematocrit and growth of immature brook trout. *Progressive Fish Culturist* 26, 59–64.

Poston, H. A. (1965). Effect of dietary vitamin E on microhematocrit, mortality and growth of immature brown trout. Cortland Hatchery Report 33 for the year 1964. Fisheries Res. Bull. 28, pp. 6–10. State of New York Conservation Department, Albany.

Poston, H. A. (1967). Effect of L-ascorbic acid on immature brook trout. Cortland Hatchery Report 35 for the year 1966. Fisheries Res. Bull. 30, pp. 46–51. State of New York Conservation Department, Albany.

Poston, H. A., Livingston, D. L., Pyle, E. A., and Phillips, A. M., Jr. (1966). The toxicity of high levels of vitamin A in the diet of brook trout. Cortland Hatchery Report 34 for the year 1965. Fisheries Res. Bull. 29, pp. 20–24. State of New York Conservation Department, Albany.

Privol'nev, T. I., and Brizinova, P. N. (1965). Melting points of fats in fishes. *Chem. Abstr.* 63, 10359h.

Pyle, E. A. (1965). Construction of a stamina testing apparatus. Cortland Hatchery Report 33 for the year 1964. Fisheries Res. Bull. 28, pp. 52–55. State of New York Conservation Department, Albany.

Pyle, E. A. (1966). A 42-week study on the growth in length of brook trout. Cortland Hatchery Report 34 for the year 1965. Fisheries Res. Bull. 29, pp. 34–39. State of New York Conservation Department, Albany.

Pyle, E. A. (1969). The effect of constant light or constant darkness on the growth and sexual maturity of brook trout. Cortland Hatchery Report 36 for the year 1967. Fisheries Res. Bull. 31, pp. 13–19. State of New York Conservation Department, Albany.

Rozin, P. N., and Mayer, J. (1961). Thermal reinforcement and thermoregulatory behavior in the goldfish, *Carassius auratus. Science* 134, 942–943.

Saito, K. (1957). Biochemical studies of fish blood. X. On the seasonal variation of serum protein components of cultured fish. *Bull. Japan. Soc. Sci. Fisheries* **22**, 768–772.

Schaeperclaus, W. (1933). Textbook of pond culture. *U.S. Fish Wildlife Serv., Fishery Leaflet* **311**, 1–240.

Schiffman, R. H. (1961). Uptake of strontium from diet and water by rainbow trout. *U.S. At. Energy Comm. Rept., H. W.* **72107**.

Schuster-Woldan, E. (1936). X-ray studies of the intestine of *Cyprinus carpio* and *Tinca vulgaris* as a contribution to the problem of the significance of the sense of smell and sight in fish in the search for food. *Z. Fischerei* **34**, 241–245.

Shabalina, A. A. (1967). Effect of cobalt chloride on the development and growth of rainbow trout. *Biol. Abstr.* **48**, 26907.

Shanks, W. E., Gahimer, G. D., and Halver, J. E. (1962). The indispensable amino acids for rainbow trout. *Progressive Fish Culturist* **24**, 68–73.

Shcherbina, M. A., Sorvachev, K. F., and Sukhoverkhov, F. M. (1964). Evaluation of fish foods on the basis of their amino acid content. *Biol. Abstr.* **45**, 85138.

Shell, E. W. (1961). Chemical composition of blood of smallmouth bass. *U.S. Fish Wildlife Serv., Res. Rept.* **57**, 1–36.

Smelova, I. V. (1962). Absorption of S^{35}-labeled compounds from water by fish. *Chem. Abstr.* **57**, 1394c.

Smith, H. W. (1930). The absorption and excretion of water and salts by marine teleosts. *Am. J. Physiol.* **93**, 480–485.

Spoor, W. A. (1946). A quantitative study of the relationship between the activity and oxygen consumption of goldfish and its application to the measurement of respiratory metabolism in fishes. *Biol. Bull.* **91**, 312–325.

Sumner, F. B., and Lanham, U. N. (1942). Studies of the respiratory metabolism of warm and cool spring fishes. *Biol. Bull.* **82**, 313–327.

Sumner, J. B. (1929). "Textbook of Biological Chemistry." Macmillan, New York.

Swift, D. R. (1964). Activity cycles in the brown trout (*Salmo trutta*). 2. Fish artificially fed. *J. Fisheries Res. Board Can.* **21**, 133–138.

Templeton, W. L., and Brown, V. M. (1963). Accumulation of calcium and strontium by brown trout from waters in the United Kingdom. *Nature* **198**, 198–200.

Thomas, A. E., Burrows, R. E., and Chenoweth, H. H. (1964). A device for stamina measurement of fingerling salmonids. *U.S. Fish Wildlife Serv., Res. Rept.* **67**, 1–15.

Tiemeier, O. W., Deyoe, C. W., and Wearden, S. (1965). Effects of growth of fingerling channel catfish of diets containing two energy and two protein levels. *Trans. Kansas Acad. Sci.* **68**, 180–186.

Togasawa, Y., and Katsumata, T. (1961). The proteinase of pyloric caeca. I. Preparation of crystalline proteinase of tunny pyloric caeca. *Chem. Abstr.* **55**, 1944h.

Togasawa, Y., Katsumata, T., Ishikawa, M. (1959). Studies on the proteinase of pyloric caeca. III. Preparation of crystalline proteinase of bonito pyloric caeca. *Bull. Japan. Soc. Sci. Fisheries* **25**, 470–472.

Tominatik, E. N., and Batyr, A. K. (1967). The role of inorganic cobalt in the nutrition of fingerling carp. *Biol. Abstr.* **48**, 26928.

Toyomizu, M., Kawasaki, K., and Tomiyasu, Y. (1963). Effect of dietary oil on the fatty acid composition of rainbow trout oil. *Bull. Japan. Soc. Sci. Fisheries* **29**, 957–961.

Tunison, A. V., Brockway, D. R., Schaffer, H. B., Maxwell, J. M., McCay, C. M., Palm, C. E., and Webster, D. A. (1944). Cortland Hatchery Report 12 for the year 1943. Fisheries Res. Bull. 5. State of New York Conservation Department, Albany.

Ushiyama, H., Fujimori, T., Shibata, T., and Yoshimura, K. (1966). Carbohydrases in the pyloric caeca of salmon (*Oncorhynchus keta*). *Chem. Abstr.* **64,** 7092h.

Vinberg, G. G., and Khartova, L. E. (1953). Intensity of metabolism in young carp. *Chem. Abstr.* **47,** 10138i.

Vincent, R. E. (1960). Some influences of domestication upon three stocks of brook trout (*Salvelinus fontinalis* Mitchell). *Trans. Am. Fisheries Soc.* **89,** 35–52.

Washbourn, R. (1936). Metabolic rates of trout fry from swift and slow running waters. *J. Exptl. Biol.* **13,** 145–147.

Wendt, C. (1964). Diet and glycogen reserves in hatchery reared Atlantic salmon during different seasons. I. Winter. *Swedish Salmon Res. Inst., RPT. LFI MEDD.*

Winberg, G. G. (1960). Rate of metabolism and food requirements of fishes. *Fisheries Res. Board Can., Transl. Ser.* **194,** 1–202.

Wohlschlag, D. E. (1957). Differences in metabolic rates of migratory and resident freshwater forms of an Artic whitefish. *Ecology* **38,** 502–510.

Wohlschlag, D. E., and Juliano, R. O. (1959). Seasonal changes in bluegill metabolism. *Limnol. Oceanog.* **4,** 195–209.

Wolf, L. E. (1951). Diet experiments with trout. *Progressive Fish Culturist* **13,** 17–24.

Wood, E. M. (1952). Methods for protein studies with applications to four selected diets. Ph.D. Thesis, Cornell University, Ithaca, New York.

Woodall, A. N., and LaRoche, G. (1964). Nutrition of salmonoid fishes. XI. Iodide requirements of chinook salmon. *J. Nutr.* **82,** 475–482.

Woodall, A. N., Ashley, L. M., Halver, J. E., Olcott, H. S., and Van Derveen, J. (1964). Nutrition of salmonoid fishes. XIII. The alpha-tocopherol requirement of chinook salmon. *J. Nutr.* **84,** 125–135.

Woynarovich, E. (1964). The oxygen consumption of fishes in the early stage of their growth in the 0.5–28° range of water temperatures. *Chem. Abstr.* **61,** 11070f.

AUTHOR INDEX

Numbers in italics refer to the pages on which the complete references are listed.

A

Abel, J., 242, 251, *286*
Acher, R., 280, *283*
Adamson, R. H., 41, 48, 49, *86*, 325, *346*
Alberts, A. W., 374, *383*
Alderdice, D. F., 296, 298, 299, 305, 306, 308, *309*
Allison, W. S., 366, *383*
Altman, P. L., 314, 322, *345*
Amberson, W. R., 22, *82*
Andrus, M. H., 45, 47, *88*
Andur, B. H., 59, 73, 75, *83*
Angerer, K. A., 43, 44, 46, *83*
Arion, W. J., 359, *383*
Armin, J., 13, *79*
Armstrong, F. H., 297, *310*
Arnheim, N., Jr., 244, *292*, 370, *383*
Arstila, A. U., 96, 224, *227, 228*
Asai, J., 377, *384*
Ashley, L. M., 408, *432*
Atkins, C. G., 392, *423*
Atkinson, D. E., 363, 379, *383*
Atwater, W. O., 393, *423*
Atz, J. W., 423, *430*
Aubert, X., 356, *383*
Audigé, J., 123, *227*
Axelrod, J., 323, *345*
Azevedo, R. L., 408, *424*

B

Babkin, B. P., 401, 405, *423*
Baggerman, B., 260, *284*, 301, *309*
Bahr, K., 102, *227*
Bailey, B. E., 404, *423*

Bailey, J., 258, *285*
Bailey, J. E., 295, *310*
Bailey, R. E., 78, *79*
Baker, J. R., 325, *345*
Baldwin, E., 335, *345*
Ball, I., 316, *345*
Ball, J., 281, *284*
Balzer, G. C., Jr., 408, 416, *429*
Bardach, J. E., 121, *233*
Barlow, J. S., 30, *80*
Barnett, H. L., 30, *82*
Barrington, E. J. W., 307, *309*
Basu, S. P., 419, *423*
Bateman, J., 268, *288*
Battle, H. I., 299, 305, *309*
Batyr, A. K., 406, *431*
Beament, J. W. L., 199, *227*
Beames, C. J., Jr., 374, *383*
Beamish, F. W. H., 414, 415, 416, 422, *423, 424*
Beams, H., 273, *288*
Beatty, D. D., 143, 145, 161, *234*
Becker, E. L., 53, 58, *80*, 191, *239*
Behrisch, H. W., 363, *383*
Belding, D. L., 416, *424*
Bellamy, D., 32, 33, 35, 63, 66, *80, 81*, 88, 150, 152, 163, 187, 207, 237, 244, 281, *284*
Benoit, G. J., 340, *345*
Bentley, P. J., 37, *80*, 103, 104, 212, *227*, 247, 280, *284, 287*, 357, 358, *383*
Berg, A., 406, *424*
Berg, L. S., 93, 119, *227*
Bergeron, J., 271, *284*
Berglund, F., 53, 58, 76, 79, *82*, 321, 322, 323, 342, *345, 347*

433

444

Shibata, T., 364, *388*, 398, *432*
Shrago, E., 358, *389*
Sindermann, C. J., 61, 66, *88*
Singhal, R. L., 358, 359, 363, *389*
Sinnhuber, R. O., 403, *427*
Skadhauge, E., 102, 143, *234*, 237, 261, *289*
Skou, J., 253, *291*
Sletten, K., 407, *426*
Slicher, A., 281, *284*
Smelova, I. V., 407, *431*
Smith, C. L., 377, *386*
Smith, D., 259, *291*
Smith, H. W., 23, 40, 42, 43, 44, 45, 46, 48, 50, 51, 52, 53, 56, 57, 76, 78, 79, *88*, 105, 106, 107, 109, 110, 111, 112, 113, 115, 116, 119, 126, 129, 133, 153, 159, 161, 169, 171, 177, 179, 186, 187, 188, 196, 203, 209, 211, 220, 229, *232*, *234*, 237, *238*, 243, 244, 247, 248, 260, 261, 287, *291*, 319, 320, 326, 327, 335, 337, 338, 339, 343, *350*, 411, *431*
Smith, L., 263, *291*
Smith, L. M., 378, *388*
Smith, W. W., 109, 112, 113, 115, 186, 188, *238*
Soberman, R. J., 11, *88*
Sokolova, M. M., 69, *89*, 259, 260, *292*
Solemdal, P., 294, 298, *311*
Solomon, A. K., 158, *239*
Solomon, S., 53, 58, *80*
Somero, G. N., 358, 368, 370, 371, 372, 380, *385*, *388*
Sorvachev, K. F., 400, *431*
Soteres, P., 202, *235*
Sperber, I., 115, 191, *231*
Spivack, S., *292*
Spoor, W. A., 419, *431*
Sprinson, D. B., 336, *346*
Srivastava, S. K., 358, 359, 363, *389*
Staedeler, G., 43, *88*
Stainer, I. M., 11, 29, 30, 72, *84*, 150, *232*
Stambaugh, R., *388*
Stanley, H. P., *238*
Stanley, J. G., 61, 66, 82, *88*, 118, 143, 150, 152, 163, 198, 205, 207, 209, *230*, *238*, 260, 281, *286*, *291*

Steen, J., 268, *291*
Sterling, K., 12, *83*, *88*
Stimpson, J. H., 360, *388*
Stolzenbach, F., 370, 371, *387*
Stone, W., Jr., 40, 41, 47, 49, 50, *80*, *82*
Stransky, E., 331, *350*
Straus, L., 273, *291*
Struempler, A., 325, *345*
Sukhoverkhov, F. M., 400, *431*
Sullivan, M. W., 407, *428*
Sullivan, M. X., 250, *291*
Sulya, L. L., 41, 53, 58, *88*
Sulze, W., 177, 188, *238*
Sumner, F. B., 414, *431*
Sumner, J. B., 412, 414, *431*
Sutherland, D. B., 421, *424*
Sutherland, L. E., 97, 100, 229, *238*
Suwa, T., 408, 409, *427*
Suyama, M., 400, *428*
Suzuki, T., 379, *389*
Sverdrup, H. U., 186, *238*
Swanson, J. R., 362, *385*
Sweet, J. G., 307, *335*
Swift, D. R., 415, *431*
Szego, C. M., 358, *388*

T

Tachino, S., 394, 396, 398, 401, 405, 406, *426*, *427*
Taggart, J. V., 115, 191, *231*, 324, *350*
Takada, N., 78, *86*, 152, 159, *236*
Takahashi, T., 401, 406, *427*
Taketa, K., 363, *388*
Talamo, B., 374, *383*
Tappel, A. L., 372, 375, 376, 377, 378, 379, 380, 381, *383*, 384, *388*
Tarr, H. L. A., 361, 362, *386*
Teller, D., 360, *384*
Templeton, W. L., 407, *431*
Theilacker, G., 297, 302, *310*
Thoesen, R. W., 396, 407, *429*
Thomas, A. E., 400, 403, 417, 419, 422, *425*, *431*
Thomas, C., 40, 47, 49, *82*
Thomas, S. F., 61, 70, *87*
Thomas, W. N., 337, *346*
Thompson, A. L., 337, *346*
Thompson, J. A., 407, *428*
Thorson, T. B., 15, 16, 17, 18, 19, 20,

SYSTEMATIC INDEX

Note: Names listed are those used by the authors of the various chapters. No attempt has been made to provide the current nomenclature where taxonomic changes have occurred.

Monkfish, 255
Mud skipper, 320, 321, *see also Peri-*
ophthalmus sobrinus
Mugil
  M. cephalus, 53, 65
  M. curema, 53
Mugilidae (Mullets), 122
Mullidae, 124
Mureana, 53, 58, 187
M. helena, 53, 58, 124, 135, 212
Muraenidae, 124, 187
Mustelus canis, 40, 41, 43, 45, 46, 47,
  49, 50, 113, 114, 115
Mycteroperca
  M. bonasi, 53
  M. tigris, 25
  M. venenosa, 53
Myoxocephalus
  M. octodecimspinosus, 126, 163, 169,
  173, 175, 187
  M. scorpius, 73, 74, 75, 163, 173, 175,
  187, 212, 221, 318, 327–329
Myxine, 99
  M. glutinosa, 32–35, 95, 97, 99, 101,
  213, 243–244, 319, 331
Myxini, 93–100, *see also* Hagfish
Myxiniformes, 4, 15, 32, 33, 39

N

Narcine brasiliensis, 40, 107
Necturus, 334
Negaprion, 325
  N. brevirostris, 18, 41, 48, 49, 115
Neoceratodus, 339
  N. forsteri, 320, 338, 339
Nerophis ophidion, 137

O

Ocyurus, 325
Oligocottus maculosus, 267
Oligophites saurus, 53
Oncocephalidae, 125
Oncocephalus, 125
Oncorhyncus, 259, 260, 280, 295, 297,
  298, 301, 303, 308
  O. gorbuscha, 23, 60, 68, 69, 125, 204,
  259, 260, 295, 298, 302, 308
  O. keta, 60, 260, 295, 298, 301, 302
  O. kisutch, 23, 62, 70, 258–260, 269,
  275–278, 298, 301, 302

O. masou, 60, 70, 71, 159
O. nerka, 23, 59, 60, 68, 69, 78, 258–
  260, 298, 302
O. tschawytscha, 60, 62, 69–71, 72,
  257–260, 278, 280, 298, 302
Ophichthyidae, 124
Ophiocephalidae, 124
Ophiodon, 361, 362
  O. elongatus, 17, 19, 22, 361, 362
Ophisurus macrorynchus, 124
Opsanus tau, 137, 181, 185, 188, 202,
  203, 209, 210, 221
Oryzias latipes, 125
Osteichthyes, 17, 20, 21, 26, 51, 78, 79,
  119, 125, 257
Ostraciidae, 125
Ostracion tuberculatus, 125

P

Paralabrax clathratus, 54, 58
Paralichthys
  P. dentatus, 205
  P. flesus, 163
  P. lethostigma, 125, 126, 137, 162–
  177, 180–181, 184–185, 187, 190–
  191, 198, 203, 205–208, 261–263
Parapristipoma trilineatum, 124
Parasiluris asotus, 124
Parophrys vetulus, 99, 101, 125, 135,
  137, 138, 140, 142, 144, 146, 148,
  155, 156, 182, 192, 194, 200, 203,
  213, 269, 297, 305, 306, 340
Pegasiformes, 93
Pempheris macrolepidotos, 124
Perca fluviatilis, 22, 165
Perch, 261, 265, 316, 323, 324, 398, 406,
  see also Perca fluviatilis, Cyprinus
  tinea
Percidae, 127
Periophthalmus, 207, 320
  P. sobrinus, 65, 320, *see also* Mud
  skipper
Petromyzon
  P. fluviatilis, see Lampetra fluviatilis
  P. marinus, 15, 16, 33, 36–39, 102,
  103, 245
  P. tridentata, 37, 38, 39
Petromyzones, 100, 141, 213, *see also*
  Lamprey
Petromyzonids, 245–247

SUBJECT INDEX

A

Acetazolamide, 45, 50
Acetoacetate, 373
4-Acetyl-4-aminoantipyrine, 11
Acetylcholine
  on gill blood flow, 268
  on rectal gland, 248
Acetyl CoA, 352, 372
Acid-base balance, 248, 263
Acid phosphatase, 201, 225, 253
Aconitase, 379
Actinomycin D, 277, 279
Active transport, 2
Acylcarnitine, 373
Adenosine diphosphate (ADP), 313, 351
Adenosine monophosphate (AMP), 313, 329, 330, 351
Adenosinetriphosphatase, 253, 255, 256
Adenosine triphosphate (ATP), 189, 255, 313, 351–374
ADP, see Adenosine diphosphate
Adrenalectomy, 67, 68
Adrenaline
  on blood glucose, 357
  on gill blood flow, 268
  on rectal gland, 248
Adrenocorticosteroids, 71, 72
Alanine, 344, 354, 375, 376
Aldactone, 66
Aldolase, 364, 365
Aldosterone, 67
Allantoicase, 333, 335, 336
Allantoin, 333, 334
Amino acids
  dietary requirements, 399, 400
  essential, 399
  in nitrogen metabolism, 315, 316, 326, 330, 333, 335, 343

p-Aminobenzoic acid, 409
p-Aminohippuric acid, 114, 115, 147, 173, 175, 324
Ammonia
  excretion, 151, 186, 327, 381
  in nitrogen metabolism, 314–319, 326, 327–331, 337–339, 343, 344
  tabulated values for blood, 34–37, 52–55
  toxicity, 316
Ammoniotelism, 314, 317, 321, 331, 332, 337, 339, 343, 344
AMP, see Adenosine monophosphate
Amylase, 397, 398
Angiotensin II, 66
Anions transport, 188–196, see also specific ions
Antipyrine, 11
Aqueous humor, 49, 50, 79
Archinephric ducts, 96, 201
Arginase, 332, 336, 339
Arginine, 212, 336, 339, 343, 399
Ascorbic acid, see Diet, vitamins
ATP, see Adenosine triphosphate
Atropine, 248
Azoreductase, 325

B

Betamethasone, 67
Bicarbonate, see also Blood chemistry
  in osmotic and ionic regulation, 248, 263, 316
  in pericardial fluid, 44
Biotin, 409
Bladder function, 202
Blastoderm, 298, 306, 307
Blood
  chemistry in Chondrichthyes, 40–42, 46–48

effect of magnesium chloride on, 183, 184
kidney function and, 91–226
of nitrogenous substances, 161–162, 196–197, 313–344
of organic acids, 115
of salt, 241–283
of urea, 196
Extracellular compartment, definition, 6
Extracellular fluid, values, 18, 24, 25, 26

F

Fat, 391, 393, 394, 395, 402–406, see also Lipid
absorption, 402
body, 403, 405
digestion, 403, 405–406
harmful effects, 403
hatchery and wild trout, 404
metabolism, 352, 372–375
requirements in trout, 403, 405
storage, 404
temperature acclimation and, 405
use in migration, 422
water content and, 411
Fatigue, 418, 421
Fatty acid metabolism, 372, 373, 374, 375, see also Lipid
FDP, see Fructose diphosphate
FDPase, see Fructose diphosphatase
Fertilization, 293, 295
Fluid, see also Blood, Urine
cerebrospinal, 45, 49
coelomic, 45
cranial, 20, 45, 49, 79
ear, 49
eye, 20, 49, 50, 79
interstitial and oncotic hydrostatic pressure of, 7
intestinal, 262
intracellular, 4, 14, 15, 16, 18, 26
pericardial, 44, 45
peritoneal, 20
perivisceral, 44, 45, 78
perivitelline, 44, 297, 305
rectal gland, 249
Fluoroacetate, 376
Folic acid, 392, 409
Food, see Nutrition, different foods
F1P, see Fructose monophosphate
F6P, see Fructose 6-phosphate

Fructose diphosphatase (FDPase), 351, 363, 364
Fructose diphosphate (FDP), 351, 362, 363, 364, 368
Fructose 6-phosphate (F6P), 351
Fructose monophosphate (F1P), 351
Fumarase, 380
Fumaric acid, 380

G

Gametes, see Eggs, Sperm
GDH, see Glutamate dehydrogenase
GDP, see Guanosine diphosphate
GFR, see Glomerular filtration rate
Gills
epithelium, 245, 246, 247, 248, 250, 251, 256
morphology, 266–276
nitrogen excretion, 314–318, 327–344
osmoregulation, 260, 264, 265, 267, 277–283, 323
Glomerular filtration rate, 97, 107–111, 141–147, 165, 167, 169–173, 179, 180–182, 205–207
Glomerulus, see also Kidney
structure, development, and evolution, 97, 106, 128, 169–171, 203, 204, 220, 221
Glucose
adrenaline effect, 357
in blood, 159, 416
digestion, 394, 396
insulin effect, 357
inulin clearance and, 145
metabolism, 353–372, 376, 411
1-phosphate (G1P), 351, 361, 362
6-phosphate (G6P), 351, 354, 358, 359, 362
reabsorption, 99, 159–161, 199, 224
renal excretion, 98, 111, 186
β-Glucuronidase, 201
Glutamate, 316, 327–331, 344, 351, 381
Glutamate dehydrogenase (GDH), 361, 327–331, 344, 351, 381–383
Glyceraldehyde 3-phosphate, 365–367
Glycerine, 405
Glycine, 400
Glycogen, 273, 359, 361, 395–396, 397
Glycogen synthetase, 361
G1P, see Glucose 1-phosphate or monophosphate